Metal Nanoparticles for Cancer Therapy

Metal Nanoparticles for Cancer Therapy

Editors

Panchanathan Manivasagan
Eue-Soon Jang

MDPI

Basel • Beijing • Wuhan • Barcelona • Belgrade • Novi Sad • Cluj • Manchester

Editors

Panchanathan Manivasagan
Department of Chemistry &
Bio-Science
Kumoh National Institute of
Technology
Gumi-si
Korea, South

Eue-Soon Jang
Department of Chemistry &
Bio-Science
Kumoh National Institute of
Technology
Gumi-si
Korea, South

Editorial Office
MDPI
St. Alban-Anlage 66
4052 Basel, Switzerland

This is a reprint of articles from the Special Issue published online in the open access journal *Pharmaceutics* (ISSN 1999-4923) (available at: www.mdpi.com/journal/pharmaceutics/special_issues/metal_therapy).

For citation purposes, cite each article independently as indicated on the article page online and as indicated below:

Lastname, A.A.; Lastname, B.B. Article Title. *Journal Name* **Year**, *Volume Number*, Page Range.

ISBN 978-3-7258-1016-1 (Hbk)
ISBN 978-3-7258-1015-4 (PDF)
doi.org/10.3390/books978-3-7258-1015-4

Contents

About the Editors

Panchanathan Manivasagan

Dr. Panchanathan Manivasagan, Ph.D., is a Research Professor in the Department of Chemistry and Bio-Science, Kumoh National Institute of Technology (KIT), Gumi, South Korea. An overarching goal of his current research is the development of multifunctional nanomaterials for multimodal imaging (in vivo fluorescence imaging, photoacoustic imaging, CT imaging, and MR imaging) and therapy (chemotherapy, phototherapy, chemodynamic therapy, electrodynamic therapy, sonodynamic therapy, immunotherapy, starvation therapy, and electrolytic ablation therapy) of cancer and microbial infections. He received his Master's and Ph.D. in Marine Biotechnology from the Faculty of Marine Sciences, Annamalai University, Tamil Nadu, India. He subsequently served as a Post-Doctoral Researcher and Visiting Research Professor at the Department of Chemistry and Department of Biomedical Engineering, Pukyong National University, Busan, South Korea. He also worked as a Research Assistant Professor in the Department of Chemical and Biological Engineering, Korea University, Seoul, South Korea. He has numerous publications (>96) in outstanding and high-impact factor journals and book chapters (>20) and has presented invited talks and presentations at international conferences. He is also a regular reviewer for many reputed international journals.

Eue-Soon Jang

Prof. Eue-Soon Jang, Ph.D., is a Professor at the Department of Chemistry and Bio-Science, Kumoh National Institute of Technology (KIT), Gumi, South Korea. He worked as a Post-Doctoral Fellow at Stanford University School of Medicine, Stanford, USA. He also worked as a Post-Doctoral Fellow at the Department of Materials Science and Engineering, POSTECH, South Korea. In addition, he worked as a Senior Researcher in the Department of Future Technology Research, Korea Telecom, South Korea. He received his B.Sc. degree from the Department of Chemistry, Chungnam National University, South Korea, and his Master's and Ph.D. degrees from the Department of Chemistry, Seoul National University, Seoul, South Korea. He won the Outstanding Paper award from the Organizing Committee of the 6th Cross Straits Symposium on Materials, Energy and Environmental Engineering, South Korea. He received a Vice President Citation award from Korea Telecom due to his outstanding contribution to the field of new technology business. He also received numerous academic awards during the last 20 years. His primary research interests are the investigation and development of nanomedicine for multimodal imaging-guided cancer therapy. His research program is mainly based on the synthesis of various metal nanoparticles and the crystal engineering of various solid nanoparticles. He developed a UV and near-infrared blocking agent for anti-aging cosmetics. To date, he has published above 85 research and review articles and has several patents (>22). He is a recipient of many research grants from various funding bodies. He actively participates in outreach and scientific dissemination for the service of the wider community.

Preface

Cancer remains a major threat to human health worldwide, causing high morbidity and mortality. Recently, owing to the rapid advancement of nanotechnology and nanoscience, the development of new multifunctional metal-based nanoparticles for cancer therapy has attracted a lot of attention because of the various advantages of the reduced rate of recurrent tumors and minimized side effects. In addition, a variety of metal-based nanomaterials with diagnostic imaging properties have been widely used to improve imaging quality, leading to more efficient and promising applications in diagnostic imaging. Combination cancer therapy also allows precise diagnosis and highly efficient tumor therapy. This reprint is a collection of efforts by various researchers engaged in the field of metal nanoparticles for cancer therapy. The presented research and review articles are valuable for the scientific community focused on metal nanoparticles for cancer therapy research.

Panchanathan Manivasagan and Eue-Soon Jang
Editors

pharmaceutics

Article

The Impact of Silver Nanoparticle-Induced Photothermal Therapy and Its Augmentation of Hyperthermia on Breast Cancer Cells Harboring Intracellular Bacteria

Sijia Liu [1], Spencer Phillips [1,2], Scott Northrup [1] and Nicole Levi [1,2,*]

1 Department of Plastic and Reconstructive Surgery, Wake Forest School of Medicine, Winston-Salem, NC 27101, USA; siliu@wakehealth.edu (S.L.); sgphilli@wakehealth.edu (S.P.); snorthru@wakehealth.edu (S.N.)
2 School of Biomedical Engineering and Sciences, Wake Forest/Virginia Tech, Winston-Salem, NC 24061, USA
* Correspondence: nlevi@wakehealth.edu

Abstract: Breast cancer can harbor intracellular bacteria, which may have an impact on metastasis and therapeutic responses. Silver nanoparticles are FDA-approved for their antimicrobial potential, plus they have pleiotropic benefits for eradicating cancer cells. In the current work we synthesized photothermal silver nanoparticles (AgNPs) with an absorption at 800 nm for heat generation when exposed to near-infrared laser irradiation. Breast cell lines MCF 10A, MCF7, and MDA MB 231 were infected with *Pseudomonas aeruginosa*, and their response to AgNPs, heat, or photothermal therapy (PTT) was evaluated. The results demonstrate that the application of a brief heating of cells treated with AgNPs offers a synergistic benefit in killing both infected and non-infected cells. Using 10 μg/mL of AgNPs plus laser stimulation induced a temperature change of 12 °C, which was sufficient for reducing non-infected breast cells by 81–94%. Infected breast cells were resistant to PTT, with only a reduction of 45–68%. In the absence of laser stimulation, 10 μg/mL of AgNPs reduced breast cell populations by 10–65% with 24 h of exposure. This concentration had no impact on the survival of planktonic bacteria with or without laser stimulation, although infected breast cells had a 42–90% reduction in intracellular bacteria. Overall, this work highlights the advantages of AgNPs for the generation of heat, and to augment the benefits of heat, in breast cancer cells harboring intracellular infection.

Keywords: silver nanoparticles; breast cancer; intracellular bacteria; photothermal therapy; hyperthermia

Citation: Liu, S.; Phillips, S.; Northrup, S.; Levi, N. The Impact of Silver Nanoparticle-Induced Photothermal Therapy and Its Augmentation of Hyperthermia on Breast Cancer Cells Harboring Intracellular Bacteria. *Pharmaceutics* **2023**, *15*, 2466. https://doi.org/10.3390/pharmaceutics15102466

Academic Editors: Manivasagan Panchanathan and Eue-Soon Jang

Received: 25 September 2023
Revised: 9 October 2023
Accepted: 11 October 2023
Published: 14 October 2023

1. Introduction

According to the American Cancer Society, breast cancer is the second leading cause of cancer death in women in the United States [1]. Previous studies have identified the presence of intracellular bacteria within breast cancer cells [2–4], which have a negative impact on therapeutic responses. A number of bacterial species have been associated with breast tumors [5–10], and intracellular bacteria have been found to enhance the survival of circulating tumor cells and promote metastasis. *Pseudomonas* has been identified in higher abundance in breast tumor tissue compared to normal breast tissue [11]. Although the exact species has not been identified, given the clinical incidence of *Pseudomonas aeruginosa*, this is a suitable culprit for the investigation of intracellular infection and response to therapies. For example, *Pseudomonas* can grow in the presence of chemotherapy, and doxorubicin stimulates *Pseudomonas* growth [11]. There is limited information on the treatment of breast cancer harboring intracellular bacteria, and few antibiotics work intracellularly, plus extended use of antibiotics can result in antibiotic resistance; therefore, there is a need to evaluate therapeutic options that can simultaneously eliminate breast cancer and bacteria.

Conventional therapies for breast cancer include surgery, radiation therapy, and chemotherapy. However, adverse side effects and therapeutic resistance limit their efficacy [12]. In recent years, photothermal therapy (PTT) has emerged as a promising strategy for the treatment of breast cancer [13–16]. By irradiating photothermal agents with deep-tissue penetrating near-infrared (NIR) light, photothermic agents can generate heat to kill cancer cells of the primary tumor, and mitigate metastasis [16–19]. Moreover, PTT has minimal side effects, such as low toxicity to healthy cells and short treatment times [20]. Gold nanoparticles of various shapes and sizes have routinely been explored as PTT agents [14,21–27], given their inert chemical nature in the body, but they have limited antimicrobial potential in the absence of light or modification [28–30].

Silver nanoparticles (AgNPs) are advantageous for breast cancer, exhibiting anti-proliferative effects on cancer cells and preventing metastasis [31–37]. In addition, AgNPs are clinically utilized as FDA-approved antibacterial agents [38–41]. Thus, AgNPs are promising for killing breast cancer cells harboring intracellular bacteria [39]. Most AgNPs used in medicine are spherical, with an absorption near 400 nm [42–45], although the optimal wavelengths for PTT are in the infrared, where body tissue is most transparent [46]. Previously, our team developed triangular AgNPs with near-infrared absorption, which induced heat generation upon exposure to near-infrared radiation [44], and demonstrated that the cellular uptake of AgNPs is not needed for beneficial effects, so long as the nanoparticles are in close proximity to the cells. In the current study, we infected three breast cell lines to investigate whether AgNPs could induce sufficient heat to destroy both non-infected and infected breast cancer cells upon exposure to laser stimulation. One of the challenges of PTT is the limited penetration depth of externally applied light [47–49]. An alternative to PTT stems from the knowledge that hyperthermia augments chemo- and radio-therapies [50–54]. AgNPs are thermally conductive, hence, we also explored their use as pleiotropic agents to augment hyperthermia (HT), kill cancer and bacteria cells directly, or serve as PTT agents. Together these data indicate the benefits of AgNPs to eliminate breast cancer and bacteria in the breast tumor microenvironment.

2. Materials and Methods

2.1. Synthesis and Characterization of AgNPs

Trisodium citrate (TSC), low molecular weight chitosan, and polyvinylpyrrolidone (PVP) were purchased from Sigma Aldrich. Ascorbic acid, sodium borohydride ($NaBH_4$), and acetic acid were purchased from Fisher Scientific (Rockingham County, NH, USA). Silver nitrate ($AgNO_3$) was purchased from EMD Chemical. Solutions were prepared using ultrapure de-ionized (DI) water. Prior to synthesis, all glassware was washed with aqua regia solution (3 parts HCl to 1 part HNO_3, and triple-rinsed with DI water).

AgNPs were prepared via a seed-mediated growth method following a reported procedure [44]. Silver seed was produced by adding 1 mL of 30 mM TSC, 2 mL of 5 mM $AgNO_3$, and 1 mL of 100 mM $NaBH_4$ (aged in the dark for 4 h before use) to 95 mL of rapidly stirred DI water. After 60 s of vigorous stirring, 1 mL of 10 mg/mL PVP was added, and the solution was stirred for an additional 30 min. The silver seed solution was stored at 4 °C in the dark until used in further synthesis.

To prepare nanoparticles using the seed, 20 mL of chitosan (1% acetic acid, 2 mg/mL low molecular weight chitosan) was vigorously stirred to which 133 mL of 116.4 mM TSC, 33 mL of 300 mM ascorbic acid, and 400 mL of silver seed were added. Then, 200 mL of 30 mM $AgNO_3$ was added dropwise to the solution, and the solution was stirred for approximately 3–7 min at room temperature while observing a color change from clear to yellow, orange, red, purple and blue. After that, 300 mL of 10 M NaOH was added to halt the reaction. The solution was centrifuged at 14,000 rpm for 10 min to isolate the nanoparticles, which were resuspended in DI water and stored at 4 °C in the dark.

The absorbance of AgNPs was measured using a Mettler Toledo UV5Nano UV/visible scanning spectrophotometer. The hydrodynamic diameter and zeta potential were determined using a Malvern Instruments Zetasizer. The shape of the AgNPs was observed using a FEI Tecnai BioTwin transmission electron microscope.

2.2. Hyperthermic Potential of AgNPs

The 800 nm light source for NIR stimulation was a K-Cube® laser (Summus Medical Laser Inc., Franklin, TN, USA), with a 1 cm beam diameter, operating under continuous wave (CW) mode and 5 W power (6.37 W/cm^2). The beam diameter completely covered a single well of a 96-well plate. To determine the heat generation, 200 µL of 0, 10, 25, or 50 µg/mL of AgNPs in water were placed into wells of a 96-well plate. Laser irradiation was applied to each well for 10, 20, 30, 40, 50, or 60 s. The initial and final temperatures were measured using a fiberoptic probe (Qualitrol Neoptix® (Fairport, NY, USA) and Nomad thermometer), and the temperature differences were plotted to generate a temperature increase curve over time.

2.3. Cell Culture and Cell Infection

Epithelial breast cells MCF 10A, and breast cancer cells MCF7 and MDA MB 231 were purchased from American Type Culture Collection (ATCC). MCF 10A cells were cultured in DMEM/F12 supplemented with 1% L-glutamine, 1% penicillin/streptomycin, 5% heat-inactivated horse serum, 20 ng/mL epidermal growth factor, 10 mg/mL insulin, 0.5 mg/mL hydrocortisone, 100 ng/mL cholera toxin, and 0.2 mg/mL gentamicin sulfate salt. MCF7 cells were cultured in EMEM supplemented with 1% L-glutamine, 1% penicillin/streptomycin, 10% fetal bovine serum, 10 mg/mL insulin, and 0.2 mg/mL gentamicin sulfate salt. MDA MB 231 cells were cultured in DMEM/F12 supplemented with 1% L-glutamine, 1% penicillin/streptomycin, 10% fetal bovine serum, and 0.2 mg/mL gentamicin sulfate salt. All cell lines were maintained at 37 °C in a 5% CO_2 humidified environment except during hyperthermia or photothermal treatments.

Based on our previously published histological results [11] demonstrating the presence of *Pseudomonas* in breast cancer cells from human patients, we utilized *P. aeruginosa* in a highly reproducible model of intracellular infection. *Pseudomonas aeruginosa* (ATCC 27853) was purchased from ATCC. Difco™ Luria–Bertani (LB) broth was purchased from Fisher Scientific. To infect the breast cells each of the cell lines was plated at a density of 200,000 cells in the media described above without penicillin/streptomycin or gentamicin sulfate (antibiotic free infection media) in two T25 culture flasks and incubated at 37 °C with 5% CO_2 overnight. One T25 flask of cells was trypsinized to count the number of breast cells, while the other flask was used for infection of the cells and experimental evaluation. *P. aeruginosa* was grown in LB broth overnight at 37 °C with shaking at approximately 160 rpm to facilitate aeration. The bacteria pellet was collected via centrifugation, resuspended in phosphate-buffered saline (PBS), and standardized using an optical density (OD) of 0.1 at 600 nm. The multiplicity of infection (MOI) of bacteria was 10:1, and bacteria were added to the cells in media without antibiotics and incubated at 37 °C for 2 h. Then, the bacteria-containing media was removed and the cells were washed twice with PBS. Media containing 0.2 mg/mL of gentamicin sulfate was added to kill any residual extracellular bacteria since gentamicin does not penetrate the eukaryotic cell membrane. After 1 h of incubation, the gentamicin-containing media was removed and cells were washed twice with PBS. Then, infected cells were cultured in their respective media containing penicillin/streptomycin and gentamicin sulfate, with the non-infected cells being cultured in the same media, although in a different room and incubator to minimize cross-contamination. All infected cell lines were maintained at 37 °C with 5% CO_2. Every week, bacterial colony-forming units (CFUs) were enumerated from the infected cells to quantify the extent of infection by lysing the breast cells for 45 min in water, serially diluting the lysate, plating onto LB agar plates, and counting the number of visible colonies present on agar plate the following day.

2.4. Cytotoxicity of AgNPs to Breast Cell Lines and P. aeruginosa

To evaluate the cytotoxicity of the AgNPs, each of the non-infected and infected cell lines was plated at a cell density of 60,000 or 200,000 cells per well in triplicate in 6-well plates. The cells were allowed to adhere for 24 h prior to incubation with AgNPs. Then, media was removed and the cells were incubated in fresh media containing 0, 10, 25, 50, 100, or 250 µg/mL of AgNPs at 37 °C for 2 or 24 h. Clinical hyperthermia treatments often use 30–120 min of elevated temperature. Even though the PTT treatment is rapid, since there are multiple plates of cells, we sought to standardize the timing of the AgNP exposure to 2 h, since this is the maximum time that cells would be exposed while also being exposed to brief laser irradiation. Cytotoxic agents are commonly evaluated following a 24 h period, or longer, of exposure. Therefore, it is prudent to evaluate cells' response to AgNPs alone, when PTT is not considered, since this time point will allow for uptake of the AgNPs and hence a potential modified cytotoxic response. The media containing AgNPs was removed and the cells were washed with PBS twice to remove extracellular AgNPs, incubated in fresh media without AgNPs for 24 h at 37 °C, then trypsinized and counted using a hemocytometer.

To evaluate AgNPs' toxicity to planktonic bacteria a single colony of *P. aeruginosa* was cultured in LB broth overnight at 37 °C and the suspended bacteria were standardized to an OD 0.1 at 600 nm. Then, 500 µL of the bacteria/ AgNPs (0, 10, 25, 50, 100 or 250 µg/mL) suspensions was added to 1 mL microcentrifuge tubes and placed on a tube revolver at 37 °C for either 2 h or 24 h, to measure the acute and prolonged exposure to AgNPs. At these timepoints 10 µL volumes were obtained, serially diluted, and plated on LB agar plates incubated at 37 °C for CFU enumeration the following day.

2.5. AgNPs to Augment Hyperthermia

As determined by the heat generation of AgNPs with 5W and 800 nm exposure, we sought to evaluate the response of the breast cells to the attained temperatures for 36 s (the same as the laser duration) in the presence of AgNPs. Starting with a baseline temperature of 37 °C, the laser alone (no AgNPs) attained 43 °C, and 10, 25, and 50 µg/mL of AgNPs attained 49, 56, and 68 °C, respectively. Each of the non-infected and infected cell lines were plated at a cell density of 250,000 cells per flask in eight T25 flasks and allowed to adhere for 24 h. Five T25 flasks cells were incubated with 4 mL of media without AgNPs and subjected to 37, 43, 49, 56, or 68 °C in a circulating water bath for 36 s. The remaining three T25 flasks were incubated with 4 mL of fresh media containing 10, 25, or 50 µg/mL of AgNPs for 2 h during which time cells in 10 µg/mL were exposed to 49 °C for 36 s, cells in 25 µg/mL were exposed to 56 °C for 36 s, and cells in 50 µg/mL were exposed to 68 °C for 36 s. After 2 h incubation, the media containing AgNPs was removed, each flask was washed twice with PBS, and the cells were incubated in media alone for 24 h at 37 °C, followed by trypsinization and counting using a hemocytometer.

Similarly, cell response to the lowest AgNPs' concentration (10 µg/mL, which is the lowest concentration needed to generate sufficient ablative temperatures with laser exposure) was evaluated at 37, 43, 49, 56, or 68 °C in a circulating water bath for 36 s. Each of the non-infected and infected cell lines was plated at a cell density of 250,000 cells per flask in five T25 flasks and allowed to adhere for 24 h. Then, media was removed and cells were incubated in 4 mL of fresh media containing 10 µg/mL of AgNPs for 2 h. During this incubation, one flask was maintained at 37 °C as a control group, and the other four flasks were treated at 43, 49, 56, or 68 °C by submersion in a circulating water bath for 36 s. After the 2 h incubation, the media was removed and each flask was washed twice with PBS. Cells were incubated in fresh media without AgNPs for 24 h at 37 °C, then were trypsinized and counted using a hemocytometer.

2.6. Photothermal Treatment of Breast Cells or P. aeruginosa

To measure the acute cellular response to photothermal therapy, each of the non-infected and infected cell lines was plated at a cell density of 5000 cells per well in 96-well plates and given 24 h to adhere. The media was removed and 200 µL of fresh media containing 0, 10, 25, or 50 µg/mL of AgNPs was added, and the plates were incubated for 2 h at 37 °C. Each well was exposed to 5 W of a 800 nm laser for 36 s once during the 2 h incubation. Following PTT the media containing AgNPs was removed, cells were washed twice with PBS, then incubated in media without AgNPs for 24 h. Twelve replicate wells per treatment group were used to have sufficient numbers for cell counting, thus these wells were trypsinized, combined, and counted using a hemocytometer. Cell viability was normalized to cells incubated without laser exposure or AgNPs.

Clonogenics assay was used to evaluate the long-term cell survivability following PTT by examining the number of colonies formed after AgNP-induced PTT. Non-infected and infected cell lines were plated at a density of 5000 cells per well in 96-well plates (24 wells for each concentration of AgNPs) and given 24 h to adhere. The media was removed and each well was incubated with 200 µL of media containing 0, 10, or 25 µg/mL of AgNPs. The plates were incubated for 2 h at 37 °C during which twelve wells were exposed to 5 W of 800 nm light for 36 s, while the remaining 12 wells were not exposed to laser light and served as controls for cells treated with AgNPs only, and not PTT. Following the 2 h incubation the media containing AgNPs was removed, cells were washed twice with PBS, trypsinized, and wells from the same treatment condition were combined, counted and seeded in triplicate at densities of 100, 500, 1000, or 5000 cells/well on 6-well plates. The cells were then cultured in fresh media for 7–14 days to allow colony formation, after which the media was removed, cells were washed twice with PBS, fixed with cold methanol, and stained with crystal violet. Breast cell colonies (>50 cells) were counted using a KEYENCE BZ-X810 Microscope. After seeding cells at certain densities on 6-well plates, all remaining infected cells were transferred into T25 flasks and allowed to expand until there was sufficient number to enumerate bacteria. Cells were trypsinized, counted, lysed, the lysate serially diluted and plated onto LB agar plates, and the number of visible colonies were counted the following day.

The response of *P. aeruginosa* to AgNP-induced PTT was evaluated using planktonic bacteria, not contained in breast cells, in an effort to understand the impact of PTT on the bacteria alone, knowing that thermal inactivation of *P. aeruginosa* is typically achieved at 60 °C [55]. A single colony of *P. aeruginosa* was cultured in LB broth overnight at 37 °C and the suspended bacteria were standardized to create treatment suspensions containing 0, 10, 25, or 50 µg/mL of AgNPs with OD 0.1 of bacteria. Then, 200 µL volumes of these solutions were added to wells of a 96-well plate, incubated at 37 °C before laser exposure, and exposed to 800 nm light at 5 W for 36 s. Immediately following PTT, 10 µL volumes were serially diluted and plated on LB agar for CFU enumeration the next day.

2.7. Statistical Analysis

All experiments were carried out in triplicate, except PTT of the breast cells, which was performed in replicates of 12, with pooling of the treated wells to harvest sufficient cells for counting using a hemocytometer. Data was normalized to control groups without AgNPs or laser exposure to obtain cell viability or surviving breast cells which formed colonies. Bacteria were grown from a single colony of *P. aeruginosa* and were standardized to OD 0.1. Bacterial CFUs/ cell were normalized to cells incubated with media alone without laser exposure or AgNPs. For the water bath hyperthermia experiments, cell viability was normalized to cells maintained in media alone at 37 °C. For cytotoxicity of the breast cells or *P. aeruginosa*, normalization was conducted with respect to cells maintained in media alone. Error bars indicate standard error of the mean. Comparison between groups was performed using one-way ANOVA to evaluate the effect of PTT, and also the effect of AgNPs plus hyperthermia delivered using a water bath. p-values less than 0.05 were deemed significant.

3. Results

3.1. Characterization of AgNPs

As shown in Figure 1A, the synthesized AgNPs exhibited a strong absorption at 800 nm, and the aqueous solution of AgNPs had a blue color. Dynamic light scattering revealed that the AgNPs had an average hydrodynamic diameter of 79 nm and zeta potential of +18 ± 2 mV, although there was a smaller population of nanoparticles with a size around 10 nm, most likely unreacted seed material (Figure 1B). Transmission electron microscopy imaging indicated the triangular shape of the AgNPs (Figure 1C). Under TEM the chitosan coating was visible under high magnification (not shown in the current image) as a fairly electron-translucent fibrous material, and the coating was thin. Hence, the TEM data correlates with the DLS sizing data. We have previously published that unreacted silver seed material can be observed in unpurified AgNPs, and the small peak in the DLS data correlated with the size of the seed material [44]. Purification of the AgNPs involves centrifugation and the small nanoparticles are not able to be removed except under ultracentrifugation. Chitosan was used as the stabilizing coat for the AgNPs, as it has been used previously, and because it is antimicrobial; however, since only a thin coating of chitosan has been observed on the AgNPs using our synthesis, the cytotoxicity of the AgNPs against bacteria is due to the release of silver ions [44].

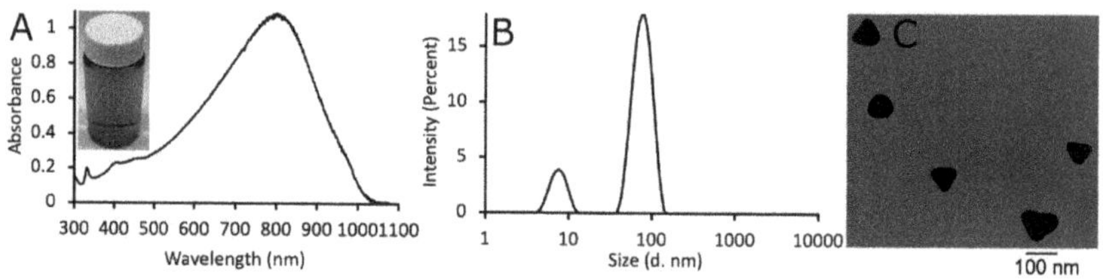

Figure 1. (**A**) Absorption spectrum of AgNPs in water. (**B**) Dynamic light scattering data of AgNPs in water. (**C**) Transmission electron microscopy image of AgNPs.

Upon exposure to the 800 nm laser, AgNPs generated heat due to their inherent plasmon resonance [44,56,57]. The temperature differences of 200 µL of 0, 10, 25, and 50 µg/mL AgNPs exposed to 5 W of 800 nm laser irradiation for 10 to 60 s were measured to generate a temperature increase curve (Figure 2). It has previously been determined that 180 joules of energy was effective for photothermal ablation experiments [44], where longer times and lower power were used. For example, 10 µg/mL AgNPs exposed to 3 W of 800 nm for 60 s resulted in a ΔT = 8 °C. For the current work we sought to apply 180 J using a more intense laser power and shorter time, since this will minimize the time for thermal transfer when the therapy is evaluated in vivo; thus, 5 W and 36 s was applied to achieve 180 J. Laser application only, without AgNPs, only induced mild hyperthermia (37 °C + 6 °C = 43 °C). For 36 s, the temperature increases for 0, 10, 25, and 50 µg/mL of AgNPs were 6, 12, 19, and 31 °C, respectively. Considering that cells and bacteria have an initial temperature of 37 °C for all experiments, then the minimal amount of AgNPs for the ablation of cancer cells is 10 µg/mL, since 37 °C + 12 °C = 49 °C, which is above the irreversible protein denaturation temperature of 45 °C [58]. The higher laser power generates a 1.5 times higher temperature than using a lower power and a longer time, even for the same concentration of AgNPs and laser fluence (180 J). Higher concentrations of AgNPs (100 and 250 µg/mL) were also considered for PTT; however, both concentrations resulted in boiling of the water, leading to inaccurate thermal measurements. Since lower concentrations of AgNPs were effective for heat generation they were used in subsequent PTT experiments against bacteria and breast cells.

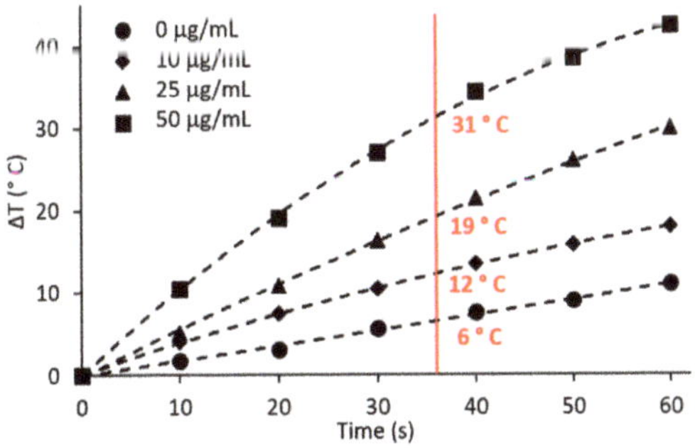

Figure 2. Temperature change versus laser irradiation (5 W, 800 nm) time for various concentrations of 200 μL of AgNPs in water.

3.2. Cytotoxicity of AgNPs

One of the unique discoveries in this work is the difference in doubling time between infected (MCF 10Ai, MCF7i, MDA MB 231i) and uninfected cells: MCF 10A and MCF 10Ai had the same doubling time, whereas MCF7 cells took 62 h, and MCF7i took only 48 h. There was a difference in MDA MB 231 cells also, although not as profound, with non-infected cells doubling in 26 h and infected cells taking 20 h. The initial intracellular infection rates before experimentation with AgNPs were 0.008 CFUs/cell for MCF 10Ai, 0.236 for MCF7i, and 0.5 for MDA MB 231i. Given the number of breast cells that were lysed to quantify the CFUs/cell, we determined that 1 out of every 10,000 non-tumorigenic cells was infected, whereas 1 out of every 100 tumorigenic cells was infected. These results indicate differences between naïve and tumorigenic cells, and also that different cell lines harbor more or less bacteria.

We first evaluated AgNPs in the absence of laser stimulation via a 2 h incubation of the cells with Ag NPs at 37 °C. Since 10, 25, and 50 μg/mL were sufficient for generating heat with laser stimulation cells or bacteriawere exposed to these concentrations for 2 h and the cell viability, or CFUs, were determined, as shown in Figure 3. We also examined the 24 h exposure of the cells or bacteria to AgNPs and the results are shown in Figure 4. Bacterial CFUs following 2 h exposure to AgNPs is shown in Figure 5A, and their exposure for 24 h is shown in Figure 5B. Figure 3 reveals the viability measured 24 h after a 2 h exposure to various concentrations of AgNPs at 37 °C. MCF 10A and MCF7 showed a progressive decrease in cell viability with increasing concentrations of AgNPs. Cells exposed to 10 μg/mL for 2 h had minimal reductions in cell viability, except MCF7 and MCF7i, which had an approximately 20% reduction. At 10 μg/mL, MDA MB 231 had no change in cell viability. All three cell lines were sensitive to 25 μg/mL of AgNPs, resulting in a 30–50% decrease in viability. The effect was more pronounced at 50 μg/mL, with a 36–75% decrease. There were minor variations between infected and non-infected cells, except MCF7, which had infected cells demonstrating more sensitivity (20% more at 25 μg/mL, or 5% more at 50 μg/mL). Even at the highest concentration, 50 μg/mL, MDA MB 231 had greater than 50% cell viability. For MDA MB 231 cells exposed to 50 μg/mL of AgNPs for 2 h, there was an additional 10.9% decrease in the viability of non-infected cells compared to infected cells. MDA MB 231 cells were resistant to the acute AgNPs exposure, with the MDA MB 231i being most resistant, and only exhibiting a 36.3% reduction in viability. An interesting observation is that MDA MB 231i cells are consistently less sensitive to AgNPs than their non-infected cellular counterpart, and also compared to normal breast epithelial, and non-triple-negative breast cancer cells, as determined in both the acute and prolonged exposure to AgNPs.

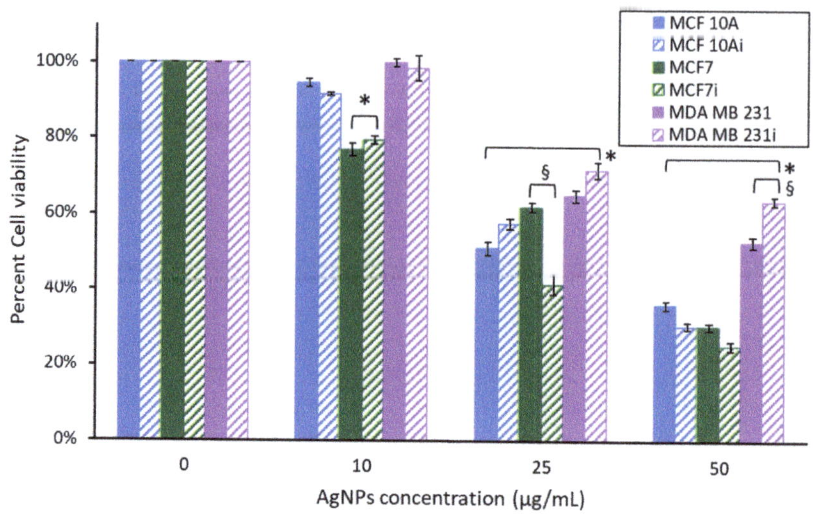

Figure 3. Cell viability of breast cell lines measured 24 h after 2 h of exposure to various concentrations of AgNPs. * indicates a statistical ($p < 0.05$) difference from the control group with 0 µg/mL of AgNPs, and § indicates a difference between non-infected and infected cells.

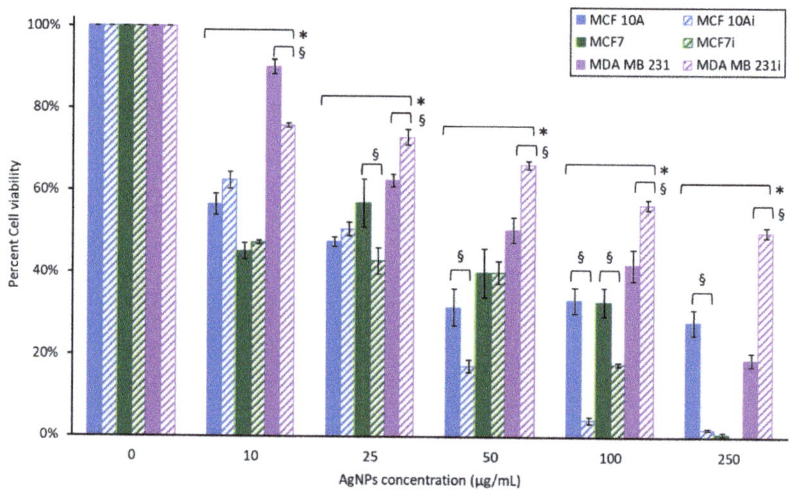

Figure 4. Cell viability of breast cell lines exposed to various concentrations of AgNPs for 24 h. * indicates a statistical ($p < 0.05$) difference from the control group with 0 µg/mL of AgNPs, and § indicates difference between non-infected and infected cells.

Figure 4 demonstrates the response of non-infected and infected breast cell lines to various concentrations of AgNPs after a 24 h exposure. The initial 2 h exposure to AgNPs was carried out at lower concentrations, which are optimal for PTT. Exposure to toxins for 24 h is a more robust approach, and we also evaluated the use of higher concentrations that might be relevant for killing intracellular bacteria in the absence of PTT. All cell lines had decreased cell viability with increasing concentrations of AgNPs. For MCF 10A, at 50 µg/mL and above, non-infected MCF 10A cells were more resistant to AgNPs than MCF 10Ai. The response of non-infected MCF7 and MCF7i were similar. MCF7 cells were the most sensitive to AgNPs at high concentrations, with nearly complete death at 250 µg/mL. MDA MB 231 was the most resistant to AgNPs, and MDA MB 231i was more resistant than non infected cells. Even at the highest concentration 250 µg/mL, and after 24 h of exposure to AgNPs, MDA MB 231i had 50% cell viability and was 2.5 times more resistant than non-infected cells. Even the low dose 10 µg/mL AgNPs generated a 38–65% reduction in viability for MCF 10A and MCF7 cells, whereas MDA MB 231 cells only had a 10–25% decrease. At lower concentrations there was no difference between infected and non-infected MCF 10A and MCF7 cells, although at higher concentrations the infected cells were more sensitive to AgNPs. At 50 µg/mL MCF 10 Ai cells were almost doubly susceptible to AgNPs. At 100 µg/mL both MCF 10Ai and MCF7i cells were four and two times more sensitive to AgNPs compared to non-infected cells, respectively. At 250 µg/mL the MCF 10Ai cells were 15 times more sensitive, while MCF7, infected or not, was completely eradicated.

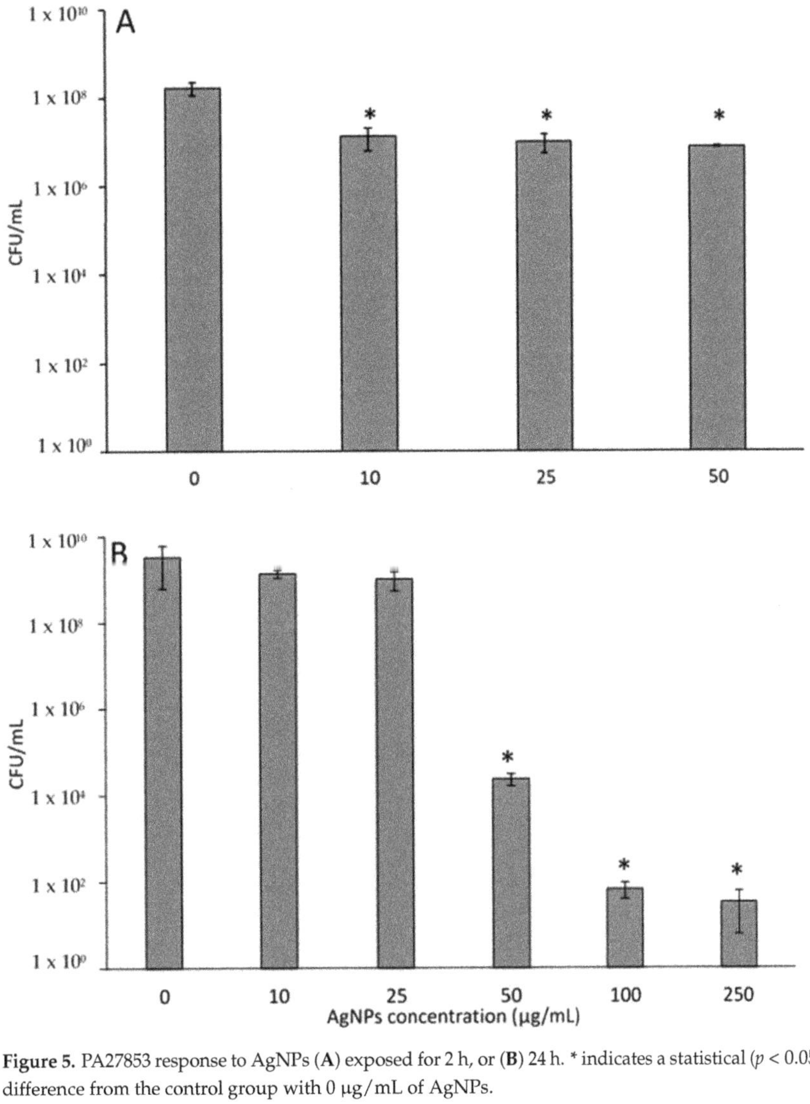

Figure 5. PA27853 response to AgNPs (**A**) exposed for 2 h, or (**B**) 24 h. * indicates a statistical ($p < 0.05$) difference from the control group with 0 µg/mL of AgNPs.

Figure 5A shows the toxic effects of AgNPs on bacteria following a 2 h exposure. Planktonic *P. aeruginosa* (not intracellular) experience minimal reductions in viability, with a maximum of about 1.5 log reduction. Interestingly, there was no difference between bacteria treated with 10, 25, or 50 μg/mL AgNPs, indicating that additional AgNPs did not increase bacterial reduction. Prolonged exposure of *P. aeruginosa* to AgNPs for 24 h leads to significant 5 to 8 log reductions, as shown in Figure 5B. High concentrations of AgNPs and a prolonged time of exposure are needed to effectively eliminate planktonic bacteria.

3.3. Cell Sensitivity to AgNPs with Hyperthermia

A unique feature of metal nanoparticles is their thermal conductivity, allowing them to facilitate heat transfer to cells even in the absence of laser stimulation. Forty-three degrees is the equivalent temperature increase of the laser alone without AgNPs. Temperatures of 49, 56, and 68 °C were obtained for AgNPs of 10, 25, and 50 μg/mL exposed to 36 s of 5 W, 800 nm light. To evaluate the sensitivity of cells to brief heating in the presence of AgNPs (no laser stimulation), infected or non-infected cells were exposed to 10 μg/mL AgNPs and heated for 36 s in a water bath (the same time of laser application needed to reach ablative temperatures). Cellular responses measured 24 h after 2 h of exposure to 10 μg/mL of AgNPs and 36 s of treatment with various temperatures are shown in Figure 6. In general, breast cells exhibited progressive reductions in cell viability with increasing temperature. Between 49–68 °C MCF 10Ai had a minimal reduction in viability, indicating that MCF 10Ai may be more resistant to hyperthermia. There were statistical differences between infected and non-infected groups at elevated temperatures with 10 μg/mL of AgNPs. Although statistically significant, there were only minor differences between MCF7 and MCF7i cells at 49, 56, and 68 °C with the addition of AgNPs. MCF7 cells are more sensitive to brief (36 s) exposure to 43 °C (Figure 6), with a 15–23% reduction, but looking back at Figure 3 this effect is indicative of their sensitivity to AgNPs alone, not heat. Non-infected MDA MB 231 cells were not susceptible, although the infected cells were, with a 22% decrease in cell viability with 43 °C and 10 μg/mL of AgNPs, which is substantially different (22% greater reduction) than treatment at 37 °C, which only had a 1.4% decrease. Examining the data on elevated temperatures and no AgNPs (from Figure 7) for cells exposed to 43 °C for 36 s, there were no differences for MCF 10A or MDA MB 231, but there was an 11% increased reduction in cell viability for MCF7 cells with 10 μg/mL AgNPs. MCF7i cells had a 6.5% increase in viability, which is not significant. Comparing brief exposure to elevated temperatures (36 s) with or without exposure to AgNPs for 2 h, there were some differences, as identified in Table 1, which shows the viability comparisons. The effect of AgNPs is variable for MCF 10A and MCF7, whether infected or not, whereas MDA MB 231 cells, infected or not, showed decreased viability when AgNPs were added to hyperthermic temperatures. Recalling from Figure 3 that MDA MB 231i cells were most resistant to AgNPs, they are more sensitive to the effects of AgNPs at elevated temperatures. The results of Table 1 indicate that AgNPs at 10 μg/mL augment hyperthermia at lower temperatures for MCF 10A, MCF 10Ai, MCF7, and MCF7i. At 68 °C though, there is no difference, or an increase in viability, for these cells, which indicates an unexpected protective effect of the nanoparticles. MDA MB 231, infected or not, had decreases in cell viability for all temperatures with the addition of 10 μg/mL AgNPs.

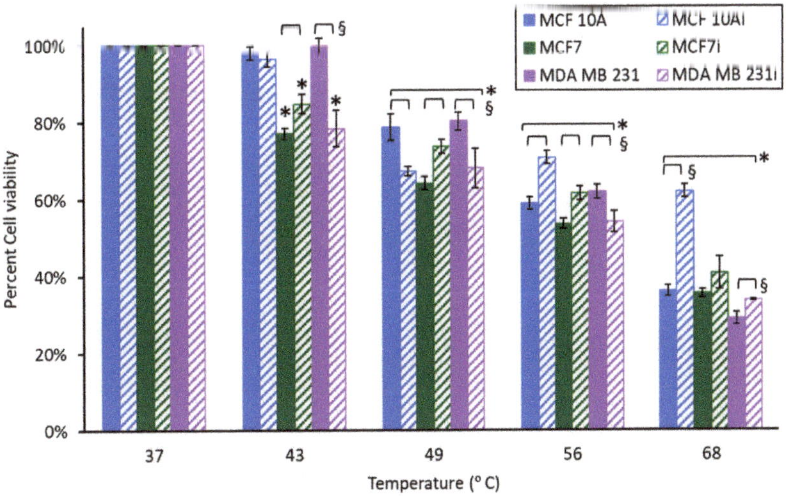

Figure 6. Cell viability of breast cell lines measured 24 h after 2 h of exposure to 10 µg/mL of AgNPs and 36 s of exposure to elevated temperatures (using a water bath). * indicates a statistical ($p < 0.05$) difference from the control group with 0 µg/mL of AgNPs, and § indicates difference between non-infected and infected cells.

Table 1. AgNPs (10 µg/mL) can augment hyperthermia. I means increase, D means decrease, and ND means no statistical difference.

	Temp (Deg. C)	Cell Viability No AgNPs	Cell Viability with AgNPs	Increase/ Decrease
MCF 10A	49	79.1	78.6	D
	56	69.8	58.8	D
	68	29.5	36.1	I
MCF 10Ai	49	77.2	67.2	D
	56	57.3	70.7	I
	68	21.7	61.8	I
MCF7	49	74	64.1	D
	56	61.6	53.5	D
	68	35.2	35.3	ND
MCFi	49	74.1	73.7	ND
	56	44.5	61.5	I
	68	27.1	40.7	I
MDA MB 231	49	101.2	80.2	D
	56	71.9	61.9	D
	68	48.9	28.8	D
MDA MB 231i	49	101.1	68	D
	56	102.5	54.1	D
	68	41.5	33.7	D

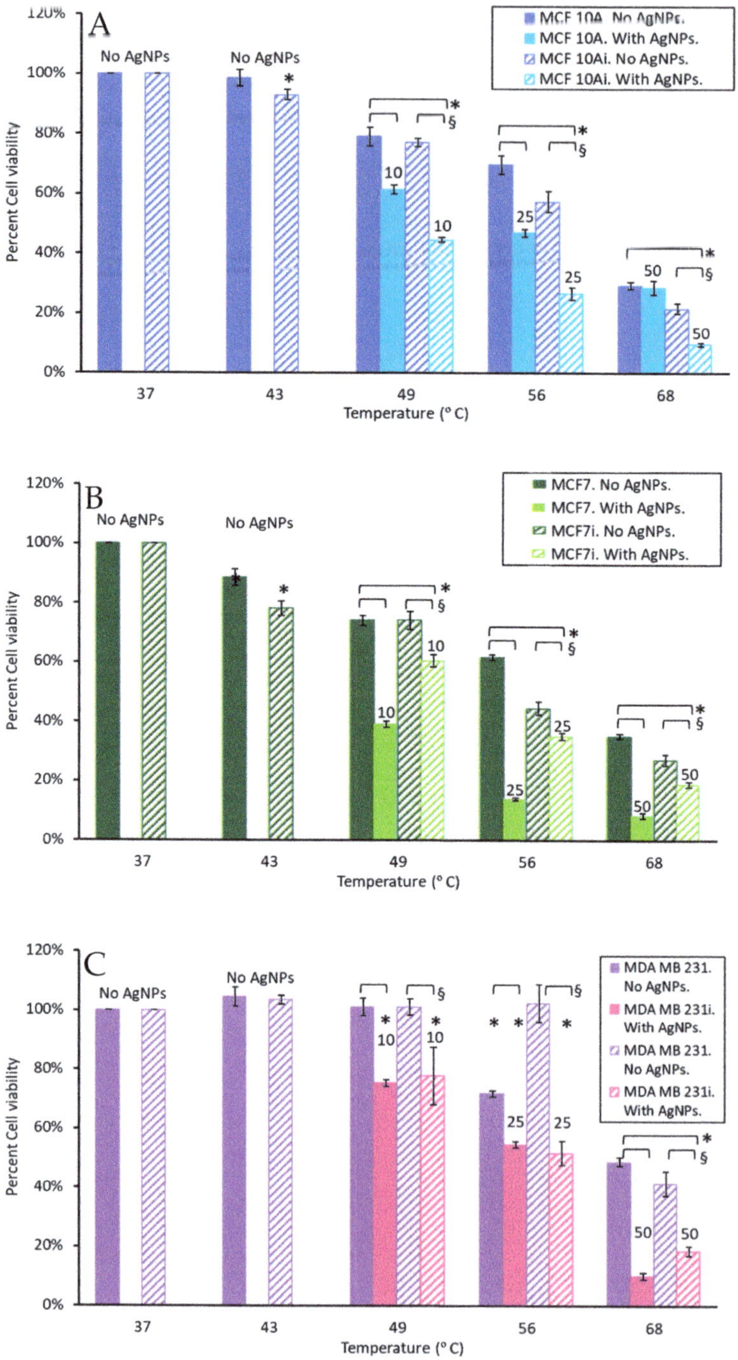

Figure 7. Cell viability of breast cell lines (**A**) MCF 10A, (**B**) MCF7, and (**C**) MDA MB 231 measured 24 h after a 36 s exposure to hyperthermia (using a water bath). * indicates a statistical ($p < 0.05$) difference from the control group with 0 µg/mL of AgNPs, and § indicates difference between non-infected and infected cells.

As shown in Figure 2, the temperature increases of 0, 10, 25, and 50 µg/mL AgNPs with 5 W of 800 nm laser irradiation for 36 s were 6 °C, 12 °C, 19 °C, and 31 °C, respectively. Accordingly, the final temperatures in the photothermal ablation experiments were 43 °C (37 °C + 6 °C), 49 °C, 56 °C, and 68 °C. To evaluate cell response to hyperthermia with AgNPs only (without laser stimulation) each of the cell lines incubated with 0, 10, 25, or 50 µg/mL AgNPs for 2 h was exposed to temperatures of 49, 56, or 68 °C for 36 s using a water bath. The cell viability measured 24 h after hyperthermia is shown in Figure 7. As expected, breast cells showed a progressive decrease in viability with increasing temperature, and cells incubated with AgNPs exhibited greater reductions in viability. Especially for MCF 10Ai (Figure 7A), non-infected MCF7 (Figure 7B), and non-infected MDA MB 231 (Figure 7C), AgNPs led to significant decreases in cell viability. From Figure 7, without AgNPs, in MCF 10A cells treated with 36 s of 49 °C there was an approximate 20% reduction in cell viability, and MCF7 had a 26% greater reduction with the addition of just 49 °C for 36 s, regardless of whether the cells were infected or not. However, MDA MB 231 cells, infected or not, had no reduction in viability. Comparing the data of Figures 3 and 7A for 10 µg/mL, a brief exposure of 36 s to 49 °C resulted in a drastic reduction in viability, where MCF 10A had a 33.3% additional reduction, and MCF 10Ai cells had an additional 47% reduction compared to cells just treated with AgNps at 37 °C. MCF7 cells had a 37.9% reduction and MCF7i cells had a 19.1% additional reduction when treated with 49 °C for 36 s plus 10 µg/mL of AgNPs compared to treatment with nanoparticles only. MDA MB 231 cells had a 24.7% additional reduction and MDA MB 231i cells had a 20.8% additional reduction. MCF 10Ai cells were most sensitive to both AgNPs and 49 °C. Although brief 49 °C exposure and AgNPs were beneficial together for non-infected cells, the MCF7i and MDA MB 231i cells were less susceptible than their non-infected counterparts. From Figure 7, brief 56 °C for 36 s reduced the viability of all the cell types by 28–55%, with the exception of MDA MB 231i cells which had no significant change in viability. MCF 10A cells had a further reduction of 4.1%, and MCF 10Ai cells had a further reduction of 30.8% with the addition of 56 °C for 36 s to 25 µg/mL of AgNPs (comparing Figures 3 and 7). MCF7 cells had a further reduction of 47.9%, and MCF7i cells had a further reduction of 6.5% with the addition of 56 °C to 25 µg/mL of AgNPs. MDA MB 231 cells had a further reduction of 10.4%, and MDA MB 231i cells had a further reduction of 10.9% with the addition of 56 °C plus 25 µg/mL of AgNPs. Brief exposure to 36 s at 68 °C reduced all cells' viability, with 70–80% for MCF 10A and 65–73% for MCF7, whether infected or not, but only 51% for MDA MB 231 and 58.5% for MDA MB 231i cells. There is a benefit to adding 68 °C to 50 µg/mL of AgNPs; it resulted in a 6.1–22% additional reduction for MCF 10A and MCF7 cells, infected or not. There is a greater benefit for MDA MB 231, and their infected cells, which had additional 42.6 and 45.1% reductions when 68 °C is combined with 50 µg/mL of AgNPs.

There are interesting observations of reduced viability between infected and non-infected cells with hyperthermia augmented by AgNPs. In Figure 7A, MCF 10A cells have 17.7, 22.9, and 0.9% additional reductions in viability and MCF 10Ai cells have 32.6, 30.6, and 12% additional reductions in viability when AgNPs at 10, 25, and 50 µg/mL are added to 49, 56, or 68 °C for 36 s. There was a 12.5% further reduction in MCF 10Ai cells treated with 56 °C and 25 µg/mL compared to MCF 10A cells, but there was only a 7.8% reduction between infected and non-infected cells when 50 µg/mL was added to 68 °C. The major benefit is that there is no statistical difference in non-infected MCF 10A cells and infected cells with hyperthermia alone, but AgNPs resulted in an additional 16.8–20.2% reduction in cell viability. This result demonstrates that AgNPs augment hyperthermia to a greater extent in the infected non-tumorigenic cells. There was no difference between MCF7 and MCF7i cells when treated with 49 °C for 36 s in the absence of AgNPs. However, there was a more profound reduction in MCF7i cells compared to non-infected cells when treated with 36 s of 56 and 68 °C, resulting in 17.1 and 8.1% increased reductions with AgNPs. As shown in Figure 7B, there were bigger reductions in non-infected MCF7 cells (34.9, 47.7, and 26.8% reductions) when 25 or 50 µg/mL of AgNPs were added to 36 s of 49, 56, and

68 °C, compared to infected cells (13.7, 9.5, and 8.2%). AgNPs seem to protect infected cells from thermal damage in this case, compared to the non-infected cells. As shown in Figure 7C, for MDA MB 231 or MDA MB 231i cells treated with 49 °C for 36 s there was no change in viability in the absence of AgNPs. There is a reduction in viability with the addition of 10 µg/mL of AgNPs due to the combination of heat and nanoparticles, although there was no difference between non-infected and infected cells. MDA MB 231 cells treated with 56 °C for 36 s had a 17.3% decrease with the addition of 25 µg/mL of AgNPs for non-infected cells compared to the infected cells, which had a 50.7% decrease. When cells were treated with 68 °C for 36 s the non-infected MDA MB 231 cells had a 38.7% reduction and the infected cells had a 22.9% reduction with the addition of 50 µg/mL of AgNPs. Infected MDA MB 231 cells either had no reduction, or an increase in viability, when 10, 25, or 50 µg/mL of AgNPs were added to 49, 56, or 68 °C. This indicates that AgNPs may not be highly beneficial for augmenting brief hyperthermia (here using a water bath) when MDA MB 231 cells are infected.

3.4. Cell Viability to Photothermal Treatment

Each of the breast cell lines incubated with 0, 10, 25, or 50 µg/mL of AgNPs for 2 h was exposed to 5 W of a 800 nm laser for 36 s to induce ablative (>45 °C) temperatures. The cellular response 24 h after photothermal treatment is presented in Figure 8. An obvious reduction in cell viability was observed at 10 µg/mL AgNPs, although infected cell lines were more resistant to AgNPs-induced PTT than non-infected cell lines, with a 22–33% enhanced viability. Nonetheless, there was a profound reduction in viability with non-infected MCF 10A, MCF7, and MDA MB231 cells having 78.2, 88.7, and 94.9% reductions with 10 µg/mL of AgNPs-induced PTT. The infected lines had 45, 66.7, and 68.1% reductions in viability for MCF 10Ai, MCF7i, and MDA MB 231i cells. PTT with AgNPs at higher concentrations resulted in nearly complete cell death for both non-infected and infected breast cells. All infected cells were more resistant to photothermal therapy with 10 µg/mL of AgNPs compared to their non-infected counterparts. There was less than 10% cell viability for all lines, infected or not, at 25 µg/mL, and no surviving cells with 50 µg/mL of AgNPs plus laser radiation.

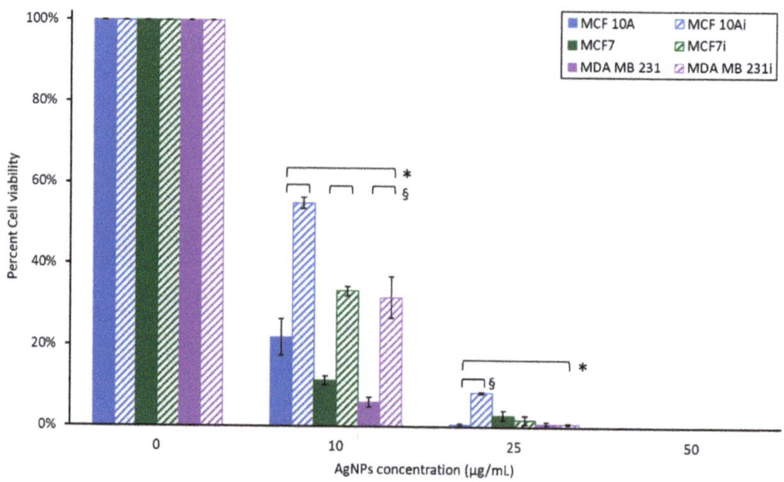

Figure 8. Cell viability of breast cell lines measured 24 h after photothermal treatment. * indicates a statistical ($p < 0.05$) difference from the control group with 0 µg/mL of AgNPs, and § indicates difference between non-infected and infected cells.

Figure 9 shows the response of planktonic *P. aeruginosa* to AgNP-induced PTT. There is a 0.3 log reduction with 10 µg/mL of AgNPs and laser stimulation, and a greater than 1 log reduction with 25 μg/mL and laser stimulation, although complete ablation of planktonic bacteria with 50 μg/mL and laser was observed. The result makes sense since temperatures above 65 °C are often needed to eradicate bacteria and PTT using 50 μg/mL resulted in a ΔT = 31 °C, so considering that bacteria had an initial temperature of 37 °C, a temperature of 68 °C in the bacterial solution would be attained. However, it was not expected that such a rapid application of hyperthermia for 36 s could induce a maximum temperature of 68 °C; an average of 52.5 °C over the course of laser exposure for 36 s would be effective [59]. This result may be due to the proximity of the AgNPs to free-floating bacteria in solution.

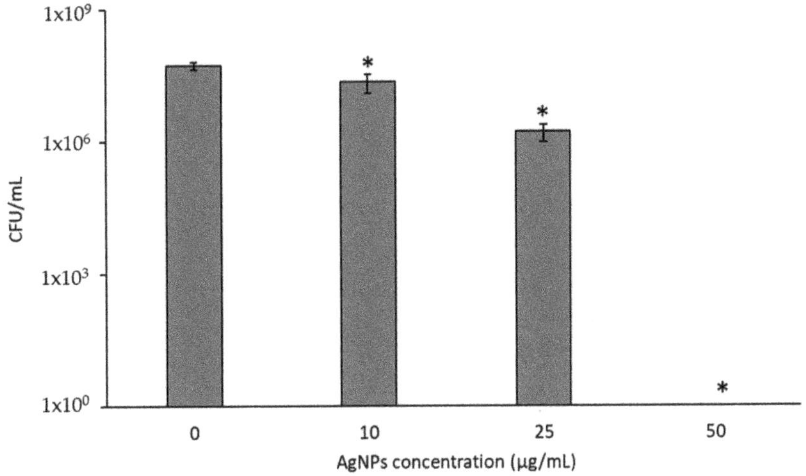

Figure 9. CFUs/ mL of *P. aeruginosa* following photothermal therapy. * indicates a statistical ($p < 0.05$) difference from the control group with 0 μg/mL of AgNPs.

Figure 10 shows the bacterial CFUs/cell for infected breast cells. MCF 10Ai cells had an infection of 0.0081 CFUs/cell with no treatment, MCF7i had 0.236 CFUs/cell, and MDA MB 231i had 0.5 CFUs/cell, which shows that breast cancer cells may harbor more intracellular bacteria than non-tumorigenic cells. For MCF 10A cells there was a statistically significant increase (80.2%) in CFUs/cell with laser only and no AgNPs, where only 43 °C was attained. When 10 μg/mL of AgNPs was used to generate a temperature of 49 °C there was no difference in CFUs/cell, with both having approximately 0.014 CFUs/cell. When 25 μg/mL of AgNPs was used with MCF 10Ai cells to generate a temperature of 56 °C there was a 100% reduction in CFUs/cell, even though there was a sufficient number of cancer cells to count since the cells treated with PTT were allowed to regrow following treatment to allow a sufficient number of cells for lysis and bacterial enumeration. This indicates that the cells that survive ablation with 25 μg/mL do not harbor bacteria. The same trend was observed in MDA MB 231i cells treated with 25 μg/mL of AgNPs and laser stimulation. For MCF7i cells treated with 36 s of 5 W and 800 nm light, without AgNPs, for a volumetric temperature change of 43 °C, there was a 39.4% reduction in CFUs/cell. This trend was augmented when 10 μg/mL of AgNPs plus laser stimulation for 36 s was applied, resulting in a 42.6% reduction in CFUs/cell. There was a slightly higher incidence of CFUs/cell when MCF7i cells were treated with 25 μg/mL, with only a 17.3% reduction in CFUs/cell with laser compared to without. MDA MB 231 cells had a 69.4% reduction in CFUs/cell with laser alone and no AgNPs, and 89.8 and 100% reductions in CFUs/cell with 10 and 25 μg/mL AgNPs plus 36 s of laser exposure to generate 49 and 56 °C. These results confirm that AgNP-induced PTT can reduce or eliminate intracellular bacteria in cells that proliferate following PTT.

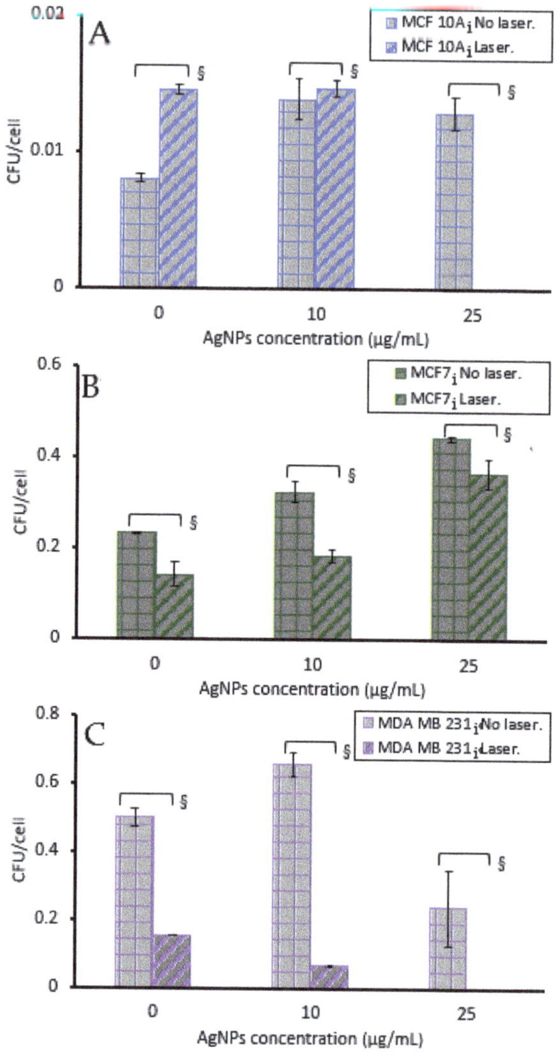

Figure 10. Bacteria colony-forming units per cell of infected (**A**) MCF 10A, (**B**) MCF7, and (**C**) MDA MB 231. Each of the cell line was incubated with various concentrations of AgNPs for 2 h and treated without/with 5 W of a 800 nm laser for 36 s. § indicates a statistical ($p < 0.05$) difference between laser and no laser.

Breast cell clonogenic results shown in Figure 11A illustrate that without laser stimulation, non-infected and infected breast cells incubated with 0, 10, or 25 µg/mL of AgNPs alone for 2 h had only slight reductions in the number of surviving cells that could form colonies. The infected cells had more breast cell colonies than their corresponding non-infected cells which supports the hyperthermia results noted earlier. The infected MCF 10A and MCF7 cells had increased survivability, with no loss in response to exposure to AgNPs for 2 h. Although the acute study results in Figure 3 show that for MCF 10A cells, infected or not, there was an approximately 45% decrease in cell viability with 25 µg/mL, the clonogenics assay demonstrates that the MCF 10A cells recover well from AgNPs exposure, and they recover better than either cancer cell line. Non-infected MCF7 cells had the greatest reduction in survivability with 25 µg/mL of AgNPs, which correlates with the acute cytotoxicity assay data of Figure 3; nonetheless, the cells recovered well from exposure to AgNPs. The acute data indicated that MDA MB 231i cells were most resistant to AgNPs, and the clonogenics assay results demonstrates these cells have minimal struggle in recovering from AgNPs exposure.

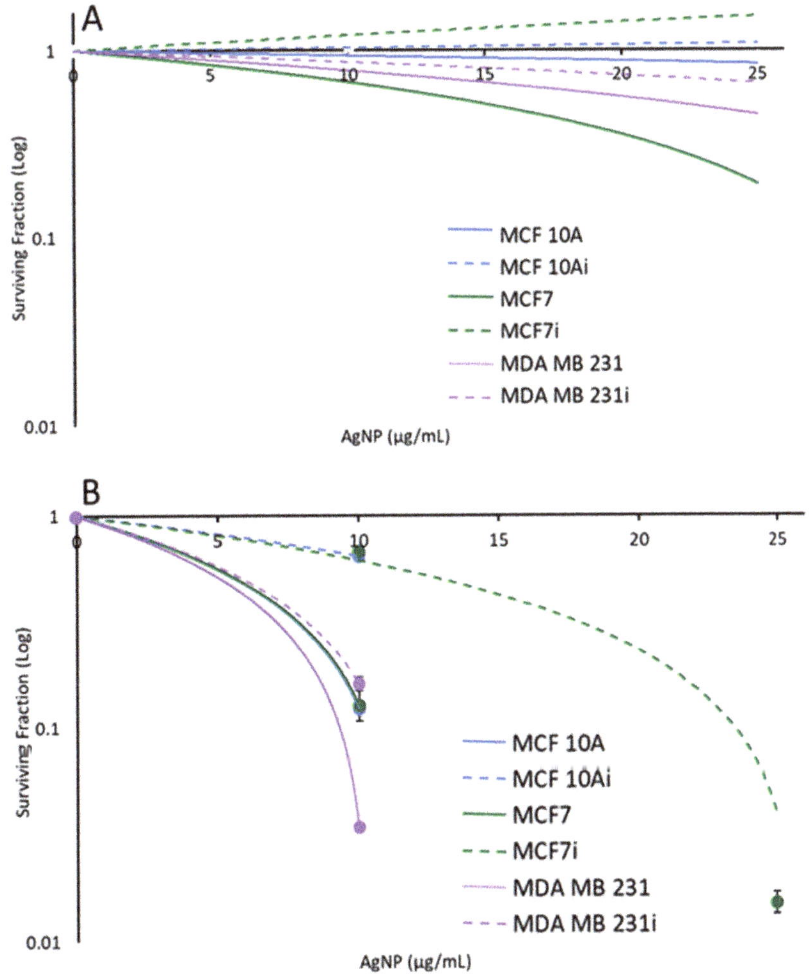

Figure 11. Semi-log plots of the survival fraction of breast cell lines treated with different concentrations of AgNPs. (**A**) In the absence of laser, cells incubated with AgNPs alone had relatively slight changes in the number of colonies formed. (**B**) Upon exposure to laser, colony numbers decreased significantly with increasing AgNPs' concentration.

As shown in Figure 11B, upon exposure to laser stimulation, breast cell colony numbers decreased with increasing AgNPs' concentration (and hence increased temperature). For MCF 10A and MCF 10Ai, MDA MB 231 and MDA MB 231i, and non-infected MCF7, 25 µg/mL AgNPs with the laser induced sufficient heat to cause no colony formation (thus no datapoints). A significant reduction in colony number was also observed for MCF7i under this condition. At 10 µg/mL, infected breast cells had a greater number of surviving cells to form colonies than their corresponding non-infected cells for all three cell lines, in agreement with the data from Figures 3 and 5, indicating that infected breast cells were more resistant to AgNPs-induced hyperthermia than non-infected breast cells. Infected and non-infected cells were all susceptible to PTT using 10 or 25 μg/mL of AgNPs. Clonogenics data confirm the same trend as observed in Figure 8 where cell viability was measured 24 h after PTT, with the infected cells being more resistant to PTT. There was a substantial decrease in survivability for all cells with 10 μg/mL of AgNP- induced PTT, with MDA MB 231 cells being most sensitive, and this confirms the results of Figure 8 where MDA MB 231 cells were the most sensitive to AgNP- induced PTT. The conclusion of the acute PTT results (Figure 8) and cell regrowth potential (Figure 11B) support the premise that although infected breast cells are resistant to lower doses of AgNPs for PTT, higher doses are most suitable for completely eliminating both infected and non-infected cancer cells.

4. Discussion

We hypothesize that the intracellular infection stresses the cells, in ways that have yet to be elucidated, and the stress makes the cells more susceptible to both heat and AgNPs. A phenomenon that has yet to be considered is that dying cancer cells may release potentially viable bacteria into the extracellular space, hence allowing for the subsequent infection of other cells; therefore, PTT may be beneficial since it can kill bacteria during the process of ablating the cancer cells. As demonstrated in our results, cells with intracellular infection can be less susceptible to AgNPs than hyperthermia, although the combination of both, and especially when AgNPs are used for PTT, can be beneficial. The results suggest that a lower incidence of intracellular infection would benefit chemotherapeutic treatments of breast cancer. Both breast cancer cell lines harbor a higher number of CFUs per cell than the non-tumorigenic cell line. Non-tumorigenic cells had 1 out of every 10,000 cells infected, but tumorigenic cells had a higher number of infected cells infected; 1 out of every 100. The current clinical literature has indicated that bacteria survive in the intracellular environment; however, this clinical incidence has not been quantified. Further information is needed to determine the number breast cancer cells that harbor intracellular bacteria in clinical samples. Our results here demonstrate that intracellular infection can be reproduced in a laboratory setting, and higher or lower levels of infection may be possible depending upon the bacterial strain used and the host cell's line. Further information is needed to determine why this trend occurs, and if all cancer is more susceptible to harboring intracellular bacteria and why. Antibiotics may not be effective against breast cells harboring intracellular infection since they can induce drug-resistant bacteria, and many antibiotics cannot enter eukaryotic cells to tackle intracellular pathogens; however, AgNPs are toxic to both cancer cells and bacteria. *P. aeruginosa* is an opportunistic pathobiont, and possibly an oncomicrobe, with the capacity to infect both non-tumorigenic and tumorigenic breast cells. A limitation of this study is that the exploratory nature of this work does not consider immune cell response to intracellular infection, and especially those cells that undergo thermal necrosis with the potential release of surviving bacteria, since both mechanisms would induce immunostimulatory effects. There are also other types of bacteria associated with breast cancer that can be used for intracellular infection, including *Fusobacterium nucleatum* and *Staphylococcus aureus* [10,60,61]. Our work uses *P. aeruginosa*, based on our previously published clinical findings from human breast cancer samples [11]. A key observation of our work is the hyperproliferation seen only in the cancer cells, with both MCF7i and MDA MB 231i cells having 22.6 and 23.1% faster doubling times compared to their non-infected counterparts. Our work supports the work of others, where pancreatic and colorectal cancer cells with intracellular infection demonstrated hyperproliferation [60,62]. The previously published works only evaluated hyperproliferation in cancer cells using colorimetric or DNA-based assays (MTS, MTT, CCK-8, live/dead, and resazurin staining); since the assays cannot differentiate between bacterial and cell viability the results may conflate the trend of hyperproliferation. Our study confirms that hyperproliferation occurs, as quantified by manual cell counting. Direct counting of viable cells is currently the only way to ensure that results between infected and non-infected cells can be compared. Furthermore, we demonstrated that non-tumorigenic MCF 10A cells can be infected at a lower level than in the breast cancer cell lines, with no observation of hyperproliferation, a result that has not been demonstrated before. We also had a 29.1 fold higher amount of CFUs/ cell for MCF7 and 61.7 fold higher amount of CFUs/ cell for MDA MB 231, respectively, compared to MCF 10A. Compared to Yu et al. [62], in which *E. coli* was used to infect colorectal cancer cells, when we used *P. aeruginosa* at the same MOI of 10:1, we found about the same level of infection: Yu et al. found 1×10^{-1} CFUs/cell, and we found 2×10^{-1} for MCF7 and 5×10^{-1} CFUs/cell for MDA MB 231.

Considering that most of the AgNPs used extracellularly for PTT will not achieve high concentrations near intracellular bacteria, it may be concluded that AgNPs do not kill bacteria via the known mechanisms of oxidative stress [63]. There might be a benefit to adding AgNPs to hyperthermia treatments to decrease cell viability in infected cells. One limitation of the current cytotoxicity studies is in determining the amount of AgNPs that are uptaken by cells, and whether intracellular infection promotes more or less uptake of the AgNPs. Acute exposure (2 h) to low dose, 10 µg/mL, AgNPs does not cause much reduction in cell viability although higher concentrations are toxic even with an acute 2 h exposure. The higher concentrations of AgNPs given to breast cells exposed for 24 h reduce cell viability profoundly for MCF 10A and MCF7 cells, which even begins at 10 µg/mL. Notably, MDA MB 231 cells were most resistant to AgNPs, which is contradictory to previous results [35,36,44]. One of the most dramatic findings is that MDA MB 231i cells were highly resistant to AgNPs, even at high concentrations (250 µg/mL) for 24 h. Lower doses of AgNPs that exhibit toxicity to breast cells (10–50 µg/mL) also cause a minor (~1 log, thus not clinically significant) reduction in planktonic *P. aeruginosa*. Prolonged exposure to AgNPs for 24 h showed that bacteria overcome the acute insult, but there is profound toxicity at concentrations higher than 50 µg/mL (5–8 log reduction).

Forty-three degrees Celsius is mild hyperthermia and it is known that cancer cells are more sensitive to heat than non-cancerous cells [51]. The reason why hyperthermia is often used against cancer is because cancer cells have been shown to be more sensitive to elevated temperatures due to their inability to overcome DNA strand breaks [64,65]. AgNPs augment hyperthermia whether cells are infected or not, but there is a greater benefit against infected cells. Given the toxicity of the AgNPs, the potential for thermal transfer, and knowledge of cell susceptibility to hyperthermia, we examined the addition of AgNPs to elevated temperatures. The temperatures that were achieved using PTT ranged from 49 to 68 °C, hence, we examined how cells with and without infection would respond to brief exposure to these temperatures with AgNPs. AgNPs can augment hyperthermia, and are especially beneficial at lower temperatures. Only the lowest dose of AgNPs (10 µg/mL) was needed to elicit such an effect, although higher doses of AgNPs also boost brief exposure to elevated temperatures. A major benefit is that AgNPs augmented hyperthermia especially in infected cells, further supporting the hypothesis that these cells may be stressed by intracellular bacteria and hence more responsive to therapeutics that increase stress responses, leading to cell death.

AgNPs, synthesized into triangular shapes for absorption at 800 nm for PTT, were synthesized and were effective for heat generation. AgNPs are inherently antimicrobial, cytotoxic at higher concentrations, and can enter cells, as multiple groups have shown [35,36,44]. They can kill extracellular bacteria, but have limited potential to kill intracellular bacteria without additional stimulation such as heat-generating PTT. Even a low dose (10 µg/mL AgNPs) and 800 nm at 5 W for 36 s could generate ΔT = 12 °C, which is sufficient for overcoming the 45 °C (with 37 + 12 °C = 49 °C) threshold of irreversible protein denaturation and hence effective for killing cancer cells. Higher concentrations of AgNPs get much hotter (up to 37 + 31 °C = 68 °C). PTT can be induced rapidly, in only 36 s, using a higher laser power that will have improved penetration depth [66]. A limitation is that PTT temperatures were measured as a volumetric change (200 µL) and temperatures that the cancer or bacterial cells experience locally at their cell surface might be much higher given the proximity of the AgNPs during laser exposure. It was expected that AgNP-induced

PTT would kill breast cells, but it was unexpected that all infected cells would be resistant to PTT. Resistance could be overcome with the use of 25 μg/mL of AgNP-induced PTT. PTT is more effective than water bath hyperthermia with AgNPs. PTT of planktonic bacteria can completely eliminate the bacteria at a concentration of 50 μg/mL, although there are also statistically significant reductions for 10 and 25 μg/mL. AgNPs at 10 μg/mL with laser stimulation did not reduce the number of intracellular bacteria in MCF 10A cells; however, there was a reduction in both breast cancer cell lines. Intracellular infection is reduced following PTT, but may not be completely eliminated depending upon the concentration of AgNPs used. Infected cells that survive PTT continue to harbor bacteria, a result that makes sense since higher temperatures are needed to kill the bacteria. Intracellular bacteria alter cell response to hyperthermia, and although PTT is effective, it is critical that high enough doses of AgNPs and heat are used to kill infected cells, since survivors could metastasize, and the bacteria could migrate with these cells to other locations.

5. Conclusions

The results of this work demonstrate that breast cells harbor intracellular pathogens, with breast cancer cells having an increased bacterial burden. The current work does not indicate that intracellular infection promotes non-tumorigenic breast cells into a cancerous phenotype, but it suggests that both tumorigenic and non-tumorigenic breast cells respond differently to heat and AgNPs, both of which induce cell stress, when bacteria are present. In the absence of the laser, breast cells incubated with AgNPs exhibited greater reductions in cell viability than cells incubated with media alone at the same temperature, indicating the potential application of AgNPs to improve the efficiency of hyperthermia treatment. Infected breast cells were more resistant to AgNP-induced PTT, and the augmentation of hyperthermia with AgNPs. In this work, 25 mg/mL of AgNPs with laser stimulation generated sufficient heat, leading to nearly complete cell death of both non-infected and infected breast cells. The lack of breast cell colony formation at the 25 mg/mL concentration demonstrated that AgNP-induced cellular damage was irreversible, indicating less potential for cell regrowth. PTT caused reductions in bacterial colony-forming units, and also in CFUs/cell for breast cells. These results highlight that triangular AgNPs could be effective for PTT against infected breast cancer cells. The overall results of this work demonstrate the benefit of using AgNPs as cytotoxic photothermal agents to kill pathogenic bacteria both inside and outside breast cells. PTT is a relatively new option under consideration to eliminate breast cancer, especially cells that harbor bacteria which may be resistant to treatment and have higher metastatic potential. Most PTT focuses on killing either cancer cells or bacteria separately; here we demonstrate the potential for killing bacteria within cancer cells, while simultaneously killing the cancer cells. The next step is to evaluate how reducing intracellular bacteria can improve responses to breast cancer treatments, where limited options exist for tackling both bacteria and cancer cells in the breast microenvironment.

Author Contributions: Conceptualization, N.L.; methodology, S.L. and N.L.; validation, S.L.; formal analysis, S.L.; investigation, S.L., S.P. and S.N.; resources, S.N.; data curation, S.L. and S.P.; writing—original draft preparation, S.L.; writing—review and editing, N.L.; visualization, S.L.; supervision, N.L.; project administration, N.L.; funding acquisition, N.L. All authors have read and agreed to the published version of the manuscript.

Funding: This research was supported by the Department of Plastic and Reconstructive Surgery at Wake Forest School of Medicine.

Institutional Review Board Statement: Not applicable.

Informed Consent Statement: Not applicable.

Data Availability Statement: The data presented in this study are held by the Department of Plastic and Reconstructive Surgery at Wake Forest School of Medicine but may be made available, subject to review, on request from the corresponding author.

Conflicts of Interest: The authors declare no conflict of interest.

References

1. Giaquinto, A.N.; Sung, H.; Miller, K.D.; Kramer, J.L.; Newman, L.A.; Minihan, A.; Jemal, A.; Siegel, R.L. Breast Cancer Statistics, 2022. *CA Cancer J. Clin.* **2022**, *72*, 524–541. [CrossRef]
2. Urbaniak, C.; Cummins, J.; Brackstone, M.; Macklaim, J.M.; Gloor, G.B.; Baban, C.K.; Scott, L.; O'Hanlon, D.M.; Burton, J.P.; Francis, K.P. Microbiota of Human Breast Tissue. *Appl. Environ. Microbiol.* **2014**, *80*, 3007–3014. [CrossRef]
3. Urbaniak, C.; Gloor, G.B.; Brackstone, M.; Scott, L.; Tangney, M.; Reid, G. The Microbiota of Breast Tissue and Its Association with Breast Cancer. *Appl. Environ. Microbiol.* **2016**, *82*, 5039–5048. [CrossRef]
4. Wang, H.; Altemus, J.; Niazi, F.; Green, H.; Calhoun, B.C.; Sturgis, C.; Grobmyer, S.R.; Eng, C. Breast Tissue, Oral and Urinary Microbiomes in Breast Cancer. *Oncotarget* **2017**, *8*, 88122. [CrossRef]
5. Fu, A.; Yao, B.; Dong, T.; Chen, Y.; Yao, J.; Liu, Y.; Li, H.; Bai, H.; Liu, X.; Zhang, Y. Tumor-Resident Intracellular Microbiota Promotes Metastatic Colonization in Breast Cancer. *Cell* **2022**, *185*, 1356–1372. [CrossRef] [PubMed]
6. Fu, A.; Yao, B.; Dong, T.; Cai, S. Emerging Roles of Intratumor Microbiota in Cancer Metastasis. *Trends Cell Biol.* **2023**, *33*, 583–593. [CrossRef]
7. Parida, S.; Drewes, J.L. Unwanted Passengers: Microbes Hitchiking in Breast Cancer Metastases. *Cell Host Microbe* **2022**, *30*, 875–877. [CrossRef] [PubMed]
8. Parida, S.; Wu, S.; Muniraj, N.; Siddharth, S.; Nagaligam, A.; Hum, C.; Mistriotis, P.; Konstantopoulos, K.; Sears, C.L.; Sharma, D. Bacteroides Fragilis: A Potential Pathogen Orchestrating EMT and Stemness in Breast Epithelial Cells via Concomitant Activation of Notch and Bcatenin Axes. *Cancer Res.* **2019**, *79*, 2834. [CrossRef]
9. Parida, S.; Wu, S.; Siddharth, S.; Wang, G.; Muniraj, N.; Nagalingam, A.; Hum, C.; Mistriotis, P.; Hao, H.; Talbot, C.C., Jr.; et al. A Procarcinogenic Colon Microbe Promotes Breast Tumorigenesis and Metastatic Progression and Concomitantly Activates Notch and β-Catenin Axes. *Cancer Discov.* **2021**, *11*, 1138–1157. [CrossRef]
10. Parhi, L.; Alon-Maimon, T.; Sol, A.; Nejman, D.; Shhadeh, A.; Fainsod-Levi, T.; Yajuk, O.; Isaacson, B.; Abed, J.; Maalouf, N. Breast Cancer Colonization by Fusobacterium Nucleatum Accelerates Tumor Growth and Metastatic Progression. *Nat. Commun.* **2020**, *11*, 3259. [CrossRef] [PubMed]
11. Chiba, A.; Bawaneh, A.; Velazquez, C.; Clear, K.Y.J.; Wilson, A.S.; Howard-McNatt, M.; Levine, E.A.; Levi-Polyachenko, N.; Yates-Alston, S.A.; Diggle, S.P.; et al. Neoadjuvant Chemotherapy Shifts Breast Tumor Microbiota Populations to Regulate Drug Responsiveness and the Development of Metastasis. *Mol. Cancer Res.* **2020**, *18*, 130–139. [CrossRef]
12. Waks, A.G.; Winer, E.P. Breast Cancer Treatment: A Review. *JAMA* **2019**, *321*, 288–300. [CrossRef] [PubMed]
13. Sun, J.; Zhao, H.; Xu, W.; Jiang, G.-Q. Recent Advances in Photothermal Therapy-Based Multifunctional Nanoplatforms for Breast Cancer. *Front. Chem.* **2022**, *10*, 1024177. [CrossRef] [PubMed]
14. Crawford, B.M.; Shammas, R.L.; Fales, A.M.; Brown, D.A.; Hollenbeck, S.T.; Vo-Dinh, T.; Devi, G.R. Photothermal Ablation of Inflammatory Breast Cancer Tumor Emboli Using Plasmonic Gold Nanostars. *Int. J. Nanomed.* **2017**, *12*, 6259. [CrossRef] [PubMed]
15. Valcourt, D.M.; Dang, M.N.; Day, E.S. IR820-Loaded PLGA Nanoparticles for Photothermal Therapy of Triple-Negative Breast Cancer. *J. Biomed. Mater. Res. Part A* **2019**, *107*, 1702–1712. [CrossRef] [PubMed]
16. Zhang, M.; Kim, H.S.; Jin, T.; Moon, W.K. Near-Infrared Photothermal Therapy Using EGFR-Targeted Gold Nanoparticles Increases Autophagic Cell Death in Breast Cancer. *J. Photochem. Photobiol. B Biol.* **2017**, *170*, 58–64. [CrossRef]
17. Xing, Y.; Kang, T.; Luo, X.; Zhu, J.; Wu, P.; Cai, C. Coral-Shaped Au Nanostructures for Selective Apoptosis Induction during Photothermal Therapy. *J. Mater. Chem. B* **2019**, *7*, 6224–6231. [CrossRef]
18. He, X.; Bao, X.; Cao, H.; Zhang, Z.; Yin, Q.; Gu, W.; Chen, L.; Yu, H.; Li, Y. Tumor-Penetrating Nanotherapeutics Loading a Near-Infrared Probe Inhibit Growth and Metastasis of Breast Cancer. *Adv. Funct. Mater.* **2015**, *25*, 2831–2839. [CrossRef]
19. Chen, Q.; Liang, C.; Wang, C.; Liu, Z. An Imagable and Photothermal "Abraxane-Like" Nanodrug for Combination Cancer Therapy to Treat Subcutaneous and Metastatic Breast Tumors. *Adv. Mater.* **2015**, *27*, 903–910. [CrossRef]
20. Kadkhoda, J.; Tarighatnia, A.; Tohidkia, M.R.; Nader, N.D.; Aghanejad, A. Photothermal Therapy-Mediated Autophagy in Breast Cancer Treatment: Progress and Trends. *Life Sci.* **2022**, *298*, 120499. [CrossRef]
21. Mackey, M.A.; Ali, M.R.K.; Austin, L.A.; Near, R.D.; El-Sayed, M.A. The Most Effective Gold Nanorod Size for Plasmonic Photothermal Therapy: Theory and In Vitro Experiments. *J. Phys. Chem. B* **2014**, *118*, 1319–1326. [CrossRef] [PubMed]
22. Cole, J.R.; Mirin, N.A.; Knight, M.W.; Goodrich, G.P.; Halas, N.J. Photothermal Efficiencies of Nanoshells and Nanorods for Clinical Therapeutic Applications. *J. Phys. Chem. C* **2009**, *113*, 12090–12094. [CrossRef]
23. Qin, Z.; Wang, Y.; Randrianalisoa, J.; Raeesi, V.; Chan, W.C.W.; Lipiński, W.; Bischof, J.C. Quantitative Comparison of Photothermal Heat Generation between Gold Nanospheres and Nanorods. *Sci. Rep.* **2016**, *6*, 29836. [CrossRef] [PubMed]
24. Ali, M.R.K.; Rahman, M.A.; Wu, Y.; Han, T.; Peng, X.; Mackey, M.A.; Wang, D.; Shin, H.J.; Chen, Z.G.; Xiao, H.; et al. Efficacy, Long-Term Toxicity, and Mechanistic Studies of Gold Nanorods Photothermal Therapy of Cancer in Xenograft Mice. *Proc. Natl. Acad. Sci. USA* **2017**, *114*, E3110–E3118. [CrossRef] [PubMed]
25. Dykman, L.A.; Khlebtsov, N.G. Gold Nanoparticles in Biology and Medicine: Recent Advances and Prospects. *Acta Naturae* **2011**, *3*, 34–55. [CrossRef]

26. Eustis, S.; El-Sayed, M.A. Why Gold Nanoparticles Are More Precious than Pretty Gold: Noble Metal Surface Plasmon Resonance and Its Enhancement of the Radiative and Nonradiative Properties of Nanocrystals of Different Shapes. *Chem. Soc. Rev.* **2006**, *35*, 209–217. [CrossRef]
27. Yin, T.; Xie, W.; Sun, J.; Yang, L.; Liu, J. Penetratin Peptide-Functionalized Gold Nanostars: Enhanced BBB Permeability and NIR Photothermal Treatment of Alzheimer's Disease Using Ultralow Irradiance. *ACS Appl. Mater. Interfaces* **2016**, *8*, 19291–19302. [CrossRef]
28. Zhang, Y.; Shareena Dasari, T.P.; Deng, H.; Yu, H. Antimicrobial Activity of Gold Nanoparticles and Ionic Gold. *J. Environ. Sci. Health Part C* **2015**, *33*, 286–327. [CrossRef] [PubMed]
29. Meeker, D.G.; Jenkins, S.V.; Miller, E.K.; Beenken, K.E.; Loughran, A.J.; Powless, A.; Muldoon, T.J.; Galanzha, E.I.; Zharov, V.P.; Smeltzer, M.S. Synergistic Photothermal and Antibiotic Killing of Biofilm-Associated Staphylococcus Aureus Using Targeted Antibiotic-Loaded Gold Nanoconstructs. *ACS Infect. Dis.* **2016**, *2*, 241–250. [CrossRef]
30. Millenbaugh, N.J.; Baskin, J.B.; DeSilva, M.N.; Elliott, W.R.; Glickman, R.D. Photothermal Killing of Staphylococcus Aureus Using Antibody-Targeted Gold Nanoparticles. *Int. J. Nanomed.* **2015**, *10*, 1953–1960. [CrossRef] [PubMed]
31. Boca, S.C.; Potara, M.; Gabudean, A.-M.; Juhem, A.; Baldeck, P.L.; Astilean, S. Chitosan-Coated Triangular Silver Nanoparticles as a Novel Class of Biocompatible, Highly Effective Photothermal Transducers for in Vitro Cancer Cell Therapy. *Cancer Lett.* **2011**, *311*, 131–140. [CrossRef]
32. Gurunathan, S.; Lee, K.-J.; Kalishwaralal, K.; Sheikpranbabu, S.; Vaidyanathan, R.; Eom, S.H. Antiangiogenic Properties of Silver Nanoparticles. *Biomaterials* **2009**, *30*, 6341–6350. [CrossRef]
33. Kalishwaralal, K.; Banumathi, E.; Pandian, S.R.K.; Deepak, V.; Muniyandi, J.; Eom, S.H.; Gurunathan, S. Silver Nanoparticles Inhibit VEGF Induced Cell Proliferation and Migration in Bovine Retinal Endothelial Cells. *Colloids Surf. B Biointerfaces* **2009**, *73*, 51–57. [CrossRef]
34. Swanner, J.; Mims, J.; Carroll, D.L.; Akman, S.A.; Furdui, C.M.; Torti, S.V.; Singh, R.N. Differential Cytotoxic and Radiosensitizing Effects of Silver Nanoparticles on Triple-Negative Breast Cancer and Non-Triple-Negative Breast Cells. *Int. J. Nanomed.* **2015**, *10*, 3937–3953. [CrossRef]
35. Swanner, J.; Fahrenholtz, C.D.; Tenvooren, I.; Bernish, B.W.; Sears, J.J.; Hooker, A.; Furdui, C.M.; Alli, E.; Li, W.; Donati, G.L.; et al. Silver Nanoparticles Selectively Treat Triple-negative Breast Cancer Cells without Affecting Non-malignant Breast Epithelial Cells in Vitro and in Vivo. *FASEB Bioadv.* **2019**, *1*, 639–660. [CrossRef] [PubMed]
36. Snyder, C.M.; Rohde, M.M.; Fahrenholtz, C.D.; Swanner, J.; Sloop, J.; Donati, G.L.; Furdui, C.M.; Singh, R. Low Doses of Silver Nanoparticles Selectively Induce Lipid Peroxidation and Proteotoxic Stress in Mesenchymal Subtypes of Triple-Negative Breast Cancer. *Cancers* **2021**, *13*, 4217. [CrossRef]
37. Kim, D.; Amatya, R.; Hwang, S.; Lee, S.; Min, K.A.; Shin, M.C. BSA-Silver Nanoparticles: A Potential Multimodal Therapeutics for Conventional and Photothermal Treatment of Skin Cancer. *Pharmaceutics* **2021**, *13*, 575. [CrossRef] [PubMed]
38. Feng, Q.L.; Wu, J.; Chen, G.Q.; Cui, F.Z.; Kim, T.N.; Kim, J.O. A Mechanistic Study of the Antibacterial Effect of Silver Ions on Escherichia Coli and Staphylococcus Aureus. *J. Biomed. Mater. Res.* **2000**, *52*, 662–668. [CrossRef] [PubMed]
39. Kang, J.; Dietz, M.J.; Hughes, K.; Xing, M.; Li, B. Silver Nanoparticles Present High Intracellular and Extracellular Killing against Staphylococcus Aureus. *J. Antimicrob. Chemother.* **2019**, *74*, 1578–1585. [CrossRef]
40. Dakal, T.C.; Kumar, A.; Majumdar, R.S.; Yadav, V. Mechanistic Basis of Antimicrobial Actions of Silver Nanoparticles. *Front. Microbiol.* **2016**, *7*, 1831. [CrossRef]
41. Jena, P.; Mohanty, S.; Mallick, R.; Jacob, B.; Sonawane, A. Toxicity and Antibacterial Assessment of Chitosancoated Silver Nanoparticles on Human Pathogens and Macrophage Cells. *Int. J. Nanomed.* **2012**, *7*, 1805–1818.
42. Pucelik, B.; Sułek, A.; Borkowski, M.; Barzowska, A.; Kobielusz, M.; Dąbrowski, J.M. Synthesis and Characterization of Size-and Charge-Tunable Silver Nanoparticles for Selective Anticancer and Antibacterial Treatment. *ACS Appl. Mater. Interfaces* **2022**, *14*, 14981–14996. [CrossRef] [PubMed]
43. Cheng, L.-C.; Huang, J.-H.; Chen, H.M.; Lai, T.-C.; Yang, K.-Y.; Liu, R.-S.; Hsiao, M.; Chen, C.-H.; Her, L.-J.; Tsai, D.P. Seedless, Silver-Induced Synthesis of Star-Shaped Gold/Silver Bimetallic Nanoparticles as High Efficiency Photothermal Therapy Reagent. *J. Mater. Chem.* **2012**, *22*, 2244–2253. [CrossRef]
44. Thompson, E.A.; Graham, E.; MacNeill, C.M.; Young, M.; Donati, G.; Wailes, E.M.; Jones, B.T.; Levi-Polyachenko, N.H. Differential Response of MCF7, MDA-MB-231, and MCF 10A Cells to Hyperthermia, Silver Nanoparticles and Silver Nanoparticle-Induced Photothermal Therapy. *Int. J. Hyperth.* **2014**, *30*, 312–323. [CrossRef] [PubMed]
45. Sears, J.; Swanner, J.; Fahrenholtz, C.D.; Snyder, C.; Rohde, M.; Levi-Polyachenko, N.; Singh, R. Combined Photothermal and Ionizing Radiation Sensitization of Triple-Negative Breast Cancer Using Triangular Silver Nanoparticles. *Int. J. Nanomed.* **2021**, *16*, 851–865. [CrossRef]
46. Weissleder, R. A Clearer Vision for in Vivo Imaging. *Nat. Biotechnol.* **2001**, *19*, 316–317. [CrossRef]
47. Wu, X.; Suo, Y.; Shi, H.; Liu, R.; Wu, F.; Wang, T.; Ma, L.; Liu, H.; Cheng, Z. Deep-Tissue Photothermal Therapy Using Laser Illumination at NIR-IIa Window. *Nano-Micro Lett.* **2020**, *12*, 38. [CrossRef]
48. Ge, X.; Fu, Q.; Bai, L.; Chen, B.; Wang, R.; Gao, S.; Song, J. Photoacoustic Imaging and Photothermal Therapy in the Second Near-Infrared Window. *New J. Chem.* **2019**, *43*, 8835–8851. [CrossRef]
49. Ash, C.; Dubec, M.; Donne, K.; Bashford, T. Effect of Wavelength and Beam Width on Penetration in Light-Tissue Interaction Using Computational Methods. *Lasers Med. Sci.* **2017**, *32*, 1909–1918. [CrossRef]

50. Ragulin, Y.; Kurpeshev, O.; Galkin, V.; Lebedeva, T.; Belokhvostova, A. Local Hyperthermia for Inflammatory Breast Cancer. *Eur. J. Cancer* **2017**, *72*, S11–S12. [CrossRef]
51. Crezee, J.; Franken, N.A.P.; Oei, A.L. Hyperthermia-Based Anti-Cancer Treatments. *Cancers* **2021**, *13*, 1240. [CrossRef] [PubMed]
52. Feyerabend, T.; Wiedemann, G.J.; Jäger, B.; Vesely, H.; Mahlmann, B.; Richter, E. Local Hyperthermia, Radiation, and Chemotherapy in Recurrent Breast Cancer Is Feasible and Effective except for Inflammatory Disease. *Int. J. Radiat. Oncol. Biol. Phys.* **2001**, *49*, 1317–1325. [CrossRef]
53. van Rhoon, G.C. Is CEM43 Still a Relevant Thermal Dose Parameter for Hyperthermia Treatment Monitoring? *Int. J. Hyperth.* **2016**, *32*, 50–62. [CrossRef] [PubMed]
54. Landon, C.D.; Park, J.-Y.; Needham, D.; Dewhirst, M.W. Nanoscale Drug Delivery and Hyperthermia: The Materials Design and Preclinical and Clinical Testing of Low Temperature-Sensitive Liposomes Used in Combination with Mild Hyperthermia in the Treatment of Local Cancer. *Open Nanomed. J.* **2011**, *3*, 38. [CrossRef]
55. Yui, S.; Karia, K.; Ali, S.; Muzslay, M.; Wilson, P. Thermal Disinfection at Suboptimal Temperature of Pseudomonas Aeruginosa Biofilm on Copper Pipe and Shower Hose Materials. *J. Hosp. Infect.* **2021**, *117*, 103–110. [CrossRef] [PubMed]
56. Shipunova, V.O.; Belova, M.M.; Kotelnikova, P.A.; Shilova, O.N.; Mirkasymov, A.B.; Danilova, N.V.; Komedchikova, E.N.; Popovtzer, R.; Deyev, S.M.; Nikitin, M.P. Photothermal Therapy with HER2-Targeted Silver Nanoparticles Leading to Cancer Remission. *Pharmaceutics* **2022**, *14*, 1013. [CrossRef] [PubMed]
57. Huang, X.; El-Sayed, M.A. Plasmonic Photo-Thermal Therapy (PPTT). *Alex. J. Med.* **2011**, *47*, 1–9. [CrossRef]
58. Lepock, J.R. Protein Denaturation during Heat Shock. In *Advances in Molecular and Cell Biology*; Elsevier: Amsterdam, The Netherlands, 1997; Volume 19, pp. 223–259, ISBN 1569-2558.
59. Cebrián, G.; Condón, S.; Mañas, P. Physiology of the Inactivation of Vegetative Bacteria by Thermal Treatments: Mode of Action, Influence of Environmental Factors and Inactivation Kinetics. *Foods* **2017**, *6*, 107. [CrossRef]
60. Udayasuryan, B.; Ahmad, R.N.; Nguyen, T.T.; Umaña, A.; Monét Roberts, L.; Sobol, P.; Jones, S.D.; Munson, J.M.; Slade, D.J.; Verbridge, S.S. Fusobacterium Nucleatum Induces Proliferation and Migration in Pancreatic Cancer Cells through Host Autocrine and Paracrine Signaling. *Sci. Signal.* **2022**, *15*, eabn4948. [CrossRef]
61. Nejman, D.; Livyatan, I.; Fuks, G.; Gavert, N.; Zwang, Y.; Geller, L.T.; Rotter-Maskowitz, A.; Weiser, R.; Mallel, G.; Gigi, E.; et al. The Human Tumor Microbiome Is Composed of Tumor Type–Specific Intracellular Bacteria. *Science* **2020**, *368*, 973–980. [CrossRef]
62. Yu, L.C.-H.; Wei, S.-C.; Li, Y.-H.; Lin, P.-Y.; Chang, X.-Y.; Weng, J.-P.; Shue, Y.-W.; Lai, L.-C.; Wang, J.-T.; Jeng, Y.-M.; et al. Invasive Pathobionts Contribute to Colon Cancer Initiation by Counterbalancing Epithelial Antimicrobial Responses. *Cell. Mol. Gastroenterol. Hepatol.* **2022**, *13*, 57–79. [CrossRef]
63. Zou, L.; Wang, J.; Gao, Y.; Ren, X.; Rottenberg, M.E.; Lu, J.; Holmgren, A. Synergistic Antibacterial Activity of Silver with Antibiotics Correlating with the Upregulation of the ROS Production. *Sci. Rep.* **2018**, *8*, 11131. [CrossRef] [PubMed]
64. Yarmolenko, P.S.; Moon, E.J.; Landon, C.; Manzoor, A.; Hochman, D.W.; Viglianti, B.L.; Dewhirst, M.W. Thresholds for Thermal Damage to Normal Tissues: An Update. *Int. J. Hyperth.* **2011**, *27*, 320–343. [CrossRef] [PubMed]
65. Dewhirst, M.W.; Viglianti, B.L.; Lora-Michiels, M.; Hanson, M.; Hoopes, P.J. Basic Principles of Thermal Dosimetry and Thermal Thresholds for Tissue Damage from Hyperthermia. *Int. J. Hyperth.* **2003**, *19*, 267–294. [CrossRef] [PubMed]
66. McCabe-Lankford, E.E.; Brown, T.L.; Levi-Polyachenko, N.H. Assessing Fluorescence Detection and Effective Photothermal Therapy of Near-Infrared Polymer Nanoparticles Using Alginate Tissue Phantoms. *Lasers Surg. Med.* **2018**, *50*, 1040–1049. [CrossRef]

pharmaceutics

Review

Metal-Based Nanoparticles for Cancer Metalloimmunotherapy

Ivan Hardianto Suliman [1], Kidong Kim [2], Weihsuan Chen [2,3], Yubin Kim [2,3], Jeong-Hyun Moon [2], Sejin Son [2,3,*] and Jutaek Nam [1,*]

[1] College of Pharmacy, Chonnam National University, Gwangju 61186, Republic of Korea
[2] Department of Biological Sciences, Inha University, Incheon 22212, Republic of Korea
[3] Department of Biological Sciences and Bioengineering, Industry-Academia Interactive R&E Center for Bioprocess Innovation, Inha University, Incheon 22212, Republic of Korea
* Correspondence: ssejin@inha.ac.kr (S.S.); namj@jnu.ac.kr (J.N.)

Abstract: Although the promise of cancer immunotherapy has been partially fulfilled with the unprecedented clinical success of several immunotherapeutic interventions, some issues, such as limited response rate and immunotoxicity, still remain. Metalloimmunotherapy offers a new form of cancer immunotherapy that utilizes the inherent immunomodulatory features of metal ions to enhance anticancer immune responses. Their versatile functionalities for a multitude of direct and indirect anticancer activities together with their inherent biocompatibility suggest that metal ions can help overcome the current issues associated with cancer immunotherapy. However, metal ions exhibit poor drug-like properties due to their intrinsic physicochemical profiles that impede in vivo pharmacological performance, thus necessitating an effective pharmaceutical formulation strategy to improve their in vivo behavior. Metal-based nanoparticles provide a promising platform technology for reshaping metal ions into more drug-like formulations with nano-enabled engineering approaches. This review provides a general overview of cancer immunotherapy, the immune system and how it works against cancer cells, and the role of metal ions in the host response and immune modulation, as well as the impact of metal ions on the process via the regulation of immune cells. The preclinical studies that have demonstrated the potential of metal-based nanoparticles for cancer metalloimmunotherapy are presented for the representative nanoparticles constructed with manganese, zinc, iron, copper, calcium, and sodium ions. Lastly, the perspectives and future directions of metal-based nanoparticles are discussed, particularly with respect to their clinical applications.

Keywords: cancer; immunotherapy; metal ion; immune cell regulation; metal-based nanoparticle; metalloimmunotherapy

Citation: Suliman, I.H.; Kim, K.; Chen, W.; Kim, Y.; Moon, J.-H.; Son, S.; Nam, J. Metal-Based Nanoparticles for Cancer Metalloimmunotherapy. *Pharmaceutics* **2023**, *15*, 2003. https://doi.org/10.3390/pharmaceutics15072003

Academic Editors: Manivasagan Panchanathan and Eue-Soon Jang

Received: 28 June 2023
Revised: 17 July 2023
Accepted: 19 July 2023
Published: 21 July 2023

1. Introduction

Cancer remains a major public health concern throughout the world. According to GLOBOCAN data, 19.3 million new cases and almost 10.0 million cancer-related deaths were recorded in 2020 [1]. Despite vast advances in the field, traditional clinical cancer therapy, such as chemotherapy, radiation therapy, and surgery, has shown limited therapeutic efficiency with considerable side effects due to a lack of specificity [2,3]. In recent years, cancer immunotherapy has emerged as a new clinical standard for many types of cancers [4], in which immune cells are trained to specifically locate cancer cells with systemic immune surveillance and generate anticancer immune responses for the elimination and long-term regression of local and metastatic tumors [5]. The therapeutic potential of cancer immunotherapy has been validated by the unprecedented bench-to-bedside clinical success of immune checkpoint inhibitors (ICIs), chimeric antigen receptor (CAR) T cell therapy, and oncolytic viruses against several tumors that have been refractory to other standard treatments [4,6]. However, cancer immunotherapy still faces challenges, such as toxicity and a low response rate [5]. For instance, most patients experience various symptoms of mild to severe immune-related adverse events, while only 20–30% benefit from ICIs [7].

Therefore, novel approaches for safe and effective cancer immunotherapy have been called for to provide more widespread clinical applications.

Metal ions are essential elements of host homeostasis and also play an important role in the regulation of the immune system [8]. They have inspired a new form of cancer immunotherapy called metalloimmunotherapy, which aims to harness the immunomodulatory features of metal ions for cancer immunotherapy [9]. However, metal ions are small charged species; thus, they exhibit physicochemical properties that present unfavorable in vivo pharmacological and drug-like activity. Therefore, effective metalloimmunotherapy requires the engineering of metal ions into suitable pharmaceutical forms that can improve the intrinsically poor in vivo performance of metal ions [9]. Nanomedicine offers promising tools to reshape metal ions into more drug-like formulations with nano-enabled engineering approaches, which can modulate the in vivo pharmacokinetic profiles of metal ions and target immune cells, reduce off-target toxicity, and improve therapeutic efficacy [10]. Accordingly, recent studies have demonstrated the potential of nanomedicine-based cancer metalloimmunotherapy. In this review, we illustrate the basic interaction between the immune system and cancer cells and the role of metal ions in the regulation of immune cells. We subsequently present recent preclinical studies that have developed and utilized metal-based nanoparticles for cancer metalloimmunotherapy and discuss the perspectives and future directions of this novel approach.

2. The Immune System and Cancer

The immune system is composed of innate and adaptive immunity that cooperate to recognize and defend the body against foreign/non-self-substances and infected/malignant cells to maintain homeostasis [11,12]. The innate immune system is the first line of defense against infection, responding to and eliminating harmful immunogens in a rapid and nonspecific manner [13,14]. Short-lived innate immunity subsequently bridges the activation of the adaptive immune system for specific and durable immune responses. Adaptive immunity can provide robust long-term protection against repeated encounters with pre-exposed antigens by the immunological memory effect, which is a cardinal feature of immunotherapy for durable and effective disease control [13,15].

Since the concept of cancer immunosurveillance was first introduced more than a century ago [12], researchers have developed the immunoediting theory to explain tumor outgrowth under active immune functions in immunocompetent hosts (Figure 1) [16]. The host immune cells of innate and adaptive arms, including natural killer (NK) cells, dendritic cells (DCs), macrophages, and T cells, are constantly searching for malignant cancer cells in systemic circulation; this is a process called immunosurveillance, which can ultimately lead to the recognition and elimination of cancer cells via anticancer immune responses [12,17,18]. This initial phase associated with cancer immunosurveillance is called elimination, as it consists of eliminating cancer cells to provide protection against tumor growth. However, the heterogeneity and genetic instability of cancer cells often prevent complete tumor eradication; the constant immune pressure results in the evolution and selection of cancer cell subclones that can resist anticancer immune responses [16,19,20]. Such cancer cell variants may survive immunosurveillance and progress toward the equilibrium phase, where cancer cells remain functionally dormant or continue to evolve into new subclones harboring immune escape mutations such as a loss of antigen presentation and an increased expression of programmed death ligand-1 (PD-L1) [16,19,20]. These immune-edited tumors can then enter the escape phase, where the immune system is unable to control their growth, and the tumors become clinically apparent [20]. Some tumors may also escape immunosurveillance by recruiting immunosuppressive leukocytes such as regulatory T cells (Tregs) and myeloid-derived suppressor cells (MDSCs), which inhibit anticancer immune cells [21]. Cancer immunotherapy aims to increase the quality and quantity of anticancer immune cells to newly prime or reinforce existing anticancer immune responses to eliminate tumors in the equilibrium or escape phase. Some key immune cells for anticancer or suppressive immune responses are described below.

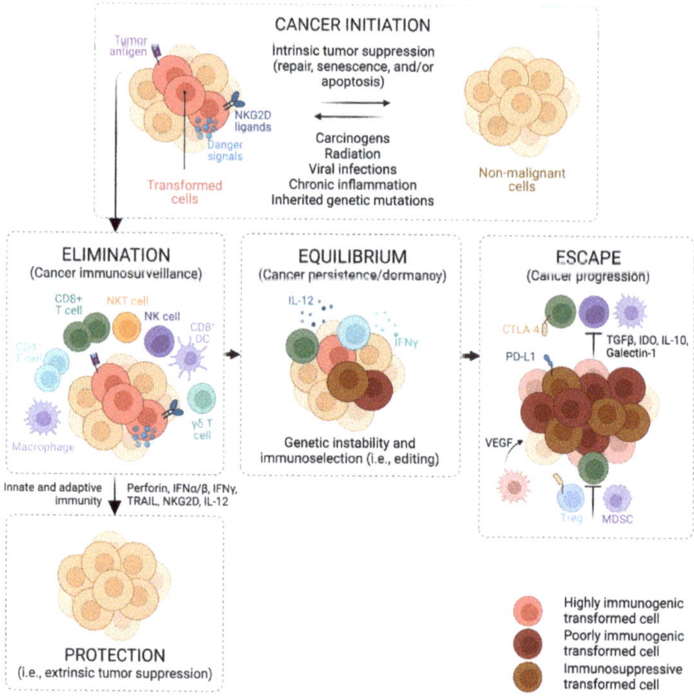

Figure 1. Cancer immunoediting. Adapted with permission from [16].

2.1. Antigen-Presenting Cells

DCs and macrophages not only play central roles in innate immune responses but they are also primary antigen-presenting cells (APCs) that induce antigen-specific T cell immunity by presenting cancer antigens in the context of major histocompatibility complex (MHC) molecules [22]. They require maturation signals such as pathogen-associated molecular patterns (PAMPs) and damage-associated molecular patterns (DAMPs) for immune activation [23]. These signals can be recognized by pattern recognition receptors (PRR), including Toll-like receptors (TLRs), retinoic acid-inducible gene I-like receptors (RLRs), NOD-like receptors (NLRs), and stimulator of interferon genes (STING) [22]. Upon maturation, DCs and macrophages exhibit an elevated expression of MHC molecules and costimulatory molecules (B7-1/B7-2) for the priming of T cells, and they secrete cytokines and chemokines that further activate T cells and enhance their migration to peripheral lymphoid organs or to the target site [24,25].

Among the major DC subsets, which include conventional DCs (cDCs), plasmacytoid DCs (pDCs), and monocyte-derived DCs (moDCs), cDCs are especially proficient at priming T cell responses [24,26]. cDCs can be subdivided into cDC1 and cDC2; cDC1s mainly induce cell immunity via $CD8^+$ T cells, whereas cDC2s induce humoral immunity via $CD4^+$ T cells [22,26]. cDC1s can also release interleukin (IL)-12 and promote interferon (IFN)-γ production and the activation of NK cells [26,27]. Furthermore, tumor-infiltrating cDC1s can produce chemokines such as the CXC chemokine ligand (CXCL) 9 and CXCL10 to attract $CD8^+$ T cells into the tumor microenvironment (TME), representing an essential role of DCs for cancer immunotherapy [22]. Conversely, macrophages are highly plastic cells capable of polarizing into different functional subtypes, including the classically activated M1 and alternatively activated M2 subtypes [28]. M1 macrophages can be activated by IFN-γ, TNF-α, and maturation signals [29,30], and they secrete pro-inflammatory cytokines such as TNF-α, IL-1α, IL-1β, IL-6, IL-12, IL-18, and IL-23 [31]. M1 macrophages are known for their tumoricidal activity and their promotion of the activation of cytotoxic T cells. In contrast, M2 macrophages are activated by IL-4, IL-10, IL-13, and glucocorticoid hormones, and they facilitate immune suppression via the anti-inflammatory response [14]. Importantly, M2 macrophages constitute the major immune cell population in the TME; tumor-associated macrophages (TAMs) support the initiation, progression, and metastasis of tumors via the down-regulation of pro-inflammatory responses and by promoting angiogenesis and connective tissue remodeling via the release of various immunosuppressive cytokines, enzymes, and chemokines [7].

2.2. Effector Immune Cells

$CD8^+$ T cells or cytotoxic T lymphocytes (CTLs) are the key effector cells for the adaptive immune response, which can directly recognize and eliminate cancer cells in an antigen-specific manner [29,32]. The multistep immune cell–cancer cell and immune cell–immune cell interactions that are involved in the killing of cancer cells via CTL are referred to as the 'cancer-immunity cycle' (Figure 2) [33]. In the first step of the cycle, cancer cell antigens released by spontaneous cell death are captured by professional APCs such as DCs (possibly accompanied by the co-release and uptake of DAMP-based maturation signals). APCs then process antigen epitopes and present them to $CD8^+$ T cells in the context of MHC I molecules in peripheral lymphoid organs [34]. Naïve T cells receive an antigenic signal via the T-cell receptor (TCR) and are primed and activated along with the co-stimulatory signals provided by APC maturation [18]. The activated T cells then migrate and infiltrate the tumor bed, recognize cancer cells via their cognate antigens bound to MHC I molecules, and exert cytotoxic activity for the induction of apoptosis [18,33]. As a consequence, cancer cells undergo immunogenic cell death, where dying cancer cells can release additional cancer cell antigens and DAMP-based maturation signals for the continued and sustained activation of the cycle [33].

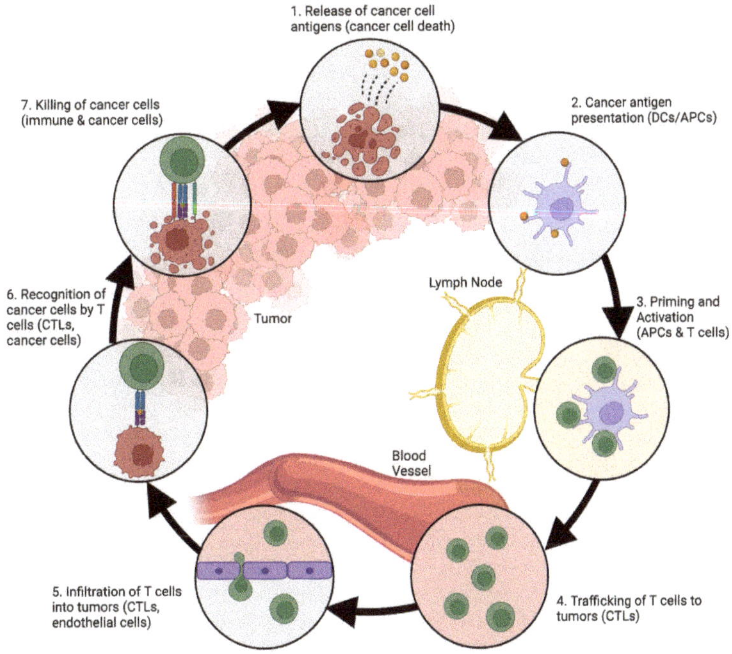

Figure 2. Cancer immunity cycle. Adapted with permission from [33].

NK cells can also directly engage and eliminate cancer cells [35,36]. Unlike CTLs, NK cells are innate immune cells and therefore do not require antigenic signals; their function is dependent on the repertoire and balance of activating and inhibitory receptor signaling [37]. For example, the inhibitory killer cell immunoglobulin-like receptor (KIR) can suppress NK cell activity after the recognition of MHC I molecules, which prevents the immune response and the subsequent collateral damage to healthy cells expressing MHC I [36,37]. Conversely, activating receptors can engage their ligands to promote NK cell activity. These include NK group 2 member D (NKG2D), such as MHC class I chain-related A and B (MICA/MICB) and several UL-16 binding proteins (ULBPs), natural cytotoxicity receptors (NCRs) like NKp30 and NKp46, and DNAX-activating molecule (DNAM1) [18,37,38]. Considering that cancer cells often down-regulate MHC I expression and up-regulate ligands for NK cell activating receptors during the immunoediting process, NK cells can effectively recognize cancer cells and subsequently undergo activation to generate robust anticancer immune responses in synergistic and/or complementary action with CTLs [18,36]. Furthermore, NK cells express Fc receptors that can bind to the common Fc region of immunoglobulins and induce antibody-dependent cellular cytotoxicity to eliminate immunoglobulin-targeted cells, thus allowing the elimination of cancer cells coated with cancer-specific antibodies produced de novo [35,38].

CTLs and NK cells can exert anticancer effector activity through the release of cytotoxic granules containing perforin and granzymes that are transmitted directly to cancer cells on contact [36]. They can also kill cancer cells via death receptor-induced cell apoptosis and the secretion of cytokines that inhibit tumor angiogenesis and further activate innate and adaptive immune responses [35,39,40].

2.3. *Immunosuppressive Cells*

Tumors in the escape phase often harbor various immunosuppressive cells that aid in immune evasion and tumor progression [41]. Among the best-characterized immunosuppressive cells are TAMs, MDSCs, and Tregs [30]. Consisting mainly of M2 macrophages, TAMs originate from monocytes that are recruited by tumor-derived cytokines, chemokines, and growth factors and then differentiate in response to IL-4, IL-10, and IL-13 [30]. They express CD68, CD163, CD204, and CD206 and produce a high level of IL-10 and arginase 1 (ARG-1), which are associated with a poor prognosis of solid tumors [42]. TAMs can negatively affect DCs and CTLs with cytokine regulation, express inhibitory ligands such as PD-L1 and B1-H4, and recruit Tregs by secreting chemokines [41]. MDSCs are immature myeloid cells that secrete various immunosuppressive substances, including ARG-1 and anti-inflammatory cytokines such as IL-10 and transforming growth factor-β (TGF-β), which upregulate PD-L1 expression and expand Tregs [10,30,43]. As a result, MDSCs disrupt T cell activation, DC antigen presentation, M1 macrophage polarization, and NK cell cytotoxicity [30]. The presence of MDSCs in the TME has been associated with drug resistance and metastatic progression, underscoring their pro-tumor effects [44–46]. Tregs are a subset of $CD4^+$ T cells with critical immunosuppressive functions for maintaining normal immune homeostasis and self-tolerance [47]. They are attracted to hypoxic environments intrinsic to tumors, where they suppress anticancer immune cells; therefore, they are considered a significant barrier to cancer immunotherapy [48,49]. Tregs produce immunosuppressive cytokines such as IL-10, IL-35, and TGF-β; deplete IL-2; and express negative co-stimulatory molecules that suppress T cells [30,41]. Other immunosuppressive cells implicated in the tumor immune escape include N2 neutrophils and regulatory DCs (DCregs) [41]. N2 neutrophils suppress $CD8^+$ T cells by releasing ARG-1, which facilitates the degradation of extracellular arginine essential for T cell function [41]. DCregs can induce T cells to undergo cell cycle arrest or apoptosis and promote Treg induction through the secretion of indoleamine-2,3-dioxygenase (IDO) [50]. Furthermore, they can also hinder T cell cytotoxicity through the production of TGF-β and IL-10 and the expression of PD-L1 and PD-L2 [51].

3. Metal Ions in the Host Response and Immune Modulation

Metal ions play an important role in the regulation of host homeostasis [52,53]. Some transition metal ions such as Fe^{2+}, Mn^{2+}, and Zn^{2+} are also essential elements in bacteria that serve as a cofactor to catalyze, stabilize, and activate many proteins and enzymes involved in vital physiological processes such as DNA replication/transcription and central metabolism [54]. Intriguingly, they can trigger host responses that are potentially associated with the inhibition of bacterial pathogens; the innate defense system not only actively exocytose these ions and thus sequester from intracellular pathogens but also accumulate them in the phagolysosomes to kill the bacteria, possibly by overdose toxicity or reactive oxygen species (ROS) production via a redox reaction [54,55]. Accordingly, studies have suggested that Cu ions accumulate in macrophage phagolysosomes via transporter proteins and induce oxidative damage to disrupt pathogens and clear infection [54].

The impact of metal ions on the activation of innate immunity has been demonstrated with inflammasome-mediated responses. The inflammasome is an oligomeric protein complex that activates caspase-1 for the maturation of IL-1β and IL-18β and subsequent pro-inflammatory responses [56]. In particular, inflammasomes that contain the nod-like receptor (NLR) family member NLRP3 can respond to several metal ions such as Ca^{2+}, Na^{+}, and K^{+}. An increase in extracellular Ca^{2+} levels has been observed at sites of infection and ischemic necrosis to activate the NLRP3 inflammasome, suggesting that Ca^{2+} is a DAMP signal released by damaged cells [57]. Na^{+} and K^{+} can also trigger the formation of the NLRP3 inflammasome by inducing an ionic imbalance within cells [56,58]. Intracellular K^{+} depletion has been shown to be necessary and sufficient to activate NLRP3 inflammasomes, which can be caused by active K^{+} efflux and the formation of membrane pores permeable to K^{+} [56]. Conversely, Na^{+} influx can dysregulate the Na^{+}/K^{+} gradient and promote K^{+} efflux, thus leading to the activation of the NLRP3 inflammasome [58]. In addition, metal ions can also directly stimulate or potentiate the cyclic GMP-AMP synthase (cGAS)-STING pathway. As a critical innate defense mechanism against viral infection, cGAS can detect cytosolic dsDNA, a hallmark of the virus, and subsequently catalyze the synthesis of cyclic GMP-AMP (cGAMP), which then binds to STING and triggers the downstream signaling pathway, stimulating the pro-inflammatory responses and production of type I IFNs [59]. Mn^{2+} has been shown to enhance sensitivity to DNA and the enzymatic activity of cGAS while also promoting the binding of cGAMP to STING, thus potentiating multiple factors in the cGAS-STING pathway [60]. Zn^{2+} has also been shown to increase cGAS activity by facilitating the phase transition of cGAS and stabilizing the cGAS-dsDNA complex [59].

Metal ions can also regulate the pivotal immune cells in the adaptive immune system. Zn^{2+} is essential for the development and function of T cells and B cells, where its deficiency causes thymic atrophy and subsequent T cell lymphopenia, disrupts T cell subtype differentiation potentially causing immune dysfunction, impairs T cell activation and expansion, and reduces B cell maturation and antibody production [55]. In addition, Zn^{2+} exhibits various immunological functions related to the priming of adaptive immunity, including polymorphonuclear cell recruitment and phagocytic activation, the maturation and differentiation of DCs and NK cells, and pro-inflammatory cytokine production [55]. For the activity of CTLs, Mg^{2+} promotes the up-regulation of lymphocyte function-associated antigen 1 (LFA-1) and costimulatory T cell receptors to enhance their cytotoxicity [4]. TCR signaling is a key step in initiating adaptive immune responses by T cells, whereby Ca^{2+} plays a critical role by manipulating phospholipids and signaling proteins [61]. Specifically, Ca^{2+} can neutralize anionic phospholipids and subsequently increase the accessibility of immunoreceptor tyrosine-based activation motif (ITAM), leading to enhanced TCR sensitivity to antigenic signals by the amplification of downstream signaling [61]. Additionally, T cells in the immune synapse promote the influx of Ca^{2+} to further regulate the phosphorylation of cytoplasmic signaling proteins and amplify signaling cascades associated with the activation and effector function of T cells [61]. In contrast, K^{+} serves as an ionic checkpoint that attenuates T cell activity; elevated intracellular K^{+} concentration inhibits the activation of effector T cells while enriching Tregs [59]. In vitro studies with human cells and in vivo studies with tumor-bearing mice have shown that the depletion of excessive K^+ can restore T cell function for anticancer immune responses [59]. Conversely, high levels of K^+ induce a state of caloric restriction by limiting the uptake and metabolism of nutrients, which triggers a starvation response that promotes metabolic and genetic reprogramming toward T cell memory formation and stemness [62]. The stemness of T cells is characterized by their ability to persist, self-renew, clonally repopulate, and exhibit pluripotency for their progeny to acquire effector functions, which is beneficial for the tumoricidal capacity of T cells [62]. This suggests that K^+ has a dual effect on T cell regulation, which should be appropriately balanced for effective cancer immunotherapy.

4. Metal-Based Nanoparticles for Cancer Metalloimmunotherapy

Nanotechnology has established a variety of nanoparticle platforms that can positively modulate the in vivo pharmacological profile of drug molecules with nano-enabled engineering approaches [63,64]. Metal-based nanoparticles have also been developed and applied to cancer imaging and therapy based on their intrinsic electromagnetic properties and direct cytotoxic activity in cancer cells [3]. Recently, the emergence of metalloimmunotherapy has spurred renewed interest in the application of metal-based nanoparticles for cancer immunotherapy with intrinsic or extrinsic anticancer and immunomodulatory features of metal ions. Below, we introduce representative preclinical studies that have utilized metal-based nanoparticles for cancer metalloimmunotherapy (Table 1). We categorized metal-based nanoparticles by their principal cellular and molecular mechanisms associated with the induction of antitumor immune responses.

4.1. Manganese- and Zinc-Based Nanoparticles for cGAS-STING Activation

Manganese-containing nanoparticles can stimulate innate and adaptive immunity through the cGAS-STING pathway [65,66]. Sun et al. showed that Mn^{2+} can potentiate the activity of the cyclic dinucleotide (CDN)-based STING agonist by an order of magnitude in multiple human STING haplotypes (Figure 3) [9]. The authors designed the self-assembled CDN-Mn^{2+} nanoparticle (CMP) to take advantage of a nanoparticle formulation to improve CDN and Mn^{2+} metabolic stability, cell permeability, intracellular activity, and in vivo performance. CMP was demonstrated to trigger anticancer immune responses through NF-kB and IRF3, the downstream signaling of the cGAS-STING pathway. It induced DC maturation, cytokine production, and M1 macrophage polarization, leading to the activation of $CD8^+$ T cells and inhibition of immunosuppressive MDSCs for remarkable antitumor efficacy in multiple difficult-to-treat murine tumor models. Liu et al. developed biomineralized manganese oxide nanoparticles (Bio-MnO_2 NPs) in combination with radiotherapy (RT), showing that it could induce apoptosis and the DNA release of cancer cells to activate the cGAS-STING pathway [67]. Bio-MnO_2 NPs not only sensitized RT to stimulate cGAS-STING signaling but also relieved tumor hypoxia and generated ROS with the release of Mn^{2+} and oxidation of hydrogen peroxide into oxygen. As a result, RT plus Bio-MnO_2 NPs elicited robust anticancer immunity while alleviating the immunosuppressive TME, significantly inhibiting tumor growth in a murine non-small-cell lung cancer (NSCLC) model. Similarly, Hou et al. constructed doxorubicin (DOX)-loaded, phospholipid (PL)-coated amorphous manganese phosphate nanoparticles for synergistic immune stimulation; DOX could trigger DNA damage that activated cGAS-STING signaling, which, in turn, was potentiated by Mn^{2+} [68]. The hybrid nanoparticles effectively inhibited primary growth, relapse, and the distant metastasis of murine 4T1 breast tumors with strong local and systemic immune responses, leading to improved survival among tumor-bearing mice [68].

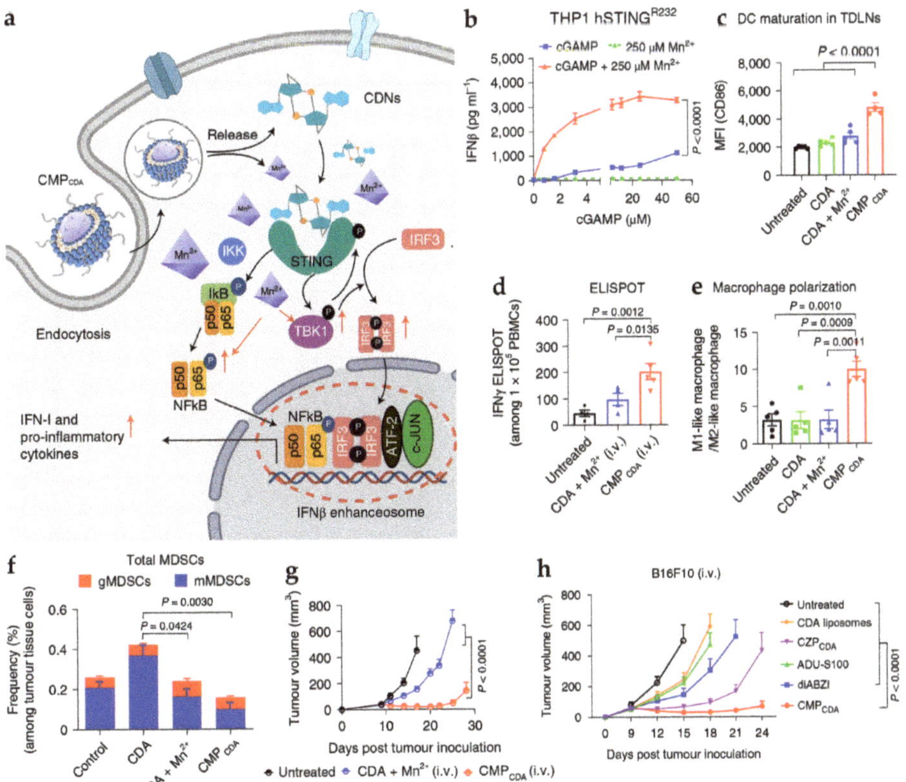

Figure 3. Antitumor activity of CMP_{CDA}. (**a**) Schematic diagram of the activation of cGAS-STING pathway by CMP_{CDA}. (**b**) STING activation in THP1 cells expressing $hSTING^{R232}$. (**c–f**) The levels of immune cells and (**g**) antitumor efficacy of CMP_{CDA} in CT26 tumor-bearing mice. (**h**) Antitumor efficacy of CMP_{CDA} in B16F10 tumor-bearing mice. Reprinted with permission from [9].

Zinc ions can also directly and indirectly stimulate cGAS-STING activation and anticancer immune responses [69]. Wang et al. showed that zinc oxide nanoparticles (ZnO NPs) could promote cancer cell apoptosis with the production of ROS [70]. ZnO NPs enhanced cellular uptake and the tumor spheroid penetration of DOX by inhibiting P-glycoprotein (Pgp)-mediated drug efflux and three-dimensional (3D) spheroid architecture-induced drug resistance, leading to the efficient killing of cancer and cancer stem cells. In addition, ZnO NPs activated macrophages and boosted anticancer immunity while decreasing the toxicity of DOX against macrophages. Cen et al. synthesized pH-responsive ZnS@BSA (bovine serum albumin) nanoclusters using a diffusive self-assembly approach that can release zinc and sulfur ions in an acidic TME (Figure 4) [69]. The released Zn^{2+} activated cGAS/STING signaling, whereas S^{2-} formed H_2S gas that inhibited catalase and cooperated with zinc ions to generate ROS in hepatocellular carcinoma (HCC) cells. The intravenous administration of ZnS@BSA significantly inhibited HCC tumor growth through anticancer immune responses that could be potentiated by the PD-L1 antibody. It also conferred long-term protection against tumor recurrence, indicating the induction of robust and durable anticancer immunity. Another study by Zhang et al. constructed Zn^{2+} into a nano-formulation using a layered double hydroxide (LDH) that could neutralize acidity in tumor tissues and induce cell apoptosis [71]. Zn^{2+} doping in LDH promoted ROS production to accelerate mitochondrial damage and immunogenic cell death, leading to the activation of cGAS-STING and down-regulation of 'don't eat me' signals (i.e., PD-L1 and CD47). As a result, Zn-LDH stimulated anticancer immune cells (M1 TAM, CTL, and NK cells) and efficiently inhibited tumor growth and the metastasis of murine models of melanoma and breast cancer [71].

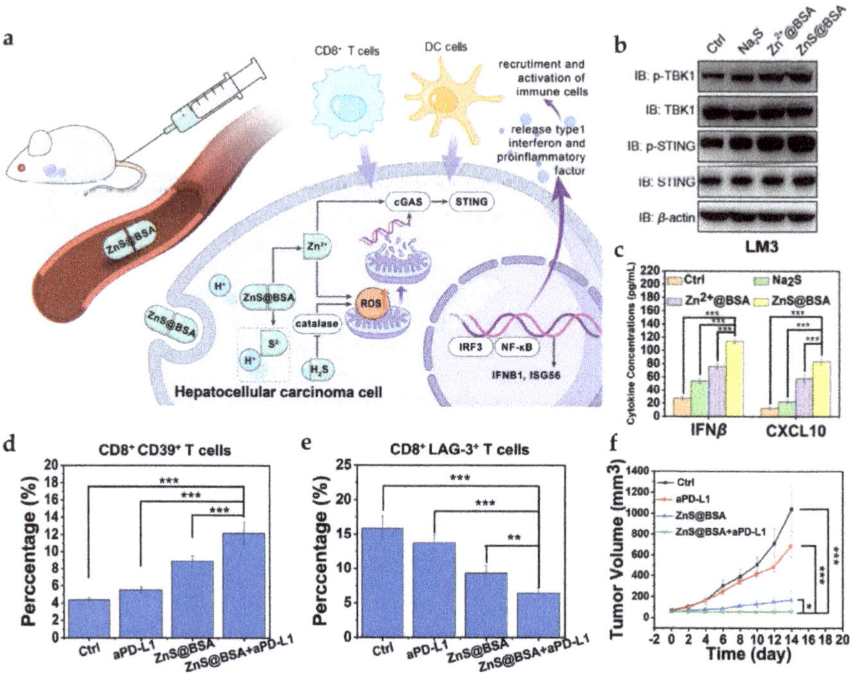

Figure 4. Antitumor activity of ZnS@BSA. (**a**) Schematic diagram of ZnS@BSA function in a HCC cell. (**b**) Western blotting analysis of STING marker expression in LM3 cells. (**c**–**e**) The levels of cytokines and immune cells in randomly selected tumor samples and (**f**) antitumor efficacy in HCC tumor-bearing mice. Reprinted with permission from [69].

4.2. Iron- and Copper-Based Nanoparticles for M1 Macrophage Polarization

Iron-based nanoparticles are mainly oxidation products that include magnetite (Fe_3O_4), maghemite (γ-Fe_2O_3), and mixed ferrites (MFe_2O_4 where M = Co, Mn, Ni, or Zn) [72]. Such iron oxide nanoparticles have proved effective for drug delivery and cancer theranostics due to their biological and electromagnetic properties [73–75]. Recently, it has also been demonstrated that iron oxide nanoparticles can elicit anticancer immune responses. Ferumoxytol, an FDA-approved iron oxide nanoparticle for the treatment of anemia, has been shown to promote the polarization of macrophages into the M1 phenotype [76]. The polarization of M1 is characterized by an increase in pro-M1 genes (TNFα, CD86), a decrease in pro-M2 genes (IL10, CD206) in vitro, and an up-regulation of CD80 expression in vivo by macrophages. Furthermore, ferumoxytol could catalyze ROS generation through the Fenton reaction with hydrogen peroxide secreted by M1 macrophages, leading to the apoptosis of cancer cells accompanied by the expression of caspase 3 [77]. In vivo mice models have demonstrated that ferumoxytol suppresses the tumor growth of early mammary cancers and inhibits the liver and lung metastasis of small-cell lung cancer. In another study, Wu et al. used hollow Fe_3O_4 nanoparticles for the loading of L-arginine (L-Arg), which was followed by poly(acrylic acid) (PAA) sealing to provide pH-responsive release in the acidic TME (Figure 5) [78]. M1 macrophages can efficiently convert L-Arg into NO with elevated levels and catalytic activity of inducible nitric oxide synthase (iNOS), allowing for NO-based gas therapy. Therefore, L-Arg could synergistically enhance the anticancer efficacy of iron oxide nanoparticles by utilizing their M2-to-M1 reprogramming capability. Chen et al. developed iron oxide nanoparticles to boost cancer vaccine activity [75]. They synthesized iron oxide-mesoporous organosilica core–shell nanospheres and then loaded protein antigen in the large-pore organosilica shell layer for vaccine application. The study showed that the hybrid nanoparticles could induce M2-to-M1 macrophage polarization and antigen-specific T cell immunity, leading to strong anticancer immune responses and the complete prevention of tumor development. Recent studies have also demonstrated that copper ions can induce M1 macrophage polarization [79,80]. Xu et al. showed that copper sulfide (CuS) nanoparticles can trigger ROS generation via mitochondrial fission and the Fenton reaction, leading to the activation and M1 polarization of bone marrow-derived macrophages (BMDMs) through the classical IKK-dependent NF-κB pathway [81]. The adoptive transfer of the ex vivo CuS nanoparticle-stimulated BMDMs prolonged the median survival time of melanoma-bearing mice, enhancing the phagocytic and digestive ability of BMDMs to the cancer cells. Another study by Ge et al. produced CuS nanoparticles using ovalbumin proteins (OVA) as a template [82]. The authors demonstrated that CuS@OVA nanoparticles can not only promote M1 macrophage polarization in the acidic TME with the release of copper ions but also induce DC activation and maturation for the priming of T cells. Furthermore, CuS@OVA allowed for mild-temperature photothermal therapy, which, together with the administration of the anti-programmed death-1 antibody (aPD-1), efficiently suppressed the primary growth and distant metastasis of B16-OVA melanoma in mice.

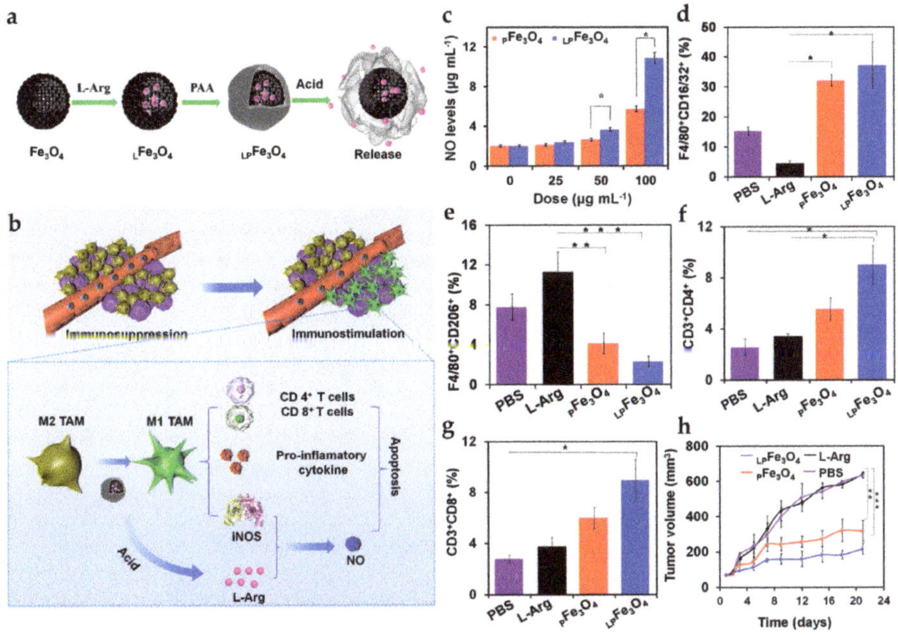

Figure 5. Antitumor activity of $_{LP}Fe_3O_4$ NPs. Schematic diagram of (**a**) L-Arg loading and pH-responsive release and (**b**) M2-M1 macrophage polarization and NO production by $_{LP}Fe_3O_4$ NPs. (**c**) NO levels of M2 macrophages treated with different formulations. (**d**,**e**) TAM phenotype (CD16/32 for M1 and CD206 for M2, respectively), (**f**,**g**) $CD4^+$ and $CD8^+$ T cells, and (**h**) antitumor efficacy in 4T1 tumor-bearing mice. Reprinted with permission from [78]. Copyright 2021 American Chemical Society.

4.3. Calcium- and Sodium-Based Nanoparticles for Ion Overloading Effect

Calcium-based nanoparticles can exhibit various anticancer and immunomodulatory effects caused by Ca^{2+} overloading. An et al. developed honeycomb OVA@$CaCO_3$ nanoparticles (HOCNs) that can be decomposed in an acidic TME and exert several functionalities associated with the release of Ca^{2+}; they alleviate tumor acidity, support autophagy, and promote DAMP secretion from cancer cells [83]. When employed with the chemotherapeutic drug mitoxantrone, HOCN induced the maturation and antigen presentation of DCs and subsequently activated CTLs and effector memory T cells, leading to robust antitumor immunity for the regression of both primary and distant tumors. Recently, Zheng et al. reported curcumin-loaded $CaCO_3$ nanoparticles for pyroptosis-based cancer immunotherapy [84]. Curcumin could stimulate the release of Ca^{2+} from the endoplasmic reticulum into the cytoplasm, which could later accumulate and overload in the mitochondria. The surges of Ca^{2+} could break the dynamic Ca^{2+} equilibrium in the mitochondria, leading to ROS induction, cytochrome C release, caspase 3 activation, gasdermin E cleavage, and eventually cell blebbing and pyroptosis. Ca^{2+} overloading-mediated pyroptosis elicited robust anticancer immune responses (marked by DC maturation and T cell activation), thereby efficiently inhibiting tumor growth. Calcium-based nanoparticles can also re-educate TAM via "Ca^{2+} interference", allowing for M2-to-M1 phenotype switching. An et al. prepared and functionalized calcium peroxide nanoparticles (CaNPs) with a circular aptamer-DNAzyme conjugate (cAD) and a PEG shell (Figure 6) [85]. "Ca^{2+} interference" by CaNPs was found to trigger p38 phosphorylation, NF-κB p65 nuclear translocation, and NLRP3-inflammasome activation, resulting in M2-to-M1 polarization and adaptive immune responses. Furthermore, CaNP induced immunogenic cancer cell death and the release of DAMPs, and cAD mimicked endonuclease activity to block the translation of PD-L1 mRNA using Ca^{2+} as a cofactor. Together, the composite nanoparticles effectively inhibited primary tumor growth, prevented lung metastasis, and developed long-term immunological memory against tumor rechallenge.

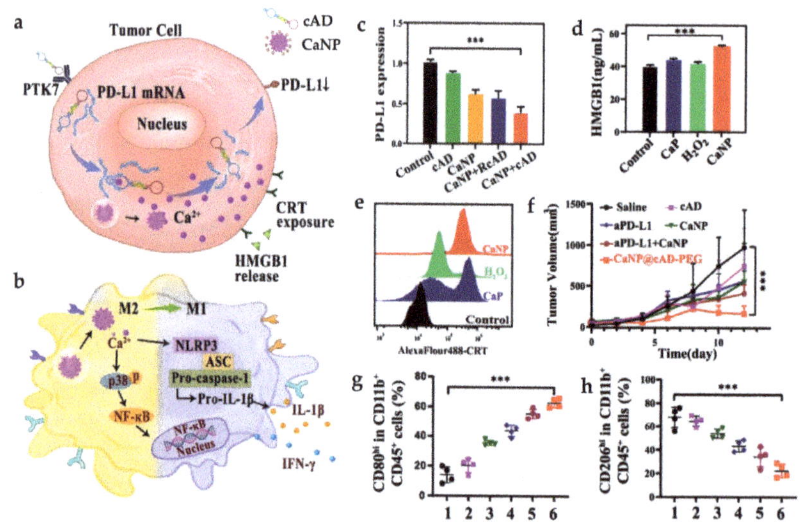

Figure 6. Antitumor activity of CaNP@cAD-PEG. Schematic diagram of (**a**) CaNP function in a tumor cell and (**b**) TAM reprogramming via Ca^{2+} interference. The expression of (**c**) PD-L1 and (**d**,**e**) DAMPs (CRT and HMGB1) in B16 cells. (**f**) Antitumor efficacy in B16 tumor-bearing mice. The levels of (**g**) M1 and (**h**) M2 macrophages. Reprinted with permission from [85].

A recent study by Tang et al. described the preparation of sodium-stabilized dendritic mesoporous aluminosilicate nanoparticles (Na-^{IV}Al-DMSN) as DC pyroptosis modulators and antigen carriers for cancer vaccine applications (Figure 7) [86]. Once internalized by DCs, Na-^{IV}Al-DMSN promoted H^+/Na^+ exchange in acidic lysosomes in a pH-responsive manner, leading to lysosomal rupture and subsequent Na^+ influx/K^+ efflux in cells. The intracellular ion perturbation induced caspase-1-dependent DC pyroptosis and inflammasome activation, which triggers the hyperactivation of bystander DCs, a superior state in priming T cells. Consequently, CT26 cancer cell lysate-loaded Na-^{IV}Al-DMSN boosted immune responses in vivo (marked by the high level of activated DCs), cytotoxic NK cells, CTLs, and effector memory T cells, showing strong anticancer efficacy when administered in a prophylactic or therapeutic setting.

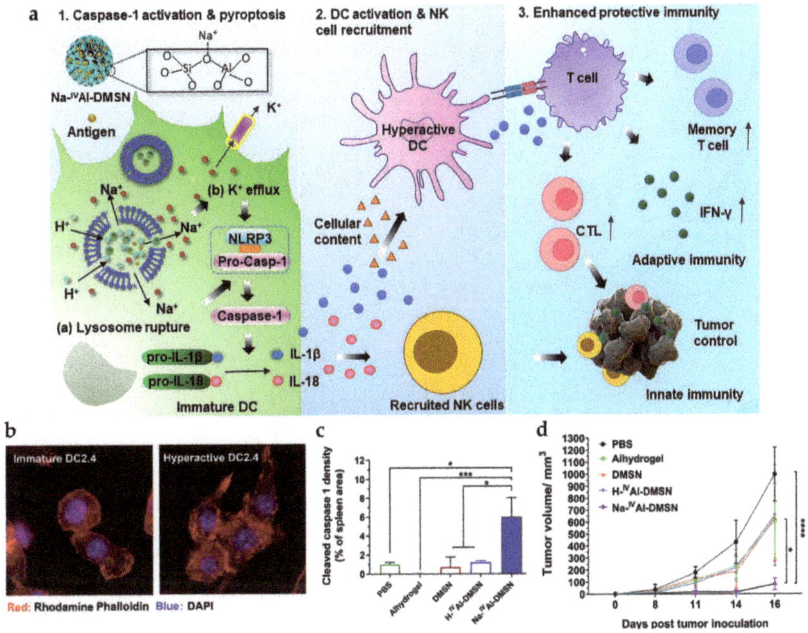

Figure 7. Antitumor activity of Na-^{IV}Al-DMSN. (**a**) Schematic diagram of Na-^{IV}Al-DMSN function in a DC. (**b**) Morphological changes in DC2.4 cells after hyperactivation. (**c**) Caspase-1 expression in the spleen of CT26 tumor-bearing mice. (**d**) In vivo tumor growth in a prophylactic tumor model. Reprinted with permission from [86].

Table 1. Summary of metal-based nanoparticles and their preclinical applications for cancer metalloimmunotherapy.

Metal	Nanoparticle Formulation	Surface Modification and Drug Loading	Nanoparticle Characteristics	Target Cells	Route and Dose (In Vivo)	Properties and Outcomes	References
Mn	CMP_{CDA}	SM: PEG-lipid bilayer	PS: 118 ± 41 nm PDI: 0.107 ZP: −2.49 ± 6.42 mV	In vitro DC: BMDCs In vitro tumor cells: THP1 In vivo tumor model: CT26, B16F10, NOOC1 mice	i.t.: CDA 5 μg, Mn^{2+} 2.5 μg i.v.: CDA 20 μg, Mn^{2+} 10 μg	Deliver STING agonist to immune cells and enhances its activity through the activation of the cGAS-STING pathway	[9]
	Bio-MnO_2 NPs + RT	-	PS: 101 nm PDI: 0.122	In vitro tumor cells: A549, PC9, H520 In vivo tumor model: LLC mice	i.t.: 2 mg/kg + IR: 8 Gy	Relieve tumor hypoxia, activate the cGAS-STING pathway, and sensitize RT	[67]
	PL/APMP-DOX NPs	SM: PL layer DL: DOX	PS: ~180 nm PDI: 0.232 ZP: −23.8 ± 0.6 mV Pore Size: 3.88 nm	In vivo tumor model: 4T1 mice	i.v.: APMP 2.50 mg/kg, DOX 2.50 mg/kg	Activate the cGAS-STING pathway and carry DOX to induce DNA damage	[68]
Zn	ZnO/DOX	DL: DOX	In DMEM/10%FBS medium: PS: 62.02 ± 0.98 nm PDI: 0.31 ± 0.03 ZP: −16.06 ± 0.09 mV	In vitro tumor cells: DOX-sensitive (MDA-MB-231 and HeLa), DOX-resistant (NCI/ADR-RES and MES-SA/Dx5) In vitro macrophage: RAW264.7	-	Generate ROS and activate caspase 3/7, polarize macrophages into M1 phenotype, downregulate the CD44 expression of stem-like cancer cells, and carry DOX to kill cancer cells synergistically	[70]
	ZnS@BSA	-	PS ≈ 100 nm	In vitro tumor cells: HCC (LM3 and Hepa1-6) In vivo tumor model: Hepa1-6 mice	i.v.	Cooperate with H_2S to accumulate ROS and activate the cGAS-STING pathway	[69]
	Zn-LDH	-	-	In vitro tumor cells: B16F10 and 4T1 In vitro DC: DC2.4 In vitro macrophage: RAW264.7 In vivo tumor model: B16F10 and 4T1 mice	PT: 1 mg	Activate the cGAS-STING pathway, block the autophagy pathway, and neutralize TME acidity	[71]

Table 1. *Cont.*

Metal	Nanoparticle Formulation	Surface Modification and Drug Loading	Nanoparticle Characteristics	Target Cells	Route and Dose (In Vivo)	Properties and Outcomes	References
Fe	Ferumoxytol (Feraheme)	-	PS 30 nm	In vitro tumor cells: MMTV-PyMT In vitro macrophage: RAW264.7/BMDMs In vivo tumor model: MMTV-PyMT mice In vivo metastatic model: KP1 mice	MFP injection: 100 µL at a conc. of 2.73 mg Fe/mL i.v.: 10 mg Fe/kg	Polarize macrophages into M1 phenotype	[76]
	Positively and Negatively Charged SPIONs	-	For S+: PS: 19.4 ± 0.8 nm ZP: +44.72 mV For S-: PS: 21.3 ± 1.[illegible] nm ZP: −27.31 mV	In vitro tumor cells: HT1080 In vitro macrophage: RAW264.7 In vivo tumor model: HT1080 mice	i.t.	Polarize macrophages into M1 phenotype	[87]
	$_{LP}Fe_3O_4$ NPs	SM: PAA layer DL: L-Arg	PS: 229.4 ± 5.0 nm ZP: −31.1 ± 0.6 mV Pore Size: 3.5 nm	In vitro tumor cells: 4T1 In vitro macrophage: RAW264.7/BMDMs In vivo tumor model: 4T1 mice	i.v.: 20 mg Fe_3O_4/kg	Polarize macrophages into M1 phenotype and carry L-Arg to induce NO production	[78]
	IO-LPMONs-OVA vaccine + IO-LPMONs	-	PS: 360 nm Pore Size: 6.3 nm	In vitro tumor cells: SCC7 In vitro macrophage: RAW264.7 In vivo tumor model: EG7-OVA mice	s.c.: IO-LPMONs-OVA vaccine (100 µL, conc. 1 mg/mL) + i.t.: IO-LPMONs (100 µL, conc. 0.2 mg/mL)	Polarize macrophages into M1 phenotype and carry OVA for vaccination	[7[illegible]]
Cu	CuS NPs	SM: PEG	PS ≈ 17 nm	In vivo tumor model: B16F10 mice	i.t.	Polarize BMDMs ex vivo into M1 phenotype for adoptive macrophage therapy	[8[illegible]]
Cu	CuS@OVA	DL: OVA	PS: 16.3 ± 0.2 nm	In vitro tumor cells: B16-OVA In vitro DC: BMDCs In vitro macrophage: BMDMs In vivo tumor model: B16-OVA and orthotopic uveal B16-Luc mice	i.t.: CuS-OVA (10 or 2 µL, conc. 1 mg/mL) + i.p.: aPD-1 10 mg/kg	Polarize macrophages into M1 phenotype and activate DCs, serve as a photosensitizer for PTT, and augment the antitumor efficacy of aPD-1	[82]

Table 1. *Cont.*

Metal	Nanoparticle Formulation	Surface Modification and Drug Loading	Nanoparticle Characteristics	Target Cells	Route and Dose (In Vivo)	Properties and Outcomes	References
Ca	MTX + HOCN	-	PS: 250 nm ZP: −7.6 mV	In vitro tumor cells: CT26 In vitro DC: DC2.4 In vivo tumor model: CT26 mice	i.v.: 40 mg/kg	Overcome the multiple barriers of DCs by disrupting autophagy inhibition, attenuating TME acidity, and releasing DAMPs	[83]
	CaNMs	DL: CUR	-	In vitro tumor cells: 4T1 In vivo tumor model: 4T1 mice	i.t.	Induce Ca^{2+} overloading-mediated pyroptosis to stimulate antitumor immune responses	[84]
	CaNP@cAD-PEG	SM: DSPE-PEG2000 DL: cAD	PS: 180 nm	In vitro tumor cells: B16 In vitro macrophage: BMDMs In vivo tumor model: B16 and CT26 mice	i.v.	Induce TAM re-education through multiple inflammation-related signaling pathways as well as NLRP3-inflammasome, promote cancer antigen release, and suppress PD-L1 expression	[85]
Na	Na-^{IV}Al-DMSN	-	PS: ~240 nm	In vitro DC: DC2.4 In vivo tumor model: CT26 mice	s.c.	Induce intracellular ion perturbation for DC pyroptosis and hyperactivation, provoking enhanced NK cell-mediated innate immunity and both cellular and humoral adaptive immune responses	[86]

4T1, murine mammary carcinoma cell; A549, human lung adenocarcinoma cell; aPD-1, anti-programmed cell death-1 antibody; B16F10, murine melanoma cell; B16-Luc, murine melanoma cell labelled with luciferase; B16-OVA, murine melanoma cell expressing OVA; Bio-MnO_2 NP, biomineralized MnO_2 nanoparticle; BMDC, bone-marrow-derived dendritic cell; BMDM, bone-marrow-derived macrophage; cAD, circular aptamer-DNAzyme conjugate; CaNM, Ca^{2+} nanomodulator; CaNP@cAD-PEG, cAD-loaded DSPE-PEG2000-coated CaO_2 nanoparticle; CDA, cyclic di-AMP; cGAS, cyclic GMP-AMP synthase; CMP_{CDA}, CDA-Mn^{2+} particle; CT26, murine colorectal carcinoma cell; CUR, curcumin; DAMP, damage-associated molecular pattern; DC, dendritic cell; DL, drug loading; DMEM/10%FBS, Dulbecco's modified Eagle's medium complete medium containing 10% fetal bovine serum; DNA, deoxyribonucleic acid; DOX, doxorubicin; DSPE-PEG2000, 1,2-distearoyl-sn-glycero-3-phosphoethanolamine-N-PEG-2000; EG7-OVA, murine thymic lymphoma cell transfected with OVA DNA; H520, human lung squamous carcinoma cell; HCC, hepatocellular carcinoma; HeLa, human cervical cancer cell; HOCN, honeycomb OVA@$CaCO_3$ nanoparticle; HT1080, human fibrosarcoma cell; IO-LPMON, Fe_3O_4-embedded large-pore mesoporous organosilica nanosphere; i.p., intraperitoneal; IR, irradiation; i.t., intratumoral; i.v., intravenous; KFI, small-cell lung cancer cell; L-Arg, L-Arginine; LLC, Lewis lung carcinoma cell; LM3, murine mammary adenocarcinoma cell; L-Fe_3O_4 NP, L-Arg-loaded PAA-coated hollow Fe_3O_4 nanoparticle; MDA-MB-231, human triple-negative breast adenocarcinoma cell; MES-SA/Dx5: multi-drug-resistant human uterine sarcoma cell; MFP, mammary fat pad; PTT, photothermal therapy; MMTV-PyMT, mouse mammary tumor virus-polyoma middle T antigen cell; MTX, mitoxantrone; Na-^{IV}Al-DMSN, sodium-stabilized dendritic mesoporous aluminosilicate nanoparticle; NCI/ADR-RES, multi-drug-resistant human ovarian cancer cell; NLRP3, nod-like receptor family pyrin domain containing 3; NO, nitric oxide; NOOC1, immune checkpoint blockade-resistant murine tobacco-associated tumor cell; OVA, ovalbumin; PAA, poly(acrylic acid); PC9, human lung adenocarcinoma cell; PDI, poly-dispersity index; PD-L1, programmed death-ligand 1; PEG, poly(ethylene glycol); PL, phospholipid; PL/APMP-DOX NP, DOX-loaded PL-coated amorphous porous manganese phosphate nanoparticle; PS, particle size; PT, peritumoral; ROS, reactive oxygen species; RT, radiotherapy; s.c., subcutaneous; SCC7, murine squamous cell carcinoma cell; SM, surface modification; SPION, super-paramagnetic Fe_3O_4 nanoparticle; STING, stimulator of interferon genes; TAM, tumor-associated macrophage; THP1, human leukemia monocytic cell; TME, tumor microenvironment; Zn-LDH, Zn^{2+}-doped layered double hydroxide nanoparticle; ZnO/DOX, DOX-loaded ZnO nanoparticle; ZnS@BSA, ZnS@bovine serum albumin nanocluster; ZP, zeta potential.

5. Conclusions

Metalloimmunotherapy offers a promising novel cancer treatment that involves utilizing metal ions to increase anticancer immune responses and anticancer efficacy. Metal-based nanoparticles can improve the pharmacological and drug-like properties of metal ions along with versatile nanoparticle engineering for the desired in vivo activity and functionality. The preclinical studies summarized in this review have demonstrated early promise, calling for further research on the clinical impact of these metal-based nanoparticles. The clinical development of metal-based nanoparticles should also consider cancer immunotherapy barriers and nanomedicine design principles to fulfil the full therapeutic potential of metalloimmunotherapy. In addition, combination therapy could be a promising approach to increse the utility of metal-based nanoparticles and improve their therapeutic outcomes with the rational selection of the combination counterparts.

6. Perspectives and Future Directions

Immunotherapy has proven to be effective in a range of cancers, including those refractory to other standard treatments. However, issues that significantly hamper clinical efficacy and ultimately lead to low response rates have emerged [88]. In particular, the TME poses a significant barrier for cancer immunotherapy. It harbors not only cancer cells but also immune cells, fibroblasts, and the extracellular matrix (ECM) in a compact organized structure, elements that are programmed by cancer cells to support tumor growth, metastasis, and migration [1]. The dense ECM layers in the periphery, together with high interstitial pressure, exclude anticancer immune cells from the tumor core (where cancer cells reside), thus preventing them from exerting effector function [4]. In addition, the TME operates abnormal biological cues that result in hypoxia, increased acidity, limited nutrient availability, and dysregulated ion concentrations, which induce the dysfunction and tolerance of anticancer immune cells while promoting immunosuppressive cells [2,3]. As a result, the TME impairs the activation, proliferation, and anticancer responses of effector cells and other anticancer immune cells crucial for cancer immunotherapy. Furthermore, cancer cells can evade anticancer immune responses through a variety of mechanisms, such as the expression of inhibitory receptors and ligands, loss of MHC and antigens, and secretion of suppressive cytokines and metabolites in the TME [2]. As described in this review, certain metal ions have the potential to stimulate innate and adaptive immune cells for robust anticancer immune responses [1]. In addition, they can directly kill cancer cells by producing ROS and impairing the equilibrium of ion levels in cells. Importantly, metal-based nanoparticles have demonstrated the ability to neutralize acidity, generate oxygen, and reprogram pro-tumor M2 macrophages into antitumor M1 macrophages in the TME, which can transform the immunosuppressive TME into an immune-responsive milieu. These findings suggest that metal-based nanoparticles can potentially address the current efficacy issues of cancer immunotherapy with multipotent anticancer and immunomodulatory effects that are intrinsic to metal ions or generated by the on-demand design and fabrication of nanoparticles. Studies have also suggested that nanoparticle formulations are clearly beneficial in maximizing the potential of metalloimmunotherapy by improving the in vivo pharmacological profile of metal ions. The promise of metal-based nanoparticles can be fulfilled through the continuous development of metal ions and nanoparticle formulations to effectively regulate the TME in order to overcome the physical and immunobiological barriers and counteract immune evasion and the resistance of cancer cells.

The optimal formulation and clinical applications of metal-based nanoparticles will need to adapt to the design principles and general frameworks of nanomedicine. Precisely targeted delivery is a fundamental requirement of nanomedicine to increase the therapeutic window. Physiochemical properties such as composition, size, shape, surface charge, and functionality are key elements that regulate in vivo behavior and the targeted delivery of nanoparticles; thus, they must constitute the primary design parameters for the construction of novel nanomedicine formulations [89]. In addition, the on-target release of drug molecules and other functional nanoparticles in their active forms is essential for maximizing the drug like properties of a nanomedicine. Furthermore, nanoparticles can be designed to decompose in response to specific signals such as high acidity, hypoxia, and enzymes that are exclusive or overexpressed in tumor tissues to allow for site-selective drug action that can decrease off-target toxicity and increase dose response [90]. The toxicity issue has been a critical hurdle for the clinical translation of nanomedicine; therefore, it must be considered in order to develop metal-based nanoparticles. The metal-based nanoparticles covered in this paper have been proven to be biocompatible when assessed according to body weight change, hematological marker levels, and/or histopathological analysis. Although preclinical studies have generally demonstrated the biocompatibility and safety of nanoformulations, these preclinical observations in mouse models often do not translate to human subjects because of the large physiological and anatomical gap, meaning that toxicity concerns are a major hindrance to the clinical application of nanomedicine. Nanoparticle surface modification is a common strategy to control specific and nonspecific interactions in vivo, thereby minimizing the off-target side effects and maximizing on-target therapeutic efficiency. Some approaches that have employed organic surface-coating layers or naturally derived cell membranes to modulate surface charge density, hydrophobicity, and functionality have been shown to ameliorate toxicity and improve the in vivo performance of nanoparticles [91]. Surface modification can also have a direct impact on nanoparticle activity [87]. Since many metal ions are essential nutritional elements that are opted into by the host, they are generally considered biocompatible; therefore, they can be less toxic drug candidates compared to other exogenous and synthetic agents [2]. Indeed, metal ions are well tolerated by animals; for example, the median lethal doses are 2000 mg/kg for Mn and Zn ions and 7500 mg/kg for Fe ion in mice, according to the information in the material safety data sheet provided by the manufacturers. However, nano-engineering modulates the physicochemical properties of metal ions, which are closely associated with their in vivo distribution, cellular responses, and toxicity profile. In addition, colloidal stability is another important factor that determines the in vivo behavior and toxicity of nanoparticles as the aggregation can alter their physicochemical properties. The preclinical examinations in the literature have generally reported that metal-based nanoparticles have good colloidal stability in a buffer and biological medium with hydrophilic surface modifications. Nonetheless, further investigations are required to test their long-term stability under practical transportation and storage conditions. Overall, how nanoparticle engineering affects the acute and chronic toxicity of metal ions remains to be thoroughly examined, and this must be investigated and optimized for individual nanoparticles with distinct physicochemical features. Accordingly, the toxicity issue will remain at the center of clinical applications of metal-based nanoparticles, and this issue should be carefully addressed along with the delivery issues. Currently, The United States Food and Drug Administration (FDA)-approved nanomedicines are mostly organic molecule-based nanoparticles composed of lipid and polymer. Although several iron-based nanoparticles have been approved by the FDA as inorganic and metal-based nanoparticles, they are indicated only for iron replacement therapy for the treatment of iron deficiency [92,93]. The FDA has published general guidelines for the production and clinical evaluation of new drug candidates; however, detailed regulations for the clinical development of novel inorganic nanoparticles, including metal-based nanoparticles, are yet to be established specifically for cancer immunotherapy applications.

The heterogeneity of cancer cells often makes a single intervention inefficient for controlling tumor growth. For more effective cancer treatment, the combination of multiple, non-overlapping treatments has been pursued to synergistically kill cancer cells through distinct cytotoxicity pathways. Metalloimmunotherapy can also play a role in combination therapy by harnessing several metal ions with unique immunostimulatory properties; for example, cGAS-STING activation (Mn, Zn), M1 macrophage polarization (Fe, Cu), and pyroptosis-induced immune responses (Ca, Na) can be simultaneously achieved by integrating the respective metal ions (mentioned directly above in parentheses) in a single nanoparticle formulation. Alternatively, metalloimmunotherapy can be applied in conjunction with other interventions that can enhance cancer cell elimination and anticancer immune responses via distinct therapeutic mechanisms, such as chemotherapy, RT, and other types of cancer immunotherapy. Some of the studies described in this review have already demonstrated the potential of combination therapy, reporting substantially improved therapeutic outcomes compared to individual treatments. For optimal combination therapy, combination agents and interventions should be rationally selected to maximize their synergistic efficacy with complementary anticancer activity. In addition, the novel engineering of nanoparticle formulations that support combination therapy would be of great interest and further facilitate the future development of metal-based nanoparticles.

Author Contributions: Conceptualization, I.H.S. and J.N.; investigation, I.H.S., K.K., W.C., Y.K. and J.-H.M.; writing—original draft preparation, I.H.S., K.K., W.C., Y.K. and J.-H.M.; writing—review and editing, S.S. and J.N.; supervision, S.S. and J.N.; funding acquisition, S.S. and J.N. All authors have read and agreed to the published version of the manuscript.

Funding: This research was funded by the National Research Foundation of Korea (NRF) grant funded by the Korean government (MSIT) (No. 2022R1A2C1006643, 2022R1A4A3026347, RS-2023-00212229), Chonnam National University, and INHA UNIVERSITY Research Grant.

Data Availability Statement: No new data were created or analyzed in this study. Data sharing is not applicable to this article.

Conflicts of Interest: The authors declare no conflict of interest.

References

1. Sung, H.; Ferlay, J.; Siegel, R.L.; Laversanne, M.; Soerjomataram, I.; Jemal, A.; Bray, F. Global Cancer Statistics 2020: GLOBOCAN Estimates of Incidence and Mortality Worldwide for 36 Cancers in 185 Countries. *CA Cancer J. Clin.* **2021**, *71*, 209–249. [CrossRef] [PubMed]
2. Shariatzadeh, S.; Moghimi, N.; Khalafi, F.; Shafiee, S.; Mehrabi, M.; Ilkhani, S.; Tosan, F.; Nakhaei, P.; Alizadeh, A.; Varma, R.S.; et al. Metallic Nanoparticles for the Modulation of Tumor Microenvironment; A New Horizon. *Front. Bioeng. Biotechnol.* **2022**, *10*, 847433. [CrossRef] [PubMed]
3. Mohapatra, A.; Sathiyamoorthy, P.; Park, I.-K. Metallic Nanoparticle-Mediated Immune Cell Regulation and Advanced Cancer Immunotherapy. *Pharmaceutics* **2021**, *13*, 1867. [CrossRef] [PubMed]
4. Waldman, A.D.; Fritz, J.M.; Lenardo, M.J. A Guide to Cancer Immunotherapy: From T Cell Basic Science to Clinical Practice. *Nat. Rev. Immunol.* **2020**, *20*, 651–668. [CrossRef] [PubMed]
5. Wang, D.-R.; Wu, X.-L.; Sun, Y.-L. Therapeutic Targets and Biomarkers of Tumor Immunotherapy: Response versus Non-Response. *Signal Transduct. Target. Ther.* **2022**, *7*, 331. [CrossRef]
6. Mishra, A.K.; Ali, A.; Dutta, S.; Banday, S.; Malonia, S.K. Emerging Trends in Immunotherapy for Cancer. *Diseases* **2022**, *10*, 60. [CrossRef]
7. Peterson, C.; Denlinger, N.; Yang, Y. Recent Advances and Challenges in Cancer Immunotherapy. *Cancers* **2022**, *14*, 3972. [CrossRef]
8. Wang, C.; Zhang, R.; Wei, X.; Lv, M.; Jiang, Z. Metalloimmunology: The Metal Ion-Controlled Immunity. *Adv. Immunol.* **2020**, *145*, 187–241. [CrossRef]
9. Sun, X.; Zhang, Y.; Li, J.; Park, K.S.; Han, K.; Zhou, X.; Xu, Y.; Nam, J.; Xu, J.; Shi, X.; et al. Amplifying STING Activation by Cyclic Dinucleotide–Manganese Particles for Local and Systemic Cancer Metalloimmunotherapy. *Nat. Nanotechnol.* **2021**, *16*, 1260–1270. [CrossRef]
10. Nam, J.; Son, S.; Park, K.S.; Zou, W.; Shea, L.D.; Moon, J.J. Cancer Nanomedicine for Combination Cancer Immunotherapy. *Nat. Rev. Mater.* **2019**, *4*, 398–414. [CrossRef]
11. Surendran, S.P.; Moon, M.J.; Park, R.; Jeong, Y.Y. Bioactive Nanoparticles for Cancer Immunotherapy. *Int. J. Mol. Sci.* **2018**, *19*, 3877. [CrossRef]
12. Pandya, P.H.; Murray, M.E.; Pollok, K.E.; Renbarger, J.L. The Immune System in Cancer Pathogenesis: Potential Therapeutic Approaches. *J. Immunol. Res.* **2016**, *2016*, e4273943. [CrossRef]
13. Abbas, A.K.; Lichtman, A.H.; Pillai, S.; Baker, D.L. *Cellular and Molecular Immunology*, 10th ed.; Elsevier: Philadelphia, PA, USA, 2022; ISBN 978-0-323-75748-5.
14. Chaplin, D.D. Overview of the Immune Response. *J. Allergy Clin. Immunol.* **2010**, *125*, S3–S23. [CrossRef] [PubMed]
15. Marshall, J.S.; Warrington, R.; Watson, W.; Kim, H.L. An Introduction to Immunology and Immunopathology. *Allergy Asthma Clin. Immunol.* **2018**, *14*, 49. [CrossRef] [PubMed]
16. Schreiber, R.D.; Old, L.J.; Smyth, M.J. Cancer Immunoediting: Integrating Immunity's Roles in Cancer Suppression and Promotion. *Science* **2011**, *331*, 1565–1570. [CrossRef] [PubMed]

17. Liu, Y.; Zeng, G. Cancer and Innate Immune System Interactions: Translational Potentials for Cancer Immunotherapy. *J. Immunother.* **2012**, *35*, 299–308. [CrossRef]
18. Janssen, L.M.E.; Ramsay, E.E.; Logsdon, C.D.; Overwijk, W.W. The Immune System in Cancer Metastasis: Friend or Foe? *J. Immunother. Cancer* **2017**, *5*, 79. [CrossRef] [PubMed]
19. Swann, J.B.; Smyth, M.J. Immune Surveillance of Tumors. *J. Clin. Investig.* **2007**, *117*, 1137–1146. [CrossRef] [PubMed]
20. O'Donnell, J.S.; Teng, M.W.L.; Smyth, M.J. Cancer Immunoediting and Resistance to T Cell-Based Immunotherapy. *Nat. Rev. Clin. Oncol.* **2019**, *16*, 151–167. [CrossRef]
21. Beatty, G.L.; Gladney, W.L. Immune Escape Mechanisms as a Guide for Cancer Immunotherapy. *Clin. Cancer Res.* **2015**, *21*, 687–692. [CrossRef]
22. Wculek, S.K.; Cueto, F.J.; Mujal, A.M.; Melero, I.; Krummel, M.F.; Sancho, D. Dendritic Cells in Cancer Immunology and Immunotherapy. *Nat. Rev. Immunol.* **2020**, *20*, 7–24. [CrossRef] [PubMed]
23. Gardner, A.; de Mingo Pulido, Á.; Ruffell, B. Dendritic Cells and Their Role in Immunotherapy. *Front. Immunol.* **2020**, *11*, 924. [CrossRef] [PubMed]
24. Gardner, A.; Ruffell, B. Dendritic Cells and Cancer Immunity. *Trends Immunol.* **2016**, *37*, 855–865. [CrossRef] [PubMed]
25. Katopodi, T.; Petanidis, S.; Charalampidis, C.; Chatziprodromidou, I.; Eskitzis, P.; Tsavlis, D.; Zarogoulidis, P.; Kosmidis, C.; Matthaios, D.; Porpodis, K. Tumor-Infiltrating Dendritic Cells: Decisive Roles in Cancer Immunosurveillance, Immunoediting, and Tumor T Cell Tolerance. *Cells* **2022**, *11*, 3183. [CrossRef]
26. Del Prete, A.; Sozio, F.; Barbazza, I.; Salvi, V.; Tiberio, L.; Laffranchi, M.; Gismondi, A.; Bosisio, D.; Schioppa, T.; Sozzani, S. Functional Role of Dendritic Cell Subsets in Cancer Progression and Clinical Implications. *Int. J. Mol. Sci.* **2020**, *21*, 3930. [CrossRef]
27. Noubade, R.; Majri-Morrison, S.; Tarbell, K.V. Beyond CDC1: Emerging Roles of DC Crosstalk in Cancer Immunity. *Front. Immunol.* **2019**, *10*, 1014. [CrossRef]
28. Kumari, N.; Choi, S.H. Tumor-Associated Macrophages in Cancer: Recent Advancements in Cancer Nanoimmunotherapies. *J. Exp. Clin. Cancer Res.* **2022**, *41*, 68. [CrossRef]
29. Galli, F.; Aguilera, J.V.; Palermo, B.; Markovic, S.N.; Nisticò, P.; Signore, A. Relevance of Immune Cell and Tumor Microenvironment Imaging in the New Era of Immunotherapy. *J. Exp. Clin. Cancer Res.* **2020**, *39*, 89. [CrossRef]
30. Lorenzo-Sanz, L.; Muñoz, P. Tumor-Infiltrating Immunosuppressive Cells in Cancer-Cell Plasticity, Tumor Progression and Therapy Response. *Cancer Microenviron.* **2019**, *12*, 119–132. [CrossRef]
31. Duan, Z.; Luo, Y. Targeting Macrophages in Cancer Immunotherapy. *Signal Transduct. Target. Ther.* **2021**, *6*, 127. [CrossRef]
32. Raskov, H.; Orhan, A.; Christensen, J.P.; Gögenur, I. Cytotoxic CD8+ T Cells in Cancer and Cancer Immunotherapy. *Br. J. Cancer* **2021**, *124*, 359–367. [CrossRef]
33. Chen, D.S.; Mellman, I. Oncology Meets Immunology: The Cancer-Immunity Cycle. *Immunity* **2013**, *39*, 1–10. [CrossRef] [PubMed]
34. Munhoz, R.R.; Postow, M.A. Recent Advances in Understanding Antitumor Immunity. *F1000Research* **2016**, *5*, 2545. [CrossRef]
35. Chu, J.; Gao, F.; Yan, M.; Zhao, S.; Yan, Z.; Shi, B.; Liu, Y. Natural Killer Cells: A Promising Immunotherapy for Cancer. *J. Transl. Med.* **2022**, *20*, 240. [CrossRef]
36. Liu, S.; Galat, V.; Galat4, Y.; Lee, Y.K.A.; Wainwright, D.; Wu, J. NK Cell-Based Cancer Immunotherapy: From Basic Biology to Clinical Development. *J. Hematol. Oncol.* **2021**, *14*, 7. [CrossRef]
37. Shimasaki, N.; Jain, A.; Campana, D. NK Cells for Cancer Immunotherapy. *Nat. Rev. Drug Discov.* **2020**, *19*, 200–218. [CrossRef] [PubMed]
38. Xiao, J.; Zhang, T.; Gao, F.; Zhou, Z.; Shu, G.; Zou, Y.; Yin, G. Natural Killer Cells: A Promising Kit in the Adoptive Cell Therapy Toolbox. *Cancers* **2022**, *14*, 5657. [CrossRef] [PubMed]
39. Sordo-Bahamonde, C.; Lorenzo-Herrero, S.; Payer, Á.R.; Gonzalez, S.; López-Soto, A. Mechanisms of Apoptosis Resistance to NK Cell-Mediated Cytotoxicity in Cancer. *Int. J. Mol. Sci.* **2020**, *21*, 3726. [CrossRef]
40. Paul, S.; Lal, G. The Molecular Mechanism of Natural Killer Cells Function and Its Importance in Cancer Immunotherapy. *Front. Immunol.* **2017**, *8*, 1124. [CrossRef]
41. Liu, Y.; Cao, X. Immunosuppressive Cells in Tumor Immune Escape and Metastasis. *J. Mol. Med.* **2016**, *94*, 509–522. [CrossRef]
42. Shi, L.; Gu, H. Emerging Nanoparticle Strategies for Modulating Tumor-Associated Macrophage Polarization. *Biomolecules* **2021**, *11*, 1912. [CrossRef] [PubMed]
43. Tie, Y.; Tang, F.; Wei, Y.; Wei, X. Immunosuppressive Cells in Cancer: Mechanisms and Potential Therapeutic Targets. *J. Hematol. Oncol.* **2022**, *15*, 61. [CrossRef] [PubMed]
44. Sui, H.; Dongye, S.; Liu, X.; Xu, X.; Wang, L.; Jin, C.Q.; Yao, M.; Gong, Z.; Jiang, D.; Zhang, K.; et al. Immunotherapy of Targeting MDSCs in Tumor Microenvironment. *Front. Immunol.* **2022**, *13*, 990463. [CrossRef] [PubMed]
45. Li, T.; Liu, T.; Zhu, W.; Xie, S.; Zhao, Z.; Feng, B.; Guo, H.; Yang, R. Targeting MDSC for Immune-Checkpoint Blockade in Cancer Immunotherapy: Current Progress and New Prospects. *Clin. Med. Insights Oncol.* **2021**, *15*, 11795549211035540. [CrossRef]
46. Gao, Y.; Ouyang, Z.; Yang, C.; Song, C.; Jiang, C.; Song, S.; Shen, M.; Shi, X. Overcoming T Cell Exhaustion via Immune Checkpoint Modulation with a Dendrimer-Based Hybrid Nanocomplex. *Adv. Healthc. Mater.* **2021**, *10*, 2100833. [CrossRef]

47. Verma, N.K.; Wong, B.H.S.; Poh, Z.S.; Udayakumar, A.; Verma, R.; Goh, R.K.J.; Duggan, S.P.; Shelat, V.G.; Chandy, K.G.; Grigoropoulos, N.F. Obstacles for T-Lymphocytes in the Tumour Microenvironment: Therapeutic Challenges, Advances and Opportunities beyond Immune Checkpoint. *eBioMedicine* **2022**, *83*, 104216. [CrossRef]
48. Bayati, F.; Mohammadi, M.; Valadi, M.; Jamshidi, S.; Foma, A.M.; Sharif-Paghaleh, E. The Therapeutic Potential of Regulatory T Cells: Challenges and Opportunities. *Front. Immunol.* **2021**, *11*, 585819. [CrossRef]
49. Chen, B.-J.; Zhao, J.-W.; Zhang, D.-H.; Zheng, A.-H.; Wu, G.-Q. Immunotherapy of Cancer by Targeting Regulatory T Cells. *Int. Immunopharmacol.* **2022**, *104*, 108469. [CrossRef]
50. Sabado, R.L.; Balan, S.; Bhardwaj, N. Dendritic Cell-Based Immunotherapy. *Cell Res.* **2017**, *27*, 74–95. [CrossRef]
51. Shurin, G.V.; Ma, Y.; Shurin, M.R. Immunosuppressive Mechanisms of Regulatory Dendritic Cells in Cancer. *Cancer Microenviron.* **2013**, *6*, 159–167. [CrossRef]
52. Serra, M.; Columbano, A.; Ammarah, U.; Mazzone, M.; Menga, A. Understanding Metal Dynamics Between Cancer Cells and Macrophages: Competition or Synergism? *Front. Oncol.* **2020**, *10*, 646. [CrossRef]
53. Monteith, A.J.; Skaar, E.P. The Impact of Metal Availability on Immune Function during Infection. *Trends Endocrinol. Metab.* **2021**, *32*, 916–928. [CrossRef] [PubMed]
54. Hood, M.I.; Skaar, E.P. Nutritional Immunity: Transition Metals at the Pathogen–Host Interface. *Nat. Rev. Microbiol.* **2012**, *10*, 525–537. [CrossRef]
55. Bonaventura, P.; Benedetti, G.; Albarède, F.; Miossec, P. Zinc and Its Role in Immunity and Inflammation. *Autoimmun. Rev.* **2015**, *14*, 277–285. [CrossRef]
56. Muñoz-Planillo, R.; Kuffa, P.; Martínez-Colón, G.; Smith, B.L.; Rajendiran, T.M.; Núñez, G. K^+ Efflux Is the Common Trigger of NLRP3 Inflammasome Activation by Bacterial Toxins and Particulate Matter. *Immunity* **2013**, *38*, 1142–1153. [CrossRef] [PubMed]
57. Rossol, M.; Pierer, M.; Raulien, N.; Quandt, D.; Meusch, U.; Rothe, K.; Schubert, K.; Schöneberg, T.; Schaefer, M.; Krügel, U.; et al. Extracellular Ca^{2+} Is a Danger Signal Activating the NLRP3 Inflammasome through G Protein-Coupled Calcium Sensing Receptors. *Nat. Commun.* **2012**, *3*, 1329. [CrossRef] [PubMed]
58. Scambler, T.; Jarosz-Griffiths, H.H.; Lara-Reyna, S.; Pathak, S.; Wong, C.; Holbrook, J.; Martinon, F.; Savic, S.; Peckham, D.; McDermott, M.F. ENaC-Mediated Sodium Influx Exacerbates NLRP3-Dependent Inflammation in Cystic Fibrosis. *eLife* **2019**, *8*, e49248. [CrossRef]
59. Du, M.; Chen, Z.J. DNA-Induced Liquid Phase Condensation of CGAS Activates Innate Immune Signaling. *Science* **2018**, *361*, 704–709. [CrossRef]
60. Haase, H. Innate Immune Cells Speak Manganese. *Immunity* **2018**, *48*, 616–618. [CrossRef]
61. Shi, X.; Bi, Y.; Yang, W.; Guo, X.; Jiang, Y.; Wan, C.; Li, L.; Bai, Y.; Guo, J.; Wang, Y.; et al. Ca^{2+} Regulates T-Cell Receptor Activation by Modulating the Charge Property of Lipids. *Nature* **2013**, *493*, 111–115. [CrossRef]
62. Vodnala, S.K.; Eil, R.; Kishton, R.J.; Sukumar, M.; Yamamoto, T.N.; Ha, N.-H.; Lee, P.-H.; Shin, M.; Patel, S.J.; Yu, Z.; et al. T Cell Stemness and Dysfunction in Tumors Are Triggered by a Common Mechanism. *Science* **2019**, *363*, eaau0135. [CrossRef] [PubMed]
63. Rasmussen, J.W.; Martinez, E.; Louka, P.; Wingett, D.G. Zinc Oxide Nanoparticles for Selective Destruction of Tumor Cells and Potential for Drug Delivery Applications. *Expert Opin. Drug Deliv.* **2010**, *7*, 1063–1077. [CrossRef] [PubMed]
64. Irvine, D.J.; Dane, E.L. Enhancing Cancer Immunotherapy with Nanomedicine. *Nat. Rev. Immunol.* **2020**, *20*, 321–334. [CrossRef]
65. Wang, C.; Guan, Y.; Lv, M.; Zhang, R.; Guo, Z.; Wei, X.; Du, X.; Yang, J.; Li, T.; Wan, Y.; et al. Manganese Increases the Sensitivity of the CGAS-STING Pathway for Double-Stranded DNA and Is Required for the Host Defense against DNA Viruses. *Immunity* **2018**, *48*, 675–687.e7. [CrossRef]
66. Lv, M.; Chen, M.; Zhang, R.; Zhang, W.; Wang, C.; Zhang, Y.; Wei, X.; Guan, Y.; Liu, J.; Feng, K.; et al. Manganese Is Critical for Antitumor Immune Responses via CGAS-STING and Improves the Efficacy of Clinical Immunotherapy. *Cell Res.* **2020**, *30*, 966–979. [CrossRef] [PubMed]
67. Liu, X.; Kifle, M.T.; Xie, H.; Xu, L.; Luo, M.; Li, Y.; Huang, Z.; Gong, Y.; Wu, Y.; Xie, C. Biomineralized Manganese Oxide Nanoparticles Synergistically Relieve Tumor Hypoxia and Activate Immune Response with Radiotherapy in Non-Small Cell Lung Cancer. *Nanomaterials* **2022**, *12*, 3138. [CrossRef]
68. Hou, L.; Tian, C.; Yan, Y.; Zhang, L.; Zhang, H.; Zhang, Z. Manganese-Based Nanoactivator Optimizes Cancer Immunotherapy via Enhancing Innate Immunity. *ACS Nano* **2020**, *14*, 3927–3940. [CrossRef]
69. Cen, D.; Ge, Q.; Xie, C.; Zheng, Q.; Guo, J.; Zhang, Y.; Wang, Y.; Li, X.; Gu, Z.; Cai, X. ZnS@BSA Nanoclusters Potentiate Efficacy of Cancer Immunotherapy. *Adv. Mater.* **2021**, *33*, 2104037. [CrossRef]
70. Wang, J.; Lee, J.S.; Kim, D.; Zhu, L. Exploration of Zinc Oxide Nanoparticles as a Multitarget and Multifunctional Anticancer Nanomedicine. *ACS Appl. Mater. Interfaces* **2017**, *9*, 39971–39984. [CrossRef]
71. Zhang, L.; Zhao, J.; Hu, X.; Wang, C.; Jia, Y.; Zhu, C.; Xie, S.; Lee, J.; Li, F.; Ling, D. A Peritumorally Injected Immunomodulating Adjuvant Elicits Robust and Safe Metalloimmunotherapy against Solid Tumors. *Adv. Mater.* **2022**, *34*, 2206915. [CrossRef]
72. Dadfar, S.M.; Roemhild, K.; Drude, N.I.; von Stillfried, S.; Knüchel, R.; Kiessling, F.; Lammers, T. Iron Oxide Nanoparticles: Diagnostic, Therapeutic and Theranostic Applications. *Adv. Drug Deliv. Rev.* **2019**, *138*, 302–325. [CrossRef]
73. Montiel Schneider, M.G.; Martín, M.J.; Otarola, J.; Vakarelska, E.; Simeonov, V.; Lassalle, V.; Nedyalkova, M. Biomedical Applications of Iron Oxide Nanoparticles: Current Insights Progress and Perspectives. *Pharmaceutics* **2022**, *14*, 204. [CrossRef]
74. Arias, L.; Pessan, J.; Vieira, A.; Lima, T.; Delbem, A.; Monteiro, D. Iron Oxide Nanoparticles for Biomedical Applications: A Perspective on Synthesis, Drugs, Antimicrobial Activity, and Toxicity. *Antibiotics* **2018**, *7*, 46. [CrossRef] [PubMed]

75. Chen, L.; Ma, X.; Dang, M.; Dong, H.; Hu, H.; Su, X.; Liu, W.; Wang, Q.; Mou, Y.; Teng, Z. Simultaneous T Cell Activation and Macrophage Polarization to Promote Potent Tumor Suppression by Iron Oxide-Embedded Large-Pore Mesoporous Organosilica Core–Shell Nanospheres. *Adv. Healthc. Mater.* **2019**, *8*, 1900039. [CrossRef]
76. Zanganeh, S.; Hutter, G.; Spitler, R.; Lenkov, O.; Mahmoudi, M.; Shaw, A.; Pajarinen, J.S.; Nejadnik, H.; Goodman, S.; Moseley, M.; et al. Iron Oxide Nanoparticles Inhibit Tumour Growth by Inducing Pro-Inflammatory Macrophage Polarization in Tumour Tissues. *Nat. Nanotechnol.* **2016**, *11*, 986–994. [CrossRef] [PubMed]
77. Zhang, J.; Wang, X.; Cui, W.; Wang, W.; Zhang, H.; Liu, L.; Zhang, Z.; Li, Z.; Ying, G.; Zhang, N.; et al. Visualization of Caspase-3-like Activity in Cells Using a Genetically Encoded Fluorescent Biosensor Activated by Protein Cleavage. *Nat. Commun.* **2013**, *4*, 2157. [CrossRef] [PubMed]
78. Wu, X.; Cheng, Y.; Zheng, R.; Xu, K.; Yan, J.; Song, P.; Wang, Y.; Rauf, A.; Pan, Y.; Zhang, H. Immunomodulation of Tumor Microenvironment by Arginine-Loaded Iron Oxide Nanoparticles for Gaseous Immunotherapy. *ACS Appl. Mater. Interfaces* **2021**, *13*, 19825–19835. [CrossRef] [PubMed]
79. Díez-Tercero, L.; Delgado, L.M.; Bosch-Rué, E.; Perez, R.A. Evaluation of the Immunomodulatory Effects of Cobalt, Copper and Magnesium Ions in a pro Inflammatory Environment. *Sci. Rep.* **2021**, *11*, 11707. [CrossRef] [PubMed]
80. Díez-Tercero, L.; Delgado, L.M.; Perez, R.A. Modulation of Macrophage Response by Copper and Magnesium Ions in Combination with Low Concentrations of Dexamethasone. *Biomedicines* **2022**, *10*, 764. [CrossRef]
81. Xu, J.; Zheng, B.; Zhang, S.; Liao, X.; Tong, Q.; Wei, G.; Yu, S.; Chen, G.; Wu, A.; Gao, S.; et al. Copper Sulfide Nanoparticle-Redirected Macrophages for Adoptive Transfer Therapy of Melanoma. *Adv. Funct. Mater.* **2021**, *31*, 2008022. [CrossRef]
82. Ge, Y.; Zhang, J.; Jin, K.; Ye, Z.; Wang, W.; Zhou, Z.; Ye, J. Multifunctional Nanoparticles Precisely Reprogram the Tumor Microenvironment and Potentiate Antitumor Immunotherapy after Near-Infrared-II Light-Mediated Photothermal Therapy. *Acta Biomater.* **2023**, *in press*. [CrossRef] [PubMed]
83. An, J.; Zhang, K.; Wang, B.; Wu, S.; Wang, Y.; Zhang, H.; Zhang, Z.; Liu, J.; Shi, J. Nanoenabled Disruption of Multiple Barriers in Antigen Cross-Presentation of Dendritic Cells *via* Calcium Interference for Enhanced Chemo-Immunotherapy. *ACS Nano* **2020**, *14*, 7639–7650. [CrossRef] [PubMed]
84. Zheng, P.; Ding, B.; Zhu, G.; Li, C.; Lin, J. Biodegradable Ca^{2+} Nanomodulators Activate Pyroptosis through Mitochondrial Ca^{2+} Overload for Cancer Immunotherapy. *Angew. Chem. Int. Ed.* **2022**, *61*, e202204904. [CrossRef]
85. An, J.; Liu, M.; Zhao, L.; Lu, W.; Wu, S.; Zhang, K.; Liu, J.; Zhang, Z.; Shi, J. Boosting Tumor Immunotherapy by Bioactive Nanoparticles via Ca^{2+} Interference Mediated TME Reprogramming and Specific PD-L1 Depletion. *Adv. Funct. Mater.* **2022**, *32*, 2201275. [CrossRef]
86. Tang, J.; Yang, Y.; Qu, J.; Ban, W.; Song, H.; Gu, Z.; Yang, Y.; Cai, L.; Theivendran, S.; Wang, Y.; et al. Mesoporous Sodium Four-Coordinate Aluminosilicate Nanoparticles Modulate Dendritic Cell Pyroptosis and Activate Innate and Adaptive Immunity. *Chem. Sci.* **2022**, *13*, 8507–8517. [CrossRef]
87. Zhang, W.; Cao, S.; Liang, S.; Tan, C.H.; Luo, B.; Xu, X.; Saw, P.E. Differently Charged Super-Paramagnetic Iron Oxide Nanoparticles Preferentially Induced M1-Like Phenotype of Macrophages. *Front. Bioeng. Biotechnol.* **2020**, *8*, 537. [CrossRef]
88. Tan, S.; Li, D.; Zhu, X. Cancer Immunotherapy: Pros, Cons and Beyond. *Biomed. Pharmacother.* **2020**, *124*, 109821. [CrossRef]
89. Sanità, G.; Carrese, B.; Lamberti, A. Nanoparticle Surface Functionalization: How to Improve Biocompatibility and Cellular Internalization. *Front. Mol. Biosci.* **2020**, *7*, 587012. [CrossRef]
90. Cao, Z.; Li, W.; Liu, R.; Li, X.; Li, H.; Liu, L.; Chen, Y.; Lv, C.; Liu, Y. PH- and Enzyme-Triggered Drug Release as an Important Process in the Design of Anti-Tumor Drug Delivery Systems. *Biomed. Pharmacother.* **2019**, *118*, 109340. [CrossRef]
91. Zhang, N.; Xiong, G.; Liu, Z. Toxicity of Metal-Based Nanoparticles: Challenges in the Nano Era. *Front. Bioeng. Biotechnol.* **2022**, *10*, 1001572. [CrossRef]
92. Thapa, R.K.; Kim, J.O. Nanomedicine-Based Commercial Formulations: Current Developments and Future Prospects. *J. Pharm. Investig.* **2023**, *53*, 19–33. [CrossRef] [PubMed]
93. Mitchell, M.J.; Billingsley, M.M.; Haley, R.M.; Wechsler, M.E.; Peppas, N.A.; Langer, R. Engineering Precision Nanoparticles for Drug Delivery. *Nat. Rev. Drug Discov.* **2021**, *20*, 101–124. [CrossRef] [PubMed]

pharmaceutics

Article

Biocompatible Calcium Ion-Doped Magnesium Ferrite Nanoparticles as a New Family of Photothermal Therapeutic Materials for Cancer Treatment

Panchanathan Manivasagan [1,†], Sekar Ashokkumar [2,†], Ala Manohar [3,†], Ara Joe [1], Hyo-Won Han [1], Sun-Hwa Seo [1], Thavasyappan Thambi [4], Hai-Sang Duong [5], Nagendra Kumar Kaushik [2], Ki Hyeon Kim [3], Eun Ha Choi [2,*] and Eue-Soon Jang [1,*]

1 Department of Applied Chemistry, Kumoh National Institute of Technology, Daehak-ro 61, Gumi 39177, Republic of Korea; manimaribtech@gmail.com (P.M.); jar@kumoh.ac.kr (A.J.); 20101414hyowon@kumoh.ac.kr (H.-W.H.); 20126051@kumoh.ac.kr (S.-H.S.)
2 Plasma Bioscience Research Centre, Applied Plasma Medicine Center, Department of Electrical and Biological Physics, Kwangwoon University, Seoul 01897, Republic of Korea; kumarebt@kw.ac.kr (S.A.); kaushik.nagendra@kw.ac.kr (N.K.K.)
3 Department of Physics, Yeungnam University, Gyeongsan 38541, Republic of Korea; amanohar.svu@gmail.com (A.M.); kee1@ynu.ac.kr (K.H.K.)
4 School of Chemical Engineering, Theranostic Macromolecules Research Center, Sungkyunkwan University, Suwon 16419, Republic of Korea; thambi@skku.edu
5 Faculty of Applied Sciences, Ton Duc Thang University, Ho Chi Minh City 700000, Vietnam; duonghaisang99@gmail.com
* Correspondence: ehchoi@kw.ac.kr (E.H.C.); euesoon@kumoh.ac.kr (E.-S.J.)
† These authors contributed equally to this work.

Citation: Manivasagan, P.; Ashokkumar, S.; Manohar, A.; Joe, A.; Han, H.-W.; Seo, S.-H.; Thambi, T.; Duong, H.-S.; Kaushik, N.K.; Kim, K.H.; et al. Biocompatible Calcium Ion-Doped Magnesium Ferrite Nanoparticles as a New Family of Photothermal Therapeutic Materials for Cancer Treatment. *Pharmaceutics* **2023**, *15*, 1555. https://doi.org/10.3390/pharmaceutics15051555

Academic Editor: Maria Nowakowska

Received: 7 April 2023
Revised: 15 May 2023
Accepted: 19 May 2023
Published: 21 May 2023

Abstract: Novel biocompatible and efficient photothermal (PT) therapeutic materials for cancer treatment have recently garnered significant attention, owing to their effective ablation of cancer cells, minimal invasiveness, quick recovery, and minimal damage to healthy cells. In this study, we designed and developed calcium ion-doped magnesium ferrite nanoparticles (Ca^{2+}-doped $MgFe_2O_4$ NPs) as novel and effective PT therapeutic materials for cancer treatment, owing to their good biocompatibility, biosafety, high near-infrared (NIR) absorption, easy localization, short treatment period, remote controllability, high efficiency, and high specificity. The studied Ca^{2+}-doped $MgFe_2O_4$ NPs exhibited a uniform spherical morphology with particle sizes of 14.24 ± 1.32 nm and a strong PT conversion efficiency (30.12%), making them promising for cancer photothermal therapy (PTT). In vitro experiments showed that Ca^{2+}-doped $MgFe_2O_4$ NPs had no significant cytotoxic effects on non-laser-irradiated MDA-MB-231 cells, confirming that Ca^{2+}-doped $MgFe_2O_4$ NPs exhibited high biocompatibility. More interestingly, Ca^{2+}-doped $MgFe_2O_4$ NPs exhibited superior cytotoxicity to laser-irradiated MDA-MB-231 cells, inducing significant cell death. Our study proposes novel, safe, high-efficiency, and biocompatible PT therapeutics for treating cancers, opening new vistas for the future development of cancer PTT.

Keywords: ferrite nanoparticles; NIR laser; biocompatibility; photothermal therapy; cancer

1. Introduction

Cancer remains one of the most challenging life- and health-threatening diseases worldwide [1,2]. Existing conventional cancer strategies, including chemotherapy, surgery, and radiotherapy, have strong side effects and suffer from cumulative radiation doses, high recurrence rates, complications, and low patient satisfaction [3]. Consequently, novel and patient-friendly therapeutic strategies are urgently needed to effectively treat cancers with high specificity [4]. Recently, owing to the rapid advancement of nanotechnology and nanoscience, many novel and emerging nanoparticle-based treatment strategies have been developed, such as photothermal therapy (PTT), radiation therapy, photocatalytic therapy, and ultrasound therapy [5–9]. PTT is a potential alternative to conventional cancer strategies, owing to its minimal invasiveness, short treatment period, easy localization, low cost, remote controllability, high efficiency, and high specificity [10]. During photothermal (PT) experiments, PT-conversion nanoparticles can absorb and convert near-infrared (NIR) light energy into heat energy to kill cancer cells [11]. Consequently, PTT is regarded as one of the most promising therapeutic strategies for treating cancers with minimal invasiveness [12–14].

To date, various nanoparticles, such as magnetic iron oxide nanomaterials [15,16], gold nanostructured materials [17,18], carbon nanomaterials [19], molybdenum sulfide [20], copper sulfide [21], and organic dye-based nanocarriers [22], have been developed as PT conversion materials for cancer treatment. Among these, magnetic iron oxide nanomaterials have recently attracted much interest in studies of drug delivery, hyperthermia, phototherapy, chemodynamic therapy, and magnetic resonance imaging (MRI) in various biomedical applications, owing to their excellent magnetic properties, low toxicity, and high stability [23,24]. Compared with magnetic iron oxide nanomaterials, superparamagnetic iron oxide nanoparticles such as calcium ion-doped magnesium ferrite nanoparticles (Ca^{2+}-doped $MgFe_2O_4$ NPs), have an average particle size smaller than 20 nm, which is more suitable for clinical diagnosis and therapy [25,26]. Ca^{2+}-doped $MgFe_2O_4$ NPs have been reported to exhibit excellent biocompatibility, stability, high heating efficiency, and good saturation magnetization, making them excellent candidates for hyperthermia, PTT, and chemodynamic therapy [27]. Ca^{2+}-doped $MgFe_2O_4$ NPs have been widely demonstrated for more efficient and safe cancer treatment approaches, such as PTT [12]. In this study, we report novel biocompatible Ca^{2+}-doped $MgFe_2O_4$ NPs as effective PT-ablation agents for cancer treatment.

2. Materials and Methods

2.1. Reagents

All reagents and solvents were of the highest analytical grade, were obtained from Sigma-Aldrich, and were used as received without any further purification.

2.2. Preparation of Ca^{2+}-Doped $MgFe_2O_4$ NPs

Ca^{2+}-doped $MgFe_2O_4$ NPs were synthesized based on the previously described solvothermal reflux (SR) method with slight modifications [28,29]. Briefly, benzyl ether (40 mL), oleic acid (9.0 mL), and oleylamine (11 mL) were mixed, stoichiometric metal precursors were added to the reaction solution, and the reaction solution was stirred for 10 min. The reaction solution was heated at 300 °C for 1 h and then, cooled to room temperature (RT). Finally, the reaction solution was separated by centrifugation and re-dispersed in *n*-hexane for further experiments.

2.3. Characterization

The optical absorbance spectra of the prepared Ca^{2+}-doped $MgFe_2O_4$ NPs were recorded using a UV–vis spectrophotometer (Shimadzu UV-2600). The particle size distribution of the prepared Ca^{2+}-doped $MgFe_2O_4$ NPs was measured using dynamic light scattering (DLS) (Brookhaven Ins. Cor., 90 plus particle size analyzer). Fourier-transform infrared (FTIR) spectra of the prepared Ca^{2+}-doped $MgFe_2O_4$ NPs were obtained as KBr pellets using a Thermo Scientific iD1 transmission spectrometer. Powder X-ray diffraction (XRD) analysis of the prepared Ca^{2+}-doped $MgFe_2O_4$ NPs was performed using an X-ray diffractometer (Miniflex 600, Rigaku) with a Cu-Kα radiation source. The morphologies of the prepared Ca^{2+}-doped $MgFe_2O_4$ NPs were investigated using a scanning electron microscope (SEM, JSM-7001F) with attached energy dispersive X-ray (EDX) mapping, transmission electron microscope (TEM, JEM 2100), and Cs-corrected field emission TEM (Cs-corrected FETEM, JEM-ARM200F) equipped with an EDX spectroscopy and mapping. X-ray photoelectron spectroscopy (XPS) was used to assess the compositional and chemical states (Kα, Thermo Fisher Scientific, Waltham, MA, USA). The stability of the prepared Ca^{2+}-doped $MgFe_2O_4$ NPs was assessed for 30 days at RT in deionized water (DW), phosphate-buffered saline (PBS), and 10% fetal bovine serum (FBS), and was determined using a UV–vis spectrophotometer and TEM. For colloidal stability in Roswell Park Memorial Institute (RPMI) 1640 medium, the Ca^{2+}-doped $MgFe_2O_4$ NPs were incubated in the RPMI medium supplemented with 10% FBS for various time intervals. The hydrodynamic size was measured using dynamic light scattering (DLS) (Brookhaven Ins. Cor., 90 plus particle size analyzer).

2.4. Determining the PT Effect of Ca^{2+}-Doped $MgFe_2O_4$ NPs

Aqueous solutions of Ca^{2+}-doped $MgFe_2O_4$ NPs at various concentrations (75, 100, and 125 μg/mL) were recorded using a UV–vis spectrophotometer. One milliliter of Ca^{2+}-doped $MgFe_2O_4$ NPs aqueous solution in a quartz cuvette, with different concentrations (75, 100, and 125 μg/mL), was exposed to 785-nm-wavelength laser irradiation (CNI laser, China) at power densities of 0.5 W/cm^2 and 0.8 W/cm^2 for 300 s, and the solutions' temperatures were monitored using a compact thermal camera (FLIR Cx-Series). The PT stability of the solution with the prepared Ca^{2+}-doped $MgFe_2O_4$ NPs (125 μg/mL) in a quartz cuvette was assessed by exposing it to an NIR laser (785 nm, 0.8 W/cm^2) for 300 s (NIR laser power turned on) and then allowing it to cool to RT for 900 s (NIR laser power turned off). The NIR laser power on/off cycles were applied four times to test the PT stability. The PT conversion efficiency (η) of the prepared Ca^{2+}-doped $MgFe_2O_4$ NPs (125 μg/mL) aqueous solution was calculated according to the available reference Equation (1) [30–32]. The UV–vis spectra and TEM images of the Ca^{2+}-doped $MgFe_2O_4$ NPs were observed after four laser on/off cycles.

$$\eta = \frac{hS(T_{Max} - T_{Sur}) - Q_{dis}}{I(1 - 10^{-A_{785}})} \quad (1)$$

2.5. Cell Culture

The normal human dermal fibroblast (HDF) cell line and human breast cancer cell line (MDA-MB-231 cells) were obtained from the Korean Cell Line Bank (Seoul, Republic of Korea). Cells were cultured and propagated in Dulbecco's modified Eagle's medium (DMEM) and Roswell Park Memorial Institute (RPMI) 1640 medium supplemented with 10% FBS and 1% penicillin/streptomycin at 37 °C in a humidified 5% CO_2 atmosphere.

2.6. In Vitro Cytotoxicity

In vitro cytotoxicity was assessed using a standard 3-(4,5-dimethylthiazol-2-yl)-2,5-diphenyltetrazolium bromide (MTT) assay. The HDF and MDA-MB-231 cells (5×10^4 cells/well) were cultured in 96-well flat-bottom plates in the DMEM and RPMI 1640 media. After 24-h-long incubation, Ca^{2+}-doped $MgFe_2O_4$ NPs at various concentrations (range, 0–400 μg/mL) were added to 96-well plates for an additional 24 h and 48 h. Next, 100 μL of the MTT solution (0.5 mg/mL) was added to each well for an additional 4 h. The MTT-containing medium was removed, and dimethyl sulfoxide (DMSO, 100 μL) was added to all wells to dissolve the formazan crystals for 30 min. The absorbance of each well was recorded at 570 nm using a plate reader (Bio-TEK MQX200).

2.7. In Vitro Hemolysis Assay

The in vitro hemolysis assay was performed according to the previously described method [33]. Red blood cells (RBCs) were collected from fresh mouse blood. Then, the RBCs were separated by centrifuging at 12,000 rpm for 7 min and diluted with 10 × PBS. Afterward, 0.3 μL of the RBC suspension was gently mixed with Ca^{2+}-doped $MgFe_2O_4$ NPs in PBS at various concentrations (25, 50, 75, 100, and 125 μg/mL). RBC-treated PBS served as a negative control, whereas RBC-treated DW was used as a positive control. After incubation at RT for 4 h, all samples were centrifuged at 12,000 rpm for 7 min, and the absorbance of the supernatant was recorded at 541 nm using a microplate reader.

2.8. Chick Chorioallantoic Membrane Assay

Freshly fertilized chicken eggs were obtained from chicken form (Chuncheon, Republic of Korea). Eggs were incubated at 37 °C and 60% relative humidity for 9 days. The chorioallantoic membrane (CAM) was removed on day 10 of the embryo development, to create a window in the eggshell. Then, 200 µL of PBS as a negative control, sodium hydroxide (NaOH, 0.1 M) as a positive control, and Ca^{2+}-doped $MgFe_2O_4$ NPs (125 µg/mL) were added to all eggs and incubated for 5 min. Finally, toxicity was observed in terms of blood vessel damage and hemorrhage [34].

2.9. In Vitro PTT

For the in vitro PTT of cancer cells, MDA-MB-231 cells (5×10^4 cells/well) were cultured in 96-well flat-bottom plates in the RPMI 1640 medium for 24 h. After that, Ca^{2+}-doped $MgFe_2O_4$ NPs at various concentrations (range, 0–125 µg/mL) were added to 96-well plates for an additional 10 h incubation, following which the cells were irradiated with an NIR laser (785 nm, 0.8 W/cm^2) for 300 s. After 5 h of additional incubation, cell viability was detected using the standard MTT assay, as previously described. In addition, cells were incubated with Ca^{2+}-doped $MgFe_2O_4$ NPs (125 µg/mL) for 10 h, and then irradiated using an NIR laser (785 nm) at 0.5 W/cm^2 and 0.8 W/cm^2 for 300 s. After 5 h of additional incubation, cell viability was measured using the standard MTT method.

For in vitro cell imaging, cells were cultured in 35-mm-diameter confocal dishes and incubated for 24 h at 37 °C. After that, cells were separated into four groups: control (100 µL PBS) only; control (100 µL PBS) + laser irradiation (0.8 W/cm^2, 300 s); Ca^{2+}-doped $MgFe_2O_4$ NPs (125 µg/mL) only; and Ca^{2+}-doped $MgFe_2O_4$ NPs (125 µg/mL) + laser irradiation (0.8 W/cm^2, 300 s). The cells were treated with PBS (100 µL) or Ca^{2+}-doped $MgFe_2O_4$ NPs (125 µg/mL) for an additional 10 h incubation, following which some cells were irradiated using the NIR laser (785 nm, 0.8 W/cm^2) for 300 s. After 5 h of incubation, cells were co-stained and tested using a live/dead staining assay with the fluorescent probes calcein-AM (green color) and propidium iodide (PI, red color), and also tested with trypan blue staining for 30 min. In addition, all cell nuclei and mitochondria were co-stained with Hoechst 33342 and MitoTracker Red for 30 min. Images of all groups of cells were acquired using a confocal laser scanning microscope (A1R; Nikon, Japan) in tail scan mode for live imaging. For the detection of reactive oxygen species (ROS), cells were cultured in a 35-mm-diameter culture plate and incubated for 24 h. After the incubation, all cells were divided into four groups, as previously described. All cells were stained with 2,7-dichlorofluorescein diacetate (DCFH-DA) for 30 min and imaged using a confocal fluorescence microscope in tail scan mode for live imaging.

For the apoptosis assay, cells were seeded into 6-well flat-bottom plates in the RPMI 1640 medium for 24 h. The cells were then separated into four groups, as previously described. After various treatments, all cells were collected, washed with PBS, and co-stained with an Annexin V-FITC and PI staining kit (BD Pharmingen, Canada). Finally, the cells were analyzed using flow cytometry to detect apoptotic cells (BD FACSCalibur, Canada).

Statistical analysis was performed, and all data were reported as the mean and standard error of the mean. Statistical differences were assessed using a one-way analysis of variance (ANOVA) test, using SPSS 14, with * denoting $p < 0.05$.

3. Results and Discussion

3.1. Preparation and Characterization of Ca^{2+}-Doped $MgFe_2O_4$ NPs

In this work, Ca^{2+}-doped $MgFe_2O_4$ NPs were prepared using the solvothermal reflux technique at 300 °C for 1 h, as shown in Scheme 1 [28,29]. The UV–vis absorbance spectrum of the prepared Ca^{2+}-doped $MgFe_2O_4$ NPs exhibited a broad absorbance in the ultraviolet to NIR range (200–1100 nm), indicating that Ca^{2+}-doped $MgFe_2O_4$ NPs could be used as a potential PT material for PTT (Figure 1a). Figure 1b shows the FTIR spectra of Ca^{2+}-doped $MgFe_2O_4$ NPs. Ca^{2+}-doped $MgFe_2O_4$ NPs exhibited two major peaks at 579 and 1408 cm^{-1}, which were attributed to the Fe–O and C=O bonds, while the absorption bands at 1558, 2847, and 2915 cm^{-1} were attributed to the C–H stretching vibration mode, confirming that the prepared Ca^{2+}-doped $MgFe_2O_4$ NPs were successfully synthesized [35]. The size of the prepared Ca^{2+}-doped $MgFe_2O_4$ NPs assessed using a DLS was 14.24 ± 1.32 nm, which is suitable for in vitro and in vivo therapeutic applications (Figure 1c). The Ca^{2+}-doped $MgFe_2O_4$ NPs were analyzed by XRD (Figure 1d). The XRD pattern of the prepared Ca^{2+}-doped $MgFe_2O_4$ NPs exhibited intense peaks corresponding to (220), (311), (400), (422), (511), (440), and (533), suggesting their crystalline nature [36]. All of the diffraction peaks matched well with the international standard data card Joint Committee on Powder Diffraction Standards (JCPDS) number 89-3084.

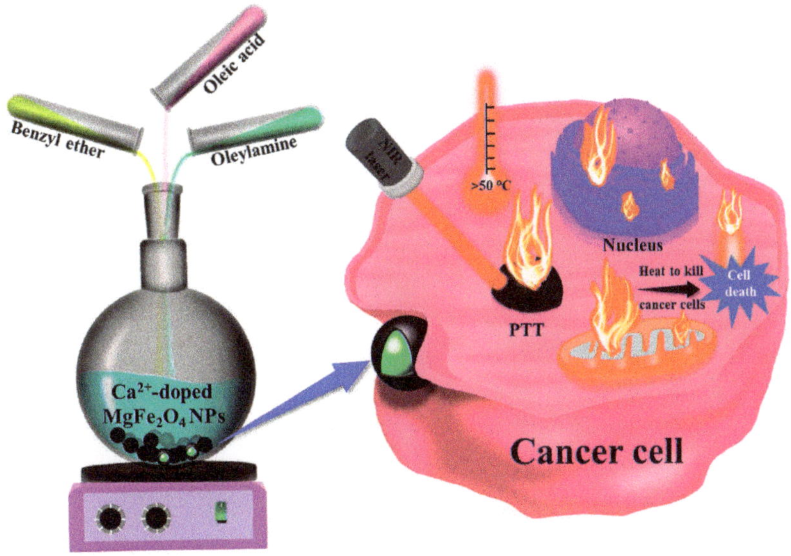

Scheme 1. Ca^{2+}-doped $MgFe_2O_4$ NPs were successfully prepared for PTT.

The morphology of the Ca^{2+}-doped $MgFe_2O_4$ NPs was investigated using TEM and Cs-corrected FETEM with EDX mapping and EDX spectroscopy. The TEM images of Ca^{2+}-doped $MgFe_2O_4$ NPs revealed that most of the particles were a nearly uniform and cube-like shape, with an average particle size of 14.24 ± 1.32 nm (Figures1 e and S1). Our TEM images are similar to those in a previous report by Heiba et al. [37] and Sivagurunathan et al. [38], which showed cube-shaped ferrite NPs. The Cs-corrected FETEM with EDX mapping (Figure S2) and EDX spectroscopy (Figure S3) of Ca^{2+}-doped $MgFe_2O_4$ NPs revealed the presence of Mg, Fe, Ca, and O. In addition, the SEM images of Ca^{2+}-doped $MgFe_2O_4$ NPs showed that the particles were almost cube shape and had uniform morphologies (Figure S4). Similar SEM images are observed by Naaz et al. [39] for the cube shape of MgFe2O4 NPs. SEM-related EDX mapping (Figure S5) and EDX spectroscopy (Figure S6) revealed the presence of Mg, Fe, Ca, and O. XPS is a versatile surface analysis technique that is frequently used to examine compositional and chemical states [40]. The goal of the XPS analysis was to learn more about the valence states of the elements on the Ca^{2+}-doped $MgFe_2O_4$ NPs surface. Survey scan spectra of the prepared nanoparticles (Ca, Mg, Fe, and O elements) are shown in Figure 2a. According to Figure 2b, the Ca 2p core is divided into 2p3/2 (binding energy = 346.91 eV) and 2p1/2 (binding energy = 350.52 eV) peaks [41]. The Mg^{2+} ions were assigned one characteristic peak at 55.59 eV and three deconvoluted peaks at 56.99, 56.36, and 55.41 eV, as shown in the Mg 2p spectra (Figure 2c).

The XPS spectra of the Fe 2P core in the Fe^{3+} valence state are shown in Figure 2d. The estimated binding energies of the Fe 2p3/2 and Fe 2p1/2 lines were 710.53 eV and 724.17 eV, respectively. Fe^{3+} was predominant, as evidenced by the satellite peaks of the Fe 2p core at 732.45 and 718.33 eV [28]. The O1s signal is shown in Figure 2e, revealing that the main peaks were located at 530.64 and 529.72 eV. The former peak was equivalent to the surface-located lattice oxygen [28]. The long-term stability of Ca^{2+}-doped $MgFe_2O_4$ NPs dispersions is essential for biomedical applications [42]. The stability test of the Ca^{2+}-doped $MgFe_2O_4$ NPs dispersed in DW, PBS, and 10% FBS for 30 days was performed using a UV–vis spectrophotometer. As shown in Figure 2f,g, no obvious changes were observed in the UV–vis absorbance spectra of the Ca^{2+}-doped $MgFe_2O_4$ NPs solution in DW, PBS, and 10% FBS for 30 days, implying that the prepared Ca^{2+}-doped $MgFe_2O_4$ NPs exhibited good stability and high biocompatibility. The TEM image of Ca^{2+}-doped $MgFe_2O_4$ NPs revealed a spherical morphology with uniform shapes in 10% FBS after 30 days (Figure 2h). For colloidal stability in the RPMI medium, the Ca^{2+}-doped $MgFe_2O_4$ NPs were incubated in the RPMI medium with 10% FBS to assess their colloidal stability in biological systems. The hydrodynamic size of the Ca^{2+}-doped $MgFe_2O_4$ NPs in RPMI medium was gradually increased because of small aggregates or precipitates in biological systems in the presence of amino acids and salts (Figure S7). Similar behavior has been reported for colloidal stability in the biological medium in the presence of NPs [43].

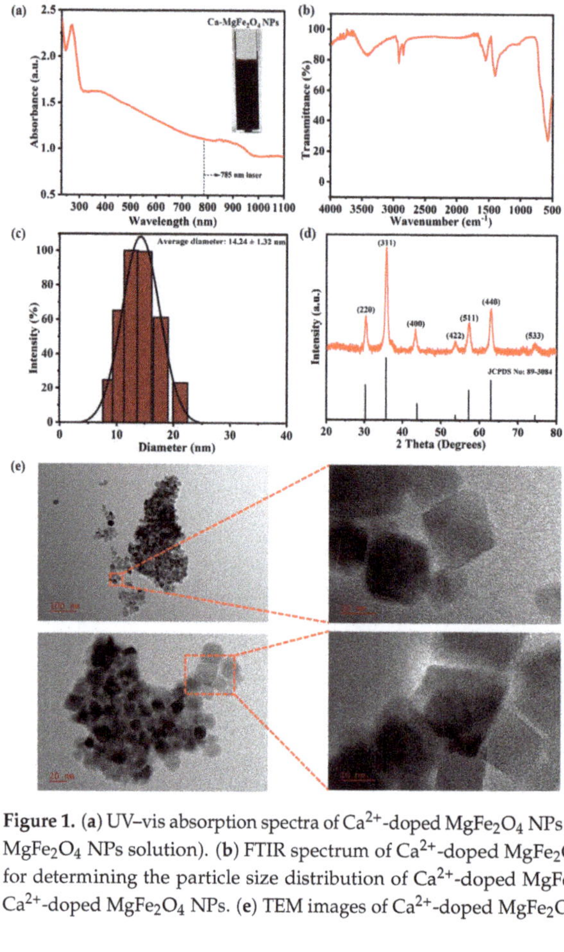

Figure 1. (**a**) UV–vis absorption spectra of Ca^{2+}-doped $MgFe_2O_4$ NPs (the inset shows the Ca^{2+}-doped $MgFe_2O_4$ NPs solution). (**b**) FTIR spectrum of Ca^{2+}-doped $MgFe_2O_4$ NPs. (**c**) DLS measurements for determining the particle size distribution of Ca^{2+}-doped $MgFe_2O_4$ NPs. (**d**) XRD analysis of Ca^{2+}-doped $MgFe_2O_4$ NPs. (**e**) TEM images of Ca^{2+}-doped $MgFe_2O_4$ NPs.

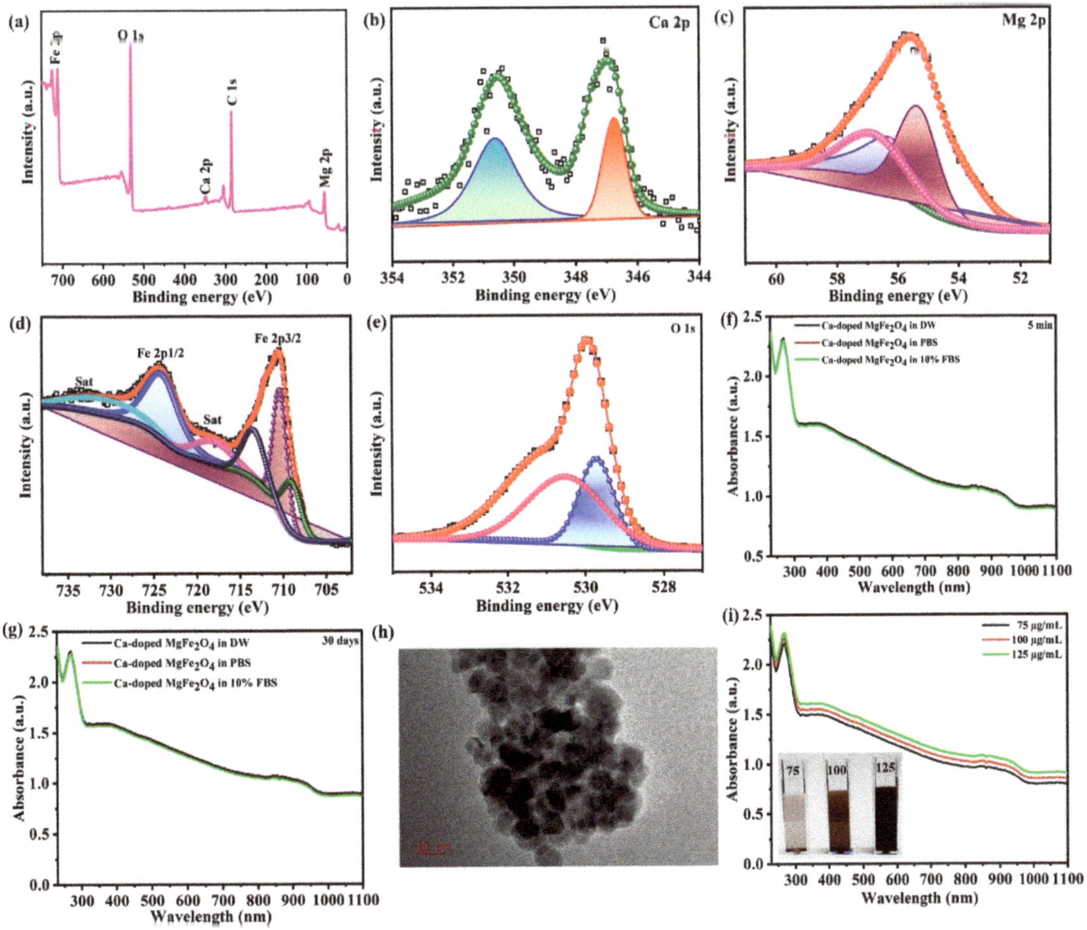

Figure 2. (**a**) XPS survey spectrum of Ca^{2+}-doped $MgFe_2O_4$ NPs. XPS spectra for Ca 2p (**b**), Mg 2p (**c**), Fe 2p (**d**), and O 1s (**e**) of Ca^{2+}-doped $MgFe_2O_4$ NPs. UV–vis absorption spectra of Ca^{2+}-doped $MgFe_2O_4$ NPs in DW, PBS, and 10% FBS for 5 min (**f**) and 30 days (**g**). (**h**) TEM images of Ca^{2+}-doped $MgFe_2O_4$ NPs in 10% FBS after 30 days. (**i**) UV–vis absorption spectra of Ca^{2+}-doped $MgFe_2O_4$ NPs at various concentrations.

3.2. Determining the PT Effect of Ca^{2+}-Doped $MgFe_2O_4$ NPs

The aqueous solutions of Ca^{2+}-doped $MgFe_2O_4$ NPs at different concentrations (75, 100, and 125 μg/mL) were observed using a UV–vis spectrophotometer, revealing good absorption in the NIR range, suggesting that Ca^{2+}-doped $MgFe_2O_4$ NPs are highly effective for PT-based cancer treatments (Figure 2i). To explore the PT effect of Ca^{2+}-doped $MgFe_2O_4$ NPs, 1-mL-aqueous solutions of Ca^{2+}-doped $MgFe_2O_4$ NPs in a quartz cuvette at various concentrations (75, 100, and 125 μg/mL) were exposed to NIR laser irradiation (785 nm) at power densities of 0.5 (Figure 3a) and 0.8 W/cm^2 (Figure 3b) for 300 s, and their temperatures were recorded using a compact thermal imaging camera. The temperature of the Ca^{2+}-doped $MgFe_2O_4$ NPs solution (125 μg/mL) increased to 51.7 °C within 300 s of the NIR laser irradiation at 0.8 W/cm^2, and real-time temperature changes of Ca^{2+}-doped $MgFe_2O_4$ NPs were captured using a compact thermal imaging camera (Figure 3c), whereas water (0 μg/mL) in the same treatment conditions only exhibited a slight temperature increase (from 27.56 °C to 29.2 °C). Therefore, Ca^{2+}-doped $MgFe_2O_4$ NPs are excellent candidates for NIR mediated PTT. The temperature change depended on the concentration, laser power density, and irradiation time [11].

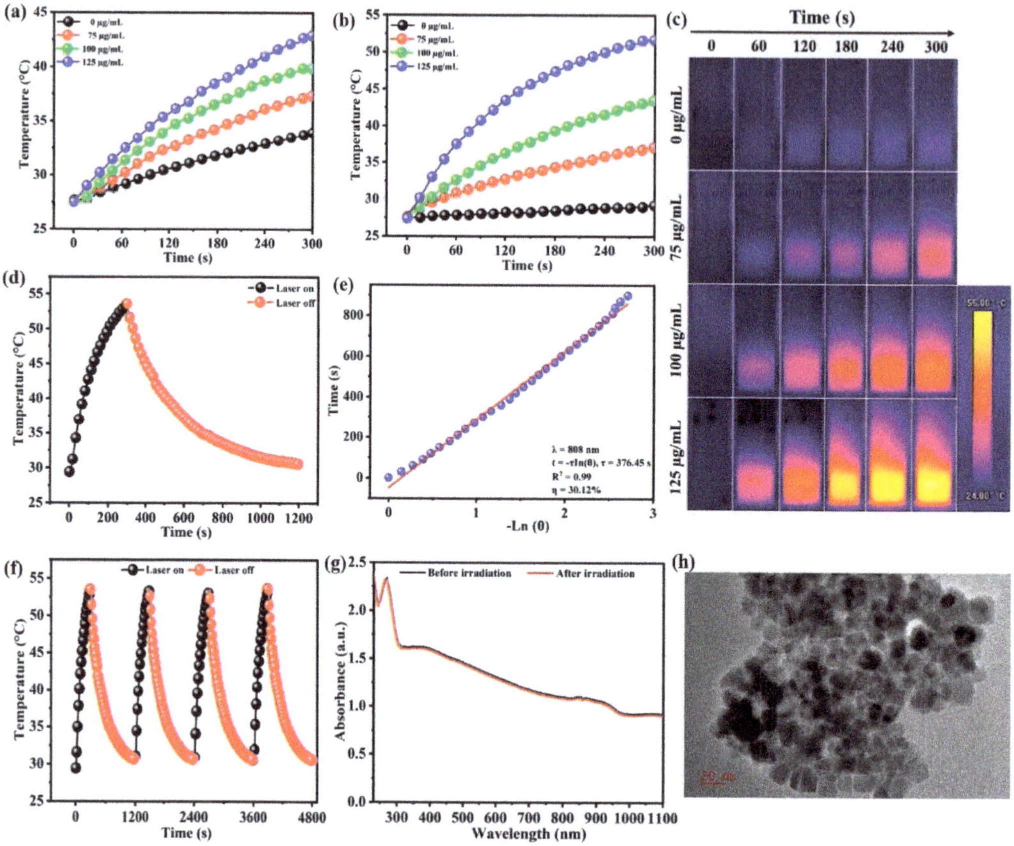

Figure 3. Temperature profiles for various concentrations of Ca^{2+}-doped $MgFe_2O_4$ NPs irradiated with a 785 nm laser at 0.5 (**a**) and 0.8 W/cm^2 (**b**) for 300 s. (**c**) Infrared thermal images for various concentrations of Ca^{2+}-doped $MgFe_2O_4$ NPs irradiated with a 785 nm laser at 0.8 W/cm^2 for 300 s. (**d**) Temperature profile of Ca^{2+}-doped $MgFe_2O_4$ NPs (125 μg/mL) irradiated with a 785 nm laser at 0.8 W/cm^2 for 300 s and then allowed to cool down to RT for 900 s. (**e**) A linear relationship between time and -Ln θ was found during the cooling period after the laser irradiation was turned off. (**f**) PT stability of Ca^{2+}-doped $MgFe_2O_4$ NPs. (**g**) UV–vis absorption spectra of Ca^{2+}-doped $MgFe_2O_4$ NPs in DW, before and after four on/off laser cycles. (**h**) TEM image of Ca^{2+}-doped $MgFe_2O_4$ NPs after four on/off laser cycles.

To investigate the PT conversion efficiency of Ca^{2+}-doped $MgFe_2O_4$ NPs, the temperature changes in the aqueous solution of Ca^{2+}-doped $MgFe_2O_4$ NPs (125 μg/mL) were monitored in response to the 5 min 785-nm-wavelength laser irradiation at 0.8 W/cm^2. After turning off the laser, the aqueous solution of Ca^{2+}-doped $MgFe_2O_4$ NPs was allowed to slowly cool down to RT for 15 min during each cycle (Figure 3d). The estimated PT conversion efficiency (η) of the solution of Ca^{2+}-doped $MgFe_2O_4$ NPs (125 μg/mL) was 30.12%, and the estimated time constant for the heat transfer of Ca^{2+}-doped $MgFe_2O_4$ NPs was 376.45 s (Figure 3e), indicating that Ca^{2+}-doped $MgFe_2O_4$ NPs can be used as an efficient PT material. In fact, the PT conversion efficiency of Ca^{2+}-doped $MgFe_2O_4$ NPs was higher than previously reported classic photothermal agents such as gold nanorods

(21%) [15], MoS_2 nanosheets (24.4%) [46], Cu2-xS nanocrystals (16.3%) [47], and SiO2/Au nanoshells (30%) [48]. To further examine the PT stability of Ca^{2+}-doped $MgFe_2O_4$ NPs, the aqueous solution of Ca^{2+}-doped $MgFe_2O_4$ NPs (125 µg/mL) was exposed to the NIR laser irradiation (785 nm, 0.8 W/cm^2) for 300 s, following which the laser was switched off for 900 s. After four on/off NIR laser cycles, the temperature profile of Ca^{2+}-doped $MgFe_2O_4$ NPs exhibited no significant changes, indicating that the prepared Ca^{2+}-doped $MgFe_2O_4$ NPs had high PT stability (Figure 3f). Furthermore, the optical properties of the aqueous solution of Ca^{2+}-doped $MgFe_2O_4$ NPs were observed by UV–vis spectrophotometry after four on/off laser cycles, and revealed no noticeable changes in the absorbance spectrum (Figure 3g). The TEM image was also clearly visible after four on/off NIR laser cycles, revealing spherical particles with uniform shapes (Figure 3h).

3.3. *In Vitro Cytotoxicity, Hemolysis, and CAM Assay*

The basic requirements for biomedical applications of NPs are high biocompatibility and safety, which are essential for industrializing nanomedical systems [49,50]. The in vitro cytotoxicity of Ca^{2+}-doped $MgFe_2O_4$ NPs at various concentrations (0–400 µg/mL) for HDF cells and MDA-MB-231 cells was assessed using the MTT assay. As shown in Figures S8 and 4a, Ca^{2+}-doped $MgFe_2O_4$ NPs exhibited no noticeable cytotoxicity after 24 h and 48 h of incubation, for concentrations in the 0–400 µg/mL range, and the cell viability remained above 95%, implying that Ca^{2+}-doped $MgFe_2O_4$ NPs had excellent biocompatibility. To investigate the hemocompatibility of Ca^{2+}-doped $MgFe_2O_4$ NPs, a hemolysis assay was performed to explore the effect of Ca^{2+}-doped $MgFe_2O_4$ NPs in PBS at various concentrations (25, 50, 75, 100, and 125 µg/mL) on RBCs. As shown in Figure 4b, the prepared Ca^{2+}-doped $MgFe_2O_4$ NPs (125 µg/mL) demonstrated an incredibly low hemolysis ratio (3.2%), which was lower than the permissible limit (5%), confirming that Ca^{2+}-doped $MgFe_2O_4$ NPs had good hemocompatibility and could be used for in vivo cancer treatments [51]. In addition, the toxicity of Ca^{2+}-doped $MgFe_2O_4$ NPs (125 µg/mL) was evaluated using the chick CAM assay (Figure 4c). Ca^{2+}-doped $MgFe_2O_4$ NP-treated chick embryos exhibited no observable toxicity in blood coagulation compared with PBS, indicating that Ca^{2+}-doped $MgFe_2O_4$ NPs were non-toxic to egg CAM [34,52].

3.4. *In Vitro PTT*

To investigate the PT effect of the prepared Ca^{2+}-doped $MgFe_2O_4$ NPs, the in vitro PT effect was evaluated using the MTT assay. Cells were incubated with Ca^{2+}-doped $MgFe_2O_4$ NPs at different concentrations (0–125 µg/mL) for 10 h and then irradiated using the NIR laser (785 nm, 0.8 W/cm^2) for 300 s. As shown in Figure 4d, the viability of non-laser-irradiated cells was 94.5% at the high concentration of Ca^{2+}-doped $MgFe_2O_4$ NPs (125 µg/mL), implying no observable effect on cancer cells. Most of the cells were killed after incubation with Ca^{2+}-doped $MgFe_2O_4$ NPs (125 µg/mL) and irradiation using the NIR laser (785 nm, 0.8 W/cm^2) for 300 s, and the cell viability decreased to 32.37%, confirming that laser-irradiated Ca^{2+}-doped $MgFe_2O_4$ NPs induced significantly more cell death. Similar results were obtained when Ca^{2+}-doped $MgFe_2O_4$ NPs were compared with various ferrite NPs (Table S1). In addition, cells were treated with Ca^{2+}-doped $MgFe_2O_4$ NPs (125 µg/mL) for 10 h and irradiated using the NIR laser (785 nm) at 0.5 and 0.8 W/cm^2 for 300 s (Figure 4e). Most of the cells were killed owing to the Ca^{2+}-doped $MgFe_2O_4$ NPs (125 µg/mL) after being NIR laser-irradiated (0.8 W/cm^2) for 300 s, demonstrating the promise of Ca^{2+}-doped $MgFe_2O_4$ NPs as effective PT materials for treating cancers.

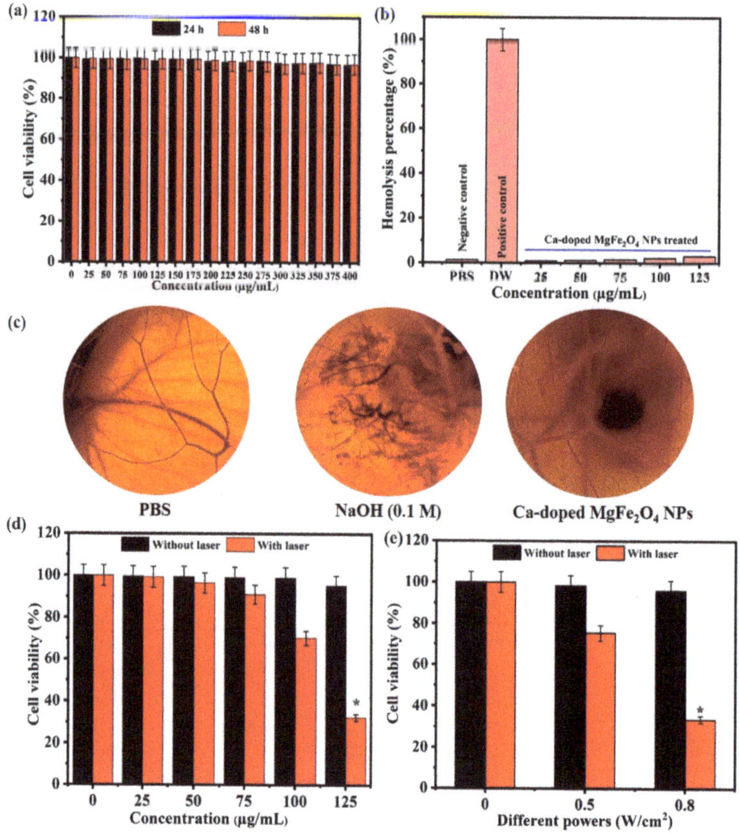

Figure 4. (**a**) In vitro biocompatibility of Ca^{2+}-doped $MgFe_2O_4$ NPs at various concentrations on MDA-MB-231 cells after 24 h and 48 h. (**b**) In vitro hemolytic activity of Ca^{2+}-doped $MgFe_2O_4$ NPs at various concentrations. (**c**) Chick CAM assay. (**d**) In vitro cell viability of Ca^{2+}-doped $MgFe_2O_4$ NPs at various concentrations, assessed on treated cells that were laser-irradiated at 0.8 W/cm^2 for 300 s. (**e**) In vitro cell viability of Ca^{2+}-doped $MgFe_2O_4$ NPs (125 μg/mL), assessed on treated cells that were laser-irradiated at 0.5 and 0.8 W/cm^2 for 300 s. The shown data are expressed as mean ± s.d., (n = 3), * $p < 0.05$.

Inspired by the above experiment, we further confirmed the results of the live/dead cell-staining calcein-AM/PI assay (Figures 5 and S9). The MDA-MB-231 cells treated with Ca^{2+}-doped $MgFe_2O_4$ NPs (125 μg/mL) were clearly evident in the presence of strong red fluorescence after the NIR laser irradiation, implying that more dead cells were detected in this group. In contrast, all the other groups exhibited intense green fluorescence and negligible cell death, indicating that more live cells were detected in the other groups. In addition, the live/dead staining assay was further confirmed by trypan blue staining, owing to their ability to distinguish between live (unstained) and dead (blue-stained) cells. As shown in Figure S10, no obvious changes were observed in the control cells with or without the NIR laser radiation, indicating that all the cells exhibited good morphology. The MDA-MB-231 cells that were treated with Ca^{2+}-doped $MgFe_2O_4$ NPs (125 μg/mL) and were not laser-irradiated exhibited negligible cell death. In contrast, cells incubated with Ca^{2+}-doped $MgFe_2O_4$ NPs (125 μg/mL) exhibited clear blue staining after the laser irradiation, indicating that more dead cells were present in this group.

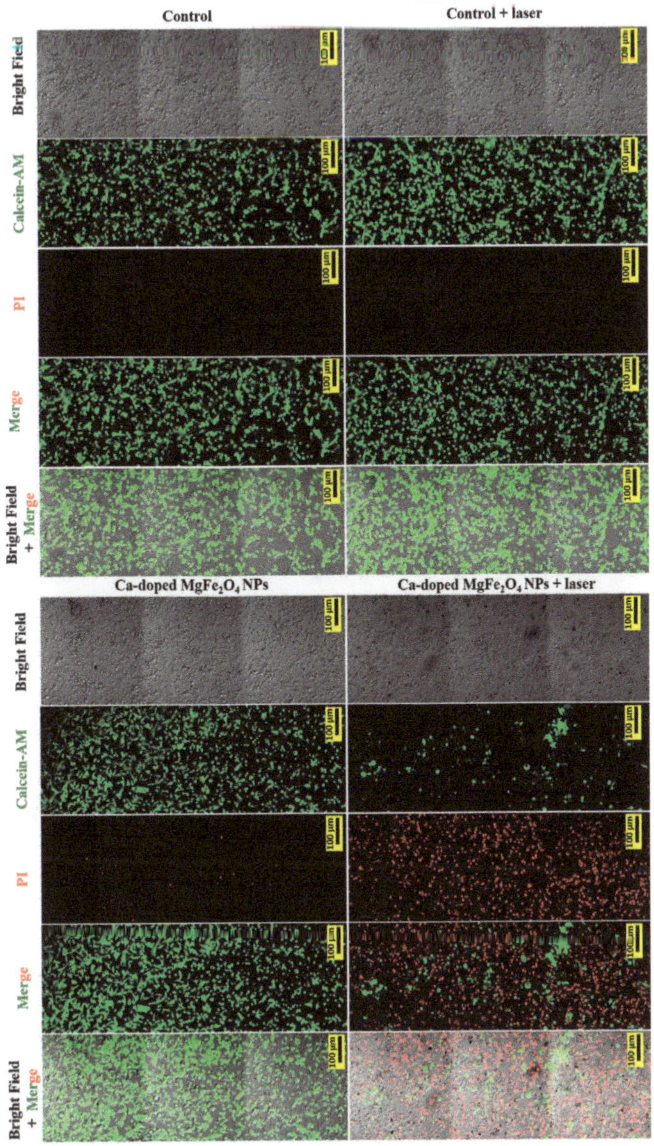

Figure 5. Confocal laser scanning microscope images of tile scans of MDA-MB-231 cells treated with PBS only, PBS + laser irradiation, Ca^{2+}-doped $MgFe_2O_4$ NPs only, and Ca^{2+}-doped $MgFe_2O_4$ NPs + laser irradiation. Calcein-AM (green) and PI (red) were used to distinguish between live and dead cells (20× magnification; scale bar: 100 μm).

In addition, each group of cells was further co-stained with Hoechst 33342 and MitoTracker Red, and evaluated using a confocal fluorescence microscope (Figure 6a). Hoechst 33342 and MitoTracker Red were used for nuclear and mitochondrial staining with blue and red fluorescence, respectively [53]. Hoechst 33342 clearly detected fragmented nuclei in the laser-irradiated Ca^{2+}-doped $MgFe_2O_4$ NPs (125 μg/mL), and MitroTracker Red was colocalized with the mitochondria-specific marker, whereas the other groups showed that the nucleus and mitochondria appeared under normal conditions. The presence of toxic ROS in the MDA-MB-231 cells was assessed using DCFH-DA as a fluorescent dye, which could be converted into a potent green fluorescent molecule DCF in the presence of toxic ROS [35]. As shown in Figures 6b and S11, the strong green fluorescence signal was almost undetectable in the control only, control + laser-irradiated, and Ca^{2+}-doped $MgFe_2O_4$ NPs (125 µg/mL), which did not affect cancer cells. In contrast, Ca^{2+}-doped $MgFe_2O_4$ NPs (125 µg/mL) caused strong fluorescence signals after the laser irradiation, which could lead to the production of ROS and damage biomolecules, mitochondria, and nuclei [54]. To evaluate the cell death mode after PT, cell apoptosis was also evaluated using flow cytometry (Figure 7). Ca^{2+}-doped $MgFe_2O_4$ NPs (125 µg/mL) induced significant apoptosis (67.2%) after the laser irradiation, compared with the other groups (1.68%, 3.31%, and 5.54%), suggesting that the PT effect of Ca^{2+}-doped $MgFe_2O_4$ NPs on MDA-MB-231 cells was enhanced by the NIR laser irradiation [55].

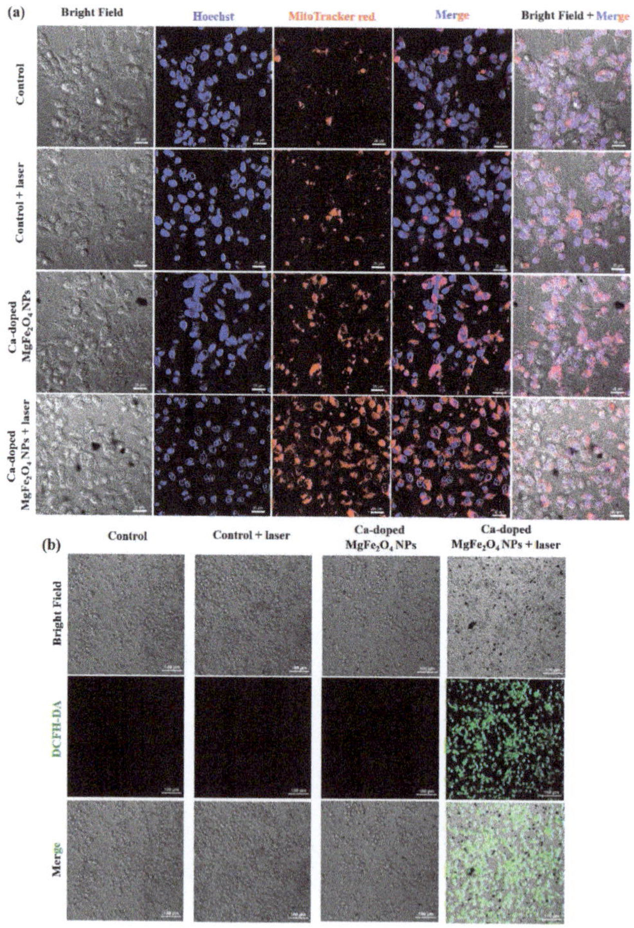

Figure 6. (**a**) Confocal laser scanning microscope images of MDA-MB-231 cells treated with PBS only, PBS + laser irradiation, Ca^{2+}-doped $MgFe_2O_4$ NPs only, and Ca^{2+}-doped $MgFe_2O_4$ NPs + laser irradiation. Hoechst 33342 and MitoTracker red were used to stain the nuclei and mitochondria (20× magnification; scale bar: 100 µm). (**b**) Confocal laser scanning microscope images of DCFH-DA-stained MDA-MB-231 cells treated with PBS only, PBS + laser irradiation, Ca^{2+}-doped $MgFe_2O_4$ NPs only, and Ca^{2+}-doped $MgFe_2O_4$ NPs + laser irradiation, for determining the intracellular ROS level (20× magnification; scale bar: 100 µm).

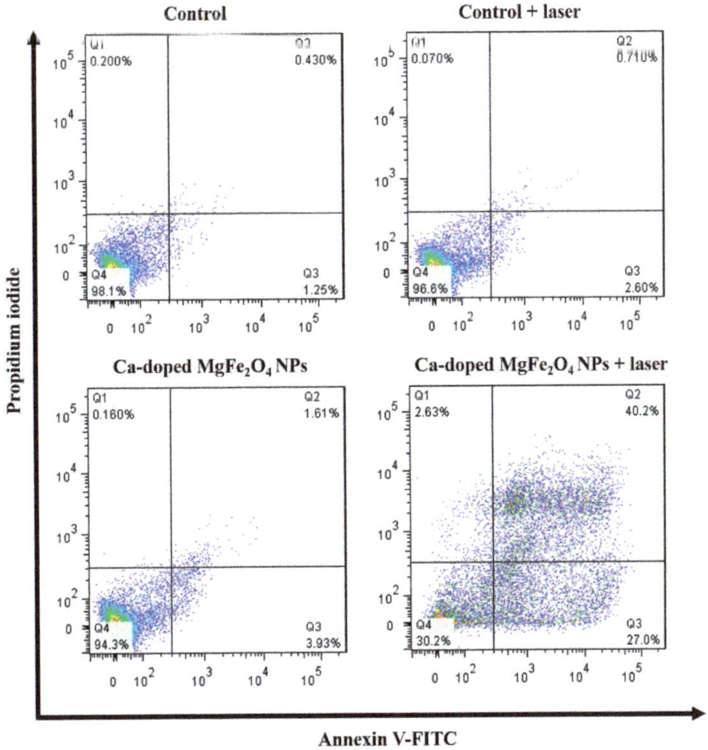

Figure 7. Flow cytometry analysis of the apoptosis of MDA-MB-231 cells, induced by PBS only, PBS + laser irradiation, Ca^{2+}-doped $MgFe_2O_4$ NPs only, and Ca^{2+}-doped $MgFe_2O_4$ NPs + laser irradiation using Annexin-V/PI staining.

4. Conclusions

In this study, we successfully designed and developed novel biocompatible PT materials for cancer therapy. Ca^{2+}-doped $MgFe_2O_4$ NPs were prepared using the solvothermal reflux method. The prepared Ca^{2+}-doped $MgFe_2O_4$ NPs had a uniform and spherical morphology, with an average particle size of 14.24 ± 1.32 nm, and exhibited a strong PT conversion efficiency (30.12%), high biocompatibility, PT stability, high NIR absorption, and biosafety. Impressively, in vitro assays indicated that Ca^{2+}-doped $MgFe_2O_4$ NPs did not exhibit any significant cytotoxicity against MDA-MB-231 cells without laser irradiation, implying that Ca^{2+}-doped $MgFe_2O_4$ NPs had excellent biocompatibility. Cells treated with Ca^{2+}-doped $MgFe_2O_4$ NPs exhibited a considerable PT ablation effect on NIR-laser-irradiated MDA-MB-231 cells, because these nanomaterials had good PT conversion efficiency, implying that laser-irradiated Ca^{2+}-doped $MgFe_2O_4$ NPs induced significantly more cell death. To the best of our knowledge, there have been no prior studies on Ca^{2+}-doped $MgFe_2O_4$ NPs as novel PT ablation materials for cancer treatment. The Ca^{2+}-doped $MgFe_2O_4$ NPs developed in this study are promising as safe and efficient PT materials for cancer therapy in the near future.

Supplementary Materials: The following supporting information can be downloaded at: https://www.mdpi.com/article/10.3390/pharmaceutics15051555/s1, Figure S1: TEM image of Ca^{2+}-doped $MgFe_2O_4$ NPs. Figure S2: Cs-corrected FETEM with EDX mapping. Figure S3: Cs-corrected FETEM with EDX spectroscopy. Figure S4: SEM images of Ca^{2+}-doped $MgFe_2O_4$ NPs. Figure S5. SEM with EDX mapping of Ca^{2+}-doped $MgFe_2O_4$ NPs. Figure S6: SEM-EDX spectra of Ca^{2+}-doped $MgFe_2O_4$ NPs. Figure S7: Hydrodynamic size of Ca^{2+}-doped $MgFe_2O_4$ NPs after incubation in RPMI medium supplemented with 10% FBS. Figure S8: In vitro biocompatibility of Ca^{2+}-doped $MgFe_2O_4$ NPs at various concentrations on HDF cells after 24 h and 48 h. Figure S9: Confocal laser scanning microscope images of tile scans of MDA-MB-231 cells treated with PBS only, PBS + laser irradiation, Ca^{2+}-doped $MgFe_2O_4$ NPs only, and Ca^{2+}-doped $MgFe_2O_4$ NPs + laser irradiation. Calcein-AM (green) and PI (red) were used to distinguish between live and dead cells (20× magnification; scale bar: 100 μm). Figure S10. Optical images of trypan blue-stained MDA-MB-231 cells treated with PBS only, PBS + laser irradiation, Ca^{2+}-doped $MgFe_2O_4$ NPs only, and Ca^{2+}-doped $MgFe_2O_4$ NPs + laser irradiation (20× magnification; scale bar: 100 μm). Figure S11. Confocal laser scanning microscopy images of tile scans of DCFH-DA-stained MDA-MB-231 cells treated with PBS only, PBS + laser irradiation, Ca^{2+}-doped $MgFe_2O_4$ NPs only, and Ca^{2+}-doped $MgFe_2O_4$ NPs + laser irradiation, for determining the intracellular ROS level (20× magnification; scale bar: 100 μm). Table S1. Comparative study of the photothermal effect of Ca^{2+}-doped $MgFe_2O_4$ NPs and various ferrite NPs. References [35,56–58] are cited in the supplementary materials.

Author Contributions: The manuscript was written through the contributions of all authors. Conceptualization, Funding acquisition, Writing—original draft., P.M.; investigation, S.A.; validation, A.M.; formal analysis, data curation, project administration, A.J., H.-W.H. and S.-H.S.; software, visualization, T.T. and H.-S.D.; methodology, resources, N.K.K.; supervision, K.H.K.; supervision, fund acquisition, E.H.C.; supervision, funding acquisition, writing—review & editing, E.-S.J. All authors have read and agreed to the published version of the manuscript.

Funding: E.-S.J is especially grateful for the research grant from the Korean Ministry of Education, Science & Technology (2016R1D1A3B0201175615) and the Grand Information Technology Research Center Program (IITP-2023-2020-0-01612) through the Institute of Information & Communications Technology Planning & Evaluation (IITP) funded by the Ministry of Science and ICT (MSIT), Korea. This research was supported by the Basic Science Research Program through the National Research Foundation of Korea (NRF) funded by the Ministry of Education (NRF-2022R1I1A1A01068693). This research was funded by the National Research Foundation (NRF) of Korea, funded by the Korean government (NRF-2021R1A6A1A03038785).

Data Availability Statement: Not applicable.

Conflicts of Interest: The authors declare no conflict of interest.

References

1. Bray, F.; Ferlay, J.; Soerjomataram, I.; Siegel, R.L.; Torre, L.A.; Jemal, A. Global cancer statistics 2018: GLOBOCAN estimates of incidence and mortality worldwide for 36 cancers in 185 countries. *CA Cancer J. Clin.* **2018**, *68*, 394–424. [CrossRef]
2. Siegel, R.L.; Miller, K.D.; Jemal, A. Cancer statistics, 2019. *CA Cancer J. Clin.* **2019**, *69*, 7–34. [CrossRef] [PubMed]
3. Zugazagoitia, J.; Guedes, C.; Ponce, S.; Ferrer, I.; Molina-Pinelo, S.; Paz-Ares, L. Current challenges in cancer treatment. *Clin. Ther.* **2016**, *38*, 1551–1566. [CrossRef]
4. Miller, K.D.; Nogueira, L.; Mariotto, A.B.; Rowland, J.H.; Yabroff, K.R.; Alfano, C.M.; Jemal, A.; Kramer, J.L.; Siegel, R.L. Cancer treatment and survivorship statistics, 2019. *CA Cancer J. Clin.* **2019**, *69*, 363–385. [CrossRef] [PubMed]
5. Nam, J.; Son, S.; Park, K.S.; Zou, W.; Shea, L.D.; Moon, J.J. Cancer nanomedicine for combination cancer immunotherapy. *Nat. Rev. Mater.* **2019**, *4*, 398–414. [CrossRef]
6. Shi, J.; Kantoff, P.W.; Wooster, R.; Farokhzad, O.C. Cancer nanomedicine: Progress, challenges and opportunities. *Nat. Rev. Cancer* **2017**, *17*, 20–37. [CrossRef] [PubMed]
7. van der Meel, R.; Sulheim, E.; Shi, Y.; Kiessling, F.; Mulder, W.J.; Lammers, T. Smart cancer nanomedicine. *Nat. Nanotechnol.* **2019**, *14*, 1007–1017. [CrossRef]
8. Wang, J.; Li, Y.; Nie, G.; Zhao, Y. Precise design of nanomedicines: Perspectives for cancer treatment. *Natl. Sci. Rev.* **2019**, *6*, 1107–1110. [CrossRef]
9. Manivasagan, P.; Joe, A.; Han, H.-W.; Thambi, T.; Selvaraj, M.; Chidambaram, K.; Kim, J.; Jang, E.-S. Recent advances in multifunctional nanomaterials for photothermal-enhanced Fenton-based chemodynamic tumor therapy. *Mater. Today Bio.* **2022**, *13*, 100197. [CrossRef] [PubMed]

10. Gao, Q.; Zhang, X.; Yin, W.; Ma, D.; Xie, C.; Zheng, L.; Dong, X.; Mei, L.; Yu, J.; Wang, C. Functionalized MoS2 nanovehicle with near infrared laser mediated nitric oxide release and photothermal activities for advanced bacteria-infected wound therapy. *Small* **2018**, *14*, 1802290. [CrossRef]
11. Manivasagan, P.; Kim, J.; Jang, E.-S. Recent progress in multifunctional conjugated polymer nanomaterial-based synergistic combination phototherapy for microbial infection theranostics. *Coord. Chem. Rev.* **2022**, *470*, 214701. [CrossRef]
12. Ding, L.; Chang, Y.; Yang, P.; Gao, W.; Sun, M.; Bie, Y.; Yang, L.; Ma, X.; Guo, Y. Facile synthesis of biocompatible L-cysteine-modified MoS2 nanospheres with high photothermal conversion efficiency for photothermal therapy of tumor. *Mater. Sci. Eng. C* **2020**, *117*, 111371. [CrossRef] [PubMed]
13. Feng, N.; Li, Q.; Bai, Q.; Xu, S.; Shi, J.; Liu, B.; Guo, J. Development of an Au-anchored Fe Single-atom nanozyme for biocatalysis and enhanced tumor photothermal therapy. *J. Colloid Interface Sci.* **2022**, *618*, 68–77. [CrossRef] [PubMed]
14. Yang, S.; Zhao, Z.; Xie, Y.; Lin, J.; Zhang, B.; Fan, J. Engineering Bio-MOF/polydopamine as a biocompatible targeted theranostic system for synergistic multi-drug chemo-photothermal therapy. *Int. J. Pharm.* **2022**, *623*, 121912. [CrossRef] [PubMed]
15. Hu, B.; Zeng, M.; Chen, J.; Zhang, Z.; Zhang, X.; Fan, Z.; Zhang, X. External Magnetic Field-Induced Targeted Delivery of Highly Sensitive Iron Oxide Nanocubes for MRI of Myocardial Infarction. *Small* **2016**, *12*, 4707–4712. [CrossRef]
16. Huang, J.; Li, Y.; Orza, A.; Lu, Q.; Guo, P.; Wang, L.; Yang, L.; Mao, H. Magnetic nanoparticle facilitated drug delivery for cancer therapy with targeted and image-guided approaches. *Adv. Funct. Mater.* **2016**, *26*, 3818–3836. [CrossRef] [PubMed]
17. Huang, X.; El-Sayed, I.H.; Qian, W.; El-Sayed, M.A. Cancer cell imaging and photothermal therapy in the near-infrared region by using gold nanorods. *J. Am. Chem. Soc.* **2006**, *128*, 2115–2120. [CrossRef] [PubMed]
18. Wang, Y.; Black, K.C.; Luehmann, H.; Li, W.; Zhang, Y.; Cai, X.; Wan, D.; Liu, S.-Y.; Li, M.; Kim, P. Comparison study of gold nanohexapods, nanorods, and nanocages for photothermal cancer treatment. *ACS Nano* **2013**, *7*, 2068–2077. [CrossRef] [PubMed]
19. Kam, N.W.S.; O'Connell, M.; Wisdom, J.A.; Dai, H. Carbon nanotubes as multifunctional biological transporters and near-infrared agents for selective cancer cell destruction. *Proc. Natl. Acad. Sci. USA* **2005**, *102*, 11600–11605. [CrossRef]
20. Liu, T.; Wang, C.; Gu, X.; Gong, H.; Cheng, L.; Shi, X.; Feng, L.; Sun, B.; Liu, Z. Drug delivery with PEGylated MoS2 nano-sheets for combined photothermal and chemotherapy of cancer. *Adv. Mater.* **2014**, *26*, 3433–3440. [CrossRef]
21. Tian, Q.; Jiang, F.; Zou, R.; Liu, Q.; Chen, Z.; Zhu, M.; Yang, S.; Wang, J.; Wang, J.; Hu, J. Hydrophilic Cu9S5 nanocrystals: A photothermal agent with a 25.7% heat conversion efficiency for photothermal ablation of cancer cells in vivo. *ACS Nano* **2011**, *5*, 9761–9771. [CrossRef] [PubMed]
22. Zhao, P.; Zheng, M.; Yue, C.; Luo, Z.; Gong, P.; Gao, G.; Sheng, Z.; Zheng, C.; Cai, L. Improving drug accumulation and photothermal efficacy in tumor depending on size of ICG loaded lipid-polymer nanoparticles. *Biomaterials* **2014**, *35*, 6037–6046. [CrossRef] [PubMed]
23. Patra, S.; Roy, E.; Karfa, P.; Kumar, S.; Madhuri, R.; Sharma, P.K. Dual-responsive polymer coated superparamagnetic nanoparticle for targeted drug delivery and hyperthermia treatment. *ACS Appl. Mater. Interfaces* **2015**, *7*, 9235–9246. [CrossRef] [PubMed]
24. Deng, K.; Chen, Y.; Li, C.; Deng, X.; Hou, Z.; Cheng, Z.; Han, Y.; Xing, B.; Lin, J. 808 nm light responsive nanotheranostic agents based on near-infrared dye functionalized manganese ferrite for magnetic-targeted and imaging-guided photodynamic/photothermal therapy. *J. Mater. Chem. B* **2017**, *5*, 1803–1814. [CrossRef]
25. Hao, R.; Xing, R.; Xu, Z.; Hou, Y.; Gao, S.; Sun, S. Synthesis, functionalization, and biomedical applications of multifunctional magnetic nanoparticles. *Adv. Mater.* **2010**, *22*, 2729–2742. [CrossRef] [PubMed]
26. Zhang, M.; Cao, Y.; Wang, L.; Ma, Y.; Tu, X.; Zhang, Z. Manganese doped iron oxide theranostic nanoparticles for combined T 1 magnetic resonance imaging and photothermal therapy. *ACS Appl. Mater. Interfaces* **2015**, *7*, 4650–4658. [CrossRef]
27. Kim, H.M.; Lee, H.; Hong, K.S.; Cho, M.Y.; Sung, M.-H.; Poo, H.; Lim, Y.T. Synthesis and high performance of magnetofluorescent polyelectrolyte nanocomposites as MR/near-infrared multimodal cellular imaging nanoprobes. *ACS Nano* **2011**, *5*, 8230–8240. [CrossRef]
28. Manohar, A.; Vijayakanth, V.; Vattikuti, S.P.; Manivasagan, P.; Jang, E.-S.; Chintagumpala, K.; Kim, K.H. Ca-Doped MgFe2O4 Nanoparticles for Magnetic Hyperthermia and Their Cytotoxicity in Normal and Cancer Cell Lines. *ACS Appl. Nano Mater.* **2022**, *5*, 5847–5856. [CrossRef]
29. Manohar, A.; Vijayakanth, V.; Vattikuti, S.P.; Manivasagan, P.; Jang, E.-S.; Kim, K.H. Electrochemical, oxygen evolution reaction and photoelectrochemical water splitting activity of Ca-doped MgFe2O4 nanoparticles. *J. Alloys Compd.* **2022**, *907*, 164566. [CrossRef]
30. Roper, D.K.; Ahn, W.; Hoepfner, M. Microscale heat transfer transduced by surface plasmon resonant gold nanoparticles. *J. Phys. Chem. C* **2007**, *111*, 3636–3641. [CrossRef]
31. Liu, X.; Li, B.; Fu, F.; Xu, K.; Zou, R.; Wang, Q.; Zhang, B.; Chen, Z.; Hu, J. Facile synthesis of biocompatible cysteine-coated CuS nanoparticles with high photothermal conversion efficiency for cancer therapy. *Dalton Trans.* **2014**, *43*, 11709–11715. [CrossRef] [PubMed]
32. Li, J.; Zhang, W.; Ji, W.; Wang, J.; Wang, N.; Wu, W.; Wu, Q.; Hou, X.; Hu, W.; Li, L. Near infrared photothermal conversion materials: Mechanism, preparation, and photothermal cancer therapy applications. *J. Mater. Chem. B* **2021**, *9*, 7909–7926. [CrossRef]
33. Liu, Y.; Guo, Z.; Li, F.; Xiao, Y.; Zhang, Y.; Bu, T.; Jia, P.; Zhe, T.; Wang, L. Multifunctional magnetic copper ferrite nanoparticles as Fenton-like reaction and near-infrared photothermal agents for synergetic antibacterial therapy. *ACS Appl. Mater. Interfaces* **2019**, *11*, 31649–31660. [CrossRef]

34. Sathivaseelan, A.; Saravanakumar, K.; Manivasagan, P.; Jeong, M.S.; Jang, E.-S.; Wang, M.-H. Folic acid conjugated chitosan encapsulated palladium nanoclusters for NIR triggered photothermal breast cancer treatment. *Carbohydr. Polym.* **2022**, *280*, 119021. [CrossRef] [PubMed]
35. Yang, C.; Chen, Y.; Guo, W.; Gao, Y.; Song, C.; Zhang, Q.; Zheng, N.; Han, X.; Guo, C. Bismuth ferrite-based nanoplatform design: An ablation mechanism study of solid tumor and NIR-triggered photothermal/photodynamic combination cancer therapy. *Adv. Funct. Mater.* **2018**, *28*, 1706827. [CrossRef]
36. Hu, B.; Cui, Y.; Yang, X.; Xu, X.; Janani, B.J.; Fakhri, A. Fabrication of novel rational Ti-Sn doped Cu-ferrite nanoparticles for robust photocatalysis reaction, magnetic resonance imaging, and chemo-magneto-photo-thermal therapy. *Surf. Interfaces* **2022**, *33*, 102226. [CrossRef]
37. Heiba, Z.K.; Mohamed, M.B.; Ellatief, A.A.; El-Denglawey, A.; Badawi, A. Structural, optical, and magnetic properties of ferrite/oxide composites MgFe2O4/(1− x) MnO–x CdO. *Appl. Phys. A Mater. Sci. Process.* **2022**, *128*, 857. [CrossRef]
38. Sivagurunathan, P.; Gibin, S. Preparation and characterization of nanosized cobalt ferrite particles by co-precipitation method with citrate as chelating agent. *J. Mater. Sci: Mater. Electron.* **2016**, *27*, 8891–8898. [CrossRef]
39. Naaz, F.; Dubey, H.K.; Kumari, C.; Lahiri, P. Structural and magnetic properties of MgFe2O4 nanopowder synthesized via co-precipitation route. *SN Appl. Sci.* **2020**, *2*, 808. [CrossRef]
40. Li, M.; Fang, H.; Li, H.; Zhao, Y.; Li, T.; Pang, H.; Tang, J.; Liu, X. Synthesis and characterization of MnZn ferrite nanoparticles with improved saturation magnetization. *J. Supercond. Nov. Magn.* **2017**, *30*, 2275–2281. [CrossRef]
41. Liu, Z.; Zhao, Z.-G.; Miyauchi, M. Efficient visible light active CaFe2O4/WO3 based composite photocatalysts: Effect of interfacial modification. *J. Phys. Chem. C* **2009**, *113*, 17132–17137. [CrossRef]
42. Phan, T.T.V.; Huynh, T.-C.; Manivasagan, P.; Mondal, S.; Oh, J. An up-to-date review on biomedical applications of palladium nanoparticles. *Nanomaterials* **2019**, *10*, 66. [CrossRef] [PubMed]
43. Yang, J.; Fan, L.; Xu, Y.; Xia, J. Iron oxide nanoparticles with different polymer coatings for photothermal therapy. *J. Nanopart. Res.* **2017**, *19*, 1–12. [CrossRef]
44. Hatamie, S.; Shih, P.-J.; Chen, B.-W.; Wang, I.-J.; Young, T.-H.; Yao, D.-J. Synergic effect of novel WS2 carriers holding spherical Cobalt Ferrite@ cubic Fe3O4 (WS2/s-CoFe2O4@ c-Fe3O4) nanocomposites in magnetic resonance imaging and photothermal therapy for ocular treatments and investigation of corneal endothelial cell migration. *Nanomaterials* **2020**, *10*, 2555. [CrossRef] [PubMed]
45. Hessel, C.M.; Pattani, V.P.; Rasch, M.; Panthani, M.G.; Koo, B.; Tunnell, J.W.; Korgel, B.A. Copper selenide nanocrystals for photothermal therapy. *Nano Lett.* **2011**, *11*, 2560–2566. [CrossRef] [PubMed]
46. Yin, W.; Yan, L.; Yu, J.; Tian, G.; Zhou, L.; Zheng, X.; Zhang, X.; Yong, Y.; Li, J.; Gu, Z. High-throughput synthesis of single-layer MoS2 nanosheets as a near-infrared photothermal-triggered drug delivery for effective cancer therapy. *ACS Nano* **2014**, *8*, 6922–6933. [CrossRef] [PubMed]
47. Wang, S.; Riedinger, A.; Li, H.; Fu, C.; Liu, H.; Li, L.; Liu, T.; Tan, L.; Barthel, M.J.; Pugliese, G. Plasmonic copper sulfide nanocrystals exhibiting near-infrared photothermal and photodynamic therapeutic effects. *ACS Nano* **2015**, *9*, 1788–1800. [CrossRef]
48. Cole, J.R.; Mirin, N.A.; Knight, M.W.; Goodrich, G.P.; Halas, N.J. Photothermal efficiencies of nanoshells and nanorods for clinical therapeutic applications. *J. Phys. Chem. C* **2009**, *113*, 12090–12094. [CrossRef]
49. Kouassi, G.K.; Irudayaraj, J. Magnetic and gold-coated magnetic nanoparticles as a DNA sensor. *Anal. Chem.* **2006**, *78*, 3234–3241. [CrossRef]
50. Gupta, A.K.; Gupta, M. Synthesis and surface engineering of iron oxide nanoparticles for biomedical applications. *Biomaterials* **2005**, *26*, 3995–4021. [CrossRef]
51. Liu, P.; Wang, Y.; An, L.; Tian, Q.; Lin, J.; Yang, S. Ultrasmall WO3–x@ γ-poly-l-glutamic acid nanoparticles as a photoacoustic imaging and effective photothermal-enhanced chemodynamic therapy agent for cancer. *ACS Appl. Mater. Interfaces* **2018**, *10*, 38833–38844. [CrossRef]
52. Unterweger, H.; Dézsi, L.; Matuszak, J.; Janko, C.; Poettler, M.; Jordan, J.; Bäuerle, T.; Szebeni, J.; Fey, T.; Boccaccini, A.R. Dextran-coated superparamagnetic iron oxide nanoparticles for magnetic resonance imaging: Evaluation of size-dependent imaging properties, storage stability and safety. *Int. J. Nanomed.* **2018**, *13*, 1899. [CrossRef] [PubMed]
53. Manivasagan, P.; Bharathiraja, S.; Santha Moorthy, M.; Mondal, S.; Nguyen, T.P.; Kim, H.; Phan, T.T.V.; Lee, K.D.; Oh, J. Biocompatible chitosan oligosaccharide modified gold nanorods as highly effective photothermal agents for ablation of breast cancer cells. *Polymers* **2018**, *10*, 232. [CrossRef] [PubMed]
54. Seo, S.-H.; Joe, A.; Han, H.-W.; Manivasagan, P.; Jang, E.-S. Methylene Blue-Loaded Mesoporous Silica-Coated Gold Nanorods on Graphene Oxide for Synergistic Photothermal and Photodynamic Therapy. *Pharmaceutics* **2022**, *14*, 2242. [CrossRef] [PubMed]
55. Manivasagan, P.; Jun, S.W.; Truong, N.T.P.; Hoang, G.; Mondal, S.; Moorthy, M.S.; Kim, H.; Phan, T.T.V.; Doan, V.H.M.; Kim, C.-S. A multifunctional near-infrared laser-triggered drug delivery system using folic acid conjugated chitosan oligosaccharide encapsulated gold nanorods for targeted chemo-photothermal therapy. *J. Mater. Chem. B* **2019**, *7*, 3811–3825. [CrossRef]
56. Wang, K.; Yang, P.; Guo, R.; Yao, X.; Yang, W. Photothermal performance of MFe2O4 nanoparticles. *Chin. Chem. Lett.* **2019**, *30*, 2013–2016. [CrossRef]

57. Yang, Y.; Shi, H.; Wang, Y.; Shi, B.; Guo, L.; Wu, D.; Yang, S.; Wu, H. Graphene oxide/manganese ferrite nanohybrids for magnetic resonance imaging, photothermal therapy and drug delivery. *J. Biomater. Appl.* **2016**, *30*, 810–822. [CrossRef]
58. Qiu, E.; Chen, X.; Yang, D.-P.; Regulacio, M.D.; Ramos, R.M.C.R.; Luo, Z.; Wu, Y.-L.; Lin, M.; Li, Z.; Loh, X.J. Fabricating Dual-Functional Plasmonic–Magnetic Au@ MgFe2O4 Nanohybrids for Photothermal Therapy and Magnetic Resonance Imaging. *ACS Omega* **2022**, *7*, 2031–2040. [CrossRef]

pharmaceutics

MDPI

Article

Impact of Plasmonic Nanoparticles on Poikilocytosis and Microrheological Properties of Erythrocytes

Tatiana Avsievich [1,*], Ruixue Zhu [1], Alexey P. Popov [2], Alexander Yatskovskiy [3], Anton A. Popov [4], Gleb Tikhonowsky [4], Andrei I. Pastukhov [5], Sergei Klimentov [4], Alexander Bykov [1], Andrei Kabashin [4,5] and Igor Meglinski [1,6,*]

1 Optoelectronics and Measurement Techniques, University of Oulu, 90570 Oulu, Finland
2 VTT Technical Research Centre of Finland, Kaitovayla 1, 90590 Oulu, Finland
3 Department of Histology, Cytology and Embryology, Institute of Clinical Medicine N.V. Sklifosovsky, I.M. Sechenov First Moscow State Medical University, Trubetskaya Street 8, 119991 Moscow, Russia
4 Institute of Engineering Physics for Biomedicine (PhysBio), National Research Nuclear University (MEPhI), Kashirskoe Shosse, 31, 115409 Moscow, Russia
5 CNRS, LP3, Aix-Marseille University, 163 Av. de Luminy, 13009 Marseille, France
6 College of Engineering and Physical Sciences, Aston University, Birmingham B4 7ET, UK
* Correspondence: tiavsievich@gmail.com (T.A.); igor.meglinski@oulu.fi (I.M.)

Abstract: Plasmonic nanoparticles (NP) possess great potential in photothermal therapy and diagnostics. However, novel NP require a detailed examination for potential toxicity and peculiarities of interaction with cells. Red blood cells (RBC) are important for NP distribution and the development of hybrid RBC-NP delivery systems. This research explored RBC alterations induced by noble (Au and Ag) and nitride-based (TiN and ZrN) laser-synthesized plasmonic NP. Optical tweezers and conventional microscopy modalities indicated the effects arising at non-hemolytic levels, such as RBC poikilocytosis, and alterations in RBC microrheological parameters, elasticity and intercellular interactions. Aggregation and deformability significantly decreased for echinocytes independently of NP type, while for intact RBC, all NP except Ag NP increased the interaction forces but had no effect on RBC deformability. RBC poikilocytosis promoted by NP at concentration 50 μg mL^{-1} was more pronounced for Au and Ag NP, compared to TiN and ZrN NP. Nitride-based NP demonstrated better biocompatibility towards RBC and higher photothermal efficiency than their noble metal counterparts.

Keywords: plasmonic nanoparticles; red blood cells; microrheology; aggregation; cell–cell interactions; poikilocytosis; optical manipulation; microscopy

Citation: Avsievich, T.; Zhu, R.; Popov, A.P.; Yatskovskiy, A.; Popov, A.A.; Tikhonowsky, G.; Pastukhov, A.I.; Klimentov, S.; Bykov, A.; Kabashin, A.; et al. Impact of Plasmonic Nanoparticles on Poikilocytosis and Microrheological Properties of Erythrocytes. *Pharmaceutics* **2023**, *15*, 1046. https://doi.org/10.3390/pharmaceutics15041046

Academic Editors: Manivasagan Panchanathan and Eue-Soon Jang

Received: 31 January 2023
Revised: 3 March 2023
Accepted: 17 March 2023
Published: 23 March 2023

1. Introduction

Plasmonic nanoparticles (NP) present one of the most important classes of nanomaterials for biomedical usage [1–3]. Owing to optical excitations of collective free electron oscillations (plasmons), these NP can exhibit a strong absorption band in optical spectra with the absorption cross-section exceeding that of conventional absorbing dyes by orders of magnitude [3]. This makes them excellent candidates as sensitizers of photo-induced therapy [2,3] and contrast agents in photoacoustic imaging [2,4], while a local enhancement in electric field promising their use as substrates for SERS-based bioidentification [5,6]. Gold (Au) and silver (Ag) nanostructures present conventional plasmonic nanomaterials and are widely used across a variety of applications. Both Au and Ag NP provide strong SERS signals [7], while Ag nanoparticles can serve as efficient antibacterial agents [8]. Au nanostructures are considered as excellent candidates for phototherapy, but one has to design complex structures (core-shells, nanorods) to shift the plasmonic peak (originally generated around 520–550 nm) toward the window of relative tissue transparency (650–900 nm) [2,3]. The plasmonic mismatch problem of conventional plasmonic materials can also be solved by the employment of alternative titanium nitride (TiN) and zirconium nitride (ZrN) NP, which generate plasmonic bands centred around 700 nm and 670 nm, respectively [9,10].

Typically, regardless of the route of entry into the human body NP are transported through the blood circulation system interacting with blood components [11,12]. Hence, hemocompatibility is a primary requirement for NP safety evaluation [13]. Red blood cells (RBC), being the most abundant cellular component of blood (40–45% volume percentage), are significant contributors to blood properties. The unique biconcave shape and high deformability of RBC enable them to squeeze through the narrowest capillaries to perform the key function of gas exchange in tissues [14]. RBC are commonly used for NP testing due to their availability and structural simplicity [15] and are actively investigated as a natural drug delivery system (DDS) [12,16], underlying the need for NP evaluation at a cellular level [17]. Many NP were found to influence RBC: fullerene NP soften RBC membrane [18], nanodiamonds may induce RBC aggregation [19], polystyrene NP adsorbed on RBC membrane or incorporated into the RBC can affect cellular deformability [17]. Exposure to NP can lead to impaired RBC properties, causing hemolysis, poikilocytosis (an increase in abnormal-shaped RBC), and alterations of deformability and aggregation. These changes can further result in complications of blood microrheology, ultimately leading to thrombosis [20–22].

NP prepared via pulsed laser ablation in liquids (LAL) present a novel class of ultra-pure nanomaterials for biomedical applications. This relatively new method is based on a natural production of nanoclusters under the action of laser radiation on a solid target in a liquid ambient (deionized water, ethanol, etc.), followed by their release into liquid ambience to form a colloidal NP solution [23–25]. The advantage of this approach is related to the possibility of performing the synthesis in an ultraclean environment in the absence of toxic by-products typical for conventional chemical synthesis. Using this approach, a series of novel nanomaterials, including plasmonic Au NP [26–28], Ag NP [29] and TiN NP [30–32], were recently synthesized to explore their use in biomedical tasks. Laser-synthesized TiN NP can provide local hyperthermia under photoexcitation in the window of relative tissue transparency [30,31] and be used as contrast agents in photoacoustic imaging [32], while Au NP can be employed as electrocatalyzers of glucose oxidation [27] and SERS probes for identification of biological molecules [33,34].

In the present study, the influence of plasmonic NP synthesized by LAL on RBC microrheology and poikilocytosis was studied on RBC in vitro. We characterized the toxicity of Ag, Au, TiN and ZrN NPs towards RBC by assessing their hemolytic activity with optical spectroscopy, NP-induced RBC morphological alterations with SEM, and RBC deformability and aggregation in autologous blood plasma via the optical tweezers (OT) method [35]. We also observed a photothermal response of NP to focused infrared (IR) laser radiation leading to hyperthermic destruction of RBC.

2. Materials and Methods

2.1. Nanoparticles Synthesis and Characterization

Colloidal solutions of plasmonic NP were synthesized by ultrafast (fs) laser ablation of the Au, Ag, and TiN solid targets in deionized water. The target was fixed vertically on the wall of a quartz cuvette filled with 15 mL of ultrapure deionized water (18.2 M ωcm at 25 °C). Radiation from a Yb:KGW laser (1030 nm wavelength, 250 fs pulse duration, up to 30 μJ pulse energy, 100 kHz repetition rate, TETA 10 model, Avesta, Moscow, Russia) was focused by a 100 mm F-theta lens on the surface of the target through a side wall of the ablation vessel. The laser beam was scanned over a 20 × 10 mm^2 area on the surface of the target, with 4000 mm/s speed with the help of a galvanometric scanner (LScanH-10-1030, AtekoTM, Moscow, Russia). The scanning was performed to avoid hole drilling in the target and to prevent the scattering of laser pulses on cavitation bubbles, which ultimately resulted in a significant NP yield increase. The thickness of the liquid layer along the incident laser beam was 4 mm. The interaction of extremely powerful fs laser pulses with liquid ambient was accompanied by energy-dependent self-focusing of the laser beam. The power of the laser pulses was 120 MW, which significantly exceeds the self-focusing threshold of water at 1030 nm (14.8 MW). Therefore, to take into account shifts of the focal plane due to the self-focusing effect, the target-lens distance was adjusted to obtain the maximum NP yield of the ablation process, measured by weighing the target before and after the ablation. Concentrations of the obtained colloidal solutions were 150 μg mL^{-1} for Ag NP, 100 μg mL^{-1} for TiN NP and 200 μg mL^{-1} for Au NP. The duration of each experiment was 15 min. To produce ZrN-based NP, the target was placed vertically in a quartz vessel filled with 25 mL of deionized water. A laser beam (Amplitude Systems, 1025 nm, 490 fs, 50 kHz) was focused on the surface of the target by a 75 mm convex lens. The energy of the incident beam was preliminarily attenuated by a half-wave plate and Brewster polarizer to 30 μJ per pulse. The calculated concentration of NP in the solution was about 110 μg mL^{-1} and was estimated by measuring the mass of the target before and after the process of ablation.

Morphology, structure and size distributions of obtained plasmonic NP were characterized by transmission electron microscopy (TEM) using a JEM-3010 microscope (JEOL, Peabody, MA, USA) at an acceleration voltage of 300 kV or scanning electron microscopy (SEM) using a MAIA 3 microscope (Tescan, Brno, Czech Republic) operating at 20 kV accelerating voltage. Samples for the electron microscopy were prepared by dropping 10 μL of the NP suspension onto a formvar-coated copper grid (200 mesh, Oxford Instruments, Oxford, UK) or a cleaned crystalline silicon substrate (for SEM), with subsequent drying at ambient conditions. The data from the TEM and SEM images were analyzed by Fiji ImageJ software (National Institutes of Health, USA). The optical extinction spectra were recorded using a UV-Vis spectrometer (UV-2600, Shimadzu, Tokyo, Japan for ZrN NP or PSI-MC 2, SOL Instruments, Minsk, Belarus) in optical glass cuvettes with a path length of 10 mm. Stock NP water solutions were centrifuged at 17,000× g (Micro Star 17, VWR, Radnor, PA, USA) for 35 min, solvents were removed and substituted with Dulbecco's Phosphate Buffered Saline (DPBS, Sigma Aldrich, pH 7.4, St. Louis, MO, USA). The estimation of PdI (polydispersity index) and zeta potential of of NP colloids was performed using Malvern Zetasizer Nano ZS (Malvern Instruments, Malvern, UK). NP absorbance spectra were also examined for NP dissolved in blood plasma with a spectrophotometer (Optronic Laboratories, Orlando, FL, USA) in the transmission mode.

2.2. Blood Samples

Whole blood specimens were obtained from two healthy volunteers by venipuncture at the Nordlab clinic (Oulu, Finland) with oral consent and ethical permission (Finnish Red Cross, No. 6095, 7/2021). Blood was collected in BD Vacutainer tubes containing EDTA (ethylenediaminetetraacetic acid) as an anticoagulant. Whole blood was centrifuged (CompactStar CS4, VWR, Radnor, PA, USA) at 2000× g (4500 RPM) for 10 min to remove formed elements. The supernatant was gently aspirated and the process was repeated at 4000× g (6500 RPM) to obtain platelet-free plasma. Healthy human RBC were freshly collected from fingertip-prick blood drops prior to each series of experiments. Blood (20 μL) was washed twice with DPBS and centrifuged at 3000× g for 10 min. NP dissolved in DPBS were prone to agglomeration, therefore prior to incubation with RBC, NP were sonicated for 1–2 min with a probe ultrasonic homogenizer (Bandelin Sonoplus HD 2070.2, Berlin, Germany) to obtain homogenous suspensions. Washed RBC were incubated with NP for 1 h at 1% hematocrit (1% Ht) to keep the constant RBC/NP ratio for all experiments. After incubation samples were centrifuged at 2000× g for 5 min. Supernatant was used for hemolytic activity examination. Intact RBC formed a dark pellet on the bottom of the tube. Sedimented RBC were diluted in plasma at 3% hematocrit for optical microscopy observation and at 0.5% hematocrit for OT. The sample preparation and experiments were conducted at room temperature 23 °C.

2.3. Hemolytic Activity

RBC were dispersed in DPBS NP solutions of 100, 50 and 10 μg mL $^{-1}$ at 1% Ht. RBC incubated in distilled water (DI) and DPBS (without NP) served as positive (+) and negative (−) controls, respectively. All mixtures were gently vortexed and incubated at room temperature (23 °C) for 1 h. After incubation, samples were centrifuged at 2000× g for 5 min, and 500 μL of supernatant was transferred into a quartz cuvette to measure the absorbance spectra (400–800 nm) of hemoglobin from lysed RBC with a spectrophotometer (Optronic Laboratories, Orlando, FL, USA) in the transmission mode. Among the 3 hemoglobin characteristic peaks at 415, 541 and 577 nm, the one at 541 nm was taken for analysis. Hemolysis percentages toward RBC in the samples were calculated using the following formula:

$$Hemolysis(\%) = \frac{(Abs(sample) - Abs(-))}{(Abs(+) - Abs(-))} \times 100\%, \quad (1)$$

where Abs(sample) is the absorbance of the treated sample, Abs(+) and Abs(−) is the absorbance of the positive and negative control, correspondingly. Data presented as the mean ± standard deviation (SD) of three independent experiments. One-way ANOVA with post hoc Dunnett's test (no matching or pairing) was performed in GraphPad Prism (GraphPad Software, San Diego, CA, USA). Statistical significance is presented for each sample against the control (* $p < 0.05$).

2.4. Optical Microscopy

RBC incubated with NP in DPBS were resuspended in autologous blood plasma at 1% Ht. Samples were gently vortexed and transferred into a measurement chamber (15 μL) made from a microscopic cover glass attached to a microscopic slide with double-sided tape. Opened edges were sealed with Vaseline to prevent drifts inside the sample due to drying. The sample was immediately placed for microscopic examination (Nikon Eclipse LV100 Upright Microscope, Tokyo, Japan) in the bright-field mode. Images were captured by supplementary NIS-Elements Advanced Research Imaging Software. The automated timelapse was recorded for each sample within 2 h with an interval of 2 or 5 min from the moment when all RBC were settled on the bottom of the cuvette. Each image was then analyzed to calculate the percentage of image area occupied by RBC. Initially, most RBC are oriented face-front to the camera, but then over time they flip on the edge while forming the aggregates, hence the whole area occupied by RBC decreases. Monochrome images (1600 × 1200 px) were then analyzed with ImageJ software (National Institutes of Health, Bethesda, MA, USA). The threshold ("Mean" method) was set to identify RBC in the stack of the images, and the image was converted to a binary image. Then RBC were analyzed automatically (command "Analyze particles; the minimum detectable particle was set to an area of one RBC (5280 px ^{2}) or the area of the edge-on oriented RBC 20 μm ^{2}. About 700–800 cells were analyzed per sample. Results displayed the relative area of RBC occupied by RBC in percentages.

2.5. Optical Tweezers

A double-channel OT setup [19,36,37] was used to measure the RBC aggregation force and deformability in autologous plasma at 0.5%Ht. Two optical traps were formed by orthogonally polarized continuous-wave laser beams from a single-mode Nd:YAG infrared laser ILML3IF-300, λ = 1064 nm, 350 mW (Leadlight Technology, Taoyuan, Taiwan) and a 100 (NA = 1.00) water-immersion objective (Olympus, LumPlanFI, Center Valley, PA, USA). One trap was set to a fixed position, while the second trap was moved by a steerable mirror. Trapped RBC were imaged in a transmission mode with a CMOS camera PL-D722MU-T (Pixelink, Gloucester, ON, Canada). Sample positioning was performed with a micro-resolution translational motorized XY stage 8MTF (Standa, Vilnius, Lithuania). OT calibration was performed using the drag force method described in our previous studies [19,36,37]. To measure the aggregation force between two individual RBC, cells were trapped and lifted at 30 μm height from the bottom of an in-house made measurement chamber (as used for optical microscopy studies described above). RBC were positioned to form a contact area with 40% overlap. The trapping power in both traps was then slowly decreased until RBC started to aggregate. The trapping power at this moment was proportional to the aggregation force, which was then retrieved using the calibration curve. The trapping force is dependent on the RBC shape, therefore calibration curves were obtained for discosytes and distorted RBC shapes—echinocytes. The deformability of RBC was estimated by subjecting the trapped RBC to viscous fluid flow (velocity 100 μm/s). The relative RBC elongation for the control sample and for samples incubated with NP was calculated as:

$$Elongation(\%) = \frac{L - L_0}{L_0} \times 100\%, \tag{2}$$

where L is the length of the stretched RBC, L_0 is the initial length of the trapped RBC. Each series of measurements was carried out on 20–30 cells. Experimental data were evaluated by one-way ANOVA with post hoc Tukey's test. Results are given as the mean ± SD. The values of * $p < 0.1$, ** $p < 0.05$ were considered as significant.

2.6. Scanning Electron Microscopy of RBC

RBC incubated with NP for 1 h in DPBS were fixed by adding 1% glutaraldehyde (Merck, Baltimore, MD, USA) for 30 min. After that, the samples were centrifuged at 2000× g for 5 min, and the supernatant was discarded. Sedimented RBC were washed with distilled water twice. After platinum sputtering, RBC smears were investigated with Zeiss Sigma FE-SEM (Carl Zeiss, Jenna, Germany) at 5 kV.

3. Results

3.1. Plasmonic Nanoparticles

Laser ablative synthesis resulted in predominantly spherical NP with a mode size of 30–40 nm as shown in Figure 1. Only ZrN NP had a significant fraction of non-spherical nanostructures, and it should be noted that only spherical NP were measured for distribution. Extinction spectra of aqueous colloidal solutions of the synthesized NP are shown in Figure 1b. Characteristic plasmonic peaks of Ag NP at 406 nm, Au NP at 524 nm, ZrN NP at 630 nm and TiN NP at 674 nm. Zeta potential measured for NP in DPBS revealed higher colloidal stability of Ag NP with ζ-potential −28.4 ± 1.6mV, while Ag, TiN and ZrN NP having originally higher ζ-potential in water (see Supplementary Materials Table S1), in DPBS became less stable with ζ-potential around −20 mV. PdI indices were below 0.5 for NP, indicating relatively narrow size distribution.

3.2. Hemolytic Activity of NP

Hemolysis describes the rupturing of RBC and the release of their contents, including hemoglobin, into the extracellular environment. Hemolytic activity of NP was assessed spectrophotometrically, results are presented in Figure 2a. According to ASTM E2524-22 (standard test method for analysis of hemolytic properties of NP) samples exhibiting hemolysis values over 5% correspond to hemolytic samples [38]. The minimal concentration of NP (10 μg mL^{-1}) was not significantly different from the control sample for all NP, 50 μg mL^{-1} caused insignificant hemolysis. Ag and Au NP were hemolytic at a concentration 100 μg mL^{-1}. NP did not shift the absorbance peaks of hemoglobin.

3.3. RBC Poikilocytosis

In total, three main RBC shapes were identified on optical images of the tested samples: a normal discocyte, a mildly perturbed echinocyte I (EI)—an irregularly contoured discocyte with up to five protrusions, and a highly perturbed echinocyte II (EII)—an oval and spherical RBC with many spicules. The relative occurrence of these fractions in each sample is presented in Figure 2b. The percentage of crenated RBC (EI and EII) increased by 5 times in samples treated with Au NP and Ag NP, from 7% in the control sample to 38% and 35%

for Au and Ag NP, respectively, a lower number was found for TiN NP 20%, and no change was found for ZrN NP.

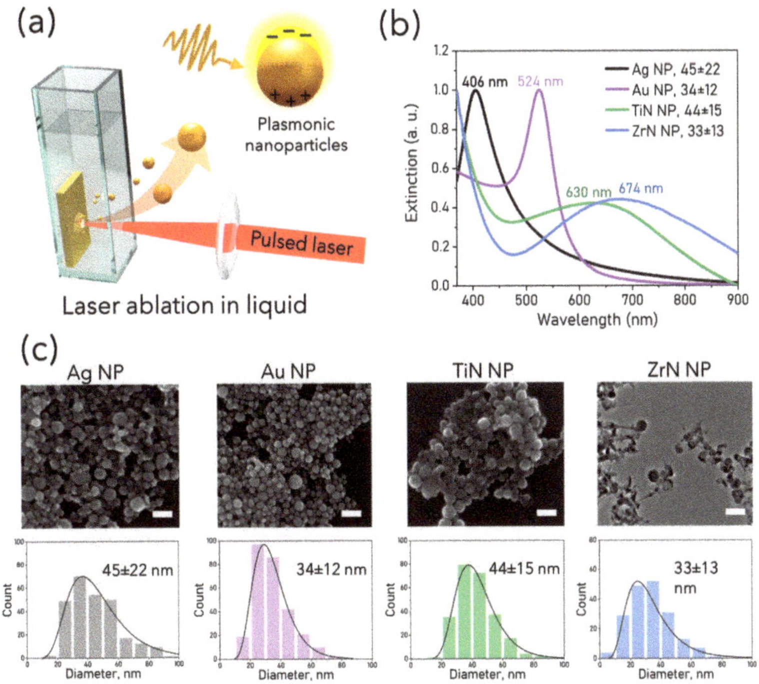

Figure 1. (**a**) Plasmonic NP synthesized by laser ablation in liquid method (LAL). (**b**) Normalized optical extinction spectra of Ag (black), Au (violet), TiN (green) and ZrN (blue) NP. (**c**) SEM images of plasmonic NP synthesized by LAL and their corresponding size distributions: Ag NP, Au NP, TiN NP, ZrN NP. Scale bar: 100 nm.

The morphology of RBC upon the NP attachment was examined via SEM (Figure 2c). For RBC treated with Ag NP besides the perturbed shapes such as teardrop-like (dacrocytes) and echinocytes, strongly spiked acanthocytes RBC were also found (<1%). The membrane perforations were found in samples treated with Au NP and ZrN NP (shown with arrows on Figure 2c, which could be a result of NP internalization into the cell. Approximately 20% of RBC incubated with Ag and Au NP were prone to exhibit a drop-like shape. More pronounced membrane transformations appeared as NP concentration increased to 100 μg mL^{-1} (Supplementary Materials, Figure S1).

The area occupied by RBC varied within 10% for the samples; however, this variation did not affect the general trend of exponential decrease, which was found repeatedly for each sample in three independent experiments. The temporal parameters of RBC aggregation were derived from the one-exponential fitting of the kinetics curves (Figure 3a). Within the first minute of observation, RBC were oriented with their face-side parallel to the glass slide of the sample chamber, occupying about 30–40% of the image area. When RBC start to aggregate, this area decreases with time due to the reorientation of RBC from "face-on" to "edge-on" position. The aggregates formed after 2 h in blood plasma in each sample are shown on microscopic images in Figure 3 for control (b), Ag NP (c), Au NP (d), TiN NP (e) and ZrN NP (f). RBC morphological changes caused by NP are shown on bright-field images (enlarged insets). Due to the higher number of single RBC at the beginning, the area occupied by RBC changed more rapidly. Aggregation kinetics retrieved from optical microscopy images revealed the difference between NP-treated samples. The time constants of aggregation kinetics on multi-cellular level for the control (23.8 ± 1.1 min), Ag and Au NP-treated samples were in the same range (29.4 ± 1.4 and 28.3 ± 1 min correspondingly), while the time constant for Zr NP sample was higher (119.34 ± 16.9 min), and significantly lower for TiN NP (13.2 ± 1.3 min) (see Supplementary Materials, Table S2).

3.4. RBC Aggregation and Deformability

RBC aggregation and deformability were examined at a non-hemolytic concentration 50 µg mL^{-1}. Aggregation force measurements were performed for intact discotic, and for crenated RBC shapes such as EI and EII. The schematic of OT measurements is presented for discocytes and EII on Figure 1a,b. The aggregation force value measured in the control sample without NP was 4.49 ± 0.86 pN. All NP except Ag NP caused an increase in RBC aggregation (Figure 4c). The aggregation force did not change with RBC shape transition into EI; however, EII demonstrated significantly reduced aggregation force compared to the control.

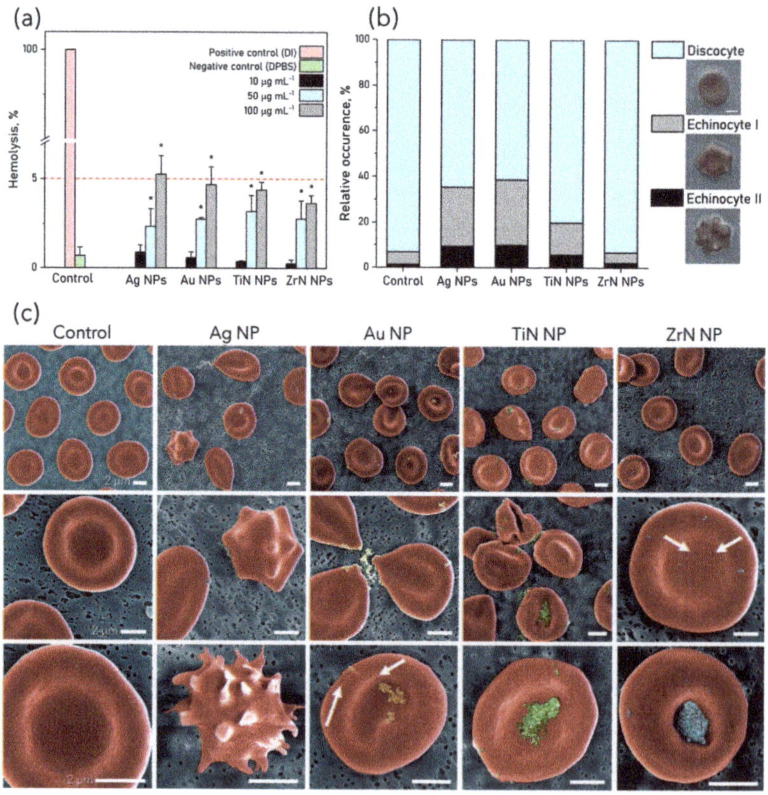

Figure 2. (**a**) In vitro hemolysis test results for RBC (1% hematocrit) exposed to NP for 1 h at concentrations 10, 50 and 100 µg mL^{-1}. The red dashed line indicates the hemolysis threshold value (5%), from which the tested NP are considered hemolytic. RBC in DPBS were used as a negative control, and RBC in DI were used as a positive control. Data are presented as a mean ± SD of three independent experiments and analyzed by one-way ANOVA (post hoc Dunnett's test). * $p < 0.05$ are significantly different from the negative control. (**b**) The relative occurrence of RBC fractions by morphology: discocyte, echinocyte I and II (EI and EII), in the samples after 1 h incubation with NP and resuspension in plasma at 50 µg mL^{-1}. Scale bar: 3 µm. (**c**) Coloured SEM images of RBC incubated for 1 h with plasmonic NP (50 µg mL^{-1}). Scale bar: 2 µm.

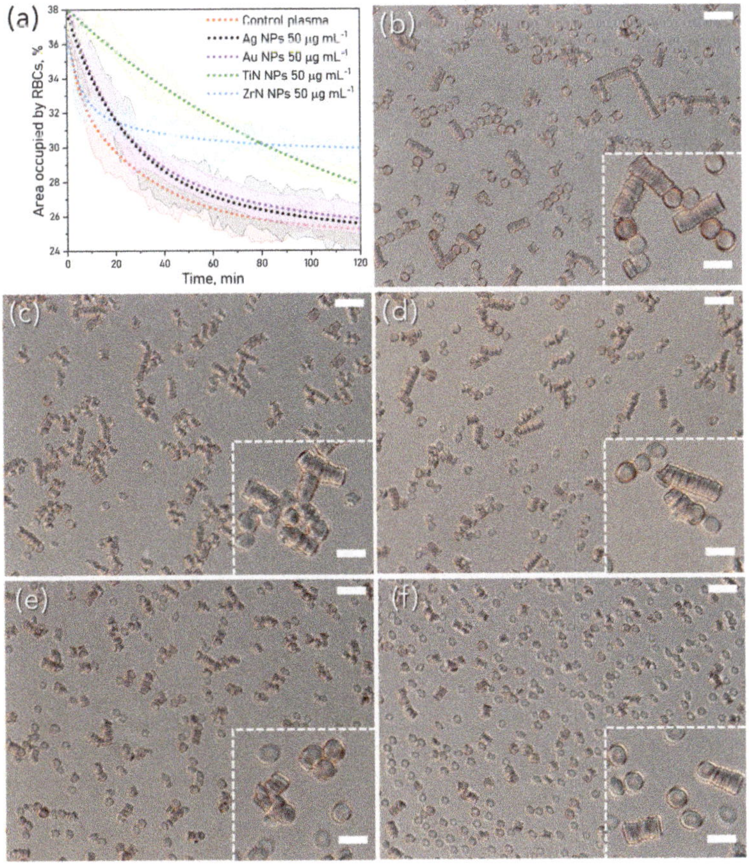

Figure 3. (**a**) Kinetics of RBC aggregation as the change of area occupied by RBC. Dotted lines are average data fitted by one-exponential decay with shaded areas representing the standard deviation. (**b–f**) Optical microscopy images of RBC incubated with NP with enlarged insets of aggregates after 2 h in autologous blood plasma: (**b**) control, (**c**) Ag NP, (**d**) Au NP, (**e**) TiN NP, and (**f**) ZrN NP at concentration 50 μg mL^{-1}. Scale: 20 μm (main images), 10 μm (insets).

Relative RBC deformation was defined by applying viscous blood plasma flow (100 μm/s) against a trapped RBC for discocytes and for EII (Figure 4d,e). Considering the occurrence of EI and EII, their deformability was also estimated. Interestingly, the behaviour of echinocytic cells was similar in all the NP-treated samples, therefore considering no significant difference among EI and EII of control and NP-treated samples, the elongation data were merged in one group for all NP types (see Figure 4f). There was no statistically significant difference between the values for RBC elongation for normal discocytes and EI RBCs after NP treatment, while for EII RBC elongation was reduced by 70%.

In samples treated with TiN and ZrN NP, many cells strongly adhered to the bottom of the glass measurement chamber while maintaining a discocyte shape, which apparently explained the remarkable difference in aggregation kinetics observed with optical microscopy. Optical trapping of 80% of RBC in these samples was not possible even at the highest power of 180 mW (55 pN). Partial adherence of RBC was visible when some RBC parts could be trapped. Attempting to move such RBC caused its stretching, where adhered parts appeared as protrusions (see Figure 5a).

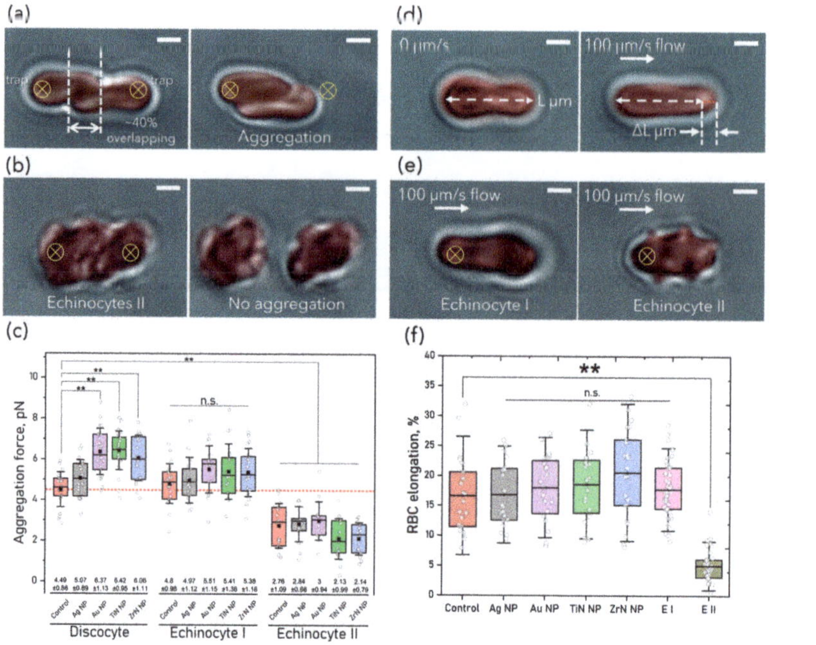

Figure 4. (**a**) OT measurements of RBC aggregation force after treatment with NP (50 µg mL^{-1}) for (**a**) discocytes and (**b**) echinocytes II. RBC relative elongation in fluid flow (100 µm/s) for (**d**) discocyte and (**e**) echinocytes I and II. Scale bar: 2 µm. (**c**) Box plot of RBC aggregation force and (**f**) scatter interval plot of relative RBC elongation in viscous flow 100 µm/s. Data are expressed as mean ± SD, n = 20–30, ** $p < 0.05$, ns: non-significant, in one-way ANOVA, followed by Tukey's post hoc test compared to control. Center lines show the medians; square dots represent the means; outliers and individual data points are represented by dots; box limits indicate the 25th and 75th percentiles; whiskers show SD.

3.5. NP-Induced Hyperthermia of RBC

The absorption spectra of NP dispersions in blood plasma compared to NP spectra in water indicate non-significant shifts of plasmonic peaks (see Figure 5b), indicating that NP preserved the optical properties in the complex biological medium. Still, 16 nm red shift observed for Ag NP could be a result of NP agglomeration in blood plasma. There was no enhanced absorption found at 1064 nm. Tightly focused laser beams of OT give rise to intensities on the order of MW/cm^2 at their focus area, which may lead to the consequent heating due to the light absorption. At 100× magnification, NP clusters of size 0.2–3 µm attached to RBC often appeared visible (see Figure 5c). Usually, these RBC cannot be optically trapped. All types of plasmonic NP showed the absorption of 1064 nm irradiation, which was strong enough to cause RBC rupture.Local radiation of the NP attached to RBC by a focused laser beam results in heat release by NP sufficient to induce fast membrane rupture. The laser power density at the focal spot (2 µm) of OT can reach about 2.5 MW/cm^2 for the trapping power 40 mW at the beam waist. The time of thermoplasmonic destruction of RBC varies in the range of 1–100 s (Figure 5d), and it largely depends on the applied laser power, the size of NP agglomerates, and the number of aggregates attached to RBC membrane due to the heterogeneous distribution of NP on the RBC surface. An example of RBC destruction resulting in RBC lysis and the formation of the ghost residue of RBC is shown in Figure 5c.

The time required for RBC to burst was almost twice as fast for TiN and ZrN NPs (up to 52 s) compared to longer times (up to 100 s) for Au and Ag NP. The mean sizes of NP agglomerates between the samples were not statistically different (Figure 5e), but hemolysis was reached faster for larger NP agglomerates (Figure 5d). The mean times (Figure 5f) are two-fold shorter for TiN and ZrN NP. Therefore, TiN and ZrN NP have higher plasmonic efficiency compared to their noble counterparts. The plasmonic peaks of TiN NPs and ZrN NP are also red-shifted compared to their counterparts, having an extended peak beyond 800 nm, closer to the trapping wavelength.

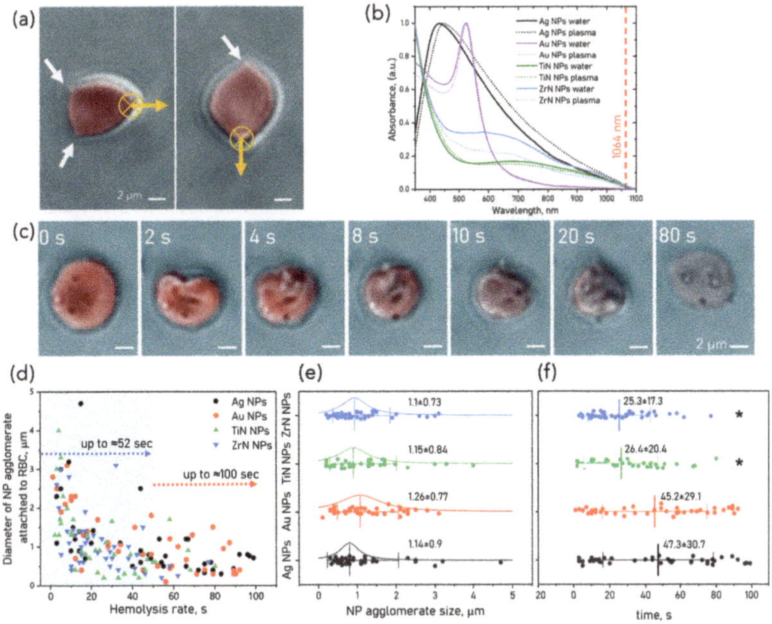

Figure 5. (**a**) RBC treated with ZrN (**left**) and TiN (**right**) NP adhered to the bottom of the measurement chamber. Regions of RBC stuck to the bottom are shown with arrows. Optical trap at maximum power 180 mW (55 pN) was unable to trap RBC. (**b**) Normalized absorption spectra of NP dispersions in water and blood plasma. (**c**) Camera snapshots during 80 s of photothermal destruction of RBC with attached Au NP clusters indicated by arrows in a laser trap (power 40 mW). RBC becomes pale as the optical contrast decreases due to hemolysis. (**d**) Scatter plot of laser trap time exposure vs. size of NP aggregates. The shadowed area indicates the maximum time required to destruct an RBC with attached TiN and ZrN NP. Scatter interval plots (one-way-ANOVA, post hoc Tukey's test, $n = 40$, comparisons with significant differences are marked for * $p < 0.05$) of mean ± SD for NP agglomerate sizes (**e**) and time of RBC destruction (**f**).

4. Discussion

LAL technique of NP synthesis offers a set of advantages as a contamination-free, fast and easily adjustable method applicable for various NP types. Since blood is an interface of utmost importance for NP biomedical applications, here we focused on NP-related effects caused by RBC-NP interactions. Recently, high safety and favourable biodistribution of laser-synthesized plasmonic NP was demonstrated in vivo on mice [28,29,31]. In vitro testing of the cellular parameters serves as a first safety assessment step of newly developed laser-synthesized plasmonic NP for biomedical applications.

NP demonstrated concentration-dependent hemolytic activity, with a lower hemolysis rate of TiN and ZrN NP compared to Ag and Au NP. Hemolytic 5% threshold indicating severe RBC damage was observed for all NP at concentrations exceeding 100 $\mu g\,mL^{-1}$. Our results align with those from a previous study that showed TiN NP to be highly biocompatible towards RBC in vitro and in vivo [31] at concentrations up to 100 $\mu g\,mL^{-1}$.

The absence of hemolysis, however, does not exclude other adverse effects, such as RBC poikilocytosis, deformability or aggregation.

In the samples treated with 50 $\mu g\,mL^{-1}$ NP, the fraction of morphologically altered EI and EII was increased in RBC population. Each detected fraction contributes to the total aggregation behaviour of RBC ensembles observed by optical microscopy. Single RBC analysis with OT is limited by the lower number of samples, which are represented by "trappable" RBC in our measurements. In turn, this allows us to measure the interaction force between the three types of RBC found in each sample (discocyte, EI, EII). The kinetics of RBC ensemble aggregation were found to be significantly different for samples treated with TiN and ZrN NP, the main reason for this difference identified with OT was a result of strong RBC adherence to the glass bottom of the chamber. A faster RBC aggregation rate was observed in ZrN NP-treated sample, while TiN NP slowed down the aggregation. Interestingly, despite the strong adherence, RBC preserved their intact shape.

The ability of RBC to deform and the degree of aggregation are critical for RBC capillary passage and oxygen delivery to biotissues. Reduced RBC deformability contributes to the pathogenesis of various hematological disorders that impair the flow and augment RBC retention in the spleen. Expectedly, the attachment of NP onto RBC membrane increases the RBC stiffness. Decreased RBC deformability was previously reported for PEGylated Au NP [39] and Ag NP [21]. The distribution of NP on RBC surface was quite uneven, including NP agglomerates. To perform an adequate NP/RBC loading ratio [17], NP must be functionalized to increase their stability. In the present study, only highly perturbed spiculated RBC (EII) demonstrated a significant loss of elasticity. Since aggregation force and relative elongation measured for echinocytes for NP-treated samples were not different from the control, the nature of this shape transition probably was not caused by the direct interaction with NP. SEM observations demonstrated how NP interaction with RBC causes local membrane distortions, which subsequently can change the RBC surface tension and result in increased rigidity.

The aggregation of intact RBC was increased for all the NP-treated samples except Ag NP at 50 $\mu g\,mL^{-1}$. NP attached to RBC can change the surface's charge distribution and either facilitate the interaction or promote the repulsive mechanisms. In the current investigation, the attachment of NP to RBC appears to be non-specific, as both the NP and the RBC surface have a negative ζ-potential (Supplementary Materials, Table S1), which suggests electrostatic repulsion. In agreement with previous study [40], we suggest that the low degree of RBC shape deterioration, corresponding to EI can be easily tolerated by the microvasculature. Significant reduction aggregation for EII could be a result of the limited elasticity and spiculated surface hindering the formation of the contact surface area large enough to form a stable aggregate, which is in vivo conditions can lead to increased blood viscosity [41].

When introduced into the bloodstream, NP are expected to be covered by plasma proteins (protein corona), which can significantly modify the intrinsic properties of NP in unpredictable ways. All NP preserved their absorption properties in blood plasma (see Figure 5b), therefore, it could be suggested that blood proteins do not affect the optical properties of NP. There was no change in absorption at 1064 nm wavelength used for the optical trapping of RBC.

Plasmonic NP can convert light into heat, which is applied in photothermal therapy (PTT), aiming for the destruction of cancer cells under NIR light. NP being in the focal area of the OT can be heated to extremely high temperatures reaching hundreds of celsius degrees [42]. The average time required for RBC to burst is clearly shorter for TiN and ZrN NP (up to 52 s) at the same trap power (40 mW, 13 pN) compared with Au and Ag NP with a time scale upper range of about 100 s (see Figure 5d). No significant difference in the size of NP aggregates was found, however, the distribution of Ag NP indicates a higher fraction of smaller agglomerates (see Figure 5e). However, the times of RBC destruction for TiN and ZrN NP were significantly shorter than in the case of Ag and Au NP (see Figure 5f). Indeed, the plasmonic peaks of TiN NP and ZrN NP are red-shifted compared to their counterparts, having a tail beyond 800 nm, which is the reason for higher photothermal conversion efficiency. In accordance with conclusions made in the study [30], we suggest that this effect is a result of the higher plasmonic efficiency of TiN and ZrN NP compared to Au NP and Ag NP. Laser heating of the RBC without structural changes is possible in the range of 37–70 °C, and spherocyte formation with following rupture can happen in the range of 70–94 °C [43]. Controlled application of laser irradiation can be used for optoporation of RBC for intracellular delivery. Observed earlier off-resonant excitation by the IR optical trap by authors of [44] probably explains the significant temperature increase enough to cause hyperthermic RBC destruction of RBC membrane.

The approach in this study includes a minimal set of hemotoxicity risk assessment to satisfy before performing in vivo studies with intravascular administration of NP. It is worth noting that interpretation of in vitro results is limited due to the extra in vivo factors such as dynamic interactions with blood contents (protein corona formation, activation of platelets and white blood cells) and shear flow of the bloodstream. Interaction of NP with immune cells and platelets is likely to initiate a cascade of immune and thrombolytic reactions. Protein corona formation and opsonization of NP will decrease the circulation time and increase their clearance from the body.

The key strategies developed in recent decades explored the advantages of using RBC as a drug delivery platform. The attachment of therapeutics to RBC by specific antibody-mediated [45] or non-specific coupling may adversely affect RBC properties [17,46]. Among the achieved benefits of therapeutic conjugation with RBC are prolonged circulation [47], enhanced activity of the enzymes which substrates circulate in plasma [48], in vivo targeting of RBC from plasma [49], and delivery to the specific site of therapeutic interest such as blood clots [50,51] and vessel wall [52,53]. Further research should be focused on the functionalization of tested NP to improve their stability and provide homogeneous loading onto RBC, and elucidation of the mechanisms behind NP-RBC interaction to ensure beneficial conjugation.

5. Conclusions

The in vitro testing of cellular parameters serves as the first safety evaluation step of newly developed laser-synthesized plasmonic NP for biomedical applications. In this study, the hemorheological properties of RBC were examined via various microscopy methods and OT after incubation with plasmonic NPs synthesized by LAL. Plasmonic NP demonstrated hemolysis for concentrations exceeding 100 $\mu g\,mL^{-1}$, while at the non-hemolytic range 50 $\mu g\,mL^{-1}$ poikilocytosis and alterations in microrheological RBC properties as aggregation and deformability were found. The kinetics of RBC aggregation for RBC ensembles were found to be significantly different for samples treated with TiN and ZrN NP. This difference is a result of strong RBC adherence to the glass base of the chamber, which was identified using OT. Except for Ag NPs, increased RBC aggregation was observed after treatment with all tested NPs. At the same time, the stiffening of RBCs was demonstrated only for the highly perturbed echinocytes. The irradiated clusters of TiN and ZrN NPs are effective photothermal heat transducers, leading to a faster RBC hyperthermal haemolysis than Au and Ag NPs. Future research towards the development of NP-RBC conjugates for clinical applications should address proper NP functionalization to minimize NP-related RBC alterations, elucidation of the molecular mechanism behind NP interaction with RBC, and effects arising in whole blood in vivo.

Supplementary Materials: The following supporting information can be downloaded at: https://www.mdpi.com/article/10.3390/pharmaceutics15041046/s1, Figure S1: Colored SEM images of RBC treated with (a) Ag NP, (b) Au NP, (c) TiN NP, (d) ZrN NP at concentration 100 $\mu g\,mL^{-1}$; Table S1: Size, zeta potential and polydispersity index (PdI) values (mean ± SD, n = 3) measured in deionized water (DI) and Dulbecco's Phosphate Buffered Saline (DPBS); Table S2: Temporal aggregation parameters derived from the kinetics curves of the area occupied by RBC change in time.

Author Contributions: Conceptualization: A.K. and I.M.; Methodology: T.A., A.P.P., A.A.P., G.T., A.I.P. and A.B.; Visualization: T.A.; Formal data analysis: T.A., R.Z. and A.Y.; Data curation: T.A., R.Z., A.K. and I.M.; Writing—original draft preparation: T.A., I.M. and A.K.; Writing—review and editing, T.A. and I.M.; Supervision: I.M, A.B, A.P.P., S.K. and A.K.; Funding acquisition: I.M., A.B., S.K. and A.K. All authors have read and agreed to the published version of the manuscript.

Funding: Current study partially supported by the Suomen Kulttuurirahasto (grant No. 00190188), China Scholarship Council (CSC No. 201706410089) and the INFOTECH and the Academy of Finland (project 325097). The authors also acknowledge the contribution of Russian Science Foundation (Project: 19-72-30012 and Project: 22-65-00096). I.M. acknowledges the support from the Leverhulme Trust and The Royal Society (Ref. no.:APX\R1\211232 APEX Awards 2021).

Informed Consent Statement: Written informed consent was obtained from all subjects involved in the study.

Data Availability Statement: Data available upon request from the authors.

Conflicts of Interest: The authors declare no conflict of interest.

Abbreviations

The following abbreviations are used in this manuscript:

NP	Nanoparticles
RBC	Red blood cells
LAL	Laser ablation in liquid
OT	Optical tweezers

References

1. Dykman, L.; Khlebtsov, N. Gold nanoparticles in biomedical applications: Recent advances and perspectives. *Chem. Soc. Rev.* **2012**, *41*, 2256–2282. [CrossRef] [PubMed]
2. Hirsch, L.R.; Stafford, R.J.; Bankson, J.; Sershen, S.R.; Rivera, B.; Price, R.; Hazle, J.D.; Halas, N.J.; West, J.L. Nanoshell-mediated near-infrared thermal therapy of tumors under magnetic resonance guidance. *Proc. Natl. Acad. Sci. USA* **2003**, *100*, 13549–13554. [CrossRef] [PubMed]
3. Jain, P.K.; Huang, X.; El-Sayed, I.H.; El-Sayed, M.A. Noble metals on the nanoscale: Optical and photothermal properties and some applications in imaging, sensing, biology, and medicine. *Acc. Chem. Res.* **2008**, *41*, 1578–1586. [CrossRef] [PubMed]
4. Gargiulo, S.; Albanese, S.; Mancini, M. State-of-the-art preclinical photoacoustic imaging in oncology: Recent advances in cancer theranostics. *Contrast Media Mol. Imaging* **2019**, *2019*, 5080267. [CrossRef] [PubMed]
5. Nie, S.; Emory, S.R. Probing single molecules and single nanoparticles by surface-enhanced Raman scattering. *Science* **1997**, *275*, 1102–1106. [CrossRef] [PubMed]
6. Kneipp, J.; Kneipp, H.; McLaughlin, M.; Brown, D.; Kneipp, K. In vivo molecular probing of cellular compartments with gold nanoparticles and nanoaggregates. *Nano Lett.* **2006**, *6*, 2225–2231. [CrossRef]
7. Sharma, B.; Frontiera, R.R.; Henry, A.I.; Ringe, E.; Van Duyne, R.P. SERS: Materials, applications, and the future. *Mater. Today* **2012**, *15*, 16–25. [CrossRef]
8. Gao, M.; Sun, L.; Wang, Z.; Zhao, Y. Controlled synthesis of Ag nanoparticles with different morphologies and their antibacterial properties. *Mater. Sci. Eng. C.* **2013**, *33*, 397–404. [CrossRef]
9. Guler, U.; Shalaev, V.M.; Boltasseva, A. Nanoparticle plasmonics: Going practical with transition metal nitrides. *Mater. Today* **2015**, *18*, 227–237. [CrossRef]
10. Lalisse, A.; Tessier, G.; Plain, J.; Baffou, G. Plasmonic efficiencies of nanoparticles made of metal nitrides (TiN, ZrN) compared with gold. *Sci. Rep.* **2016**, *6*, 38647. [CrossRef]
11. Tian, Y.; Tian, Z.; Dong, Y.; Wang, X.; Zhan, L. Current advances in nanomaterials affecting morphology, structure, and function of erythrocytes. *RSC Adv.* **2021**, *11*, 6958–6971. [CrossRef]
12. Zhu, R.; Avsievich, T.; Popov, A.; Bykov, A.; Meglinski, I. In vivo nano-biosensing element of red blood cell-mediated delivery. *Biosens. Bioelectron.* **2020**, *175*, 112845. [CrossRef]
13. Avsievich, T.; Zhu, R.; Popov, A.; Bykov, A.; Meglinski, I. Blood–nanomaterials interactions. In *Nanotechnology for Hematology, Blood Transfusion, and Artificial Blood*; Elsevier: Amsterdam, The Netherlands, 2022; pp. 1–40.
14. Danielczok, J.G.; Terriac, E.; Hertz, L.; Petkova-Kirova, P.; Lautenschläger, F.; Laschke, M.W.; Kaestner, L. Red blood cell passage of small capillaries is associated with transient Ca2+-mediated adaptations. *Front. Physiol.* **2017**, *8*, 979. [CrossRef]
15. Wadhwa, R.; Aggarwal, T.; Thapliyal, N.; Kumar, A.; Yadav, P.; Kumari, V.; Reddy, B.S.C.; Chandra, P.; Maurya, P.K. Red blood cells as an efficient in vitro model for evaluating the efficacy of metallic nanoparticles. *3 Biotech* **2019**, *9*, 279. [CrossRef] [PubMed]

16. Lomanovskaya, T.; Piavchenko, G.; Venediktov, A.; Kustavinova, E.; Kartashkina, N.; Mukhamedova, S.; Boronikhina, T.; Markov, A.; Telyshev, D.; Meglinski, I.; et al. Structural changes of erythrocyte membrane revealed by 3D confocal optical profilometer. *J. Biophoton.* **2022**, *10*, e202200222. [CrossRef] [PubMed]
17. Pan, D.C.; Myerson, J.W.; Brenner, J.S.; Patel, P.N.; Anselmo, A.C.; Mitragotri, S.; Muzykantov, V. Nanoparticle properties modulate their attachment and effect on carrier red blood cells. *Sci. Rep.* **2018**, *8*, 1615. [CrossRef] [PubMed]
18. Zhang, X.; Zhang, Y.; Zheng, Y.; Wang, B. Mechanical characteristics of human red blood cell membrane change due to C 60 nanoparticle infiltration. *Phys. Chem. Chem. Phys.* **2013**, *15*, 2473–2481. [CrossRef]
19. Avsievich, T.; Popov, A.; Bykov, A.; Meglinski, I. Mutual interaction of red blood cells influenced by nanoparticles. *Sci. Rep.* **2019**, *9*, 5147. [CrossRef]
20. Bian, Y.; Kim, K.; Ngo, T.; Kim, I.; Bae, O.N.; Lim, K.M.; Chung, J.H. Silver nanoparticles promote procoagulant activity of red blood cells: A potential risk of thrombosis in susceptible population. *Part. Fibre Toxicol.* **2019**, *16*, 9. [CrossRef]
21. Kim, M.J.; Shin, S. Toxic effects of silver nanoparticles and nanowires on erythrocyte rheology. *Food Chem. Toxicol.* **2014**, *67*, 80–86. [CrossRef]
22. He, Z.; Liu, J.; Du, L. The unexpected effect of PEGylated gold nanoparticles on the primary function of erythrocytes. *Nanoscale* **2014**, *6*, 9017–9024. [CrossRef] [PubMed]
23. Fojtik, A.; Henglein, A. Laser ablation of films and suspended particles in a solvent: formation of cluster and colloid solutions. *Ber. Bunsenges. Phys. Chem.* **1993**, *97*, 252–252.
24. Kabashin, A.V.; Delaporte, P.; Pereira, A.; Grojo, D.; Torres, R.; Sarnet, T.; Sentis, M. Nanofabrication with pulsed lasers. *Nanoscale Res. Lett.* **2010**, *5*, 454–463. [CrossRef] [PubMed]
25. Barcikowski, S.; Compagnini, G. Advanced nanoparticle generation and excitation by lasers in liquids. *Phys. Chem. Chem. Phys.* **2013**, *15*, 3022–3026. [CrossRef]
26. Maximova, K.; Aristov, A.; Sentis, M.; Kabashin, A.V. Size-controllable synthesis of bare gold nanoparticles by femtosecond laser fragmentation in water. *Nanotechnology* **2015**, *26*, 065601. [CrossRef]
27. Hebie, S.; Holade, Y.; Maximova, K.; Sentis, M.; Delaporte, P.; Kokoh, K.B.; Napporn, T.W.; Kabashin, A.V. Advanced electrocatalysts on the basis of bare Au nanomaterials for biofuel cell applications. *ACS Catal.* **2015**, *5*, 6489–6496. [CrossRef]
28. Bailly, A.L.; Correard, F.; Popov, A.; Tselikov, G.; Chaspoul, F.; Appay, R.; Al-Kattan, A.; Kabashin, A.V.; Braguer, D.; Esteve, M.A. In vivo evaluation of safety, biodistribution and pharmacokinetics of laser-synthesized gold nanoparticles. *Sci. Rep.* **2019**, *9*, 12890. [CrossRef]
29. Kabashin, A.V.; Meunier, M. Femtosecond laser ablation in aqueous solutions: A novel method to synthesize non-toxic metal colloids with controllable size. *J. Phys. Conf. Ser.* **2007**, *59*, 074. [CrossRef]
30. Popov, A.A.; Tselikov, G.; Dumas, N.; Berard, C.; Metwally, K.; Jones, N.; Al-Kattan, A.; Larrat, B.; Braguer, D.; Mensah, S.; et al. Laser-synthesized TiN nanoparticles as promising plasmonic alternative for biomedical applications. *Sci. Rep.* **2019**, *9*, 1194. [CrossRef]
31. Zelepukin, I.V.; Popov, A.A.; Shipunova, V.O.; Tikhonowski, G.V.; Mirkasymov, A.B.; Popova-Kuznetsova, E.A.; Klimentov, S.M.; Kabashin, A.V.; Deyev, S.M. Laser-synthesized TiN nanoparticles for biomedical applications: Evaluation of safety, biodistribution and pharmacokinetics. *Mater. Sci. Eng. C* **2021**, *120*, 111717. [CrossRef]
32. Maldonado, M.E.; Das, A.; Gomes, A.S.; Popov, A.A.; Klimentov, S.M.; Kabashin, A.V. Nonlinear photoacoustic response of suspensions of laser-synthesized plasmonic titanium nitride nanoparticles. *Opt. Lett.* **2020**, *45*, 6695–6698. [CrossRef]
33. Uusitalo, S.; Popov, A.; Ryabchikov, Y.V.; Bibikova, O.; Alakomi, H.L.; Juvonen, R.; Kontturi, V.; Siitonen, S.; Kabashin, A.; Meglinski, I.; et al. Surface-enhanced Raman spectroscopy for identification and discrimination of beverage spoilage yeasts using patterned substrates and gold nanoparticles. *J. Food Eng.* **2017**, *212*, 47–54. [CrossRef]
34. Kögler, M.; Ryabchikov, Y.V.; Uusitalo, S.; Popov, A.; Popov, A.; Tselikov, G.; Välimaa, A.L.; Al-Kattan, A.; Hiltunen, J.; Laitinen, R.; et al. Bare laser-synthesized Au-based nanoparticles as nondisturbing surface-enhanced Raman scattering probes for bacteria identification. *J. Biophoton.* **2018**, *11*, e201700225. [CrossRef] [PubMed]
35. Ashkin, A. Acceleration and trapping of particles by radiation pressure. *Phys. Rev. Lett.* **1970**, *24*, 156. [CrossRef]
36. Avsievich, T.; Tarakanchikova, Y.; Zhu, R.; Popov, A.; Bykov, A.; Skovorodkin, I.; Vainio, S.; Meglinski, I. Impact of nanocapsules on red blood cells interplay jointly assessed by optical tweezers and microscopy. *Micromachines* **2020**, *11*, 19. [CrossRef] [PubMed]
37. Zhu, R.; Avsievich, T.; Bykov, A.; Popov, A.; Meglinski, I. Influence of pulsed He–Ne Laser irradiation on the red blood cell interaction studied by optical tweezers. *Micromachines* **2019**, *10*, 853. [CrossRef]
38. ASTM E2524-22. Standard Test Method for Analysis of Hemolytic Properties of Nanoparticles. 2022. Available online: https://www.astm.org/e2524-22.html (accessed on 16 March 2022).
39. He, Z.; Li, C.; Zhang, X.; Zhong, R.; Wang, H.; Liu, J.; Du, L. The effects of gold nanoparticles on the human blood functions. *Artif. Cells. Nanomed. Bbiotechnol.* **2018**, *46*, 720–726. [CrossRef]
40. Piety, N.Z.; Reinhart, W.H.; Pourreau, P.H.; Abidi, R.; Shevkoplyas, S.S. Shape matters: The effect of red blood cell shape on perfusion of an artificial microvascular network. *Transfusion* **2016**, *56*, 844–851. [CrossRef]
41. Reinhart, W.; Singh-Marchetti, M.; Straub, P. The influence of erythrocyte shape on suspension viscosities. *Eur. J. Clin. Invest.* **1992**, *22*, 38–44. [CrossRef]
42. Seol, Y.; Carpenter, A.E.; Perkins, T.T. Gold nanoparticles: Enhanced optical trapping and sensitivity coupled with significant heating. *Opt. Lett.* **2006**, *31*, 2429–2431. [CrossRef]

43. Heger, M. Thrombosis versus thermal coagulum formation as a result of endovenous laser treatment: Biochemistry versus photophysics. *Phlebology* **2014**, *29*, 701–705. [CrossRef] [PubMed]
44. Urban, A.S.; Carretero-Palacios, S.; Lutich, A.A.; Lohmüller, T.; Feldmann, J.; Jäckel, F. Optical trapping and manipulation of plasmonic nanoparticles: Fundamentals, applications, and perspectives. *Nanoscale* **2014**, *6*, 4458–4474. [CrossRef] [PubMed]
45. Villa, C.H.; Pan, D.C.; Johnston, I.H.; Greineder, C.F.; Walsh, L.R.; Hood, E.D.; Cines, D.B.; Poncz, M.; Siegel, D.L.; Muzykantov, V.R. Biocompatible coupling of therapeutic fusion proteins to human erythrocytes. *Blood Adv.* **2018**, *2*, 165–176. [CrossRef]
46. Glassman, P.M.; Hood, E.D.; Ferguson, L.T.; Zhao, Z.; Siegel, D.L.; Mitragotri, S.; Brenner, J.S.; Muzykantov, V.R. Red blood cells: The metamorphosis of a neglected carrier into the natural mothership for artificial nanocarriers. *Adv. Drug Deliv. Rev.* **2021**, *178*, 113992. [CrossRef]
47. Antonelli, A.; Magnani, M. Red blood cells as carriers of iron oxide-based contrast agents for diagnostic applications. *J. Biomed. Nanotechnol.* **2014**, *10*, 1732–1750. [CrossRef]
48. RB, M.; Kirch, H. Enhanced biological activity of human recombinant interleukin 2 coupled to mouse red blood cells as evaluated using the mouse Meth A sarcoma model. *Biotechnol. Appl. Biochem.* **1996**, *23*, 29–36.
49. Glassman, P.M.; Villa, C.H.; Marcos-Contreras, O.A.; Hood, E.D.; Walsh, L.R.; Greineder, C.F.; Myerson, J.W.; Shuvaeva, T.; Puentes, L.; Brenner, J.S.; et al. Targeted In Vivo Loading of Red Blood Cells Markedly Prolongs Nanocarrier Circulation. *Bioconjug. Chem.* **2022**, *33*, 1286–1294. [CrossRef] [PubMed]
50. Zaitsev, S.; Spitzer, D.; Murciano, J.C.; Ding, B.S.; Tliba, S.; Kowalska, M.A.; Marcos-Contreras, O.A.; Kuo, A.; Stepanova, V.; Atkinson, J.P.; et al. Sustained thromboprophylaxis mediated by an RBC-targeted pro-urokinase zymogen activated at the site of clot formation. *Blood* **2010**, *115*, 5241–5248. [CrossRef]
51. Ganguly, K.; Krasik, T.; Medinilla, S.; Bdeir, K.; Cines, D.B.; Muzykantov, V.R.; Murciano, J.C. Blood clearance and activity of erythrocyte-coupled fibrinolytics. *J. Pharmacol. Exp. Ther.* **2005**, *312*, 1106–1113. [CrossRef]
52. Smirnov, V.; Domogatsky, S.; Dolgov, V.; Hvatov, V.; Klibanov, A.; Koteliansky, V.; Muzykantov, V.; Repin, V.; Samokhin, G.; Shekhonin, B. Carrier-directed targeting of liposomes and erythrocytes to denuded areas of vessel wall. *Proc. Natl. Acad. Sci. USA* **1986**, *83*, 6603–6607. [CrossRef]
53. Ferguson, L.T.; Hood, E.D.; Shuvaeva, T.; Shuvaev, V.V.; Basil, M.C.; Wang, Z.; Nong, J.; Ma, X.; Wu, J.; Myerson, J.W.; et al. Dual Affinity to RBCs and Target Cells (DART) Enhances Both Organ-and Cell Type-Targeting of Intravascular Nanocarriers. *ACS Nano* **2022**, *16*, 4666–4683. [CrossRef] [PubMed]

pharmaceutics

Article

The Role of Functionalization and Size of Gold Nanoparticles in the Response of MCF-7 Breast Cancer Cells to Ionizing Radiation Comparing 2D and 3D In Vitro Models

Marika Musielak [1,2,3,*], Agnieszka Boś-Liedke [4], Oliwia Piwocka [1,2,3], Katarzyna Kowalska [5], Roksana Markiewicz [6], Barbara Szymkowiak [7], Paweł Bakun [2,8] and Wiktoria M. Suchorska [1,3]

1 Department of Electroradiology, Poznan University of Medical Sciences, 61-866 Poznan, Poland
2 Doctoral School, Poznan University of Medical Sciences, 60-812 Poznan, Poland
3 Radiobiology Laboratory, Department of Medical Physics, Greater Poland Cancer Centre, 61-866 Poznan, Poland
4 Department of Macromolecular Physics, Faculty of Physics, Adam Mickiewicz University, 61-614 Poznan, Poland
5 Department of Histology and Embryology, Poznan University of Medical Sciences, 61-781 Poznan, Poland
6 NanoBioMedical Centre, Adam Mickiewicz University, 61-614 Poznan, Poland
7 Faculty of Physics, Adam Mickiewicz University, 61-614 Poznan, Poland
8 Chair and Department of Chemical Technology of Drugs, Poznan University of Medical Sciences, 61-781 Poznan, Poland
* Correspondence: marikamusielak@gmail.com; Tel.: +48-505-372-290

Abstract: Gold nanoparticles (AuNPs), as an agent enhancing radiosensitivity, play a key role in the potential treatment of breast cancer (BC). Assessing and understanding the kinetics of modern drug delivery systems is a crucial element that allows the implementation of AuNPs in clinical treatment. The main objective of the study was to assess the role of the properties of gold nanoparticles in the response of BC cells to ionizing radiation by comparing 2D and 3D models. In this research, four kinds of AuNPs, different in size and PEG length, were used to sensitize cells to ionizing radiation. The in vitro viability, uptake, and reactive oxygen species generation in cells were investigated in a time- and concentration-dependent manner using 2D and 3D models. Next, after the previous incubation with AuNPs, cells were irradiated with 2 Gy. The assessment of the radiation effect in combination with AuNPs was analyzed using the clonogenic assay and γH2AX level. The study highlights the role of the PEG chain in the efficiency of AuNPs in the process of sensitizing cells to ionizing radiation. The results obtained imply that AuNPs are a promising solution for combined treatment with radiotherapy.

Keywords: Au nanoparticles; metal nanoparticles; radiotherapy; radiosensitization; oncology

Citation: Musielak, M.; Boś-Liedke, A.; Piwocka, O.; Kowalska, K.; Markiewicz, R.; Szymkowiak, B.; Bakun, P.; Suchorska, W.M. The Role of Functionalization and Size of Gold Nanoparticles in the Response of MCF-7 Breast Cancer Cells to Ionizing Radiation Comparing 2D and 3D In Vitro Models. *Pharmaceutics* **2023**, *15*, 862. https://doi.org/10.3390/pharmaceutics15030862

Academic Editor: Elisa Panzarini

Received: 27 January 2023
Revised: 22 February 2023
Accepted: 6 March 2023
Published: 7 March 2023

1. Introduction

Nanotechnology is a field of science that studies structures who have at least one dimension less than or equal to 100 nm [1]. The most important feature of materials in the nanometric scale is the transition in physicochemical properties along with the change in their size [2]. Gold nanoparticles (Au nanoparticles, AuNPs) are often used to sensitize cancer cells to ionizing radiation [3]. These are characterized by a high surface to volume ratio, the possibility of surface modification, and specific optical properties [4]. The surface modification of AuNPs is highly related to the aim to be achieved (e.g., AuNPs have been chemically modified with low molecular branched polyethylenimine for the efficient delivery of gapmers targeting p53 mutant protein) [5]. Xue et al. designed a multi-functional three-dimensional (3D) DNA shell consisting of DNA bricks with sticky ends (sticky-YTDBs) and tiled them onto siRNA-packaged AuNPs [6]. Recent advances in the multi-functional design of AuNPs enable the generation of localized heat in the cancer tissues as well as the controlled and targeted delivery of multiple desired drugs. AuNPs have numerous advantages that make them suitable for photothermal or irradiation treatment [7]. Introducing AuNPs into the tumor area is a potential strategy to solve the problem of protecting healthy tissues in the beam field. Sensitizing tumor cells to radiotherapy would allow for lower radiation doses, reducing the negative effects of radiation on healthy tissues. This involves still unknown physical, chemical, and biological processes that require further research. Butterworth et al. [8] performed a time course experiment to determine the effect of AuNPs on cell survival in the absence of radiation and the optimal nanoparticle incubation period prior to irradiation. The implementation of 1.9 nm gold particles caused a variety of cell line-specific responses such as decreased clonogenic survival, increased apoptosis, and DNA damage induction, which may be mediated by the production of reactive oxygen species. It was the first study using 1.9 nm sized particles to report multiple cellular responses that influence the radiation dose modifying effect. This article emphasized the importance of thoroughly characterizing responses to gold nanoparticles when assessing dose-enhancing potential in cancer therapy. The radiosensitization effect of glucose-capped AuNPs (Glu-AuNPs) of different sizes (16 nm and 49 nm) on MDA-MB-231 cells in the presence of megavoltage X-rays was also described [9]. The authors discovered that Glu-AuNPs could increase the radiosensitivity toward BC cells, most likely by regulating the cell cycle distribution, with a higher number of cells in the G2/M phase. They highlighted that the amount of Glu-AuNPs in the cells might be related to the effect of radiation enhancement. Due to the wide range of surface functionalization possibilities of AuNPs, there is a noticeable lack of systematic analysis in the literature, which does not allow for a proper comparison of the experimental results. There is a need for research groups to focus on improving the already proposed strategies and attempts to conduct experiments in multidimensionally controlled experimental conditions.

Another important aspect is the method of evaluating the effectiveness of nanoparticles depending on the research model used. Promising results obtained in 2D conditions are often not reflected in the results of in vivo studies using animal models [10]. 2D cultures are devoid of a three-dimensional network of different cell types surrounded by an extracellular matrix. These two aspects play a crucial role in the diffusion and uptake of nanoparticles by cells. However, the evaluation of their performance remains limited as most work is carried out on two-dimensional culture systems.

The main objective of this work was to assess the role of gold nanoparticle properties in the response of breast cancer cells to ionizing radiation in 2D and 3D models. Specific objectives included (1) the synthesis, functionalization, and characterization of spherical gold nanoparticles; (2) the evaluation and comparison of internalization of nanoparticles with different properties between cells in 2D and 3D culture; and (3) the evaluation of the impact of the incubation time of gold nanoparticles with breast cancer cells in 2D and 3D culture on the response of cells after exposure to ionizing radiation.

2. Materials and Methods

2.1. Methods

2.1.1. Synthesis of AuNPs

Nanospheres were prepared using two techniques: the inverse method published by Schulz et al. [11] and the Turkevich technique presented by Wang et al. [12].

The preparation procedure of 10 nm nanospheres:

According to Schulz et al. [11], presenting the inverse method, 25.8 mg NaCl (trisodium citrate dihydrate) was dissolved in 50 mL of water. Then, 1 mL of EDTA (3.8 mg/mL, ethylenediaminetetraacetic acid) was added to the solution and boiled for 15 min before precursor addition. The precursor solution (8.49 mg/mL $HAuCl_4 \cdot 3H_2O$, gold(III) chloride) was prepared in a separate flask according to the calculated molar ratio (MR). After boiling, 1 mL of precursor solution was injected while the mixture was rapidly stirred. When the color changed to wine-red, which indicated the formation of AuNPs, the heating was switched off. The solution was removed when the temperature reached 70 °C. The solution was filtered. Using this technique, only 10 + 2 nm nanospheres could be synthesized with MR = 1. The initial concentration of $AuNPs_{10nm}$ was 0.277 mg/mL.

The preparation procedure of 30 nm nanospheres:

The 30 nm nanospheres were synthesized according to Wang et al. [12] where 25 mL of water was stirred in a flask and heated to 100 °C. Separately, 52.4 mg of $HAuCl_4$ was dissolved in 75 mL of water, and 10 mL of the solution was poured into the hot water flask. Afterward, the citrate stock was prepared by dissolving 506.3 mg of NaCt in 75 mL of water. Then, 4 mL of the stock solution was added to $HAuCl_4$ and boiled for 1.5 h. Next, the solution was left on the table until the temperature reached room temperature. The initial concentration of $AuNPs_{30nm}$ was 0.286 mg/mL.

2.1.2. Functionalization of AuNPs

For the functionalization of 10 nm and 30 nm nanospheres, the RGD (Ary–Gly–Asp) complex and polyethylene glycol (PEG) chains were used in two molecular weights of 800 and 2000. The functionalization was performed according to the protocol by Yang et al. [13]. We obtained four nanostructures: RGD-PEG_{800}-$AuNPs_{10nm}$, RGD-PEG_{2000}-$AuNPs_{10nm}$, RGD-PEG_{800}-$AuNPs_{30nm}$, and RGD-PEG_{2000}-$AuNPs_{30nm}$.

2.1.3. Characterization of AuNPs

The morphology of the synthesized AuNPs with sizes of 10 nm and 30 nm was analyzed by transmission electron microscopy (TEM). On a copper grid covered in a Formvar-carbon membrane (300 mesh, Ted Pella Inc., Redding, CA, USA)), 5–10 L of an aqueous dispersion was first applied and air dried at room temperature. Ultrastructural and morphometric analysis of the gold nanoparticles was performed using a TEM model JEM 1010 (Jeol, Tokyo, Japan) equipped with a MEGAVIEW G2 camera and the cooperating iTEM Digital Imaging Solutions program (Olympus Soft Imaging Solutions GmbH, Münster, Germany). The average diameter of the AuNPs reached about 10.5 nm +/− 1.3 nm and 30.9 nm +/− 1.8. The UV–VIS measurements were performed using a Jasco 650v (ABL&E-JASCO, Kraków, Poland).

2.2. *Cell Culture*

2.2.1. 2D Model

The breast cancer cell line MCF-7 was used in the study (ATCC). The standard conditions for cell maintenance were a 37 °C incubator (Binder, Tuttlingen, Germany) with a water vapor-saturated atmosphere that was 5% CO_2 enriched. Penicillin/streptomycin (P/S) at a final concentration of 1% (Merck Millipore Corporation, Darmstadt, Germany) and 0.01 mg/mL of insulin (Gensulin R, Bioton, Macierzysz, Poland) were added to DMEM (Biowest, Nuaillé, France) supplemented with 10% fetal bovine serum (FBS) (Biowest, France) for the MCF-7 culture medium. Cells were passaged with trypsin-EDTA when confluency reached 80–90%. Tests were executed using a laminar flow hood under aseptic conditions.

2.2.2. 3D Model

MCF-7 cells were seeded on the U-bottom 96-well in the concertation of 6000 cells per well. After 6 days, the mammospheres were ready for the experiments. The cell culture medium and cell culture conditions were the same as for the 2D model.

2.3. *Live/Dead Assay*

Cells were seeded at concentrations of 4000, 6000, and 8000 per well. The staining solution consisted of 2.5 µL calcein AM, 10 µL ethidium homodimer-1, and 5 mL of PBS. The medium was removed from the cells, and wells were washed with PBS. Then, 100 µL of the staining solution was added directly to cells and incubated for 30 min at room temperature. Afterward, the cells were imaged on an Olympus IX83 inverted fluorescence microscope (Olympus, Tokyo, Japan).

2.4. Viability Test

Cells were plated at a concentration of 15,000 cells per well (MDA-MB-231 and MCF12A) onto 96-well flat-bottomed plates. After 24 h, RGD-PEG_{800}-$AuNPs_{10nm}$, RGD-PEG_{2000}-$AuNPs_{10nm}$, RGD-PEG_{800}-$AuNPs_{30nm}$, and RGD-PEG_{2000}-$AuNPs_{30nm}$ were added at 0.0004, 0.0008, 0.0012, 0.0020, 0.0060, 0.0120, and 0.0200 mg/mL concentrations and distributed in culture media at final volumes of 100 µL per well. Tests were performed for 24 and 48 h of cell incubation. The culture media containing AuNPs were then removed, and MTT (3-(4,5-dimethylthiazol-2-yl)-2,5-diphenyltetrazolium bromide) (Affymetrix, Cleveland, OH, USA) was added at a final concentration of 0.5 mg/mL. The medium was removed after 3 h of cell culture, and 100 µL of DMSO (Thermo Scientific, Waltham, MA, USA) was then added to each well to dissolve the formazan crystals that had formed. A Multiskan plate reader (Thermo Scientific, Waltham, MA, USA) at 570/590 nm, background 655 nm, was used to read the outcome.

2.5. Reactive Oxygen Species (ROS) Assay

A total of 200,000 cells per well were used for cell seeding. In the culture medium devoid of serum, a 1:1000 solution of 2′,7′-dichlorofluorescin diacetate DCFH-DA (Merck Millipore Corporation, Germany) was made. Cells were PBS-washed once after the culture media were removed. After that, a dye solution was applied to the cells, and then incubated at 37 °C for 45 min. Following this, the dye solution was removed, and cells underwent one PBS wash. Complete culture medium mixed with each AuNP—RGD-PEG_{800}-$AuNPs_{10nm}$, RGD-PEG_{2000}-$AuNPs_{10nm}$, RGD-PEG_{800}-$AuNPs_{30nm}$ and RGD-PEG_{2000}-$AuNPs_{30nm}$—at a concentration of 0.0004, 0.0008, and 0.0012 mg/mL was added to the cells followed by incubation of 10, 20, 30, 40 min and 1, 3, 6, and 24 h. Cells were gathered, suspended in PBS, and subjected to flow cytometric analysis. Fluorescein isothiocyanate-height (FITC-H) channel measurements were applied using a flow cytometer (at 485 nm excitation wavelength and 527 nm emission wavelength).

2.6. Internalization Analysis

The uptake of the AuNPs was measured according to the method published by Park et al. [14]. Cells were seeded on the 12-well plates (200,000 cells/well). After 24 h, RGD-PEG_{800}-$AuNPs_{10nm}$, RGD-PEG_{2000}-$AuNPs_{10nm}$, RGD-PEG_{800}-$AuNPs_{30nm}$, and RGD-PEG_{2000}-$AuNPs_{30nm}$ were added in the concentration of 0.0004, 0.0008, and 0.0012 mg/mL for each type of AuNP. The internalization level of AuNPs in the cells was investigated after 0.5, 1, 3, 6, and 24 h. After incubation, cells were harvested, suspended in phosphate buffered saline (PBS) (Biowest, France) and washed once to discard excess AuNPs from the sample. A cytometric analysis was conducted using a Cytoflex Beckmann Coulter cytometer (Beckman Coulter Life Sciences, Indianapolis, IN, USA). By examining fluorescence at 611 nm, the side scatter parameter (SSC) was determined. FlowJo v10 was used for the analysis of the outcomes.

2.7. Irradiation

Cells were irradiated after incubation with RGD-PEG_{800}-$AuNPs_{10nm}$, RGD-PEG_{2000}-$AuNPs_{10nm}$, RGD-PEG_{800}-$AuNPs_{30nm}$, and RGD-PEG_{2000}-$AuNPs_{30nm}$. Three different time points (TPs) of cells incubated with AuNPs were chosen based on the previous results: 30 min, 3 h, and 24 h. First, the cells were plated on 10 cm plates. Then, a medium with one type of AuNP was added to the cells and incubated for the chosen TP. After that, the cells were gathered and exposed to radiation. Using a closed source of Cs-137 with an activity of 20.4 TBq, the radiation was administered using a Gamma Cell® 1000 Elite device (BestTheratronics Ltd., Kanata, ON, Canada) at a dose rate of 2.5 Gy/min, and 2 Gy of radiation was applied to the cells.

2.8. Clonogenic Assay

After irradiation, the optimized number of cells was plated on 6-well plates and incubated for 7 days. Clonogenic assays were closed when a colony that reached a minimum of 50 clones was observed in the control group. The medium was discarded, and cells were washed with PBS. Next, the fixation of cells was performed with denatured ethanol. Following the removal of the ethanol, the plates were stained roughly with 2 mL of Coomassie Blue solution (Merck Millipore Corporation, Germany) and left to develop for 20 min. The plates were then washed in warm water, dried, and the buffer was discarded. The ChemiDoc Touch Bio-Rad system (Hercules, Clearwater, FL, USA) was used to take pictures of the plates. The Gene Tools Syngene program was applied to complete an automatic colony counting. The clonogenic assay on the 3D model was conducted in the same conditions as for the 2 D model. The difference was that the incubation with AuNPs before irradiation was performed in the 3D model. The culture medium was replaced with the fresh medium and AuNPs in the selected concentration. Next, spheres were collected and partially resuspended for irradiation. Further procedures where the same in both models. The spheres were dissociated into single cells for plating. To assess the radiosensitivity effect of AuNPs after irradiation, the group incubated with AuNPs only was applied for the calculation as a control to the group treated with AuNPs + 2 Gy.

2.9. Flow Cytometry Analysis after Irradiation—γH2AX

After irradiation, cells were separated into samples of 200,000 cells. After 45 min, cells were washed with PBS, then fixed and permeabilized with the Fixation/Permeabilization Kit (BD Biosciences, Franklin Lakes, NJ, USA). Cells were incubated with the 3.5 µL anti-γH2AX antibody (Becton Dickinson, Franklin Lakes, NJ, USA) and PBS in the final volume of 20 µL for 30 min at 4 °C. Finally, cells were washed with PBS and suspended in 200 µL of PBS for the cytometric analysis.

2.10. Immunofluorescence

The immunofluorescence was completed for γH2AX after irradiation of the cells. Cells were seeded on 8-well chamber slides (VWR, Darmstadt, Germany) with a 50,000 cells/well density. After 24 h, a solution of the culture medium with each AuNP—RGD-PEG_{800}-$AuNPs_{10nm}$, RGD-PEG_{2000}-$AuNPs_{10nm}$, RGD-PEG_{800}-$AuNPs_{30nm}$ and RGD-PEG_{2000}-$AuNPs_{30nm}$—at a concentration of 0.0008 mg/mL was added to the cells. The 50 min TP was chosen for the incubation time of the cells with AuNPs. Following each TP incubation with AuNPs, the cells were irradiated with a dose of 2 Gy. After irradiation, the cells were incubated for 45 min at 37 °C. Next, the cells were washed with PBS, fixed in 4% paraformaldehyde for 20 min at room temperature (RT), and permeabilized with ice-cold 100% methanol at −20 °C for 20 min. Next, the blocking was performed by incubation with 0.2% Triton X-100 and 1% bovine serum albumin (BSA) (VWR, Germany) solution for 30 min at RT. After blocking, cells were washed with PBS. Next, 200 µL of a primary antibody solution γH2AX (ab22551, Abcam, Cambridge, UK) produced in mice was added into each chamber and slides were incubated overnight at 4 °C. After incubation, the cells were washed thrice with 2% BSA in PBS solution and incubated with 250 µL of secondary antibody solution (Jackson Immuno Research Labs, West Grove, PA, USA) for 1 h at 37 °C in darkness. All slides were washed thrice with 2% BSA in PBS solution, and 400 µL of DAPI (catalogue number: SAFSD8417) (VWR, Germany) solution was added. Immunofluorescence was imaged using an Olympus IX83 microscope (Boston Industries, Inc., Walpole, MA, USA). The quantification of γH2AX was performed using image deconvolution. To be able to compare the results, the parameter of the mean fluorescence intensity of γH2AX foci was used in both cell models. In the 2D model, the results were calculated as a ratio of the mean foci intensity per nuclei whereas in the 3D model, it was the mean foci intensity per area [μm^2].

2.11. Statistical Analysis

The statistical analysis was performed using PQStat Software v.1.8.2. and Microsoft® Excel® (Microsoft Office Professional Plus 2019). The normality of the observed data distribution was assessed using the Shapiro–Wilk test. The one-way ANOVA was conducted for multiple comparisons. If Levene's test indicated that the variances are not equal across the groups, the unequal variance t Test (Welch's t Test) was implemented. To calculate the differences for a complex system (more than two groups), multiple comparison procedures were used by applying Tuckey's post hoc test. The data were deemed significant at $p < 0.05$. The setting of the p-value was * $p < 0.05$, ** $p < 0.01$, *** $p < 0.001$.

3. Results

3.1. Synthesis, Functionalization, and Characterization of AuNPs

The first specific objective was to synthesize, functionalize, and characterize the spherical gold nanoparticles (Figure 1). Two sizes, 10 nm and 30 nm, of AuNPs were obtained. Nanospheres were prepared using two techniques: the inverse method published by Schulz et al. [11] for 10 nm and the Turkevich technique presented by Wang et al. [12] for 30 nm. The primary studies using the protocol by Schultz et al. focused on the most suitable NaCt (trisodium citrate dihydrate) to $HAuCl_4 \cdot 3H2O$ molar ratio MR choice. For that reason, the NaCt concentration was modified to achieve MR from 0.5 to 15 and the following synthesis was repeated several times. Using this technique only, 10 ± 2 nm nanospheres were able to be synthesized with MR = 4. The diameter and the shape of the synthesized nanostructures were checked using UV–VIS and TEM. Using the protocol by Yang et al. [13] for the functionalization oof AuNPs, we obtained four different nanostructures: RGD-PEG$_{800}$-AuNPs$_{10nm}$, RGD-PEG$_{2000}$-AuNPs$_{10nm}$, RGD-PEG$_{800}$-AuNPs$_{30nm}$, and RGD-PEG$_{2000}$-AuNPs$_{30nm}$.

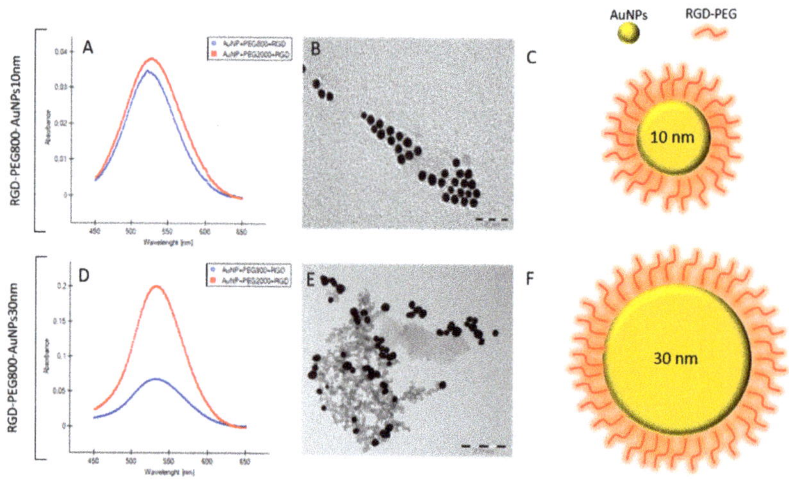

Figure 1. Characterization and morphology of the synthesized AuNPs. (**A**) UV–VIS spectrum for RGD-PEG$_{800}$-AuNPs$_{10nm}$ and RGD-PEG$_{2000}$-AuNPs$_{10nm}$. (**B**) TEM image of 10 nm AuNPs, based on the TEM measurements the average size of AuNPs was 10.5 nm +/− 1.3 nm. (**C**) The scheme of RGD-PEG$_{800/2000}$-AuNPs$_{10nm}$. (**D**) UV–VIS spectrum for RGD-PEG$_{800}$-AuNPs$_{30nm}$ and RGD-PEG$_{2000}$-AuNPs$_{30nm}$. (**E**) TEM image of 30 nm AuNPs, based on the TEM measurements, the average size of AuNPs was 30.9 nm +/− 1.8 nm. (**F**) The scheme of RGD-PEG$_{800/2000}$-AuNPs$_{30nm}$.

3.2. Cell Culture—2D and 3D Models

For the experiments, the MCF-7 breast cancer cell line was used. This cell line is characterized by the Luminal A subtype expressing estrogen receptor (ER) and progesterone receptor (PR) [15]. The luminal-A is the most common subtype and represents 50–60% of all breast cancers [16]. For the 3D model optimization, the live/dead assay was used to compare three numbers of seeded cells: 4000, 6000, and 8000 cells and times of formation of 5 and 7 days. The middle number of cells was chosen, and 7 days was the most effective scheme considering different incubation times in further experiments and the maximum size of the sphere, enabling us to imitate the minimum physiology of breast cancer tumors. After 7 days of forming, the breast cancer spheres were ready to use in experiments (Figure 2).

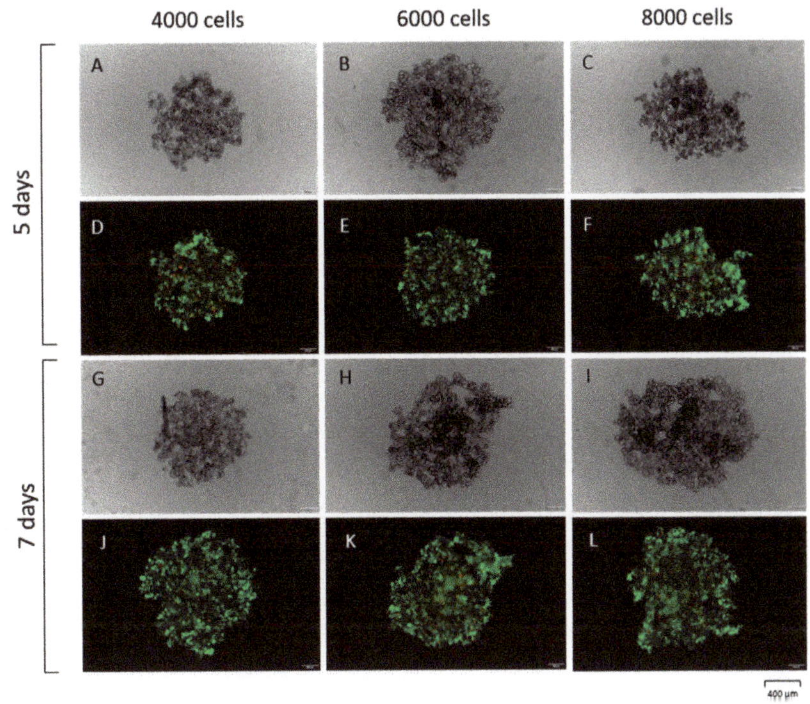

Figure 2. Live/dead assay of the MCF-7 3D model. BC spheres were observed after 5 days of forming in a bright field containing 4000 cells (**A**), 6000 cells (**B**), and 8000 cells (**C**). The identical spheres were tested with the live/dead assay (**D–F**). After 7 days of formation, identical BC spheres were photographed using the bright field (**G–I**) and live/dead assay (**J–L**). The FITC channel was used to present the green alive cells, and the Texas Red channel for the red, dead ones.

3.3. Viability Assay

The MTT assay was evaluated for assessing the MCF-7 cell viability after incubation with AuNPs. In this test, RGD-PEG$_{800}$-AuNPs$_{10nm}$, RGD-PEG$_{2000}$-AuNPs$_{10nm}$, RGD-PEG$_{800}$-AuNPs$_{30nm}$, and RGD-PEG$_{2000}$-AuNPs$_{30nm}$ were used in a concentration of 0.0004, 0.0008, 0.0012, 0.0020, 0.0060, 0.0120, and 0.0200 mg/mL. To observe a wide range of effects, the TPs of 24, 48, and 72 h of the cell incubation time with AuNPs were chosen. All experiments were performed in triplicate in both models; 2D and 3D cell cultures. In the 2D model, the highest heterogeneity in results was observed in the smallest concentrations of AuNPs (Figure 3). The stability trend was noticed with a higher concentration of AuNPs. The lowest viability level was about 65% after 48 h of incubation with AuNPs of 10 nm in size. Statistically significant differences were not observed when comparing the viability results of all four nanostructures. The viability maintained a stable trend, reaching about 80–90% in all incubation times. The MTT assay was also performed after incubation with AuNPs and BC spheres (Figure 4). In this model, the differences at the lowest concentration were observed, but they were higher than the 2D model. The highest peak reached 170% viability after 72 h of incubation with the 30 nm AuNPs. Some similarities between 2D and 3D models were observed. The 48 h TP again became the most toxic for cells using RGD-PEG$_{2000}$-AuNPs$_{10nm}$. Moreover, the mean viability level in each TP and kind of AuNP was again 80–90%. Analyzing the viability level from each type of AuNP alone, some stability trends were observed. Still, comparing AuNPs of all sizes and functionalization, some differences were noticed between those stabilities, especially at the 48 and 72 h TPs. Considering the differences in the distribution level of the AuNPs because of the used models, a diversity in results was observed. The incorporation in the 3D model was uncontrolled, thus, greater differences in viability level were detected. The viability results of the 2D (Table 1) and 3D (Table 2) models were collected as a resume of values for each condition tested.

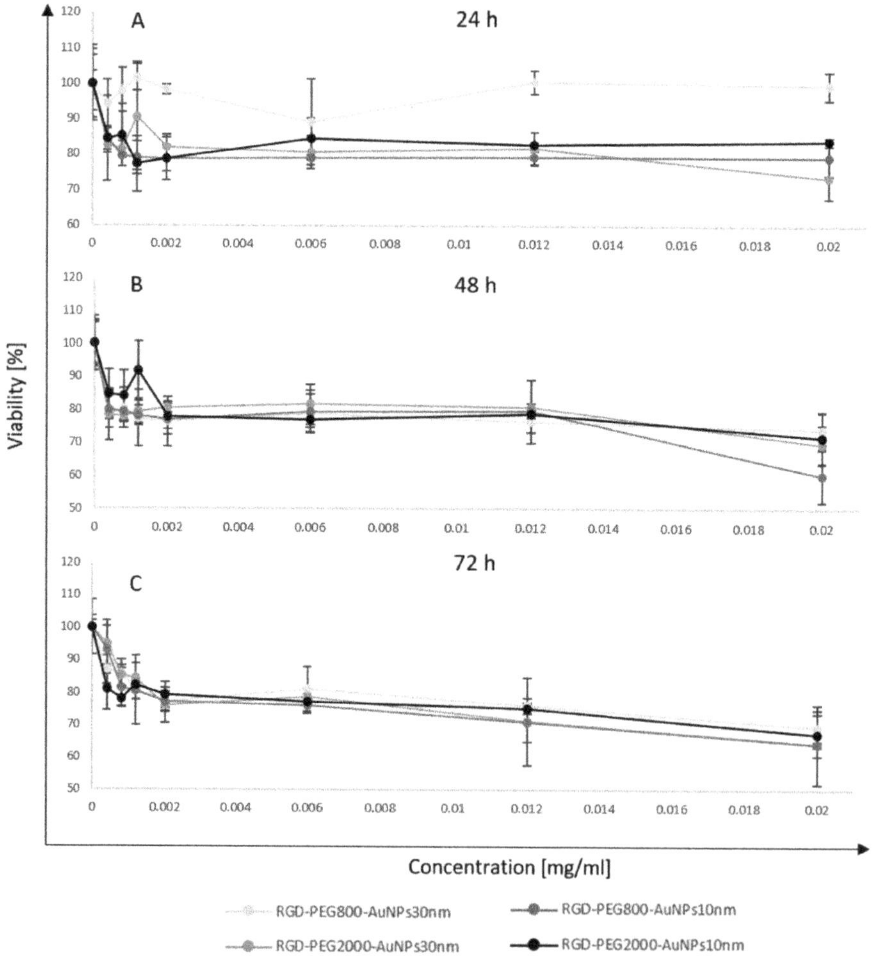

Figure 3. The viability level of the MCF-7 2D model after incubation with AuNPs. In this test, RGD-PEG$_{800}$-AuNPs$_{10nm}$, RGD-PEG$_{2000}$-AuNPs$_{10nm}$, RGD-PEG$_{800}$-AuNPs$_{30nm}$, and RGD-PEG$_{2000}$-AuNPs$_{30nm}$ were used in concentrations of 0.0004, 0.0008, 0.0012, 0.0020, 0.0060, 0.0120 and 0.0200 mg/mL. The time points of incubation time with AuNPs were 24 h (**A**), 48 h (**B**), and 72 h (**C**).

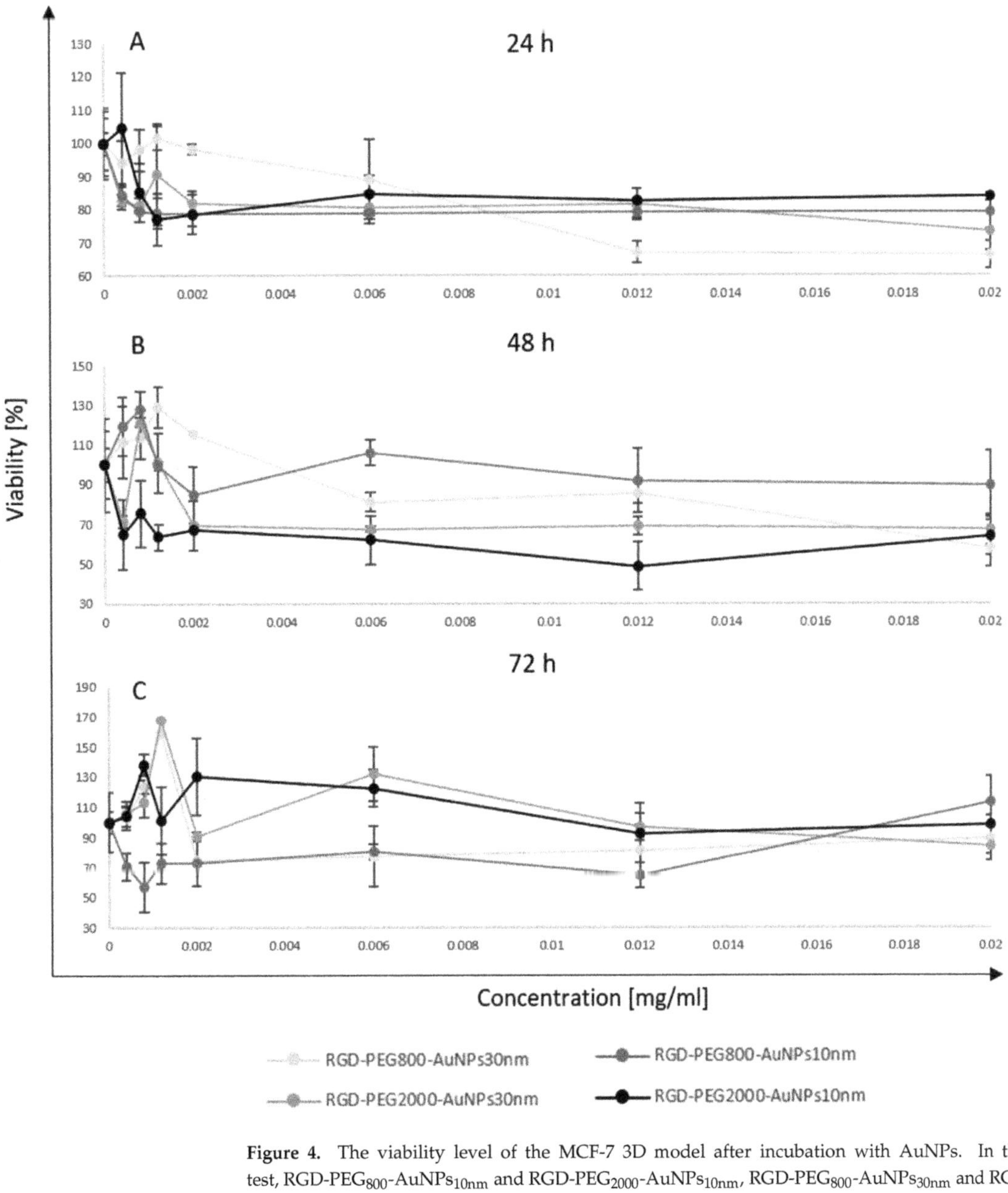

Figure 4. The viability level of the MCF-7 3D model after incubation with AuNPs. In this test, RGD-PEG_{800}-$AuNPs_{10nm}$ and RGD-PEG_{2000}-$AuNPs_{10nm}$, RGD-PEG_{800}-$AuNPs_{30nm}$ and RGD-PEG_{2000}-$AuNPs_{30nm}$ were used in concentrations of 0.0004, 0.0008, 0.0012, 0.0020, 0.0060, 0.0120 and 0.0200 mg/mL. The time points of incubation time with AuNPs were 24 h (**A**), 48 h (**B**), and 72 h (**C**).

Table 1. The resume of the viability level of the MCF-7 2D model after incubation with AuNPs. In this test, RGD-PEG$_{800}$-AuNPs$_{10nm}$, RGD-PEG$_{2000}$-AuNPs$_{10nm}$, RGD-PEG$_{800}$-AuNPs$_{30nm}$, and RGD-PEG$_{2000}$-AuNPs$_{30nm}$ were used in concentrations of 0.0004, 0.0008, 0.0012, 0.0020, 0.0060, 0.0120, and 0.0200 mg/mL. Abbreviations: SD—standard deviation.

The Cell Viability of the MCF-7 2D Model									
	Concentration mg/mL	RGD-PEG800-AuNPs30nm		RGD-PEG2000-AuNPs30nm		RGD-PEG800-AuNPs10nm		RGD-PEG2000-AuNPs10nm	
		Mean [%]	SD [%]	Mean [%]	SD [%]	Mean [%]	SD [%]	Mean [%]	SD [%]
24 h	0.0000	100.00	9.65	100.00	10.73	100.00	3.42	100.00	7.82
	0.0004	93.95	6.90	82.49	2.23	84.41	3.20	84.37	12.00
	0.0008	98.08	6.20	81.29	2.28	79.51	0.46	85.23	8.90
	0.0012	101.65	3.64	90.63	15.31	78.87	4.59	77.11	7.87
	0.0020	98.30	1.42	81.91	3.58	78.72	3.81	78.62	6.03
	0.0060	89.17	12.16	80.45	4.71	78.95	0.66	84.47	5.75
	0.0120	100.44	3.25	81.56	4.66	79.15	1.80	82.68	0.62
	0.0200	99.48	4.09	73.46	5.78	79.21	5.07	83.79	1.23
48 h	0.0000	100.00	7.13	100.00	6.21	100.00	6.67	66.67	8.23
	0.0004	78.42	7.75	78.65	1.51	80.05	5.68	84.82	7.28
	0.0008	80.33	6.02	77.94	1.79	79.53	0.29	84.20	7.54
	0.0012	77.41	8.56	79.51	3.80	78.19	3.03	91.69	9.00
	0.0020	76.31	7.52	80.63	1.60	77.18	4.82	77.86	3.82
	0.0060	78.80	6.00	81.65	6.14	79.52	6.24	76.95	2.43
	0.0120	76.35	3.15	80.99	0.93	79.59	9.58	78.75	1.15
	0.0200	73.95	5.83	69.71	5.84	60.27	8.07	71.66	7.75
72 h	0.0000	100.00	8.43	100.00	1.96	100.00	3.49	100.00	8.34
	0.0004	86.93	4.36	94.99	7.10	92.92	7.33	80.93	6.42
	0.0008	85.70	4.29	85.16	3.17	81.25	5.93	78.11	2.51
	0.0012	80.80	3.09	84.26	4.42	80.49	10.64	82.18	0.77
	0.0020	77.00	1.94	75.82	5.46	77.06	3.18	79.22	5.24
	0.0060	81.01	7.05	78.62	2.76	75.97	2.35	76.96	1.19
	0.0120	75.80	2.38	71.07	13.47	70.68	6.00	75.09	3.90
	0.0200	69.23	4.21	64.10	12.24	64.07	3.54	67.06	3.70

3.4. Flow Cytometry Analysis

3.4.1. The Reactive Oxygen Species Generation and Time Optimization

Based on the viability level, three concentrations of AuNPs were chosen for further investigation: 0.0004, 0.0008, and 0.0012 mg/mL. In this study, the crucial aim was to check the various conditions of the planned experiments related to the concentration, size, and functionalization of the AuNPs and the incubation time. At first, a wide range of TPs was tested in a 2D model to observe which areas changed in terms of the ROS generation that occurred. The experiment was performed singly at the TPs of 30 min, 2, 4, and 6 h (Figure 5A–D). The increase that was detected at the 30 min TP reached about a three times higher level of ROS compared to the control group. Thus, in the next step, shorter incubation times were investigated. The range of 1 h was investigated, hence, the ROS level was checked every 10 min (Figure 5E–H). The highest level of ROS induction was reached using a 0.0008 mg/mL concentration in all kinds of AuNPs. Only, in the case of RGD-PEG$_{800}$-AuNPs$_{30nm}$, the same results were observed for the smaller and middle-used concentrations. Analyzing shorter periods, a strong trend in the increasing ROS level was detected at the TP of 50 min. This case was recognized in all AuNPs. Due to this, the 30, 50 min, 2, and 4 h TPs of incubation with AuNPs were chosen for further analysis and comparison between the 2D and 3D models.

Table 2. The resume of the viability level of the MCF-7 3D model after incubation with AuNPs. In this test, RGD-PEG$_{800}$-AuNPs$_{10nm}$ and RGD-PEG$_{2000}$-AuNPs$_{10nm}$, RGD-PEG$_{800}$-AuNPs$_{30nm}$ and RGD-PEG$_{2000}$-AuNPs$_{30nm}$ were used in concentrations of 0.0004, 0.0008, 0.0012, 0.0020, 0.0060, 0.0120 and 0.0200 mg/mL. Abbreviations: SD—Standard Deviation.

The Cell Viability of the MCF-7 3D Model									
	Concentration mg/mL	RGD-PEG800-AuNPs30nm		RGD-PEG2000-AuNPs30nm		RGD-PEG800-AuNPs10nm		RGD-PEG2000-AuNPs10nm	
		Mean [%]	SD [%]	Mean [%]	SD [%]	Mean [%]	SD [%]	Mean [%]	SD [%]
24 h	0.0000	100.00	9.65	100.00	10.73	100.00	3.42	100.00	7.82
	0.0004	93.95	6.90	82.49	2.23	84.41	3.20	104.62	16.64
	0.0008	98.08	6.20	81.29	2.28	79.51	0.46	85.23	8.90
	0.0012	101.65	3.64	90.63	15.31	78.87	4.59	77.11	7.87
	0.0020	98.30	1.42	81.91	3.58	78.72	3.81	78.62	6.03
	0.0060	89.17	12.16	80.45	4.71	78.95	0.66	84.47	5.75
	0.0120	66.96	3.25	81.56	4.66	79.15	1.80	82.68	0.62
	0.0200	66.32	4.09	73.46	5.78	79.21	5.07	83.79	1.23
48 h	0.0000	100.00	17.18	100.00	17.08	100.00	23.54	100.00	8.57
	0.0004	111.37	18.37	71.28	3.15	119.50	14.85	64.92	17.77
	0.0008	113.53	10.50	120.94	7.03	127.93	8.85	75.55	16.67
	0.0012	128.89	10.16	100.91	15.26	99.53	2.11	63.60	6.63
	0.0020	115.23	0.00	69.59	12.38	84.87	14.07	67.11	0.00
	0.0060	80.96	4.75	67.30	1.46	105.96	6.49	61.87	12.04
	0.0120	85.51	5.26	68.92	4.70	91.65	16.09	48.64	12.29
	0.0200	57.79	9.51	66.91	7.74	89.27	17.51	63.91	9.71
72 h	0.0000	100.00	7.08	100.00	3.12	100.00	19.87	100.00	1.90
	0.0004	102.58	7.41	105.76	8.34	70.79	9.03	104.09	6.85
	0.0008	123.63	4.43	113.27	10.07	57.40	16.49	138.32	6.98
	0.0012	159.96	0.00	167.70	0.00	72.83	13.38	101.07	22.41
	0.0020	74.65	0.00	90.88	3.17	73.01	15.02	130.81	25.41
	0.0060	77.42	20.40	132.04	17.81	80.46	5.41	122.58	12.57
	0.0120	81.29	16.83	96.79	9.07	64.67	8.47	92.52	19.87
	0.0200	89.06	14.93	83.97	4.98	113.61	17.04	98.60	14.70

To compare the 2D and 3D models, the same concentrations of AuNPs were used. In the 2D model (Figure 6A–D), the same concentration of 0.0008 mg/mL appeared to be the most effective in the ROS induction, in which an about three times higher ROS level was observed compared to the control group. The effect at 30 min was similar to that at 50 min in most cases, but the differences between these two TPs increased after incubation with 10 nm AuNPs (Figure 6C,D). These might be due to dependency—the smaller the size of the AuNP, the deeper the incorporation into the cells. As in the previous preliminary experiments, no effect was detected at 2 and 4 h of incubation. Consequently, the TP of 50 min and 0.0008 mg/mL was chosen as the most effective in generating ROS in the 2D model using all nanostructures.

More homogenous results were presented in the 3D model (Figure 6E–H). In all kinds of AuNPs, the increase at the 30 min TP was observed. There was a considerable similarity between effects taking into consideration the used concentration of AuNPs, so it was not possible to choose the most effective one. Comparing the results of the AuNPs of different sizes, a diversification was noticed. The bigger AuNPs caused a higher ROS level at the 30 min TP, reaching about 3.5–4.5 higher values in contrast to the 10 nm AuNPs, where the effect was about 1.5–3.0 times higher compared to the control.

Moreover, when analyzing the diversity in PEG length inside the size of the AuNPs, one trend observed was that the higher the ROS level, the shorter the PEG chain. It occurred in both sizes of AuNPs in the 3D model. At 2 and 4 h TPs, the same situation was discovered; no effect of ROS generation was observed.

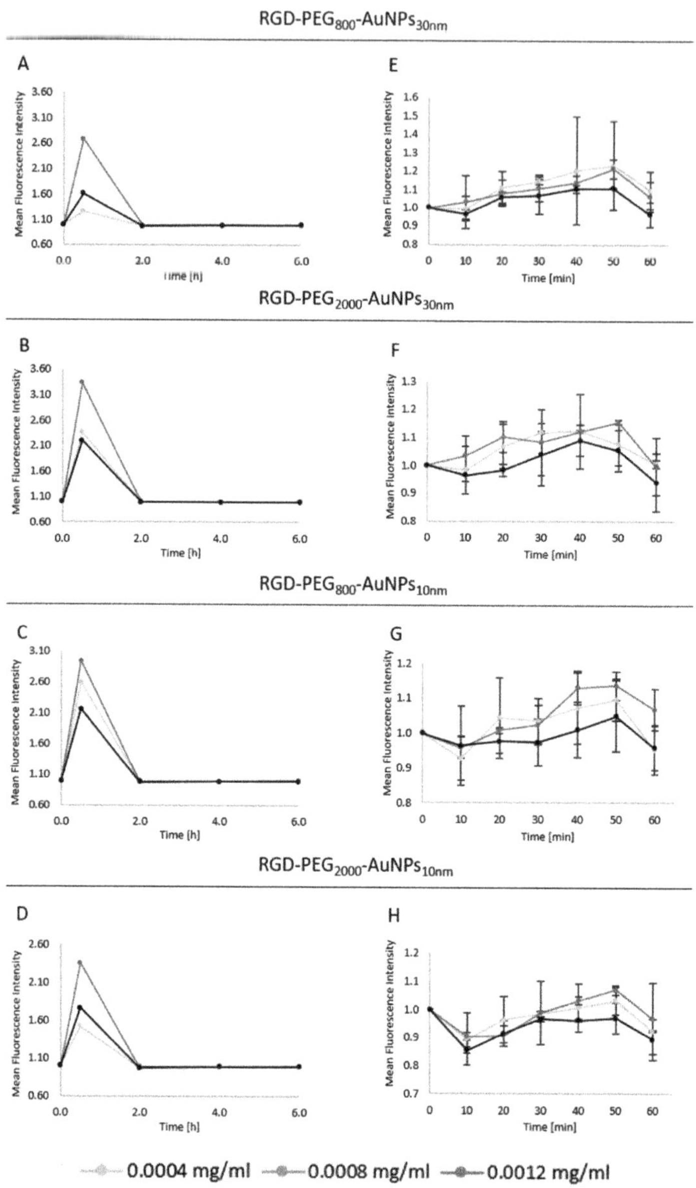

Figure 5. The ROS induction—optimization of the incubation time and AuNP concentration. The preliminary experiments were performed to optimize the TPs of the highest level of ROS generation. The experiment was performed singly ats TP of 30 min, 2, 4, and 6 h incubation with RGD-PEG$_{800}$-AuNPs$_{10nm}$ (**A**), RGD-PEG$_{2000}$-AuNPs$_{10nm}$ (**B**), RGD-PEG$_{800}$-AuNPs$_{30nm}$ (**C**). and RGD-PEG$_{2000}$-AuNPs$_{30nm}$ (**D**). The range of 1 h was investigated; thus, the ROS level was checked every 10 min using RGD-PEG$_{800}$-AuNPs$_{10nm}$ (**E**), RGD-PEG$_{2000}$-AuNPs$_{10nm}$ (**F**), RGD-PEG$_{800}$-AuNPs$_{30nm}$ (**G**), and RGD-PEG$_{2000}$-AuNPs$_{30nm}$ (**H**).

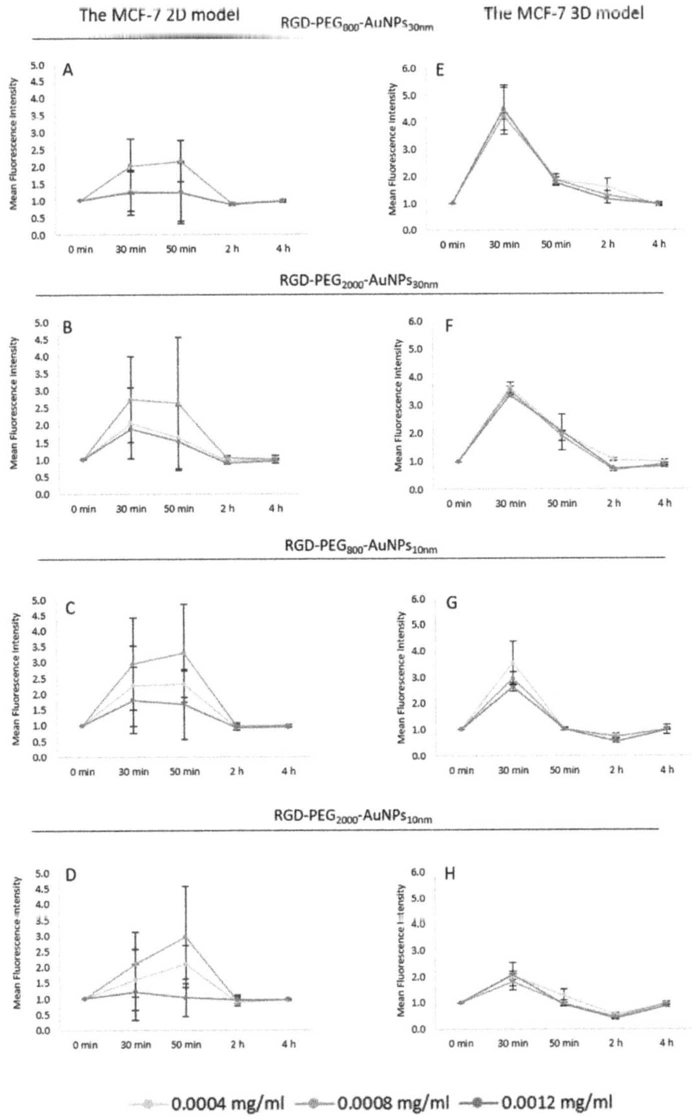

Figure 6. The comparison of the ROS induction level between the 2D and 3D models. In the 2D model, the experiment was performed in triplicate at the TPs of 30, 50 min, 2, and 4 h incubation with RGD-PEG$_{800}$-AuNPs$_{10nm}$ (**A**), RGD-PEG$_{2000}$-AuNPs$_{10nm}$ (**B**), RGD-PEG$_{800}$-AuNPs$_{30nm}$ (**C**), and RGD-PEG$_{2000}$-AuNPs$_{30nm}$ (**D**). In the 3D model, the experiment was performed in triplicate at the TPs of 30, 50 min, 2, and 4 h incubation with RGD-PEG$_{800}$-AuNPs$_{10nm}$ (**E**), RGD-PEG$_{2000}$-AuNPs$_{10nm}$ (**F**), RGD-PEG$_{800}$-AuNPs$_{30nm}$ (**G**), and RGD-PEG$_{2000}$-AuNPs$_{30nm}$ (**H**). The concentrations of all AuNPs were 0.0004, 0.0008, and 0.0012 mg/mL.

3.4.2. Uptake of AuNPs

Based on previous work [17], the evaluation of AuNP uptake was established. The experiment assumed that the higher granularity of the cells, the more nanoparticles they incorporated. This parameter was checked by SSC-H analysis using flow cytometry; higher side scatter levels inform on the higher granularity of the cells. For this analysis, the same concentration and TPs were used as in the experiments for the ROS induction level. After incubation with RGD-PEG_{800}-$AuNPs_{10nm}$, RGD-PEG_{2000}-$AuNPs_{10nm}$, RGD-PEG_{800}-$AuNPs_{30nm}$, and RGD-PEG_{2000}-$AuNPs_{30nm}$, the cells were collected and analyzed using the SSC-H parameter in both cell culture models. The results appeared to be highly heterogeneous. The SSC-H level was too low to assess the uptake of cells in the 2D model (Figure 7A–D) and 3D model (Figure 7E–H). It is likely that this is correlated to the size of the cells and used AuNPs. Considering the level of the standard deviations and SSC-H values, no apparent effect was observed. Only, in the case of the 3D model, the smallest size and PEG chain AuNPs, RGD-PEG_{800}-$AuNPs_{10nm}$ changed the granularity of MCF-7 cells, reaching 1.7 of the relative SSC-H level. Based on these results, there was no information about the TP or concentration changing the granularity of the cells, however, this is not identical to the situation in that the cells did not absorb the AuNPs.

3.5. Irradiation

3.5.1. Clonogenic Assay

An important part of this research was related to the effect of induction radiosensitivity by AuNPs of MCF-7 cells and comparing those results between the 2D and 3D models. Based on previous optimization in time and AuNP concentration, the TP of 50 min and 0.0008 mg/mL of AuNPs was used for the irradiation experiments. Although the highest level of ROS was detected at 30 min in the 3D model and in the 2D model at 50 min, the second TP was chosen. There was no intention that the irradiation should be implemented in the highest level of ROS, but after ROS induction and thus, the molecular effect they caused. Due to this, the longer TP was used in both models and the middle concentration, which was the most effective in generating ROS, especially in the 2D model. In contrast, similar results were observed in all AuNP concentrations in the 3D model. Cells were irradiated using the dose of 2 Gy after 50 min incubation with RGD-PEG_{800}-$AuNPs_{10nm}$, RGD-PEG_{2000}-$AuNPs_{10nm}$, RGD-PEG_{800}-$AuNPs_{30nm}$, and RGD-PEG_{2000}-$AuNPs_{30nm}$ in the concentration of 0.0008 mg/mL.

Many statistically significant differences were discovered in the 2D model (Figure 8A). The surviving fraction (SF) of 2 Gy was compared to the SF of 2 Gy in addition to the previous incubation with AuNPs. The highly significant differences occurred between SF of 2 Gy and SF of 2 Gy + RGD-PEG_{2000}-$AuNPs_{30nm}$ ($p = 0.000275$) and 2 Gy + RGD-PEG_{2000}-$AuNPs_{10nm}$ ($p = 0.00016$). Moreover, there were also statistically significant differences between SF of 2 Gy and SF of 2 Gy + RGD-PEG_{800}-$AuNPs_{30nm}$ ($p = 0.018624$) and 2 Gy + RGD-PEG_{800}-$AuNPs_{10nm}$ ($p = 0.025877$). In this case, there might be a correlation between the used PEG chain because higher statistical differences appeared between previously non-treated and treaded cells with AuNPs functionalized with PEG of 2000 MW.

Moreover, there were statistically significant differences in analyzing the PEG chains or sizes between AuNPs. Considering AuNPs at the size of 30 nm used for previous incubation with cells before irradiation, a statistical difference was observed ($p = 0.036042$) where the SF of 2 Gy + RGD-PEG_{2000}-$AuNPs_{30nm}$ compared to the SF of 2 Gy + RGD-PEG_{800}-$AuNPs_{30nm}$ decreased. A greater difference was also detected between the SF value of 2 Gy + 2 Gy + RGD-PEG_{2000}-$AuNPs_{10nm}$ SF and 2 Gy + RGD-PEG_{800}-$AuNPs_{30nm}$ ($p = 0.002943$). Another exciting occurrence was between the smallest and biggest AuNPs. The SF of 2 Gy + RGD-PEG_{2000}-$AuNPs_{30nm}$ was lower than that of 2 Gy + RGD-PEG_{800}-$AuNPs_{10nm}$ ($p = 0.025877$). The last difference in SF was between AuNPs at the size of 10 nm used before irradiation. Again, AuNPs with the longer PEG chain showed that the SF value decreased using the previously incubated AuNPs functionalized with the longer chain ($p = 0.002215$). The 3D model is inherently more resistant to any exposure. In this model, there was no statistically significant difference between the SF of 2 Gy and SF of 2 Gy + AuNPs. The statistical analysis aiming to compare all variants showed one difference between the cells irradiated and previously incubated with the same size of AuNPs of 30 nm, but with diversity in the PEG chain ($p = 0.03396$). Concerning the objective of this study, this difference had no investigational sense because the SF of 2 Gy + RGD-PEG$_{800}$-AuNPs$_{30nm}$ was higher than the SF value for 2 Gy alone (Figure 8B).

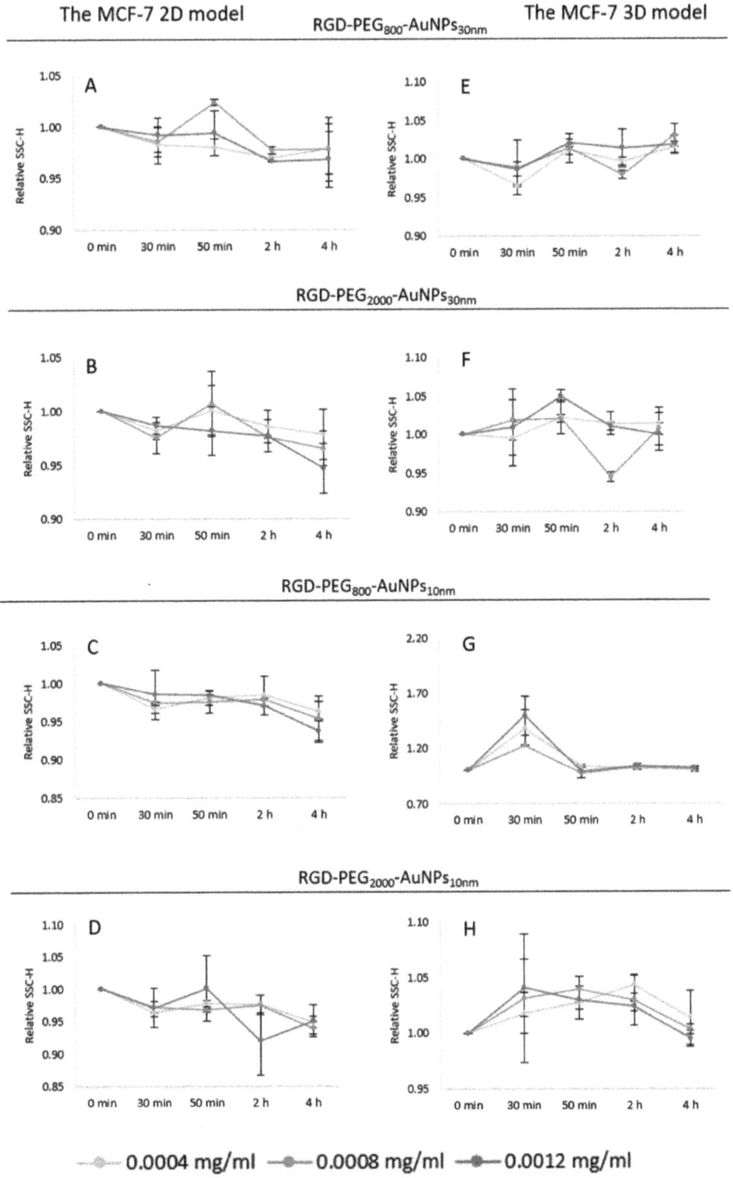

Figure 7. The relative SSC-H parameter level of MCF-7 cells in the 2D and 3D models. In the 2D model, the experiment was performed in triplicate at the TPs of 30, 50 min, 2, and 4 h incubation with RGD-PEG$_{800}$-AuNPs$_{10nm}$ (**A**), RGD-PEG$_{2000}$-AuNPs$_{10nm}$ (**B**), RGD-PEG$_{800}$-AuNPs$_{30nm}$ (**C**), and RGD-PEG$_{2000}$-AuNPs$_{30nm}$ (**D**). In the 3D model, the experiment was performed in triplicate at the TPs of 30, 50 min, 2, and 4 h incubation with RGD-PEG$_{800}$-AuNPs$_{10nm}$ (**E**), RGD-PEG$_{2000}$-AuNPs$_{10nm}$ (**F**), RGD-PEG$_{800}$-AuNPs$_{30nm}$ (**G**), and RGD-PEG$_{2000}$-AuNPs$_{30nm}$ (**H**). All of the concentrations of the AuNPs were 0.0004, 0.0008, and 0.0012 mg/mL.

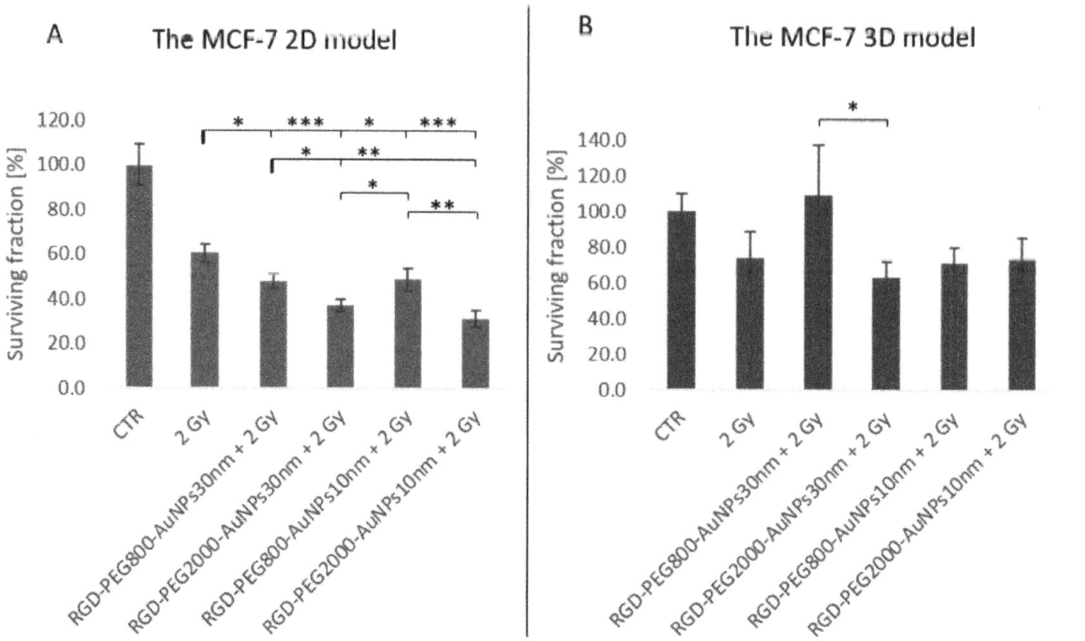

Figure 8. The survival fraction of MCF-7 cells after exposure to a dose of 2 Gy and dose of 2 Gy with the previous 50 min incubation with RGD-PEG$_{800}$-AuNPs$_{10nm}$, and RGD-PEG$_{2000}$-AuNPs$_{10nm}$, RGD-PEG$_{800}$-AuNPs$_{30nm}$, and RGD-PEG$_{2000}$-AuNPs$_{30nm}$ in the concentration of 0.0008 mg/mL. (**A**) Results for the MCF-7 2D model and (**B**) for the MCF-7 3D model. The setting of the p-value * $p < 0.05$, ** $p < 0.01$, *** $p < 0.001$.

3.5.2. γH2AX Analysis—Flow Cytometry and Immunofluorescence

After incubation with AuNPs and irradiation with 2 Gy, the γH2AX analysis was performed. At first, using flow cytometry, the γH2AX relative level was checked. The results after treating cells with the previous incubation with AuNPs and 2 Gy of irradiation were compared to the results of 2 Gy alone. Considering the γH2AX relative level in the 2D model (Figure 9A) between variants of 2 Gy and 2 Gy + AuNPs, the most significant difference was between 2 Gy and 2 Gy + RGD-PEG$_{2000}$-AuNPs$_{30nm}$ (p = 0.000501). Moreover, similar statistical differences compared the γH2AX relative level of 2 Gy to the γH2AX relative level of 2 Gy + RGD-PEG$_{800}$-AuNPs$_{10nm}$ (p = 0.02105) and 2 Gy + RGD-PEG$_{2000}$-AuNPs$_{10nm}$ (p = 0.015123). As in the case of SF, there was a difference in the γH2AX relative level between cells treated with 2 Gy and the same AuNP size of 30 nm. Using 30 nm AuNPs with a longer PEG chain turned out to be more effective in the induction of γH2AX compared to the AuNPs with a shorter PEG chain (p = 0.020192). In the 3D model (Figure 9B), a similar tendency in results was observed, whereas only one difference was statistically significant. This occurred between the γH2AX relative level after the exposure of cells to 2 Gy alone and 2 Gy + RGD-PEG$_{2000}$-AuNPs$_{30nm}$ (p = 0.013027). In other cases, the trend of increasing γH2AX relative level was noticed, but it was not enough for a significant difference.

The immunofluorescence of γH2AX was investigated (Figures 10 and 11). After the previous 50 min incubation time of cells with AuNPs, the irradiation in the dose of 2 Gy was implemented. Using the image deconvolution, the quantification of the γH2AX foci mean fluorescence intensity (MFI) was performed (Figure 12). The increase in the γH2AX foci MFI was observed in the MCF-7 model where the RGD-PEG$_{2000}$-AuNPs$_{30nm}$ were added.

In the other variants, a decrease in the γH2AX foci MFI was detected when comparing both the 2D and 3D models, although lower values were noticed using MCF-7 spheres.

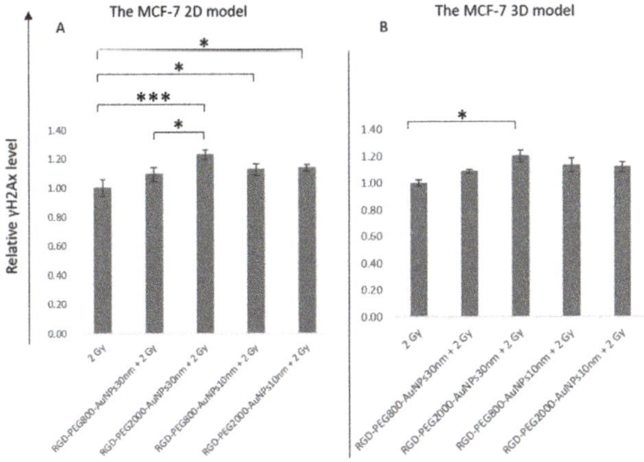

Figure 9. The γH2AX relative level of the MCF-7 cells after exposure to a dose of 2 Gy and dose of 2 Gy with the previous 50 min incubation with RGD-PEG$_{800}$-AuNPs$_{10nm}$, RGD-PEG$_{2000}$-AuNPs$_{10nm}$, RGD-PEG$_{800}$-AuNPs$_{30nm}$, and RGD-PEG$_{2000}$-AuNPs$_{30nm}$ in the concentration of 0.0008 mg/mL. (**A**) Results for the MCF-7 2D model and (**B**) for the MCF-7 3D model. The setting of the *p*-value * $p < 0.05$, *** $p < 0.001$.

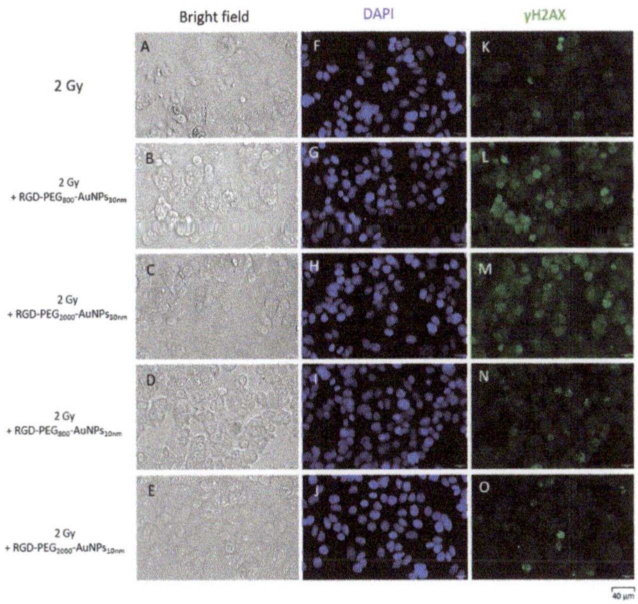

Figure 10. The immunofluorescence of γH2AX after cell exposure to 2 Gy and dose of 2 Gy with the previous 50 min incubation with AuNPs in the 2D model. The morphology of cells was presented in the bright field (**A–E**). The nuclei of cells were dyed with the DAPI solution (**F–J**). The γH2AX foci were dyed using the solution in green (488 nm) (**K–O**).

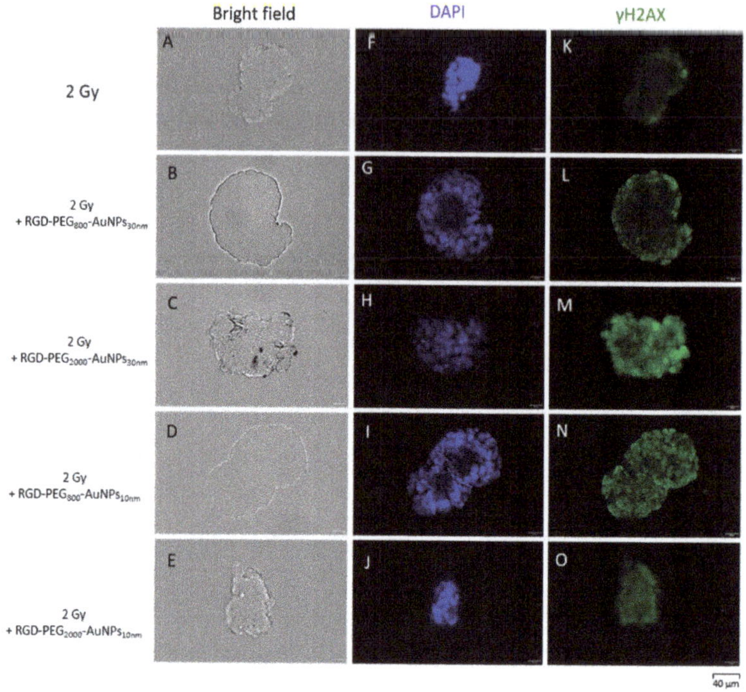

Figure 11. The immunofluorescence of γH2AX after cell exposure to 2 Gy and dose of 2 Gy with the previous 50 min incubation with AuNPs in the 3D model. The morphology of cells was presented in the bright field (**A–E**). The nuclei of cells were dyed with DAPI solution (**F–J**). The γH2AX foci were dyed using the solution in green (488 nm) (**K–O**).

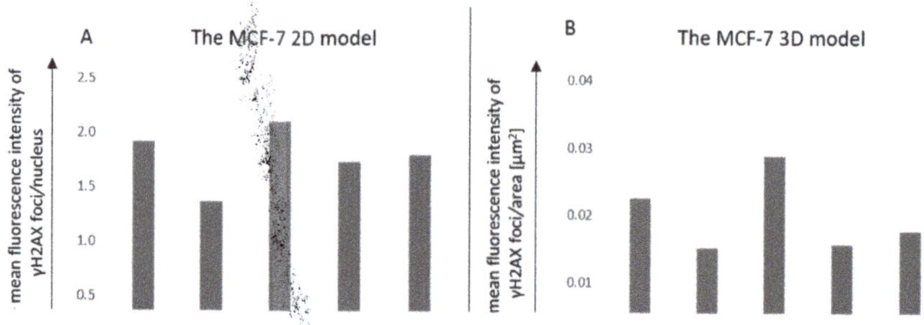

Figure 12. The quantification of the mean fluorescence intensity of the γH2AX foci after cell exposure to 2 Gy and dose of 2 Gy with the previous 50 min incubation with AuNPs in the 2D model (**A**) and 3D model (**B**).

4. Discussion

Introducing the functionalized AuNPs into the tumor area before IR could change the treatment outcome. It is crucial to obtain a high efficiency of their active or passive intracellular transport and the results they could cause in combination with radiotherapy. The radiosensitivity effect of AuNPs is known, but there is a knowledge gap in the case of choosing the most effective experimental conditions related to not only the size of the kind of functionalization, but also to the time and concentration dependent manner of the implementation of AuNPs. The main aim of the study was to assess how the radiosensitivity effect of AuNPs depends on the diversity of the properties of AuNPs and how the effects of different AuNP properties acted in 2D and 3D models.

AuNPs were synthesized, with the result of 10 mm and 30 nm sized AuNPs and then functionalized with PEG and the RGD peptide. PEG is a polyether compound derived from petroleum with applications in biomedicine [18]. This kind of AuNP functionalization is biocompatible and can protect gold surfaces from aggregation in vitro [19]. Moreover, various types of peptides can be used to allow biomaterials to cross cell membranes, aiming for cancer cell targeting [20]. When examining the penetration efficiency of AuNPs, the different lengths of the attached PEG chain should be taken into account. It was observed [21] that a longer PEG chain resulted in a more significant reduction in nanoparticle toxicity, but at the same time, limited internalization due to increased nanostructure size. Achieving better intracellular transport and avoiding degradation in cell structures is a significant challenge and a rarely undertaken research topic. According to earlier studies, the in vivo blood, circulation time, and clearance rate of AuNPs are reportedly impacted by their surface functionalization. For instance, Cho et al. [22] (examined the biodistribution of PEG-encapsulated AuNPs in tumor-bearing mice. According to their research, within 7 days of intravenous injection, AuNPs (4 and 13 nm) were dispersed throughout the organs including the liver and spleen. Their concentration peaked in these organs, named the reticuloendothelial system (RES) organs. RES organs were disseminated by 100 nm PEG-AuNPs, which peaked 30 min after intravenous injection and remained at high levels for six months. This discovery is supported by numerous investigations that have demonstrated that greater PEG chain lengths aided in the longer blood circulation times oof NPs [23].

Another important aspect is the method of evaluating the effectiveness of nanoparticles depending on the research model used. Promising results obtained in 2D conditions are often not reflected in the results of in vivo studies using animal models. 2D cultures are devoid of a 3D network of different cell types surrounded by an extracellular matrix. These two aspects play pivotal roles in the diffusion and uptake of nanoparticles by cells. However, the evaluation of their performance remains limited, as most work is carried out on 2D culture systems [24]. To extend the existing reports, research using 3D systems should be implemented. Spheroids can be formed by single cells that can form cell aggregates. During proliferation, intercellular communication is established, and cells form a specific microenvironment. The findings suggest that mammosphere culture of metastatic cells could be an excellent model for studying the sensitivity of tumorigenic stem cells to therapeutic agents and characterizing the tumor-inducing subpopulation of breast cancer cells [25]. Yousefnia et al. [26] demonstrated the stemness phenotypes of mammospheres generated for further applications in therapeutic approaches and used the MCF-7, MDA-MB-231, and SKBR3 BC cell lines. MCF-7 had the highest mammosphere formation efficiency. Mammospheres generated from all three cell lines had significantly higher proliferation, migration rate, and drug resistance. Moreover, the deposition of the extracellular matrix prompts spheroids to study the ability of nanoparticles to penetrate and diffuse in solid tumors.

At first, the toxicity of the AuNPs was evaluated. The viability of cells slightly decreased after 24, 48, and 72 h compared to the control, and reached the lowest AuNP concentrations of about 90%. Based on these results, the lowest three concentrations were chosen for the following experiments, aiming to define the granularity of the cells and ROS level after incubation with AuNPs. Analysis of the results in the 2D and 3D models at a concentration of 0.002 mg/ML indicated heterogenous viability using different kinds of AUNPs. Concentrations in the range of 0.002–0.02 mg/ML in the 2D model induced viability at 80%, whereas in the 3D model, viability was more varied, indicating that the AuNP distribution is related to different cell morphology. Interestingly, in 3D culture, the viability seemed to increase at the lowest concentration of 30 nm AuNPs, reaching about 170% of the viability. In fact, some studies that investigated the biological effects of nanoparticles found a hormetic dose-response [27]. As a result, it appears clear that future research should focus on this issue by investigating the potential adverse health effects caused by low-level nanoparticle exposure [28]. Xia et al. [29] investigated the toxicity levels of 5 nm AuNPs in the HepG2 cancer cells. They suggested that smaller AuNPs are more likely to accumulate in the cell after entering the cytoplasm and were confined within the endosome or lysosome. Acidic pH in the endosomes and lysosomes may induce the release of toxic ions, indicating ion-specific toxicity. Rostami et al. [30] demonstrated that MCF-7 cells treated with glucose-coated AuNPs decreased the cell viability by 13.2%.

One of the main mechanisms claimed to be the potential reason for radiosensitivity is ROS induction after AuNPs are introduced to the BC cells. It is the one widely recognized biological pathway of radiosensitization. ROS may generate cell damage directly by interacting with biomolecules including cellular DNA or cause cell death indirectly by the oxidation of lipids, proteins, and mitochondrial dysfunction [31]. In this study, to implement the time–concentration manner of using AuNPs, three AuNP concentrations of 0.0004, 0.0008, and 0.0012 mg/mL were used at four TPs of 30 min, 2, 4, and 6 h. The enhancement in ROS generation was noticed at 30 min of incubation. Due to this, the ROS level was analyzed in shorter periods, from point "0" to 60 min per every 10 min in the real-time analysis. The most effective concentration, causing the highest ROS level, was 0.0008 mg/mL at 50 min of the cell incubation with AuNPs. The next step was implementing the optimized results to compare the 2D and 3D models. The same concentrations of AuNPs were used, but the TPs were 30 min, 50 min, 2, and 4 h. The main difference between the models was that in the 3D model, the highest point of ROS level was reached at 30 min, in contrast to the 2D model where the ROS induction was similar at 30 min and 50 min, but the trend of highest ROS peak was at 50 min of the experiment. As in the optimization, performed using a 2D model, the increasing ROS level was noticed with 0.0008 mg/mL of AuNPs. In the 3D model, all concentrations achieved similar results of ROS level enhancement. Considering the 2D and 3D models, differences in ROS generation between results gained with the 30 nm AuNPs were observed. Values in the 3D model were significantly higher than the quantification detected in the 2D model. Analyzing the PEG role in the ROS induction, in the 2D model, there was no indication of a difference between the AuNPs whereas in the 3D model, the enhancement of ROS generation was noticed using AuNPs with the shorter PEG chain. This suggests that ROS induction is associated more with the presence of AuNPs, but not with the used surface modification. In addition, the enhancement of ROS in a short treatment time may lead to a cytotoxic effect due to the triggering of oncogenic signaling pathways by ROS [32]. Kowsalya et al. [33] measured the intracellular ROS production in MCF-7 cells after incubation with AuNPs. They indicated that ROS production increased significantly after using a higher concentration of AuNPs (40.63 +/− 1.69 μg/mL). This did not correlate with our results, which might be associated with another synthesis method. They synthesized AuNPs using green synthesis. Moreover, the AuNPs in our study were functionalized to increase the uptake, thus a lower AuNP concentration was needed for the induction of ROS.

Furthermore, AuNPs altered the cellular redox state in the MCF-7 cells by generating intracellular ROS, thereby inducing ROS-mediated cell death. Another group [34] investigated whether the chitosan functionalized AuNPs, in sizes of 3–10 nm, could induce ROS production in MCF-7 and HeLa cells. The level of ROS generation similarly reached 35% in both cell lines compared to the control group.

In our previous work [17] and the article published by Wu et al. [35], the granularity of cells was analyzed. The uptake after the cells' incubation with AuNPs can be established using the SSC-H signal, as the presence of AuNPs raised the SSC-H response. Using the AuNP functionalization of PEG and RGD, we tried to enhance the AuNP internalization process. Functionalization with targeting biomolecules highly induces the tracking behavior and indicates the localization of AuNP accumulation. In our study, pegylated AuNPs were conjugated with Arg–Gly–Asp peptides to target BC cells. It is usually used for targeting cells expressing RGD-binding integrins such as αV-integrins [36]. The SSC-H parameter indicating the cellular uptake did not alter enough during selected TPs to indicate changing granularity. The probable reason could be based on the size of the AuNP cells and MCF-7 cell morphology, as the 10 and 30 nm AuNPs may be too small to generate differences in the SSC-H parameter. Moreover, the value of the MCF-7 cells could reach about 3375 to 16,873 μm^3, which makes them one of the bigger cells, compared to, for example, MCF10A cells [37]. Gaining the highest uptake might be undetectable in this kind of cell. Despite the knowledge regarding a diverse morphology of cells in different culture models, there were no statistical differences between the 2D and 3D models analyzing the SSC-H parameter. Since the evaluation of uptake was not conclusive in this experiment, the following step could be functionalized AuNPs with a fluorescent probe to better detect their localization as well as to perform experiments with chemical inhibitors of endocytosis to assess the molecular mechanisms. The diffusion of AuNPs into the cancerous spheres plays a crucial role in nanostructure optimization as it promotes the ability of AuNPs to penetrate and enter the hypoxic and necrotic cores of tumors [38]. The researchers in [39] studied 15 nm AuNPs and 50 nm AuNPs in the CAL-27 and HeLa cell lines using the 2D and 3D culture models. The results of the depth of the penetration tests, an essential property that enables the targeting of deep-set tumor cells, showed that 15 nm AuNPs penetrated more successfully than the 50 nm AuNPs.

The main aim of the study was to assess the radiosensitivity effect of MCF-7 cells after irradiation and the previous incubation with AuNPs as well as its dependence on the properties of the AuNPs, and a comparison of the obtained results between the 2D and 3D models. Radiotherapy is an irreplaceable treatment for the effective control of local tumors, mostly connected with chemotherapy and surgical therapy [38]. Numerous studies have shown that using nanoplatforms to deliver therapeutic compounds to tumor tissues in a targeted manner increases the bioavailability of cytotoxic medicines while reducing the risk of damage to healthy tissues. Additionally, research has indicated that one of the most crucial elements impacting radiation effectiveness is the biodistribution of AuNPs [39]. They cause high local ionization in tumor tissues, which minimizes the treatment time and radiation doses, providing that they have been dispersed preferentially in the tumor area and it follows the same energy absorption mode as that of the surrounding healthy tissues.

Based on the viability, ROS, and uptake tests, the point of implementing the IR in the dose of 2 Gy was after 50 min of incubation with 0.0008 mg/mL of all kinds of AuNPs. Using the clonogenic assay, the SF was analyzed by comparing the SF of the cells irradiated with 2 Gy alone to cells previously incubated with AuNPs and irradiated with an IR dose. In the 2D model, statically significant differences were noticed. AuNPs combined with 2 Gy caused a significant SF reduction. There was a similar tendency to analyze values inside the AuNP sizes. In each case, AuNPs conjugated with PEG in MW of 2000 showed a decreased SF value compared to PEG_{800}. This probably suggests that scientists should consider longer PEG chains during the AuNP decoration. Results established in the 3D models were not so promising. There was no significant difference between the cells treated with the AuNPs and IR combination and the cells irradiated alone, although the lowest results were caused using RGD-PEG_{2000}-$AuNPs_{30nm}$. The dose of 2 Gy was chosen because the aim was to apply the scheme of a patient treatment. In clinical conditions, the administration of AuNPs could be performed daily as the radiotherapy fractions, but it depends highly on the time distribution of AuNPs in the blood circulation. The observed radiosensitivity effect might also be caused by ROS generation after incubation with AuNPs. The "energy powerhouse of cells" and amplifiers of ROS generation are mitochondria and is one of the crucial targets that must not be overlooked during the radiosensitization process [40]. Apoptosis may result from oxidative stress-induced mitochondrial malfunction, which includes damage to mitochondrial DNA (mtDNA), the mobilization of cytochrome C, and other biological repercussions [41]. Additionally, it has been proposed that AuNP-induced oxidative stress may be caused by the suppression of proteins necessary for cellular oxidative homeostasis such as thioredoxin reductase TrxR1 [42]. Further research is required to fully understand the mechanism of AuNP-induced oxidative stress, which is currently thought to originate from mitochondrial malfunction brought on by high intracellular ROS levels.

The nanoparticles used in this study may be employed to deliver bioactive compounds to enhance the therapeutic effect of ionizing radiation. To minimize the side effects and increase the effectiveness of the therapy, one proposal is use a drug delivery system containing gold nanoparticles incorporated into liposomes with bioactive substance or as a factor changing the biological response of cells to ionizing radiation. The combination of nanoparticles with a new delivery system could increase the internalization of the nanomaterial, especially in the 3D model, and the bioactive substance would improve the response of cancer cells to combined radiotherapy. According to some research results [43], AuNPs decreased the proliferation, migration, and invasion processes of BCPAP and TPC-1 papillary thyroid cancer cells. The group also showed that AuNPs reduced CCT3 mRNA expression in the papillary thyroid cancer cells, which further demonstrated their antitumor effects. These discoveries might result in the creation of a new potential method for combating various cancer types.

Comparison of 2D and 3D cell culture models by Fontoura et al. [10] indicated that tumor cells grown on 3D models are more resistant to chemotherapy drugs and have similar morphologies resembling in vivo tumors. This could be related to our results that the 3D model reacted differently than the 2D model. The clinical promise of this novel therapy was further supported by a study by Yang et al. [44], who showed the therapeutic benefit of the combination therapy on MDA-MB-231 cells in the presence of MV radiation using RGD-modified AuNPs in conjunction with the chemotherapy medication cisplatin. The survival rate of cells treated with AuNP-RGD:Cis and radiation was reduced noticeably, even at low concentrations, at values substantially lower than the survival rate of cells treated with cisplatin and radiation alone. As a result, they anticipate that this AuNP-mediated chemoradiation will soon be incorporated into the treatment of cancer. Raitanen et al. [45] published an article describing the radiation response comparison in 2D and 3D cell cultures of various cancer cell lines (PC-3, LNCaP, and T-47D) irradiated with 1, 2, 4, 6, 8, or 20 Gy doses of X-ray beams. The findings support the existence of significant diversity in radiobiological response to X-rays in the 2D and 3D cell culture models. They used the same 3D cell culture method as in our study. The results showed that the radiobiological response to X-rays measured in 2D was not reflected in 3D. When compared to the respective monolayers, the spheroid model demonstrated higher radioresistance in all cell lines. To evaluate the radiosensitization effect of AuNPs, Hebert et al. [46] created 5 nm AuNPs coated with the gadolinium chelating agent dithiolated diethylenetriaminepentaacetic gadolinium (Au@DTDTPA:Gd) for the in vitro and in tumor-bearing mice (MC7-L1) analysis. Although AU@DTDPA:Gd indicated no radiosensitizing effect in vivo, in vitro experiments revealed that cell death was induced at 5 mM, a concentration 100 times lower than that detected in the tumors. One proposed reason for this toxicity was the free Gd distribution in the tumor as a result of the surface instability of the AuNPs. Despite promising preclinical confirmation in vitro and a small number of in vivo experiments as well as approaches in radiobiological understanding, AuNPs have not yet entered the clinic. This could be due to a mismatch between the expected levels of radiosensitization based on the experimental conditions and analyzed radiobiological response as well as a limited mechanistic explanation [47]. When designing experiments, many factors can be taken into account to accelerate the introduction of AuNPs into clinical treatment. The use of megavolt energy as well as cell lines isolated from patients are one of many examples that will produce nanoparticles used in the treatment of patients in the near future.

Single-strand breaks (SSBs), double-strand breaks (DSBs), DNA-protein cross-links, and DNA base alterations are only a few types of DNA damage that radiotherapy can cause. Among them, DSBs are the most destructive kind of radiation-induced damage and are connected to the destruction of clonogenic cells. Cell death can occur in a number of ways if the genomic stability is compromised by the inability to repair DNA DSBs. The earliest sensitive markers are thought to be p53-binding protein 1 (53BP1) and phosphorylated histone variant γH2AX [48]. In our study, γH2AX, after exposure of the cells to a dose of 2 Gy and a dose of 2 Gy with the previous 50 min incubation with AuNPs was analyzed. As in the previous clonogenic results, the results established in the 2D model presented higher statistically significant differences. There was a noticeable increase in every relative γH2AX level in the cells treated previously with AuNPs. The highest value was reached using RGD-PEG_{2000}-$AuNPs_{30nm}$, again indicating that those AuNPs are the most effective in inducing the radiosensitivity effect of BC cells in both models. Based on the quantification results of the γH2AX foci MFI, RGD-PEG_{2000}-$AuNPs_{30nm}$ induced the increase compared to the control group (2 Gy alone). On one hand, this result is associated with the flow cytometry outcome. On the other hand, analyzing the rest of the values, the opposite results were detected. Using flow cytometry, in all variants, the γH2AX relative level increased, whereas using the γH2AX foci MFI based on the immunofluorescence, a reduction was observed compared to cells irradiated with 2 Gy alone. This might be related to the number of cell used in both analysis; for flow cytometry, 200,000 cells were used while for immunofluorescence quantification, it was a fragment (image) of 30,000 cells used in the staining. In addition, the increase in the γH2AX relative level correlates with the obtained SF value. Therefore, DNA repair inhibition appears to be a crucial physiologic process underlying AuNP radiosensitization. However, other researchers contend that AuNPs have little impact on the kinetics of DNA repair [49].

5. Conclusions

Implementing gold nanoparticles into the tumor area is a potential strategy to solve the problem of protecting healthy tissues in the beam field. Sensitizing tumor cells to radiotherapy would allow for lower radiation doses, reducing the negative effects of radiation on healthy tissues. This involves still unknown physical, chemical, and biological processes. The results of this research suggest that scientists might pay more attention to the conditions when planning experiments depending on the effects they wish to obtain.

The size and functionalization of nanoparticles have a very significant impact on the way they penetrate the cell. The study highlighted the role of the PEG chain in the efficiency of the AuNPs in the process of sensitizing cells to ionizing radiation. In both sizes of AuNPs, the 2000 MW PEG led to greater statistically significant differences than the AuNPs functionalized with a PEG of 800 MW when analyzing the 2D model. The most effective AuNPs in inducing the sensitivity for IR were the biggest in size and functionalization AuNPs used in both cell culture models.

In the future, scientists should consider a time-dependent manner using AuNP planning experiments, aiming to decrease SF after exposure to AuNPs and IR as well as the size of PEG in functionalization. Results obtained in the 2D model were satisfactory enough to claim that AuNPs are the appropriate solution for combined treatment with RT. In the 3D model, the results after IR suggest that further research is needed to gain a higher efficiency of AuNPs in the sensitizing of BC cells. The future clinical success of the use of nanoparticles is predicted based on a more detailed understanding of the mechanisms by which their physicochemical properties influence the cellular radiobiological response.

Author Contributions: Conceptualization, M.M.; Methodology, M.M., A.B.-L. and R.M.; Formal analysis, M.M.; Investigation, M.M., O.P., B.S., K.K. and P.B.; Writing—original draft preparation, M.M.; Writing—review and editing, M.M. and W.M.S.; Visualization, M.M.; Supervision, W.M.S.; Project administration, M.M. and A.B.-L. All authors have read and agreed to the published version of the manuscript.

Funding: This research was financed from the large research grant no. 502-14-24125270-45086 from statutory funding for young researchers—doctoral students for 2021.

Institutional Review Board Statement: Not applicable.

Informed Consent Statement: Not applicable.

Data Availability Statement: Not applicable.

Conflicts of Interest: The authors declare no conflict of interest.

References

1. Singh, A.; Amiji, M.M. Application of Nanotechnology in Medical Diagnosis and Imaging. *Curr. Opin. Biotechnol.* **2022**, *74*, 241–246. [CrossRef] [PubMed]
2. Musielak, M.; Potoczny, J.; Boś-Liedke, A.; Kozak, M. The Combination of Liposomes and Metallic Nanoparticles as Multifunctional Nanostructures in the Therapy and Medical Imaging—A Review. *IJMS* **2021**, *22*, 6229. [CrossRef] [PubMed]
3. Her, S.; Jaffray, D.A.; Allen, C. Gold Nanoparticles for Applications in Cancer Radiotherapy: Mechanisms and Recent Advancements. *Adv. Drug Deliv. Rev.* **2017**, *109*, 84–101. [CrossRef] [PubMed]
4. Milan, J.; Niemczyk, K.; Kus-Liśkiewicz, M. Treasure on the Earth—Gold Nanoparticles and Their Biomedical Applications. *Materials* **2022**, *15*, 3355. [CrossRef]
5. García-Garrido, E.; Cordani, M.; Somoza, Á. Modified Gold Nanoparticles to Overcome the Chemoresistance to Gemcitabine in Mutant P53 Cancer Cells. *Pharmaceutics* **2021**, *13*, 2067. [CrossRef]
6. Xue, C.; Hu, S.; Gao, Z.-H.; Wang, L.; Luo, M.-X.; Yu, X.; Li, B.-F.; Shen, Z.; Wu, Z.-S. Programmably Tiling Rigidified DNA Brick on Gold Nanoparticle as Multi-Functional Shell for Cancer-Targeted Delivery of SiRNAs. *Nat. Commun.* **2021**, *12*, 2928. [CrossRef]
7. Vines, J.B.; Yoon, J.-H.; Ryu, N.-E.; Lim, D.-J.; Park, H. Gold Nanoparticles for Photothermal Cancer Therapy. *Front. Chem.* **2019**, *7*, 167. [CrossRef]
8. Butterworth, K.T.; Coulter, J.A.; Jain, S.; Forker, J.; McMahon, S.J.; Schettino, G.; Prise, K.M.; Currell, F.J.; Hirst, D.G. Evaluation of Cytotoxicity and Radiation Enhancement Using 1.9 Nm Gold Particles: Potential Application for Cancer Therapy. *Nanotechnology* **2010**, *21*, 295101. [CrossRef]
9. Wang, C.; Jiang, Y.; Li, X.; Hu, L. Thioglucose-Bound Gold Nanoparticles Increase the Radiosensitivity of a Triple-Negative Breast Cancer Cell Line (MDA-MB-231). *Breast Cancer* **2015**, *22*, 413–420. [CrossRef]
10. Fontoura, J.C.; Viezzer, C.; dos Santos, F.G.; Ligabue, R.A.; Weinlich, R.; Puga, R.D.; Antonow, D.; Severino, P.; Bonorino, C. Comparison of 2D and 3D Cell Culture Models for Cell Growth, Gene Expression and Drug Resistance. *Mater. Sci. Eng. C* **2020**, *107*, 110264. [CrossRef]
11. Schulz, F.; Homolka, T.; Bastús, N.G.; Puntes, V.; Weller, H.; Vossmeyer, T. Little Adjustments Significantly Improve the Turkevich Synthesis of Gold Nanoparticles. *Langmuir* **2014**, *30*, 10779–10784. [CrossRef] [PubMed]
12. Wang, A.; Ng, H.P.; Xu, Y.; Li, Y.; Zheng, Y.; Yu, J.; Han, F.; Peng, F.; Fu, L. Gold Nanoparticles: Synthesis, Stability Test, and Application for the Rice Growth. *J. Nanomater.* **2014**, *2014*, 1–6. [CrossRef]
13. Yang, C.; Bromma, K.; Chithrani, D. Peptide Mediated In Vivo Tumor Targeting of Nanoparticles through Optimization in Single and Multilayer In Vitro Cell Models. *Cancers* **2018**, *10*, 84. [CrossRef] [PubMed]
14. Park, J.; Ha, M.K.; Yang, N.; Yoon, T.H. Flow Cytometry-Based Quantification of Cellular Au Nanoparticles. *Anal. Chem.* **2017**, *89*, 2449–2456. [CrossRef] [PubMed]
15. Moon, H.; Ospina-Muñoz, N.; Noe-Kim, V.; Yang, Y.; Elzey, B.D.; Konieczny, S.F.; Han, B. Subtype-Specific Characterization of Breast Cancer Invasion Using a Microfluidic Tumor Platform. *PLoS ONE* **2020**, *15*, e0234012. [CrossRef]
16. Yersal, O. Biological Subtypes of Breast Cancer: Prognostic and Therapeutic Implications. *WJCO* **2014**, *5*, 412. [CrossRef] [PubMed]
17. Musielak, M.; Boś-Liedke, A.; Piotrowski, I.; Kozak, M.; Suchorska, W. The Role of Gold Nanorods in the Response of Prostate Cancer and Normal Prostate Cells to Ionizing Radiation—In Vitro Model. *IJMS* **2020**, *22*, 16. [CrossRef]
18. Wang, S.; Cheng, K.; Chen, K.; Xu, C.; Ma, P.; Dang, G.; Yang, Y.; Lei, Q.; Huang, H.; Yu, Y.; et al. Nanoparticle-Based Medicines in Clinical Cancer Therapy. *Nano Today* **2022**, *45*, 101512. [CrossRef]
19. Boisselier, E.; Astruc, D. Gold Nanoparticles in Nanomedicine: Preparations, Imaging, Diagnostics, Therapies and Toxicity. *Chem. Soc. Rev.* **2009**, *38*, 1759. [CrossRef]
20. Zong, J.; Cobb, S.L.; Cameron, N.R. Peptide-Functionalized Gold Nanoparticles: Versatile Biomaterials for Diagnostic and Therapeutic Applications. *Biomater. Sci.* **2017**, *5*, 872–886. [CrossRef]
21. Attia, M.F.; Anton, N.; Wallyn, J.; Omran, Z.; Vandamme, T.F. An Overview of Active and Passive Targeting Strategies to Improve the Nanocarriers Efficiency to Tumour Sites. *J. Pharm. Pharmacol.* **2019**, *71*, 1185–1198. [CrossRef]
22. Cho, W.-S.; Cho, M.; Jeong, J.; Choi, M.; Han, B.S.; Shin, H.-S.; Hong, J.; Chung, B.H.; Jeong, J.; Cho, M.-H. Size-Dependent Tissue Kinetics of PEG-Coated Gold Nanoparticles. *Toxicol. Appl. Pharmacol.* **2010**, *245*, 116–123. [CrossRef] [PubMed]
23. Schuemann, J.; Berbeco, R.; Chithrani, D.B.; Cho, S.H.; Kumar, R.; McMahon, S.J.; Sridhar, S.; Krishnan, S. Roadmap to Clinical Use of Gold Nanoparticles for Radiation Sensitization. *Int. J. Radiat. Oncol.*Biol.*Phys.* **2016**, *94*, 189–205. [CrossRef] [PubMed]
24. Van Zundert, I.; Fortuni, B.; Rocha, S. From 2D to 3D Cancer Cell Models—The Enigmas of Drug Delivery Research. *Nanomaterials* **2020**, *10*, 2236. [CrossRef]
25. Grimshaw, M.J.; Cooper, L.; Papazisis, K.; Coleman, J.A.; Bohnenkamp, H.R.; Chiapero-Stanke, L.; Taylor-Papadimitriou, J.; Burchell, J.M. Mammosphere Culture of Metastatic Breast Cancer Cells Enriches for Tumorigenic Breast Cancer Cells. *Breast Cancer Res.* **2008**, *10*, R52. [CrossRef] [PubMed]
26. Yousefnia, S.; Ghaedi, K.; Seyed Forootan, F.; Nasr Esfahani, M.H. Characterization of the Stemness Potency of *Mammospheres* Isolated from the Breast Cancer Cell Lines. *Tumour Biol.* **2019**, *41*, 101042831986910. [CrossRef]
27. Iavicoli, I.; Calabrese, E.J.; Nascarella, M.A. Exposure to Nanoparticles and Hormesis. *Dose-Response* **2010**, *8*, 501–507. [CrossRef]
28. Sthijns, M.M.J.P.E.; Thongkam, W.; Albrecht, C.; Hellack, B.; Bast, A.; Haenen, G.R.M.M.; Schins, R.P.F. Silver Nanoparticles Induce Hormesis in A549 Human Epithelial Cells. *Toxicol. Vitro* **2017**, *40*, 223–233. [CrossRef]

29. Xia, Q.; Li, H.; Liu, Y.; Zhang, S.; Feng, Q.; Xiao, K. The Effect of Particle Size on the Genotoxicity of Gold Nanoparticles: The Effect Of Particle Size On The Genotoxicity Of Gold Nanoparticles. *J. Biomed. Mater. Res.* **2017**, *105*, 710–719. [CrossRef]
30. Rostami, A.; Toossi, M.T.B.; Sazgarnia, A.; Soleymanifard, S. The Effect of Glucose-Coated Gold Nanoparticles on Radiation Bystander Effect Induced in MCF-7 and QUDB Cell Lines. *Radiat. Environ. Biophys.* **2016**, *55*, 461–466. [CrossRef]
31. Chen, Y.; Yang, J.; Fu, S.; Wu, J. Gold Nanoparticles as Radiosensitizers in Cancer Radiotherapy. *IJN* **2020**, *15*, 9407–9430. [CrossRef] [PubMed]
32. Nakamura, H.; Takada, K. Reactive Oxygen Species in Cancer: Current Findings and Future Directions. *Cancer Sci.* **2021**, *112*, 3945–3952. [CrossRef] [PubMed]
33. Kowsalya, E.; MosaChristas, K.; Jaquline, C.R.I.; Balashanmugam, P.; Devasena, T. Gold Nanoparticles Induced Apoptosis via Oxidative Stress and Mitochondrial Dysfunctions in MCF-7 Breast Cancer Cells. *Appl. Organomet. Chem.* **2021**, *35*, e6071. [CrossRef]
34. Martínez-Torres, A.C.; Zarate-Triviño, D.; Lorenzo-Anota, H.Y.; Ávila-Ávila, A.; Rodríguez-Abrego, C.; Rodríguez-Padilla, C. Chitosan Gold Nanoparticles Induce Cell Death in HeLa and MCF-7 Cells through Reactive Oxygen Species Production. *IJN* **2018**, *13*, 3235–3250. [CrossRef] [PubMed]
35. Wu, Y.; Ali, M.R.K.; Dansby, K.; El-Sayed, M.A. Improving the Flow Cytometry-Based Detection of the Cellular Uptake of Gold Nanoparticles. *Anal. Chem.* **2019**, *91*, 14261–14267. [CrossRef]
36. Wu, P.-H.; Onodera, Y.; Ichikawa, Y.; Rankin, E.; Giaccia, A.; Watanabe, Y.; Qian, W.; Hashimoto, T.; Shirato, H.; Nam, J.-M. Targeting Integrins with RGD-Conjugated Gold Nanoparticles in Radiotherapy Decreases the Invasive Activity of Breast Cancer Cells. *IJN* **2017**, *12*, 5069–5085. [CrossRef]
37. Geltmeier, A.; Rinner, B.; Bade, D.; Meditz, K.; Witt, R.; Bicker, U.; Bludszuweit-Philipp, C.; Maier, P. Characterization of Dynamic Behaviour of MCF7 and MCF10A Cells in Ultrasonic Field Using Modal and Harmonic Analyses. *PLoS ONE* **2015**, *10*, e0134999. [CrossRef]
38. Zhang, Y.; Huang, F.; Ren, C.; Liu, J.; Yang, L.; Chen, S.; Chang, J.; Yang, C.; Wang, W.; Zhang, C.; et al. Enhanced Radiosensitization by Gold Nanoparticles with Acid-Triggered Aggregation in Cancer Radiotherapy. *Adv. Sci.* **2019**, *6*, 1801806. [CrossRef]
39. Jain, S.; Coulter, J.A.; Butterworth, K.T.; Hounsell, A.R.; McMahon, S.J.; Hyland, W.B.; Muir, M.F.; Dickson, G.R.; Prise, K.M.; Currell, F.J.; et al. Gold Nanoparticle Cellular Uptake, Toxicity and Radiosensitisation in Hypoxic Conditions. *Radiother. Oncol.* **2014**, *110*, 342–347. [CrossRef]
40. Mkandawire, M.M.; Lakatos, M.; Springer, A.; Clemens, A.; Appelhans, D.; Krause-Buchholz, U.; Pompe, W.; Rödel, G.; Mkandawire, M. Induction of Apoptosis in Human Cancer Cells by Targeting Mitochondria with Gold Nanoparticles. *Nanoscale* **2015**, *7*, 10634–10640. [CrossRef]
41. Decrock, E.; Hoorelbeke, D.; Ramadan, R.; Delvaeye, T.; De Bock, M.; Wang, N.; Krysko, D.V.; Baatout, S.; Bultynck, G.; Aerts, A.; et al. Calcium, Oxidative Stress and Connexin Channels, a Harmonious Orchestra Directing the Response to Radiotherapy Treatment? *Biochim. Biophys. Acta (BBA)—Mol. Cell Res.* **2017**, *1864*, 1099–1120. [CrossRef]
42. Liu, R.; Wang, Y.; Yuan, Q.; An, D.; Li, J.; Gao, X. The Au Clusters Induce Tumor Cell Apoptosis via Specifically Targeting Thioredoxin Reductase 1 (TrxR1) and Suppressing Its Activity. *Chem. Commun.* **2014**, *50*, 10687. [CrossRef] [PubMed]
43. Liu, F.; Ma, D.; Chen, W.; Chen, X.; Qian, Y.; Zhao, Y.; Hu, T.; Yin, R.; Zhu, Y.; Zhang, Y.; et al. Gold Nanoparticles Sup-pressed Proliferation, Migration, and Invasion in Papillary Thyroid Carcinoma Cells via Downregulation of CCT3. *J. Nanomater.* **2019**, *2019*, 1687340. [CrossRef]
44. Yang, J.C.; B Chithrani, D. Nuclear Targeting of Gold Nanoparticles for Improved Therapeutics. *CTMC* **2015**, *16*, 271–280. [CrossRef] [PubMed]
45. Raitanen, J.; Barta, B.; Hacker, M.; Georg, D.; Balber, T.; Mitterhauser, M. Comparison of Radiation Response between 2D and 3D Cell Culture Models of Different Human Cancer Cell Lines. *Cells* **2023**, *12*, 360. [CrossRef]
46. Hébert, E.M.; Debouttière, P.-J.; Lepage, M.; Sanche, L.; Hunting, D.J. Preferential Tumour Accumulation of Gold Nanoparticles, Visualised by Magnetic Resonance Imaging: Radiosensitisation Studies in Vivo and in Vitro. *Int. J. Radiat. Biol.* **2010**, *86*, 692–700. [CrossRef]
47. Rosa, S.; Connolly, C.; Schettino, G.; Butterworth, K.T.; Prise, K.M. Biological Mechanisms of Gold Nanoparticle Radiosensitization. *Cancer Nano* **2017**, *8*, 2. [CrossRef]
48. Djuzenova, C.S.; Elsner, I.; Katzer, A.; Worschech, E.; Distel, L.V.; Flentje, M.; Polat, B. Radiosensitivity in Breast Cancer Assessed by the Histone γ-H2AX and 53BP1 Foci. *Radiat. Oncol.* **2013**, *8*, 98. [CrossRef]
49. Jain, S.; Coulter, J.A.; Hounsell, A.R.; Butterworth, K.T.; McMahon, S.J.; Hyland, W.B.; Muir, M.F.; Dickson, G.R.; Prise, K.M.; Currell, F.J.; et al. Cell-Specific Radiosensitization by Gold Nanoparticles at Megavoltage Radiation Energies. *Int. J. Radiat. Oncol. Biol. Phys.* **2011**, *79*, 531–539. [CrossRef]

pharmaceutics

MDPI

Article

Thermosensitive Polymer-Modified Mesoporous Silica for pH and Temperature-Responsive Drug Delivery

Kokila Thirupathi [1,†], Madhappan Santhamoorthy [2,†], Sivaprakasam Radhakrishnan [3], Selvakumari Ulagesan [4], Taek-Jeong Nam [5], Thi Tuong Vy Phan [6,7,*] and Seong-Cheol Kim [2,*]

1 Department of Physics, Government Arts and Science College for Women, Karimangalam, Dharmapuri 635111, Tamil Nadu, India
2 School of Chemical Engineering, Yeungnam University, Gyeongsan 38541, Republic of Korea
3 Department of Organic Materials and Fiber Engineering, Jeonbuk National University, 567 Baekje-daero, Deokjin-gu, Jeonju-si 54896, Republic of Korea
4 Division of Fisheries Life Sciences, Pukyong National University, Nam-gu, Busan 48513, Republic of Korea
5 Institute of Fisheries Sciences, Pukyong National University, Gijang-gun, Busan 46041, Republic of Korea
6 Center for Advanced Chemistry, Institute of Research and Development, Duy Tan University, 03 Quang Trung, Hai Chau, Danang 550000, Vietnam
7 Faculty of Environmental and Chemical Engineering, Duy Tan University, 03 Quang Trung, Hai Chau, Danang 550000, Vietnam
* Correspondence: phanttuongvy4@duytan.edu.vn (T.T.V.P.); sckim07@yu.ac.kr (S.-C.K.)
† These authors contributed equally to this work.

Citation: Thirupathi, K.; Santhamoorthy, M.; Radhakrishnan, S.; Ulagesan, S.; Nam, T.-J.; Phan, T.T.V.; Kim, S.-C. Thermosensitive Polymer-Modified Mesoporous Silica for pH and Temperature-Responsive Drug Delivery. *Pharmaceutics* **2023**, *15*, 795. https://doi.org/10.3390/pharmaceutics15030795

Academic Editors: Manivasagan Panchanathan and Eue-Soon Jang

Received: 28 January 2023
Revised: 24 February 2023
Accepted: 27 February 2023
Published: 28 February 2023

Abstract: A mesoporous silica-based drug delivery system (MS@PNIPAm-PAAm NPs) was synthesized by conjugating the PNIPAm-PAAm copolymer onto the mesoporous silica (MS) surface as a gatekeeper that responds to temperature and pH changes. The drug delivery studies are carried out in vitro at different pH (7.4, 6.5, and 5.0) and temperatures (such as 25 °C and 42 °C, respectively). The surface conjugated copolymer (PNIPAm-PAAm) acts as a gatekeeper below the lower critical solution temperature (LCST) (<32 °C) and as a collapsed globule structure above LCST (>32 °C), resulting in controlled drug delivery from the MS@PNIPAm-PAAm system. Furthermore, the 3-(4,5-dimethylthiazol-2-yl)-2,5-diphenyltetrazolium bromide (MTT) assay and cellular internalization results support the prepared MS@PNIPAm-PAAm NPs being biocompatible and readily taken up by MDA-MB-231 cells. The prepared MS@PNIPAm-PAAm NPs, with their pH-responsive drug release behavior and good biocompatibility, could be used as a drug delivery vehicle where sustained drug release at higher temperatures is required.

Keywords: mesoporous silica; drug delivery; thermosensitive copolymer; pH-stimuli; biocompatibility

1. Introduction

Cancer has attracted a lot of research focus in the last few decades because of its relationship to the environment and genetics, and it is still one of the major causes of mortality in the twenty-second century [1–3]. Chemotherapy has become the mainstay of cancer treatment among traditional cancer treatments (surgical resection, radiation therapy) due to its high efficiency in comparison to other treatments [4]. However, because most chemotherapeutic drugs lack specificity, they kill tumor cells while causing toxic side effects in normal tissues, severely limiting their clinical application. The stimulus-triggered drug carrier system is combined with the stimulus-responsive components and drug, allowing for timed and targeted drug delivery in response to internal/external stimuli [5–8]. It has the potential to improve drug bioavailability while decreasing the undesirable side effects of chemotherapy drugs [9–11].

Stimulus-responsive drug delivery systems are classified as either internal stimuli-responsive (such as pH, temperature, enzyme) or external stimuli-responsive (such as light and magnetic field) [12–16]. However, because the microenvironment of disease sites differs greatly from that of normal tissues, a single stimulus-responsive drug delivery system is unable to meet the demands of modern drug therapy. Multiple stimulus-response materials can help the lesion site adapt to its complex environment and allow for controlled drug release. It has emerged as a critical area of growth for the drug delivery system.

To enrich solid tumors, mesoporous silica nanoparticles could take advantage of the enhanced permeability and retention (EPR) effect. Mesoporous silica materials (MS) are regarded as an ideal candidate for drug delivery applications due to a variety of distinct characteristics when compared to other nanoparticles, including easy synthesis steps, controlled pores, excellent mechanical stability, and good cytocompatibility [17]. MSs' high surface areas and pore sizes make it easier to load and distribute a significant amount of drug to tumors in a controlled manner. By altering the pore blocking agent, the drug can be placed into the mesopore channel and shielded from premature release. An efficient drug carrier must prevent premature drug leakage before reaching the target sites and release the drugs rapidly at the specific sites [18].

Poly(N-isopropyl acrylamide) (PNIPAM) is a typical example of a temperature-sensitive polymer, with dramatic volume changes at temperatures below the lower critical solution temperature (LCST) of 32 °C [19]. The polymer is in a highly swollen state when the temperature is below LCST. When the temperature rises above LCST, the polymer transforms into a dehydrated, collapsed state, which promotes drug release [20]. Although the LCST of PNIPAM is lower than the physiological temperature of humans, the LCST can be increased by adding hydrophilic copolymer monomers [21]. PAAm, a linear-chain structure polymer with amide substituents on alternating carbon atoms, has long been utilized in contact lenses, and its hydrophilic and bioinert properties make it appealing for biomedical applications [22,23]. As a result, the PNIPAm-PAAm copolymer could be used as a coating in biological applications. To control drug release in response to environmental stimuli, a variety of polymers have been used to coat silica materials [24–27]. The application of a PNIPAm-PAAm copolymer to the outer surface of MS is anticipated to act as a shell that is pH-responsive to regulate the drug release under the desired pH, which could be useful in controlling the release of chemotherapeutic drugs.

The pH- and temperature-responsive drug delivery system (MS@PNIPAm-PAAm NPs) was created in this study. The temperature-responsive copolymer PNIPAm-PAAm was modified on the MS surface by an epoxy-ring opening reaction approach. To examine the drug release properties, doxorubicin (Dox) was utilized as a model drug. The loaded Dox was successfully protected inside the mesopore channels, allowing it to be released under temperature-stimulated intracellular pH conditions. The cytocompatibility and cell uptake of the prepared MS@PNIPAm-PAAm NPs were assessed using a human breast cancer (MDA-MB-231) cell line.

2. Materials and Methods

2.1. Chemicals and Reagents

Tetraethyl orthosilicate (TEOS, 99%), cetyltrimethylammonium bromide (CTAB, 99%), N-isopropyl acrylamide (NIPAm, 97%), acrylamide (AAm, 98%), 3-glycidoxypropyl trimethoxysilane (GPTMS, 98%) azobisisobutyronitrile (AIBN, 98%), and doxorubicin hydrochloride (Dox), tetrahydrofuron (THF, 99%), toluene (98%), hydrochloric acid (HCl, 36%), sodium hydroxide (NaOH, 98%), ammonium nitrate (NH_4NO_3, 98%) were purchased from Aldrich Chemical Co., Saint Louis, MO, USA. All the chemicals were used as received. Human breast cancer cells, MDA-MB-231 (Korea cell line bank, Seoul, Republic of Korea), were purchased and used for this study.

2.2. Synthesis of PNIPAm-Acrylamide (PNIPAm-PAAm) Copolymers

In a typical synthesis of the PNIPAm-PAAm copolymer, NIPAm (1 g, 0.0088 mol) was dissolved in 10 mL THF in a three-neck flask and degassed with nitrogen. To this, acrylamide (0.62 g, 0.0088 mol) was added, and the obtained monomer solution was continuously purged with nitrogen for 45 min before AIBN was introduced to initiate the reaction. Next, the reaction flask was heated to 70 °C for 24 h. After the specified time, the mixture was concentrated, and the resulting sticky, viscous mass was then dissolved in chloroform and reprecipitated from cold hexane [28,29]. The reprecipitation process was done thrice. The obtained copolymer product was dried in a vacuum. We chose NIPAm monomer to incorporate thermoresponsive property and AAm for chemical conjugation onto mesoporous silica nanoparticles through GPTMS linking agent. The synthesized copolymer obtained was a random copolymer with an average molecular mass of about 16,500 g/mol.

2.3. Synthesis of Mesoporous Silica (MS) NPs

To dissolve CTAB completely, 1.0 g in 450 mL deionized water was used and allowed to stir at 35 °C. To this, about 5.0 mL of NaOH solution (2.0 M) was added and stirred at 35 °C. The mixture was then stirred at 90 °C after the gradual addition of 5.0 mL of TEOS solution. Finally, the resultant white precipitate was centrifuged, washed several times with water, and dried at 60 °C under vacuum for 24 h. The as-made sample obtained was named MS@CTAB NPs (Scheme 1, Step 1) [30].

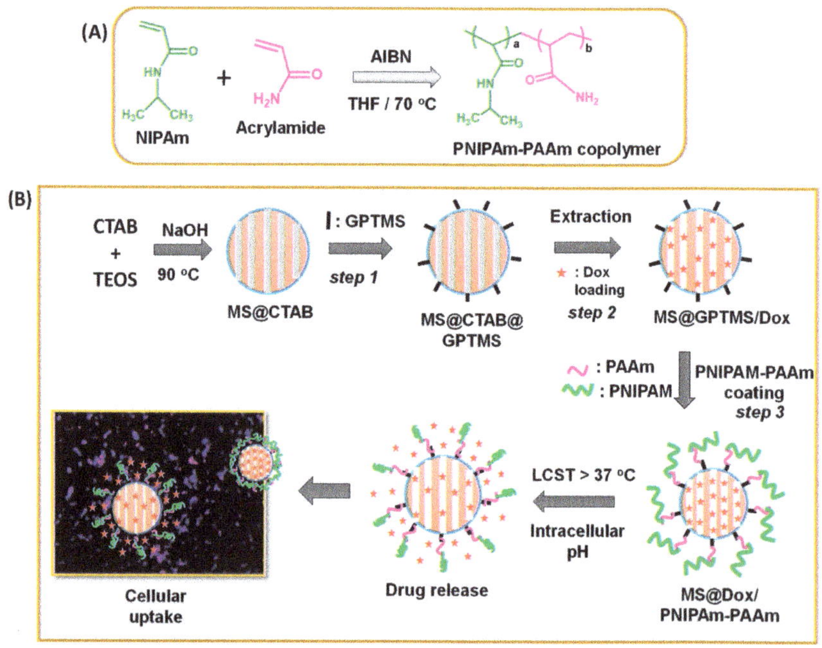

Scheme 1. Schematic representation of the (**A**) synthesis of the PNIPAm-PAAm copolymer and (**B**) the MS@Dox/PNIPAm-PAAm NPs' synthesis, surface modifications, drug loading, and release behavior, respectively.

2.4. GPTMS Modification onto the MS@CTAB NPs

To perform this process, MS@CTAB NPs (0.3 g) were suspended in 60 mL of dry toluene. To this, GPTMS (1 mL) was dropped, and the suspension was refluxed at 80 °C for 8 h under inert conditions. The resultant product was centrifuged and washed with toluene (3 × 10 mL each) and subsequently washed with ethanol (3 × 10 mL each) to remove the toluene, and then dried at 60 °C [31]. The obtained GPTMS-modified sample was named MS@CTAB@GPTMS NPs (Scheme 1, Step 2). In the next step, the occluded CTAB was extracted by dispersing MS@CTAB@GPTMS NPs (0.2 g) in a 100 mL ethanol solution containing ammonium nitrate (0.3 g) and refluxing for 6 h at 60 °C [32]. The extraction steps were performed thrice and then separated, rinsed with ethanol, and dried at 60 °C. The obtained material was named MS@GPTMS NPs (Scheme 1, Step 2).

2.5. Dox Loading into the MS@GPTMS NPs

For Dox loading, MS@GPTMS NPs (0.1 g) were dispersed in toluene (10 mL). Next, 5 mL (1 mg/mL) of Dox was introduced and stirred for 24 h. Then, the resultant suspension was centrifuged and rinsed with 1 mL of toluene and a minimum amount of ethanol (1 mL) to eliminate the physisorbed drugs and toluene solvent, and then vacuum dried at 60 °C. The Dox-encapsulated sample was named MS@Dox/GPTMS NPs. The encapsulated Dox was estimated at 498 nm using a UV-Vis spectrometer as follows:

$$\text{Drug loading (wt\%)} = \frac{\text{Mass of the drug in sample}}{\text{Mass of the drug in feed}} \times 100\%$$

A small amount of drug released during the surface modification process was collected, combined with the final drug solution after drug loading, and used to determine the amount of drug loading into the nanoparticles. From the initial and final drug solutions, the loaded Dox was determined to be ~72% (Scheme 1, Step 2).

2.6. Synthesis of PNIPAM-PAAm Coated MS@Dox/PNIPAm-PAAm NPs

The PNIPAm-PAAm copolymer coated MS@Dox/PNIPAm-PAAm NPs were prepared as follows. To perform this process, MS@GPTMS NPs (0.1 g) were suspended in 50 mL of dry THF. Next, about 2 mL of PNIPAm-PAAm copolymer solution was added to the suspension in the presence of a trimethylamine catalyst and allowed to react at 40–70 °C for 24 h. During this reaction process, the amine (-NH_2) groups of PAAm can interact with the hydrogen groups of the opened epoxy ring via hydrogen bonding interactions between the amine part of the PAAm segment and the GPTMS units that were modified onto the outer surface of the mesoporous silica nanoparticles (Scheme 1, Step 2). Generally, this ring opening reaction requires a high temperature and catalytic conditions to perform a complete reaction [33,34]. However, in this work, an appropriate temperature condition was used to achieve a partial reaction to avoid the crowding accumulation of PNIPAm-PAAm copolymer onto the silica surfaces. The polymer-modified silica sample was collected by filtration, rinsed with THF, and vacuum dried at RT. The obtained PNIPAm-PAAm copolymer modified material was named as MS@Dox/PNIPAm-PAAm NPs.

2.7. Stimuli-Responsive In Vitro Drug Release Study

Phosphate buffer saline (PBS) solution was used for the drug release study after being adjusted to the appropriate pH by using 0.1 M HCl or 0.1 M NaOH solution. In vitro Dox release from the MS@Dox/PNIPAm-PAAm NPs was examined under different conditions: (i) different pH (5.0, 6.5, and 7.4); (ii) different temperature (25 °C and 42 °C); (iii) the combined pH and temperature (7.4/25 °C, 5.0/25 °C, and 7.4/42 °C, 5.0/42 °C), respectively. In a dialysis bag (molecular weight cutoff of 10,000), 10 mg of MS@Dox/PNIPAm-PAAm NPs were suspended in PBS (5 mL) at different pHs (7.4, 6.5, and 5.0, respectively). The dialysis bag was immersed in 20 mL of phosphate buffer saline (PBS) medium maintained at the appropriate temperature using a water bath. The samples were then placed in a beaker equipped with a magnetic stirrer. At the predetermined time, 1 mL of the release medium was extracted and analyzed at 498 nm using a UV-Vis spectrophotometer. The control sample MS@Dox/PNIPAm-PAAm NPs was prepared without Dox loading.

2.8. Biocompatibility (MTT Assay) Analysis

The in vitro biocompatibility of the generated MS@PNIPAm-PAAm NPs without Dox loading and Dox-loaded MS@Dox/PNIPAm-PAAm NPs, as well as the same concentration of pure Dox, were evaluated on the MDA-MB-231 cell line. Firstly, the cells were cultured in a 96-well plate for 24 h at 37 °C and 5% CO_2. Secondly, different concentrations of MS@PNIPAm-PAAm NPs, MS@Dox/PNIPAm-PAAm NPs, and free Dox, respectively, were added (5 replicate wells for each sample) and cultured for 24 h. Finally, 20 μL of MTT solution was added and incubated for another 4 h. Next, 200 μL of DMSO was added to each well and gently shaken to replace the existing medium. For each well, the absorbance at 570 nm was measured. The percentage of cell viability was calculated by measuring the absorbance at 570 nm.

$$\text{Cell viability } (\%) = \frac{A_{sample} - A_{blank}}{A_{control} - A_{blank}} \times 100$$

where $OD_{treated}$: cells treated with MS@PNIPAm-PAAm NPs without drug, and MS@Dox/PNIPAm-PAAm NPs, respectively. $OD_{control}$: control cells.

2.9. Fluorescence Microscopy Analysis

Fluorescence microscopic examination was used to investigate the intracellular internalization of MS@Dox/PNIPAm-PAAm NPs. For this, the MDA-MB-231 cells were cultured in a 6-well plate in DMEM medium for 24 h. The cultured plate was then incubated for a further 6 h after being treated with MS@Dox/PNIPAm-PAAm NPs at a concentration of 10 μg/mL. Following incubation, the sample-exposed cells were fixed using 4% paraformaldehyde, the existing medium was withdrawn, and the plate was rinsed thrice using fresh PBS buffer to eliminate any adhered samples. Using a fluorescence microscope, fluorescence pictures of the cells were captured.

2.10. Instrumental Characterization

Low-angle X-ray diffraction (XRD) investigation was carried out in the low-angle range ($2\theta = 0.5°$ to $6°$) utilizing a Bruker AXN analyzer. N_2 sorption analysis was performed by using the Nova 4000e surface area instrument. Fourier transform infrared (FTIR; JASCO FTIR 4100) spectra with a frequency ranging from 4000 to 400 cm^{-1}, were recorded using the KBr pellet technique. The surface morphology was studied by scanning electron microscopy (SEM, S4800). Transmission electron microscopy (TEM) images were collected on a JEOL 2010 operating at an acceleration voltage of 200 kV. Scanning electron microscopy (SEM) images were recorded with a JEOL 6400 microscope operating at 20 kV. Thermogravimetric analysis (TGA) was performed on a Perkin-Elmer Pyris Diamond thermogravimetric analyzer with a heating rate of 10 °C/min under N_2 atm. Particle size and zeta potential measurements were performed on the Malvern Zetasizer Nano-ZS. UV-vis spectral analysis was performed using an Agilent Inc. UV-Vis spectrophotometer.

3. Results and Discussion

3.1. Synthesis of PNIPAm-PAAm Copolymers, Surface Modification, and Characterization of PNIPAm-PAAm Modified MS@PNIPAm-PAAm NPs

The PNIPAm-PAAm copolymer was synthesized via free radical copolymerization of NIPAm and AAm as shown in Scheme 1A, and the obtained copolymer product was characterized by ^1H NMR and FTIR analysis (Figure 1). Figure 1A displays the ^1H NMR spectrum of the PNIPAm-PAAm copolymer. As shown in Figure 1A, the proton signals around 1.1 ppm attributed to the methyl ($-CH_3$) groups of NIPAm units, and the peaks at 3.7–4.1 ppm are allocated to the $-CH_2$-CH_2O- protons of NIPAm amide groups. The signals in the range of 1.9 to 2.2 ppm and 1.2 to 1.8 ppm are assigned to secondary and tertiary carbon and amine protons in the copolymer backbone, respectively. Figure 1B shows the FTIR spectrum of the synthesized PNIPAm-PAAm copolymer. The strong band appeared at 3425 cm^{-1} which is ascribed to the bending of amine ($-NH_2$) groups. The vibration bands at 1637 cm^{-1} and 1715 cm^{-1} are indicating the presence of carbonyl (–C=O) units of acrylamide and NIPAm, respectively. Moreover, the bending vibration band at 2781 cm^{-1}, 2922 cm^{-1} indicates the presence of alkyl C-H groups in the PNIPAm in the copolymer backbone. These FTIR absorption peaks demonstrated the successful synthesis of the PNIPAm-PAAm copolymer.

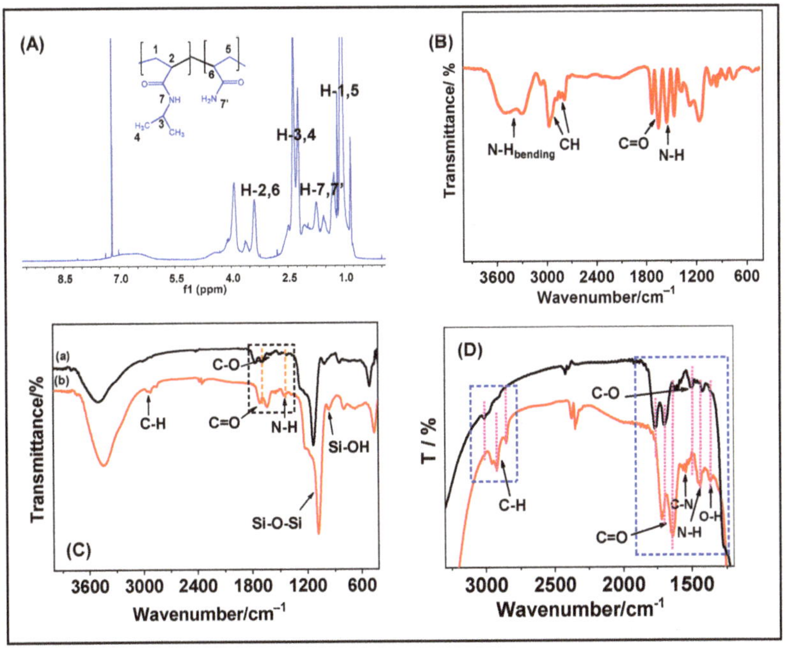

Figure 1. (**A**) ^{1}H NMR; (**B**) FTIR spectrum of PNIPAm-PAAm copolymer; and (**C**,**D**) FTIR spectra of (a) MS@GPTMS NPs and (b) copolymer modified MS@PNIPAm-PAAm NPs, respectively.

Figure 1C,D depicts the FTIR spectra of MS@GPTMS NPs and MS@PNIPAm-PAAm NPs, respectively. Figure 1C(a) shows that the existence of a propyl carbon chain was indicated by a -C–H band at 2863 cm^{-1}, 2936 cm^{-1}, and a stretching band at 1389 cm^{-1} was attributed to the C–O vibrations of the epoxy moiety of the GPTMS groups. In addition, the characteristic bands at 963 cm^{-1}, 1092 cm^{-1} were identified as vibration peaks of Si-OH and Si-O-Si stretching of MS@GPTMS NPs, respectively. As shown in Figure 1D, after PNIPAm-PAAm copolymer modification onto the MS@GPTMS NPs, the new intense absorption peak of MS@PNIPAm-PAAm NPs appeared at 1639 cm^{-1}, 1449 cm^{-1} could be assigned to the amide (-C=O) and (-C-N, -N-H bending) in NIPAm and acrylamide, respectively. As compared to the sample MS@GPTMS NPs, the copolymer-modified MS@PNIPAm-PAAm NPs showed a noticeable peak shift of ~45 cm^{-1} for the C-O peak, indicating the epoxy ring opening by the reaction of epoxy groups with the amine part of the PAAm segments. Moreover, the peak intensity of the alkyl C-H groups was considerably increased, and new peaks appeared for O-H at 1364 cm^{-1}, C-N for 1542 cm^{-1}, and 1637 cm^{-1} for C=O groups, respectively, which implies that the copolymer has been modified onto the MS@PNIPAm-PAAm NPs. From the FTIR analysis data, the surface modification of the PNIPAm-PAAm copolymer onto the MS@GPTMS NPs was confirmed [35].

The size and morphology of MS@GPTMS NPs and MS@PNIPAm-PAAm NPs were observed by SEM and TEM (Figure 2). As illustrated in Figure 2A, the SEM images of MS@GPTMS NPs have a particle size of about ~175 nm and a uniform pore structure. As displayed in Figure 2b, the size of the particle increased slightly after surface modification, and the particles were slightly aggregated due to the surface modification of the PNIPAm-PAAm polymer onto the MS@GPTMS NPs surface. Meanwhile, the mesopore channels were completely covered by the surface-coated PNIPAm-PAAm copolymer on the MS@PNIPAM-PAAm NPs and are visible and distinguishable when compared to MS@GPTMS NPs, indicating that they were successfully coated with copolymer (Figure 2c,d).

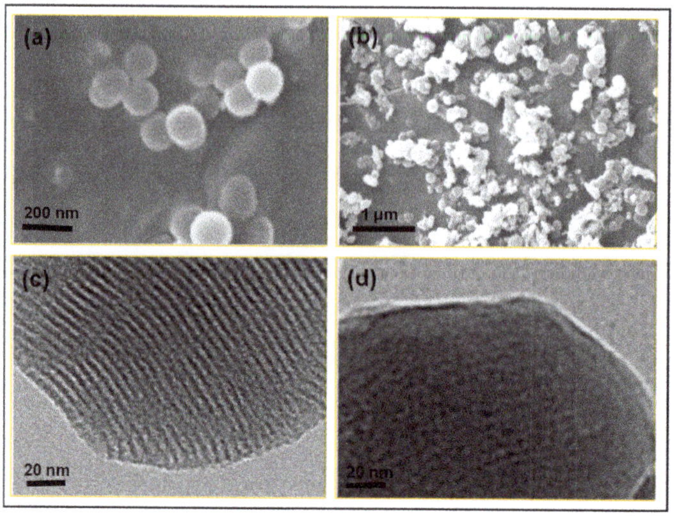

Figure 2. SEM images of (**a**) MS@GPTMS NPs and (**b**) MS@PNIPAm-PAAm NPs, without Dox loading, respectively. TEM images of (**c**) MS@GPTMS NPs and (**d**) MS@PNIPAm-PAAm NPs, respectively.

Furthermore, the particle size distribution of MS@GPTMS NPs and MS@PNIPAm-PAAm NPs was determined using a dynamic light scattering (DLS) analysis (Figure 3A). The average particle size of the MS@GPTMS NPs was determined to be between ~75–300 nm, with a polydispersity index (PDI) value of about 0.18, as illustrated in Figure 3A(a). The particle size distribution of the PNIPAm-PAAm copolymer-modified MS@PNIPAm-PAAm NPs, on the other hand, is about ~100 to ~350 nm with a determined polydispersity index (PDI) value of about 0.25 (Figure 3A(b)). The polydispersity value was slightly increased for the MS@PNIPAm-PAAm NPs as compared to the MS@GPTMS NPs, indicating that the particles were monodispersed in an aqueous medium. The presence of a surface-modified PNIPAm-PAAm copolymer on the MS@PNIPAm-PAAm NPs could account for the increased particle size of around ~25–40 nm. Moreover, no considerable small or large particles were observed below ~75 nm and above ~350 nm. However, when increasing the medium temperature to 42 °C, the average particle size was slightly decreased to about ~150 nm due to the shrinkage of surface-modified PNIPAM-PAAm copolymers at higher temperatures. Figure 3B(a,b)) shows the surface charge of MS@GPTMS and MS@PNIPAm-PAAm NPs, respectively. As observed in Figure 3B(a), the MS@GPTMS NPs had a negative zeta potential value of approximately −2.6 mV at pH 5 and ~−32 mV at pH 10. A slight negative zeta potential value was observed for the MS@GPTMS NPs at an acidic pH (pH 5) due to the presence of surface silanol (Si-OH) groups of silica nanoparticles. In contrast, the PNIPAm-PAAm modified MS@PNIPAm-PAAm NPs had more positive values of approximately +24 mV at pH 5 and a negative value of about ~ −30 mV at pH 10 (Figure 3B(b)). As compared to the zeta potential value of MS@GPTMS NPs, the PNIPAm-PAAm copolymer modified MS@PNIPAm-PAAm NPs showed an increased positive zeta potential value due to the presence of -CO-NH- groups in the PNIPAm and PAAm segments which undergo protonation (-CO-NH^+) under an acidic pH condition. Therefore, the overall zeta potential value for MS@PNIPAm-PAAm NPs showed a positive charge under the low pH condition. These results indicate that the PNIPAm-PAAm copolymer has been modified onto the mesoporous silica MS@GPTMS NPs.

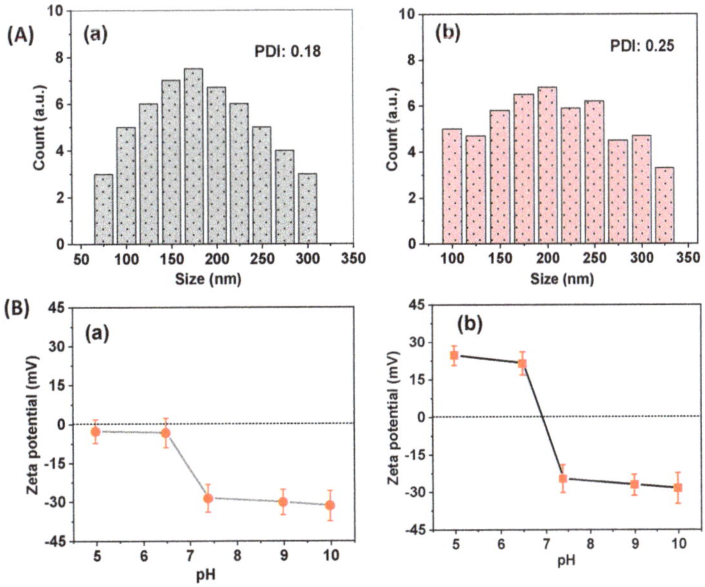

Figure 3. (**A**) Particle size distribution; and (**B**) Zeta potential analysis of (**a**) MS@GPTMS NPs and (**b**) MS@PNIPAm-PAAm NPs, respectively.

The XRD patterns of the MS@GPTMS NPs showed three crystal diffraction peaks at 2θ = 2.13°, 3.62°, and 4.19°, which were assigned to (100), (110), and (200), respectively, indicating the formation of ordered mesoporous materials (Figure 4A(a)). However, after PNIPAm-PAAm copolymer modification, the diffraction crystal plane at 2θ = 1.75° (100) shifted toward a lower angle and the peak intensity decreased, and the peaks at 2θ = 3.62° and 4.13° (110) and (200) are not visible. This might be the ratio of ordered material: the total sample decreased, and therefore, the diffraction planes became less visible due to the presence of surface-modified PNIPAm-PAAm copolymers on the MS@PNIPAm-PAAm NPs. This result indicates that the PNIPAm-PAAm copolymer was covered onto the mesopore channels of the MS@PNIPAm-PAAm NPs [36].

N_2 adsorption-desorption analysis was used to examine the surface and mesopore properties of the MS@GPTMS and MS@PNIPAm-PAAm NPs. The specific surface area was estimated by the Brunauer–Emmet–Teller (BET) method. The pore size distribution curve was obtained from an analysis of the adsorption branch using the Barret–Joyner–Halenda (BJH) method. Both samples had type IV isotherms, as illustrated in Figure 4B(a,b), indicating that the presence of mesoporous structures in the prepared mesoporous silica nanoparicles. The MS@GPTMS NPs' surface area, mesopore size, and pore volume were calculated to be about ~658 m^2/g, 2.3 nm, and 0.98 cm^3/g, respectively. The MS@PNIPAm-PAAm NPs, on the other hand, had a significantly reduced surface area of ~152 m^2/g after PNIPAm-PAAm copolymer modification (Figure 4B). The presence of surface coated PNIPAm-PAAm copolymers on the MS@PNIPAm-PAAm NPs resulted in a significant reduction in surface area and pore size (Figure 4C). These results also supported the finding that the PNIPAm-PAAm copolymer was successfully modified on the surface of MS@GPTMS NPs.

Figure 5 depicts the TGA curves for the GPTMS modified sample, MS@GPTMS NPs, MS@GPTMS NPs without the CTAB extraction process, and PNIPAM-PAAm copolymer coated MS@PNIPAm-PAAm NPs, respectively. The MS@GPTMS NPs showed about 11.3% indicating the degradation of surface-modified GPTMS functional units. Further, the MS@GPTMS-CTAB sample showed ~18.5% weight loss due to the degradation of occluded CTAB molecules. The MS@PNIPAm-PAAm NPs had an initial weight loss of about ~2.5 wt.% between 30 °C and 100 °C due to the evaporation of physisorbed water or moisture. Moreover, the PNIPAM-PAAm copolymer-coated MS@PNIPAm-PAAm NPs showed a gradual weight loss of about ~24.5 wt.% in mass, which was implied to be due to the thermal degradation of surface-modified PNIPAm-PAAm copolymer from the surface of the mesoporous silica nanoparticles.

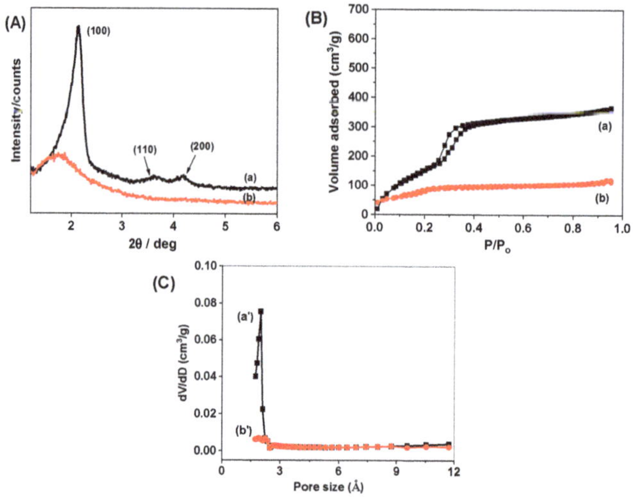

Figure 4. (**A**) XRD patterns; (**B**) N_2 adsorption-desorption of (a) MS@GPTMS NPs and (b) MS@PNIPAm-PAAm NPs; and (**C**) Pore size distributions of (a′) MS@GPTMS NPs and (b′) MS@PNIPAm-PAAm NPs, respectively.

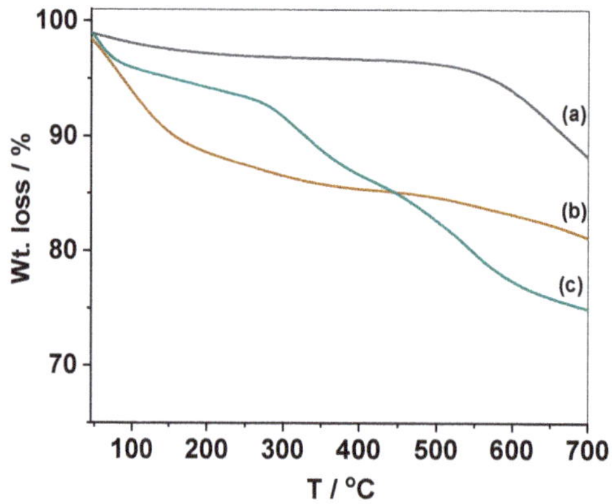

Figure 5. Thermogravimetric analysis of (a) MS@GPTMS NPs, (b) MS@GPTMS-CTAB NPs, and (c) PNIPAm-PAAm copolymer modified MS@PNIPAm-PAAm NPs, respectively.

3.2. *Temperature-Responsive Dox Release from MS@Dox/PNIPAM-PAAm NPs*

PNIPAm is a prominent thermoresponsive polymer composed of hydrophilic amide and hydrophobic alkyl groups, with an LCST of around 32 °C, which is somewhat lower than the body temperature (37 °C). As a result, PNIPAm solution has a sol state at ambient temperature and a gel state when heated to body temperature. This distinct characteristic enables PNIPAm to be used in biological applications. Similarly, the PAAm polymer features amine (-NH) groups in the side chain that can serve as drug binding sites to hold the drugs under physiological pH circumstances and release the loaded drugs under acidic pH conditions owing to protonation of -NH- sites. The combination of thermoresponsive and pH-responsive characteristics of the synthesized PNIPAm-PAAm copolymer make it an excellent candidate for the modification of silica nanoparticles for drug delivery applications. Temperature-responsive drug release behavior of the Dox-loaded MS@Dox/PNIPAm-PAAm NPs system was evaluated at different temperatures of 25 °C and 42 °C, respectively (Figure 6a). As shown in Figure 6a, only a small amount of Dox release, ~17.6%, was seen in the 48-h release period. These results revealed that the H-bonding interactions between the -C=O and -N-H groups present in the PNIPAm-PAAm copolymer are more stable at pH 7.4 and 25 °C, implying that the PNIPAm-PAAm copolymer on the MS@Dox/PNIPAm-PAAm NPs effectively retains the drugs from premature release at pH 7.4 and at 25 °C conditions.

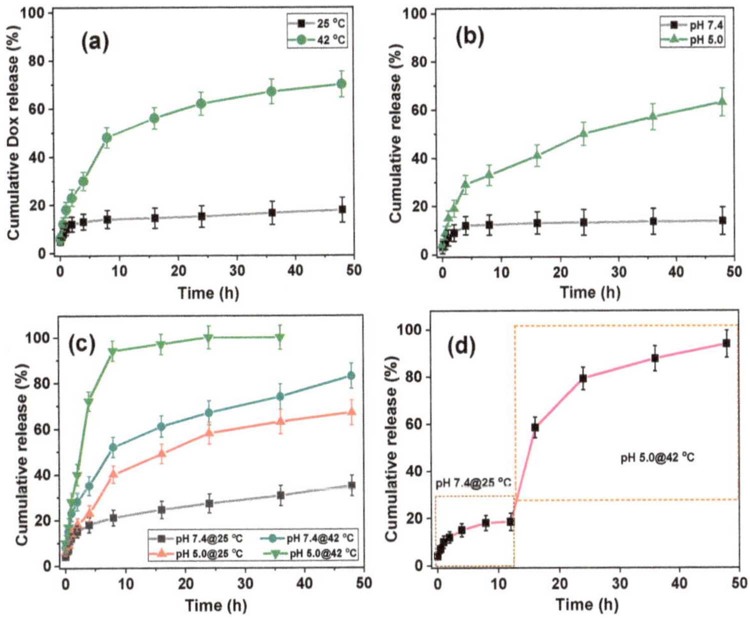

Figure 6. In vitro Dox release from the MS@Dox/PNIPAm-PAAm NPs at (**a**) different temperatures, (**b**) different pH, and (**c**) the combined pH and temperature, respectively. (**d**) pH- and temperature-dependent in vitro Dox release at the combined pH and temperature conditions.

In contrast, due to the temperature-responsive shrinkage of PNIPAm segments at 42 °C, the Dox release was increased to ~69.8% even at pH 7.4 (Table 1). As a result, the pore channels were opened, and the loaded drugs began to release from the mesopore channels. As shown in Figure 6a, as the temperature increased, more drugs were released from MS@Dox/PNIPAm-PAAm NPs, indicating that the prepared MS@Dox/PNIPAm-PAAm NPs is temperature responsive. The main reason was that when the temperature fell below LCST (<32 °C), the copolymer layer on the drug carrier's surface swelled, and therefore, the Dox was blocked in the pore channel, reducing the release. When the temperature exceeded LCST (>32 °C), the surface-modified PNIPAm-PAAm polymer layer on the carrier surface collapsed, exposing the pore channel and promoting the release of large amounts of drug from MS@Dox/PNIPAM-PAAm NPs [25,37].

Table 1. Dox release efficiency from the MS@Dox/PNIPAm-PAAm system at different temperatures.

Sample	LCST (°C)	Release Efficiency (%)	
		25 °C	42 °C
MS@Dox/PNIPAM-PAAm	~40	17.6	64.8

3.3. Stimuli-Responsive Dox Release from MS@Dox/PNIPAm-PAAm NPs at Different pH Conditions

Drug release behavior of the MS@Dox/PNIPAm-PAAm NPs was accessed under different pH (pH 7.4 and 5.0) stimuli conditions, respectively. Figure 6b shows the cumulative Dox release at pH 7.4 was 18.5% after 48 h, while at pH 5.0, the cumulative release was increased to ~63.5% after 48 h. At pH 7.4, only a small amount of Dox (about ~18.5%) was released during a 48-h period, indicating that surface-coated PNIPAm-PAAm copolymers retain the drug molecules inside mesoporous carriers due to the existence of H-bonding interactions between the –C=O and -NH groups of PNIPAm and PAAm, resulting in less drug release. These phenomena could be explained by using zeta potential measurements. As shown in Figure 3B, at pH 7.4, the MS@PNIPAm-PAAm NPs had a higher negative charge, indicating that the loaded Dox molecules interacted significantly with drug binding sites such as hydroxyl and amine sites. As a result, it demonstrated a significantly lower level of Dox release induced by the release of physisorbed drug molecules. However, when the pH was decreased to pH 5.0, the -NH and -OH groups, as well as the surface hydroxyl groups, became protonated under the reduced pH (pH 5.0) conditions. According to the zeta potential measurements, the MS@PNIPAM-PAAm NPs became more positively charged under the acidic pH conditions (pH 5.0), and thus the amine groups in the PNIPAm and PAAm copolymer and hydroxyl groups in silica surfaces were protonated to -NH$^+$ and -OH$^+$, resulting in a higher cumulative release of Dox at pH 5.0 than the release of Dox at pH 6.5 due to an increased electrostatic repulsive force between protonated Dox molecules and the protonated drug binding sites. With decreasing pH, the protonation degree of -NH$^+$ increased, resulting in greater electrostatic repulsion between the copolymer chains. As a result, after 48 h, the cumulative Dox drug release at pH 5.0 was ~63.5% and about 3.5 times higher than at physiological pH (7.4), confirming the MS@Dox/PNIPAm-PAAm NP's excellent pH responsiveness (Figure 6b and Table 2) [38].

Table 2. Dox release efficiency from the MS@Dox/PNIPAm-PAAm system at different pH values.

pH	Release Efficiency (%)
7.4	18.5
6.5	52.4
5.0	63.5
7.4/42 °C	67
5.0/42 °C	100

3.4. Combined pH and Temperature-Responsive Dox Release from MS@Dox/PNIPAm-PAAm NPs

The Dox release from the MS@Dox/PNIPAm-PAAm NPs at pH 7.4/25 °C, 7.4/42 °C, and 5.0/25 °C and 5.0/42 °C, respectively, are shown in Figure 6c. At pH 7.4/25 °C, approximately ~35% of the Dox was released, and at pH 7.4/42 °C, Dox release increased to about ~81.5% as shown in Figure 6c. In contrast, at lower pH such as pH 5.0/25 °C and pH 5.0/42 °C, the MS@Dox/PNIPAm-PAAm system released ~67% and complete release of Dox, respectively (Figure 6c and Table 2). In comparison to Dox release only under acidic pH or temperature circumstances, the proportion of Dox released under the combined acidic pH and temperature (42 °C) stimulus conditions was significantly higher. This suggests that the combination of pH and temperature is the key driving factor in the release of drugs from the mesopore channels. Protonation of -NH groups and dissociation of H-bonding interactions between PNIPAm-PAAm copolymer chains can be caused by acidic pH stimuli. Furthermore, at temperatures above LCST, the PNIPAm segments shorten and open pores, and the loaded drugs are released quickly from the mesopore channels. As a result, Dox release was significantly higher at pH 5.0/42 °C than at either pH or temperature alone [39]. In this study, it was found that at pH 7.4, only ~18.5% of the drug was released in 24 h, implying that the loaded drug was firmly confined inside the mesopores under these pH 7.4 conditions. However, at higher temperatures (25 °C and 42 °C), around ~17.6 and ~64.8% were released in 24 h due to slow dissolution and diffusion out of the mesopore channels, respectively. The drug carrier system requires additional time to achieve complete release (over 24 h). In contrast, when the acidic pH (pH 5.0) and temperature (42 °C) stimuli were combined, an enhanced drug release and approximately complete drug release were achieved in 24 h. This might be attributed to the acidic pH-induced electrostatic repulsion as well as the temperature-induced drug solubility and fast kinetics, which all play important roles in boosting drug release from the drug carrier system, and therefore nearly complete release was achieved in 24 h.

The release of Dox can be studied by altering the condition from pH 7.4/25 °C to pH 5.0/42 °C to access the MS@Dox/PNIPAm-PAAm NPs. For the first 12 h, the pH was kept at 7.4 at 25 °C, and the Dox release was studied. Then, the pH and temperature were changed to pH 5.0 after 12 h, and the released drug was measured for 48 h. Only a small percentage (~19.5%) of Dox was released over the first 12 h, as shown in Figure 6d. When the pH was raised to 5.0/42 °C, however, it resulted in a ~95% increase in Dox release (Figure 6d). In this study, a slightly increased ~32% of Dox release was observed at pH 5.0 at 42 °C compared to only at pH 5.0 (Table 2). The combined pH and temperature stimulation has been shown to significantly increase drug release behavior, and therefore the obtained results evidence the dual pH and temperature responsiveness of the prepared MS@PNIPAm-PAAm NPs system [40].

3.5. Cytotoxicity of MS@PNIPAm-PAAm NPs

Biocompatible polymers or functional groups of modified silica materials showed considerable cytocompatibility [41,42]. To determine the in vitro biocompatibility, the PNIPAm-PAAm copolymer modified MS@PNIPAm-PAAm NPs were tested at the sample doses of 0, 25, 50, 100, 150, and 200 µg/mL, respectively, without or with Dox loading using MDA-MB-231 cells. The cell viability of MS@PNIPAm-PAAm NPs without Dox loading had ~90% cell viability in all the tested concentrations, as shown in Figure 7A, indicating that MS@PNIPAm-PAAm NPs have favorable biocompatibility. Further, the cytotoxicity of free Dox and Dox-loaded MS@Dox/PNIPAm-PAAm NPs, respectively, was investigated. All of the samples displayed concentration-dependent cell toxicity to MDA-MB-231 cells, as shown in Figure 7B. Specifically, the Dox-loaded MS@Dox/PNIPAm-PAAm NPs treated cells showed considerably enhanced cell toxicity of ~>95% to the MDA-MB-231 cells treated with the MS@Dox/PNIPAm-PAAm NPs concentration of about ~200 µg/mL after being cultured for 24 h at 37 °C (Figure 7B(b)).

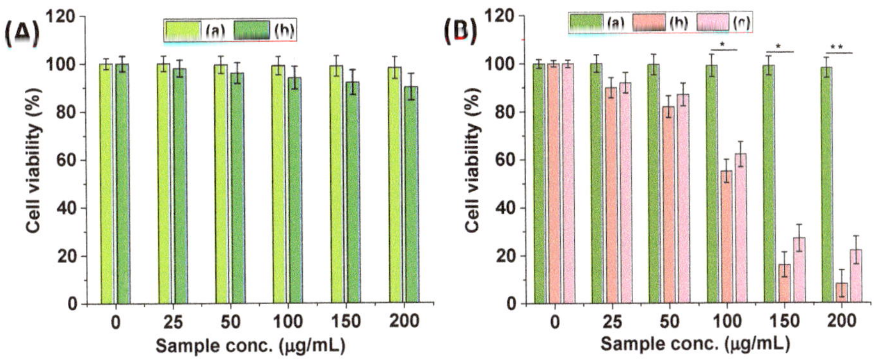

Figure 7. (**A**) In vitro biocompatibility of (a) control MDA-MB-231 cells; (b) MS@PNIPAm-PAAm NPs without Dox loading, at different concentrations. (**B**) In vitro cytotoxicity of (a) control MDA-MB-231 cells, (b) MS@Dox/PNIPAm-PAAm NPs, and (c) a pure Dox sample, respectively, at different sample concentrations. Statistical significance to the cell toxicity with different samples (*, significant $p < 0.05$; **, highly significant $p < 0.01$).

As shown in Figure 7B(b), there was a significant increase in toxicity at the sample concentration of 200 µg/mL as compared to the sample concentration of 25 µg/mL or

50 µg/mL, indicating that more drug molecules were released from the MS@Dox/PNIPAm-PAAm NPs system due to the increased cellular uptake of the nanoparticles and the intracellular pH conditions allowing the mesopores to open and allow the drugs to be released from the mesoporous silica carriers. MS@Dox/PNIPAm-PAAm NPs showed considerably more toxicity to the MDA-MB-231 cells than the cells treated with the same concentration of free DOX (Figure 7B(c)). This study's results indicate that the prepared MS@PNIPAm-PAAm NPs are biocompatible, and Dox-loaded MS@Dox/PNIPAm-PAAm NPs are more easily endocytosed by cells, resulting in increased intracellular Dox concentration and increased tumor cell killing efficiency.

3.6. Cell Uptake Study

To assess intracellular uptake, MS@Dox/PNIPAm-PAAm NPs were treated for 5 h with MDA-MB-231 cells at a sample dose of 10 µg/mL at 37 °C. Figure 8 shows the cellular uptake and consequent internalization of the MS@Dox/PNIPAm-PAAm NPs into the MDA-MB-231 cells are observed by the Dox fluorescence, demonstrating that the MS@Dox/PNIPAm-PAAm NPs can be engulfed by MDA-MB-231 cells in a short incubation time (Figure 8b,c). This could be due to the surface-coated PNIPAm-PAAm copolymer, which promotes cellular internalization of MS@Dox/PNIPAm-PAAm NPs by MDA-MB-231 cells and thus facilitates drug release in a controlled manner (Figure 8b,c). The intracellular uptake behavior of the MDA-MB-231 cells was also examined at different time (1, 2, 4, and 8 h) intervals. According to Figure 8d, increased fluorescence intensity of Dox in the cells with respect to time indicates effective cellular internalization of MS@Dox/PNIPAm-PAAm NPs inside the cells via the intracellular internalization process [43–45]. The fluorescence image results provide direct evidence for the pH-responsive delivery performance of Dox from the MS@Dox/PNIPAm-PAAm NPs.

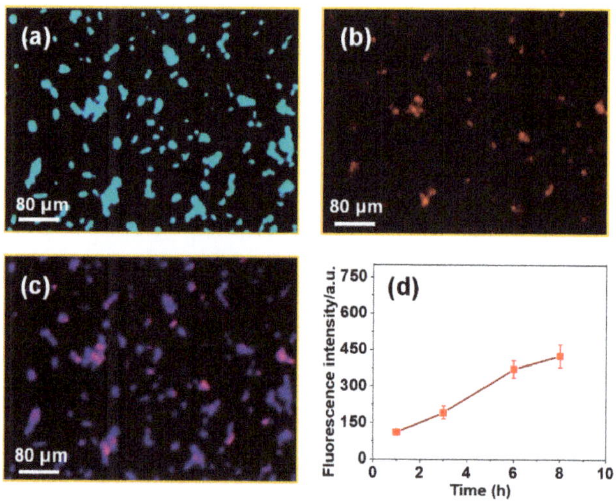

Figure 8. Fluorescence microscopy images of MDA-MB-231 cells treated with (**a**) a nucleus stain; (**b**) MS@/Dox/PNIPAm-PAAm NPs treated cells; and (**c**) merged images of MS@/Dox/PNIPAm-PAAm NPs treated cells. The blue signal is given by the nuclei, and the red signal is given by Dox molecules present in the MS@Dox/PNIPAm-PAAm NPs. Scale bar = 80 µm. (**d**) Fluorescence intensity of released Dox at different times.

4. Conclusions

In this study, PNIPAm-PAAm copolymer conjugated mesoporous silica MS@PNIPAm-PAAm NPs were synthesized for pH- and temperature-responsive in vitro drug delivery applications. The surface-conjugated PNIPAm-PAAm copolymer acts as a gatekeeper to prevent the drugs from impulsive release under physiological conditions, while the mesoporous silica serves as a drug reservoir. The MS@Dox/PNIPAm-PAAm NPs demonstrated improved drug release performance under pH and temperature stimuli due to the pH-responsive collapse and temperature-responsive shrinkage of the surface-conjugated PNIPAm-PAAm copolymer above the LCST. The MS@PNIPAm-PAAm system is biocompatible and readily engulfed by MDA-MB-231 cells, according to the in vitro biocompatibility (MTT assay) results. All the experimental results support the idea that the MS@PNIPAm-PAAm NPs could be used for pH- and temperature-responsive drug delivery that requires the target sites to be exposed to sustained drug release at higher temperatures.

Author Contributions: Conceptualization, methodology, and formal analysis, K.T. and M.S.; validation, S.U. and S.R.; data curation, K.T.; writing—original draft preparation, K.T., M.S., T.T.V.P. and S.R.; writing—review and editing; supervision, project administration, funding acquisition, T.-J.N. and S.-C.K. All authors have read and agreed to the published version of the manuscript.

Funding: This research was supported by the Basic Science Research Program through the National Research Foundation of Korea (NRF), funded by the Ministry of Education (2020R1I1A3052258). In addition, the work was also, supported by the Technology Development Program (S3060516) funded by the Ministry of SMEs and Startups (MSS, Republic of Korea) in 2021.

Institutional Review Board Statement: Not applicable.

Informed Consent Statement: Not applicable.

Data Availability Statement: Not applicable.

Conflicts of Interest: The authors have no conflicts of interest to declare.

References

1. Sonnenschein, C.; Soto, A.M. Theories of carcinogenesis: An emerging perspective. *Semin. Cancer Biol.* **2008**, *18*, 372–377. [CrossRef] [PubMed]
2. Parsa, N. Environmental Factors Inducing Human Cancers. *Iran. J. Public Health* **2012**, *41*, 1–9. [PubMed]
3. Bagot, R.C.; Meaney, M.J. Epigenetics and the Biological Basis of Gene × Environment Interactions. *J. Am. Acad. Child Adolesc. Psychiatry* **2010**, *49*, 752–771. [CrossRef]
4. Arruebo, M.; Vilaboa, N.; Sáez-Gutierrez, B.; Lambea, J.; Tres, A.; Valladares, M.; González-Fernández, Á. Assessment of the Evolution of Cancer Treatment Therapies. *Cancers* **2011**, *3*, 3279–3330. [CrossRef]
5. Akbarian, M.; Gholinejad, M.; Mohammadi-Saamni, S.; Farjadian, F. Theranostic mesoporous silica nanoparticles made of multi-nuclear gold or carbon quantum dots particles serving as pH responsive drug delivery system. *Microporous Mesoporous Mater.* **2022**, *329*, 111512. [CrossRef]
6. Kong, J.; Park, S.S.; Ha, C.S. pH-Sensitive Polyacrylic Acid-Gated Mesoporous Silica Nanocarrier Incorporated with Calcium Ions for Controlled Drug Release. *Materials* **2022**, *15*, 5926. [CrossRef] [PubMed]
7. Li, Z.; Yang, Y.; Wei, H.; Shan, X.; Wang, X.; Ou, M.; Liu, Q.; Gao, N.; Chen, H.; Mei, L.; et al. Charge-reversal biodegradable MSNs for tumor synergetic chemo/photothermal and visualized therapy. *J. Control. Release* **2021**, *338*, 719–730. [CrossRef] [PubMed]
8. Huang, P.; Lian, D.; Ma, H.; Gao, N.; Zhao, L.; Luan, P.; Zeng, X. New advances in gated materials of mesoporous silica for drug controlled release. *Chin. Chem. Lett.* **2021**, *32*, 3696. [CrossRef]
9. Taghizadeh, B.; Taranejoo, S.; Monemian, S.A.; Moghaddam, Z.S.; Daliri, K.; Derakhshankhah, H.; Derakhshani, Z. Classification of stimuli-responsive polymers as anticancer drug delivery systems. *Drug Deliv.* **2015**, *22*, 145–155. [CrossRef]
10. Sheng, Y.; Hu, J.; Shi, J.; Lee, L.J. Stimuli-responsive Carriers for Controlled Intracellular Drug Release. *Curr. Med. Chem.* **2019**, *26*, 2377–2388. [CrossRef]
11. Zhang, P.; Li, M.; Xiao, C.; Chen, X. Stimuli-responsive polypeptides for controlled drug delivery. *Chem. Commun.* **2021**, *57*, 9489–9503. [CrossRef] [PubMed]
12. Mi, P. Stimuli-responsive nanocarriers for drug delivery, tumor imaging, therapy and theranostics. *Theranostics* **2020**, *10*, 4557–4588. [CrossRef]
13. Zhao, J.; Wang, M.; Ying, H.; Su, D.; Zhang, H.; Lu, G.; Chen, J. Extracellular Matrix Component Shelled Nanoparticles as Dual Enzyme-Responsive Drug Delivery Vehicles for Cancer Therapy. *ACS Biomater. Sci. Eng.* **2018**, *4*, 2404–2411. [CrossRef]
14. Qian, B.; Zhao, Q.; Ye, X. Ultrasound and Magnetic Responsive Drug Delivery Systems for Cardiovascular Application. *J. Cardiovasc. Pharm.* **2020**, *76*, 414–425. [CrossRef] [PubMed]
15. Nastyshyn, S.; Stetsyshyn, Y.; Raczkowska, J.; Nastishin, Y.; Melnyk, Y.; Panchenko, Y.; Budkowski, A. Temperature-Responsive Polymer Brush Coatings for Advanced Biomedical Applications. *Polymers* **2022**, *14*, 4245. [CrossRef]

16. Stetsyshyn, Y.; Raczkowska, J.; Harhay, K.; Gajos, K.; Melnyk, Y.; Dabczynski, P.; Shevtsova, T.; Budkowski, A. Temperature-responsive and multi-responsive grafted polymer brushes with transitions based on critical solution temperature: Synthesis, properties, and applications. *Colloid Polym. Sci.* **2021**, *299*, 363. [CrossRef]
17. Zhang, J.; Yuan, Z.-F.; Wang, Y.; Chen, W.H.; Luo, G.-F.; Cheng, S.-X.; Zhuo, R.-X.; Zhang, X.-Z. Multifunctional Envelope-Type Mesoporous Silica Nanoparticles for Tumor-Triggered Targeting Drug Delivery. *J. Am. Chem. Soc.* **2013**, *135*, 5068–5073. [CrossRef] [PubMed]
18. LaBauve, A.E.; Rinker, T.E.; Noureddine, A.; Serda, R.E.; Howe, J.Y.; Sherman, M.B.; Rasley, A.; Brinker, C.J.; Sasaki, D.Y.; Negrete, O.A. Lipid-Coated Mesoporous Silica Nanoparticles for the Delivery of the ML336 Antiviral to Inhibit Encephalitic Alphavirus Infection. *Sci. Rep.* **2018**, *8*, 13990. [CrossRef] [PubMed]
19. Hajebi, S.; Abdollahi, A.; Roghani-Mamaqani, H.; Salami-Kalajahi, M. Temperature-Responsive Poly(N-Isopropylacrylamide) Nanogels: The Role of Hollow Cavities and Different Shell Cross-Linking Densities on Doxorubicin Loading and Release. *Langmuir* **2020**, *36*, 2683–2694. [CrossRef] [PubMed]
20. Sánchez-Moreno, P.; de Vicente, J.; Nardecchia, S.; Marchal, J.A.; Boulaiz, H. Thermo-Sensitive Nanomaterials: Recent Advance in Synthesis and Biomedical Applications. *Nanomaterials* **2018**, *8*, 935. [CrossRef]
21. Lanzalaco, S.; Armelin, E. Poly(N-isopropylacrylamide) and Copolymers: A Review on Recent Progresses in Biomedical Applications. *Gels* **2017**, *3*, 36. [CrossRef]
22. Fernández, E.; López, D.; López-Cabarcos, E.; Mijangos, C. Viscoelastic and swelling properties of glSucose oxidase loaded polyacrylamide hydrogels and the evaluation of their properties as glucose sensors. *Polymer* **2005**, *46*, 2211–2217. [CrossRef]
23. Rosiak, J.; Burozak, K.; Pękala, W. Polyacrylamide hydrogels as sustained release drug delivery dressing materials. *Radiat. Phys. Chem.* **1977**, *22*, 907–915. [CrossRef]
24. Begines, B.; Ortiz, T.; Pérez-Aranda, M.; Martínez, G.; Merinero, M.; Argüelles-Arias, F.; Alcudia, A. Polymeric Nanoparticles for Drug Delivery: Recent Developments and Future Prospects. *Nanomaterials* **2020**, *10*, 1403. [CrossRef] [PubMed]
25. Zhang, Y.; Ang, C.Y.; Li, M.; Tan, S.Y.; Qu, Q.; Luo, Z.; Zhao, Y. Polymer-Coated Hollow Mesoporous Silica Nanoparticles for Triple-Responsive Drug Delivery. *ACS Appl. Mater. Interfaces* **2015**, *7*, 18179–18187. [CrossRef]
26. Popova, M.; Trendafilova, I.; Zgureva, D.; Kalvachev, Y.; Boycheva, S.; Novak, N.; Szegedi, T.A. Polymer-coated mesoporous silica nanoparticles for controlled release of the prodrug sulfasalazine. *J. Drug Deliv. Sci. Technol.* **2018**, *44*, 415–420. [CrossRef]
27. Xing, L.; Zheng, H.; Cao, Y.; Che, S. Coordination Polymer Coated Mesoporous Silica Nanoparticles for pH-Responsive Drug Release. *Adv. Mater.* **2012**, *24*, 6433–6437. [CrossRef] [PubMed]
28. Feng, C.; Shen, Z.; Li, Y.; Gu, L.; Zhang, Y.; Lu, G.; Huang, X. PNIPAM-b-(PEA-g-PDMAEA) double-hydrophilic graft copolymer: Synthesis and its application for preparation of gold nanoparticles in aqueous media. *J. Polym. Chem. Part A* **2009**, *47*, 1811–1824. [CrossRef]
29. Feng, C.; Shen, Z.; Gu, L.; Zhang, S.; Li, L.; Lu, G.; Huang, X. Synthesis and characterization of PNIPAM-b-(PEA-g-PDEA) double hydrophilic graft copolymer. *J. Polym. Chem. Part A* **2008**, *15*, 5638–5651. [CrossRef]
30. Tang, F.; Li, L.; Chen, D. Mesoporous silica nanoparticles: Synthesis, biocompatibility and drug delivery. *Adv. Mater.* **2012**, *24*, 1504–1534. [CrossRef] [PubMed]
31. Mousavi, M.; Fini, E. Silanization Mechanism of Silica Nanoparticles in Bitumen Using 3-Aminopropyl Triethoxysilane (APTES) and 3-Glycidyloxypropyl Trimethoxysilane (GPTMS). ACS Sustain. *Chem. Eng.* **2020**, *8*, 3231–3240.
32. Ghaedi, H.; Zhao, M. Review on Template Removal Techniques for Synthesis of Mesoporous Silica Materials. *Energy Fuels* **2022**, *36*, 2424–2446. [CrossRef]
33. Chang, K.-C.; Lin, C.-Y.; Lin, H.-F.; Chiou, S.-C.; Huang, W.-C.; Yeh, J.-M.; Yang, J.-C. Thermally and mechanically enhanced epoxy resin-silica hybrid materials containing primary amine-modified silica nanoparticles. *J. Appl. Polym. Sci.* **2008**, *108*, 1629–1635. [CrossRef]
34. Wach, A.; Drozdek, M.; Dudek, B.; Biazik, M.; Łątka, P.; Michalik, M.; Kustrowski, P. Differences in Catalytic Activity of Poly(vinylamine) Introduced on Surface of Mesoporous SBA-15 by Grafting from and Grafting onto Methods in Knoevenagel Condensation. *J. Phys. Chem. C* **2015**, *119*, 19954–19966. [CrossRef]
35. Yegane, M.M.; Hashemi, F.; Vercauteren, F.; Meulendijks, N.; Gharbi, R.; Boukany, P.E.; Zitha, P. Rheological response of a modified polyacrylamide–silica nanoparticles hybrid at high salinity and temperature. *Soft Matter* **2020**, *16*, 10198–10210. [CrossRef] [PubMed]
36. Moorthy, M.S.; Subramanian, B.; Panchanathan, M.; Mondal, S.; Kim, H.; Lee, K.D.; Oh, J. Fucoidan-coated core–shell magnetic mesoporous silica nanoparticles for chemotherapy and magnetic hyperthermia-based thermal therapy applications. *New J. Chem.* **2017**, *41*, 15334–15346. [CrossRef]
37. Peralta, M.E.; Jadhav, S.A.; Magnacc, G.; Scalarone, D.; Mártire, D.O.; Carlos, M.E.P.L. Synthesis and in vitro testing of thermoresponsive polymer-grafted core-shell magnetic mesoporous silica nanoparticles for efficient controlled and targeted drug delivery. *J. Colloid Interface Sci.* **2019**, *544*, 198–205. [CrossRef]
38. Porrang, S.; Rahemia, N.; Davaran, S.; Mahdavi, M.; Hassanzadeh, B. Synthesis of temperature/pH dual-responsive mesoporous silica nanoparticles by surface modification and radical polymerization for anti-cancer drug delivery. Colloid. Surf. A: Physicochem. *Eng. Asp.* **2021**, *623*, 126719–127630. [CrossRef]

39. Jin, X.; Wang, Q.; Sun, J.; Panezai, H.; Bai, S.; Wu, X. Dual (pH- and temperature-) stimuli responsive nanocarrier with bimodal mesoporous silica nanoparticles core and copolymer shell for controlled ibuprofen-releasing: Fractal feature and diffusion mechanism. *Microporous Mesoporous Mater.* **2017**, *254*, 77–85. [CrossRef]
40. Wu, X.; Wang, Z.; Zhu, D.; Zong, S.; Yang, L.; Zhong, Y.; Cui, Y. pH and Thermo Dual-Stimuli-Responsive Drug Carrier Based on Mesoporous Silica Nanoparticles Encapsulated in a Copolymer–Lipid Bilayer. *ACS Appl. Mater. Interfaces* **2013**, *5*, 10895–10903. [CrossRef]
41. Ravi Kumar, M.N.; Gameti, M.; Mohapatra, S.S. Cationic Silica Nanoparticles as Gene Carriers: Synthesis, Characterization and Transfection Efficiency In vitro and In vivo. *J. Nanosci. Nanotechnol.* **2004**, *4*, 876–881. [CrossRef] [PubMed]
42. Park, S.S.; Moorthy, M.S.; Ha, C.-S. Periodic mesoporous organosilicas for advanced applications. *NPG Asia Mater.* **2014**, *6*, e96. [CrossRef]
43. Moorthy, M.S.; Park, J.-H.; Bae, J.-H.; Kim, S.-H.; Ha, C.-S. Mesoporous organosilica hybrids with a tunable amphoteric framework for controlled drug delivery. *J. Mater. Chem. B* **2014**, *2*, 6487–6499. [CrossRef] [PubMed]
44. Moorthy, M.S.; Bae, J.-H.; Kim, M.-J.; Kim, S.-H.; Ha, C.-S. Design of a Novel Mesoporous Organosilica Hybrid Microcarrier: A pH Stimuli-Responsive Dual-Drug-Delivery Vehicle for Intracellular Delivery of Anticancer Agents. *Part. Part. Syst. Charact.* **2013**, *30*, 1044–1055. [CrossRef]
45. Moorthy, M.S.; Thirupathi, K.; Periyasamy, T.; Ramkumar, V.; Kim, S.C. Ethidium bromide-bridged mesoporous silica hybrid nanocarriers for fluorescence cell imaging and drug delivery applications. *New J. Chem.* **2021**, *45*, 20641–20648.

pharmaceutics

MDPI

Review

Recent Advances in Polymer-Based Nanomaterials for Non-Invasive Photothermal Therapy of Arthritis

Muktika Tekade [1,*], Prashant Pingale [2], Rachna Gupta [3], Bhakti Pawar [3], Rakesh Kumar Tekade [3,*] and Mukesh Chandra Sharma [1]

[1] School of Pharmacy, Devi Ahilya Vishwavidyalaya, Takshila Campus, Khandwa Road, Indore 452001, Madhya Pradesh, India
[2] Department of Pharmaceutics, Sir Dr. M.S. Gosavi College of Pharmaceutical Education and Research, Nashik 422005, Maharashtra, India
[3] National Institute of Pharmaceutical Education and Research (NIPER), Ahmedabad, Palaj, Opp. Air Force Station, Gandhinagar 382355, Gujarat, India
* Correspondence: muktikarakeshtekade@gmail.com (M.T.); rakeshtekade@gmail.com (R.K.T.)

Abstract: To date, nanomaterials have been widely used for the treatment and diagnosis of rheumatoid arthritis. Amongst various nanomaterials, polymer-based nanomaterials are becoming increasingly popular in nanomedicine due to their functionalised fabrication and easy synthesis, making them biocompatible, cost-effective, biodegradable, and efficient nanocarriers for the delivery of drugs to a specific target cell. They act as photothermal reagents with high absorption in the near-infrared region that can transform near-infrared light into localised heat with fewer side effects, provide easier integration with existing therapies, and offer increased effectiveness. They have been combined with photothermal therapy to understand the chemical and physical activities behind the stimuli-responsiveness of polymer nanomaterials. In this review article, we provide detailed information regarding the recent advances in polymer nanomaterials for the non-invasive photothermal treatment of arthritis. The synergistic effect of polymer nanomaterials and photothermal therapy has enhanced the treatment and diagnosis of arthritis and reduced the side effects of drugs in the joint cavity. In addition, further novel challenges and future perspectives must be resolved to advance polymer nanomaterials for the photothermal therapy of arthritis.

Keywords: polymer; nanomaterials; photothermal; arthritis; near infra-red; inflammation; drugs

Citation: Tekade, M.; Pingale, P.; Gupta, R.; Pawar, B.; Tekade, R.K.; Sharma, M.C. Recent Advances in Polymer-Based Nanomaterials for Non-Invasive Photothermal Therapy of Arthritis. *Pharmaceutics* **2023**, *15*, 735. https://doi.org/10.3390/pharmaceutics15030735

Academic Editors: Avi Domb, Manivasagan Panchanathan and Eue-Soon Jang

Received: 7 December 2022
Revised: 25 January 2023
Accepted: 8 February 2023
Published: 22 February 2023

1. Introduction

Rheumatoid arthritis (RA) is a chronic autoimmune condition that is correlated with a high death rate and affects nearly 1% of the population worldwide. It is distinguished by the infiltration of inflammatory cells, synovial hyperplasia, pannus development, persistent cartilage, and bone deterioration [1]. The inflammatory response, predominantly in RA, deals with immune cells such as macrophages, neutrophils, and T cells, which secrete a large number of pro-inflammatory cytokines and also generate reactive oxygen species (ROS) that cause structural damage to cartilage and weaken the functional ability in the arthritic area (Figure 1). Due to the breakdown of bone and cartilage, joint swelling, and discomfort, this chronic autoimmune inflammatory disease limits movement and lowers quality of life. The relevance of RA diagnosis and therapy is critical because obtaining an accurate, early diagnosis of RA can be challenging, while the lack thereof can be detrimental to RA treatment. Many disease-modifying anti-rheumatic drugs (DMARDs) such as methotrexate (MTX), dexamethasone (Dex), tocilizumab (TCZ), and others have been used to reduce joint inflammation and deformity. Even so, these drugs have not been proven helpful for patients due to their adverse effects following long-term use [2].

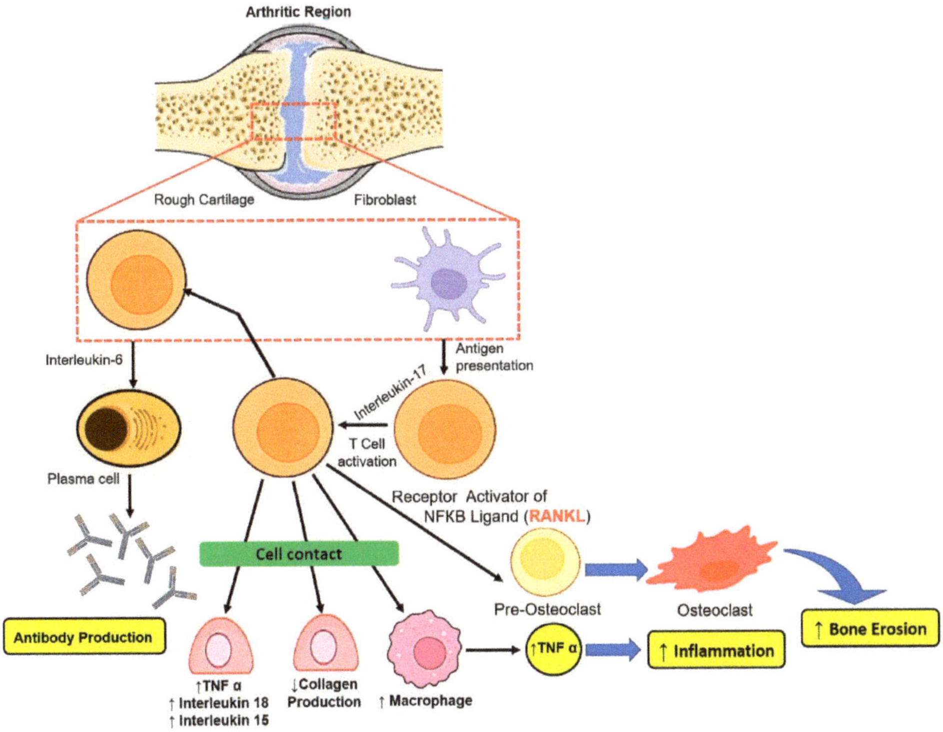

Figure 1. Pathophysiology of Arthritis.

The quality of life for many patients has increased thanks to recent developments and novel treatment modalities that have significantly weakened disease progression [3]. The advancement of nanotechnology opens fresh perspectives for the study and treatment of RA. Using nanomaterials (NMs) as nanocarriers for drugs has enhanced their bioavailability and bioactivities and upgraded the selective targeting of inflamed joints. In addition, NMs' small size and considerable surface activity have enabled the controlled release of drugs that reduces their dose-dependent side effects. Combining diagnosis and treatment is the most effective strategy for managing the condition of RA patients. Recent research has demonstrated improved arthritis treatment via the employment of the combined effect of nanomaterial and photothermal therapy (PTT). Among various NMs, polymer-based NMs act as vital photothermal reagents with high absorption in the near-infrared (NIR) region that can transform near-infrared light into localised heat, thus offering fewer side effects, easier integration with existing therapies, and increased effectiveness [4]. NMs with PTT can easily penetrate joint tissues without causing damage and facilitate imaging for an arthritis diagnosis. Recently, manifold works have been carried out that have revealed the efficiency of the combinatory effect of polymer-based NMs and PTT with respect to the treatment and diagnosis of RA [5]. This review article will discuss the advancement of polymer-based NMs for non-invasive PTT in arthritis.

Healthy bone is made up of two neighbouring bones that are covered in cartilage. The articular cavity is the area between the ends that carries synovial fluid and is delineated by the synovial membrane on both sides. The synovial membrane is a thin layer of cells made up of two types of synovial cells: type A, which resemble macrophages, and type B, which resemble fibroblasts. Due to the porous structure of the synovial membrane, nutrients in the serum can diffuse cartilage while also producing synovial fluid. The development of rheumatoid arthritis and immunological reactions are complicated and involve interaction between elements of innate and adaptive immune systems. Numerous inflammatory cells are present in the synovial membrane, where they interact to promote joint degeneration [6]. The existence of dendritic cells, a significant class of antigen-presenting cells (APC) that produces several cytokines, suggests the significance of the adaptive immune pathway with respect to RA. The mechanism of T-cell stimulation includes dendritic cells, which also provide antigens to T cells already existing in the synovium [7]. There are two signals needed for T-cell activation: the presentation of an antigen to the T-cell receptor is the preliminary signal, while the CD80/86 surface protein present on the dendritic cell must connect with the CD28 protein of the T cell for the second signal, which is known as the costimulatory signal [8]. The activation of T-cells cause macrophages and dendritic cells to secrete different TGF-β and interleukins (IL-1,6,17, 21, and 23) and reduces the generation of regulatory T-cells, hence leading to inflammation in the synovial fluid. In addition to this, the activation of T-cells also causes the release of IL-17, which stimulates fibroblasts and macrophages such as synovial cells for the activation of IL-26, in turn leading to the further production of IL-6, TNF-α, and IL-1β, thereby causing inflammation [9].

Another mechanism involved in RA is an antigen-nonspecific pathway that requires cell-to-cell interaction between fibroblasts and macrophages with activated T-cells [10]. Adaptive humoral immunity is also a significant component of the aetiology of RA. Multiple potential strategies may be used to mediate the role of B cells in autoimmune disorders. Additionally, B cells can produce anti- and pro-inflammatory cytokines, which may be involved in the generation of RA [11]. An overview of the abovementioned arthritis pathogenesis is provided in Figure 1.

2. Fundamentals of Photothermal Therapy

PTT uses the near infra-red wavelength (NIR; 780 nm to 2500 nm) of the electromagnetic spectrum to treat diseases such as cancers and arthritis. Upon exposure to NIR, the tissue temperature increases due to the conversion of light energy to heat, thereby killing cancerous cells or tissue. Three photophysical steps are involved in the conversion of NIR energy into heat. In the first and second steps, upon absorbing light, NMs convert into hot lattices, resulting in charge separation particle oscillation. Once light has been absorbed and transformed into heat, electron–photon relaxation occurs. The final step is cooling, during which heat is emitted from metal NMs into the surrounding environment. (Figure 2). Thus, such energy is utilised in arthritis treatment that causes cell demolition and apoptosis.

Owing to the small size of NMs, the restriction of electronic motion differs from that of bulk material. When light energy is incident on the NMs, the electrons on the NMs' surface become excited, and an electron cloud forms over this surface. Here, the free electrons transfer from the valence band to the conduction band, which leads to the enhancement of the resonance peak. NMs, including graphene, carbon nanotubes, and gold nanoparticles (NPs), are the subject of preclinical investigation as possible agents for the photothermal treatment of cancer, particularly tumours. However, there are still obstacles to be overcome to improve light conversion efficiency and decrease damage to healthy cells and tissues. A new class of drugs called biological response modifiers targets inflammatory mediators in the treatment of RA. At optical frequencies, metal NPs effectively scatter and absorb visible light. Recent advancements in the synthesis, conjugation, and fabrication of NPs have prompted a greater interest in employing plasmon-resonant NPs. Additionally, their optical and photonic usage for biological purposes has been developed (Tank ND, 2017). To control biological autoimmune systems and treat diseases, a novel therapeutic strategy has recently arisen due to its potential to minimize side effects.

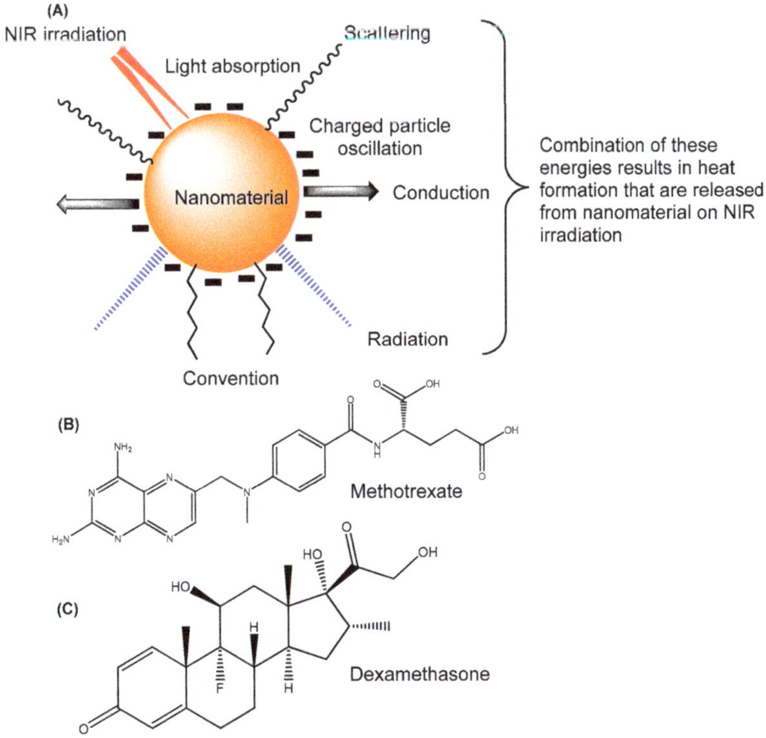

Figure 2. (**A**) Photothermal conversion of nanomaterial. In this process, heat is released in the form of radiation, scattering, charged particle oscillation, conduction, and convection, which are used for RA therapy and cause cell apoptosis; structures of (**B**) Methotrexate and (**C**) Dexamethasone.

3. Various Polymer-Based Nanomaterials for Non-Invasive Photothermal Therapy of Arthritis

3.1. Polymer-Based Nanomaterials

In recent decades, conjugated, polymer-based nanomaterials have been used in photothermal therapy [12]. Conjugated polymers possess recurring chemical units, delocalised electron systems, and conjugated bonding. These polymer-based nanomaterials (NMs) are biocompatible, cost-effective, biodegradable, possess high photothermal conversion efficiency, and are efficient nanocarriers for the delivery of drugs to a target cell. Furthermore, they can easily be surface-modified and functionalised with numerous functional groups and can encapsulate therapeutic agents. Examples of such polymers include polypyrrole, polyaniline, and polythiophene, which have been extensively used in biomedical applications.

So far, numerous switches have been used to control a drug's release from the formulation. Amongst them all, the use of a light-triggered sequential release system stands out from the rest of the stimuli. The combination of the use of conjugated, polymer-based nanomaterial with laser therapy can achieve controlled drug release. The release of an anti-rheumatic drug incorporated into the polymeric material upon laser irradiation is depicted in Figure 3. The combined effect of PTT and polymer NMs have improved the treatment and diagnosis of RA and reduced the side effects of drugs in the joint cavity. Various reports exhibiting the efficacy of polymeric-based NMs for RA treatment are available.

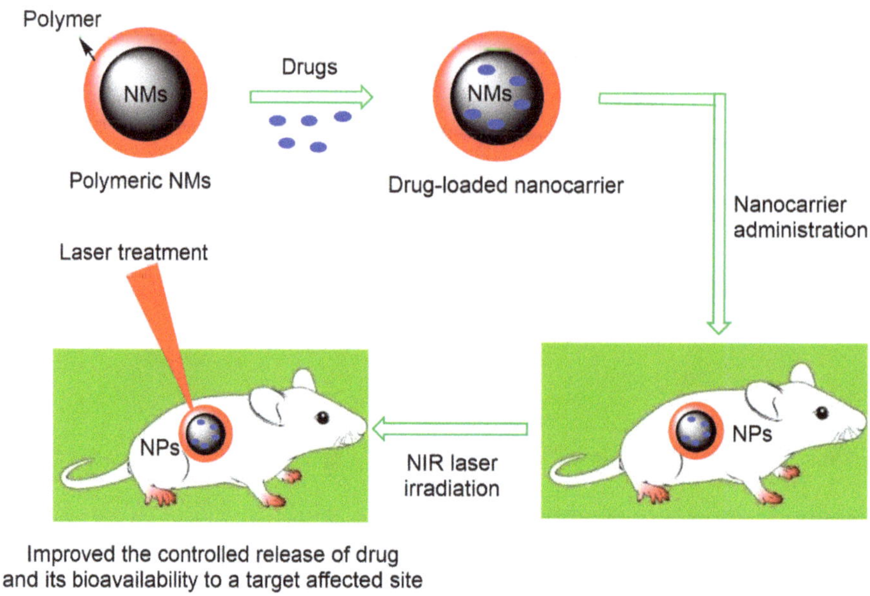

Figure 3. Scheme representing the use of polymeric nanomaterials as photothermal agents for arthritis therapy. Such polymeric nanomaterials enhanced DMARD's bioavailability and sustained release and proved to be an effective RA therapy.

For instance, polyethylene glycol (PEG), primarily, is used for the surface modification of NMs. In one study, Quindeel and co-workers prepared a polycaprolactone–polyethylene glycol–polycaprolactone (PCL-PEG-PCL) triblock copolymer via a ring-opening copolymerization reaction and utilised such polymer NMs as nano vehicles for the delivery of the drug MTX. Their results revealed the excellent efficacy of NMs with respect to an RA mouse model compared to those models of free MTX. These fabricated NMs decrease the expression of the inflammatory factors and recover the behavioural response capability of mice [13].

Another study by Chen and colleagues [14] demonstrated the efficacy of photothermally triggered nitric oxide (NO) nanogenerators via combining photothermal agents and NO within polymeric NMs. NIR light within 650–900 nm is an excellent trigger-responsive system for the sustained release of NO. The authors synthesised NO-Hb@siRNA@PLGA-PEG (NHsPP), wherein hemoglobin (Hb) acts as a carrier for NO that absorbs NIR at 650 nm, and siRNA was also used in combination for the inhibition of the inflammatory response through the Notch signalling pathway. Their results exhibited controlled NO release in RAW 265.7 cells and collagen-induced arthritis (CIA) mice. Moreover, by combining NO, siRNA, and PTT, an excellent macrophage inflammatory response was shown that could be used for other inflammatory diseases [14]. A few of the other studies are discussed in Table 1.

Another study by Lee et al. [15] revealed the effect of MTX-loaded PLGA NMs on RA treatment by conjugating them with NIR irradiation. Their nanocomposite showed an excellent therapeutic outcome compared with MTX alone in CIA mice and complete Freund's adjuvant-induced arthritic mice (Figure 4). Their results also improved the controlled release of MTX and its bioavailability to the specific arthritic site using the developed functional polymer-NIR bioconjugate nanocomposite [15].

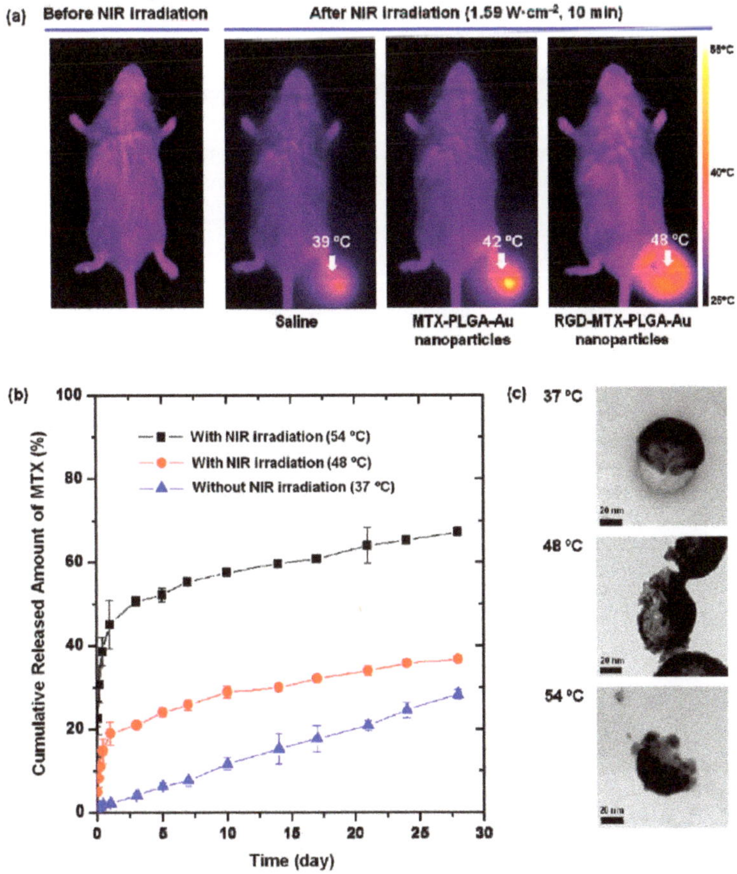

Figure 4. (**a**) Thermal images of CIA mice treated with saline, MTX-PLGA-Au nanoparticles (150 µL, 1 mg/mL dispersed in PBS), or RGD-MTX-PLGA-Au nanoparticles (150 µL, 1 mg/mL dispersed in PBS) before and after NIR exposure (1.59 W/cm^2, 10 min) of the right paw. (**b**) Profiles of MTX release from RGD-MTX-PLGA-Au nanoparticles with and without NIR irradiation of 0.38 (48 °C) or 0.53 (54 °C) W/cm^2 for 10 min at the initial time. Data represent mean values for n = 3, and the error bars represent the standard deviation of the means. (**c**) TEM images of RGD-MTX-PLGA-Au nanoparticles measured after MTX release experiments without (top) or with NIR irradiation of 0.38 (48 C, middle) or 0.53 (54 C, bottom) W/cm^2 for 10 min at the initial time. Adapted with permission from reference [15]. Copyright 2022 ACS publisher.

3.2. Polymer-NIR780 Bioconjugate

IR780, a lipophilic, cationic, heptamethine dye with an emission wavelength of 780 nm, generates heat upon its irradiation with NIR light. NIR780 is a NIR bioactive dye that is usually conjugated with a polymer for photothermal therapy [16]. PTT can target specific arthritic sites and produce heat due to interaction of such NIR active dye with light. The combinatory effect of PTT and polymer NMs has enhanced absorption in the NIR region (700–950 nm) [17]. Using a polymer–NIR 780 bioconjugate, light irradiation triggers the heat generation response of the material that serves as a mediator to activate structural polymer modifications, and this process, in turn, offers novel material functionalities such as the switching of drug delivery on/off or signals providing the basis for imaging. Together with PTT, such polymer bioconjugate materials can be used for the treatment and diagnosis of arthritis issues.

For example, in one report, researchers developed a nanogold core multifunctional dendrimer nanocomposite that was conjugated with the drug MTX, wherein the latter acts as a targeting ligand for receptors in arthritic tissues. In addition, NIR-780 was also loaded with this nanocomposite to provide PTT via light irradiation. Their results showed the sustained release of MTX and the enhanced anti-inflammatory efficacy of the nanocomposite via its combination with PTT towards mouse macrophage RAW 265.7 cells and CIA mice model. Hence, their multifunctional, targeted nanocomposite proved to be a potent therapeutic strategy for improved RA therapy [16].

3.3. Polymer-Based Gold Nanomaterials

Gold nanoparticles number among the FDA-approved metallic nanoparticles. They are widely used for diverse biomedical applications owing to the simplicity of their synthesis, their biocompatible nature, the ease with which their surfaces can be modified, and their remarkable optical activities [18]. The tuning of the morphological and physical attributes of AuNPs can be used to easily adjust their optical activities from the visible to the NIR spectrum. In the NIR region, materials exhibit low absorption due to the more profound penetration abilities of NIR light compared to visible light. However, a combination of AuNPs with PTT has improved RA therapy via reducing the levels of inflammatory cytokines [16]. Li and colleagues developed hydrogels of hyaluronic, acid-modified, Triptolide-loaded AuNPs to treat RA in a CIA mice model [19].

Triptolide (TP) is a disease-modifying anti-rheumatic drug (DMARD). Owing to the NIR irradiation of AuNPs, TP can be released to specific inflammatory joint tissues and improve the treatment of RA in vivo. Moreover, these modified materials successfully minimised the invasion and migration of RA fibroblast-like synoviocytes in vitro. NIR irradiation increased the temperature and generated heat that accelerated the release rate of TP from the hybrid hydrogels (Figure 5). This investigation suggests that the TP/Au/HA hybrid hydrogel combined with PTT is more effective in terms of RA therapy than conventional treatment with TP alone. Moreover, their results also revealed minimal dose-related side effects.

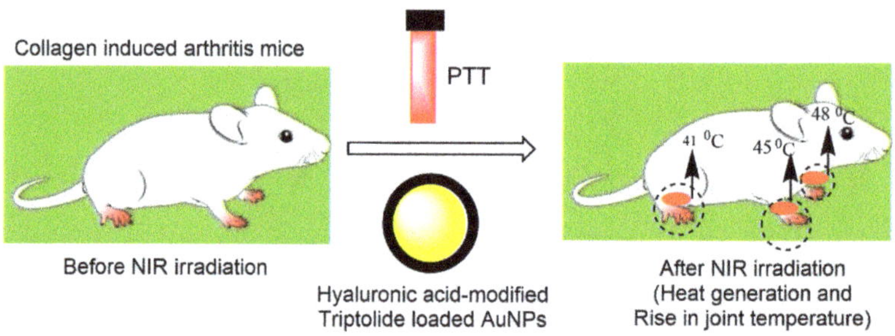

Figure 5. Indicates the synergistic activity of PTT and hyaluronic-acid-modified, Triptolide-loaded AuNPs for the treatment of CIA mice. The combined effect of PTT and drug-loaded polymer AuNPs has improved arthritis treatment via heat generation, which causes a temperature increment.

Another study demonstrated the combined effect of PTT and MTX-loaded PLGA Au/Fe/Au NPs conjugated with arginine-glycine-aspartic acid (RGD) for RA therapy applied to CIA mice [20]. This study also showed the enhanced release of the employed drug, namely, MTX, upon NIR irradiation, thus allowing the AuNPs to generate heat and offering sustained drug release at the inflammation sites of mice. In this case, Fe supports the NPs' delivery to inflamed joints by applying an external magnetic field and enhancing their residence time in the joints. Moreover, this investigation suggests that the combination of targeted PTT and an external magnetic field improves MTX's release and therapeutic effect in vivo (Table 1).

3.4. Polymer-Based Copper Nanomaterials

Like AuNPs, Cu-based NPs also play a vital role in various biomedical applications because of their superior biocompatibility, ease of synthesis, and PTT alteration efficacy. In addition, Cu aids osteogenic differentiation and bone development and improves chondrocyte proliferation [21]. Therefore, Cu is a promising candidate for repairing the RA-induced breakage of bone and cartilage.

The combination of PTT and photodynamic therapy (PDT) with $Cu_{7.2}S_4$ NPs under the influence of NIR reduced the intrusion of synovial liquid and enhanced anti-inflammatory effects in the arthritic joints of CIA rats. This also inhibited C-reactive protein and tumour necrosis factor (TNF-α). As the concentration of NPs increases, the solution temperature rises from 41 °C (125 mg/mL) to 55 °C (500 mg/mL). Singlet oxygen and reactive oxygen species generation were investigated to study their photodynamic effects.

Moreover, a comparative study between the impact of (NaCl + NIR) and ($Cu_{7.2}S_4$ NPs + NIR) on the inflamed paws of rats was reported. In this study, the researchers revealed the significantly greater therapeutic effect of ($Cu_{7.2}S_4$ NPs + NIR) compared to (NaCl + NIR). In another approach, Huang and associates designed a CuS shell and Au nanorod core (Au NR@CuS) together with PTT and PDT for RA treatment to reduce the hyperplasia of the synovium [9,22]. The localised surface plasmon resonance of both Au and CuS in the Au NR@CuS yolk-shell NPs was sufficient to enhance the efficacy of PTT, which may ultimately prove fruitful for the development of a cure for RA and the prevention of normal cells from being damaged.

Moreover, this significant core–shell structure was surface-modified with hyaluronic acid (HA) and vasoactive intestinal peptide (VIP) for the loading of MTX and its release in a controlled manner to a specific arthritic site. VIP-HA-Au NR@CuS-MTX demonstrated excellent therapeutic effects in vivo and higher NIR absorption than Au NR@CuS. Furthermore, both HA and VIP played a significant part in controlling or regulating the release of MTX. Hence, this investigation suggests that the synergistic activity of NPs and PTT improved RA treatment (Table 1).

Table 1. Polymer-based nanomaterials used in photothermal combined therapy of arthritis.

Photothermal Therapy	Objective of Investigation	Method of Synthesis	Drug Used	In Vitro Model Used	Animal Model Investigated	Outcome of Investigation	Ref
Polymer-based nanomaterials	Establish the controlled release of drug at specific arthritic sites using PCL-PEG micelles	Film dispersion	Dexamethasone	Murine macrophage Raw264.7 cell line and human umbilical vein endothelial cell	Rats with adjuvant-induced arthritis	Reduced joint swelling, bone erosion, and inflammatory cytokine expression in both joint tissue and serum	[12]
	Methotrexate-loaded PCL–PEG–PCL) triblock copolymer against RA	Copolymerization reaction and precipitation	Methotrexate	Fresh human blood and non-activated and lipopolysaccharide-activated macrophages	Mouse with RA	MTX-loaded nanomicelles proved to be a promising agent against RA	[13]
	To evaluate the influential role of PLGA NPs coated with anti-cyclooxygenase-2 (COX2) siRNA in arthritis treatment	Water-in-oil-in-water solvent evaporation technique	Dexamethasone	Human chondrocyte cell line (C28/I2)	-	Synergistic action of dexamethasone and COX-2 siRNA treatment reduced the expression of inflammatory and apoptosis-related factors produced in C28/I2 cells	[23]
	Evaluation of QRu-PLGA-RES-DS NPs) for effective arthritis therapy	Precipitation and sonication technique	Resveratrol	RAW 264.7 and HUVECs cells	CIA mice	QRu-PLGA-RES-DS NPs effectively treated RA after eliminating the inflammatory response	[24]
	NO-Hb@siRNA@PLGA-PEG (NHsPP) was utilised for osteoarthritis therapy	Minor medication, precipitation, and conjugation processes	Nitric oxide	RAW 264.7 cells and HEK 293T)	Mice	The therapeutic effect of the NHsPP NPs was significantly enhanced compared to the treatment groups using only NO, siRNA, or PTT.	[14]
Polymer-NIR780 bioconjugate	To investigate nanogold-core multifunctional dendrimer to establish photothermal therapy of RA	Citrate reduction method and conjugation process	Methotrexate	Mouse Macrophage RAW264.7 cells	-	Multifunctional targeted NPs proved to be potential therapeutics for the improved treatment of RA	[16]
Gold nanomaterial	Development of MTX-loaded poly(DL-lactic-co-glycolic acid) Au half-shell NPs (MTX-PLGA-Au) for arthritis treatment	Turkevich process for Au synthesis followed by MTX-PLGA conjugation	Methotrexate	Joint tissues were extracted for histological study	CIA Mice	This drug delivery system proved to be effective and minimised dosage-related MTX side effects in the treatment of RA	[25]
	Hyaluronate–gold nanoparticle/Tocilizumab (HA-AuNP/TCZ) complex was prepared for RA	Thiolated HA (HA-SH) was synthesised by reductive amination and conjugated with Au prepared using the Turkevich method	Tocilizumab	HUVECs cells	CIA Mice	Ha/Au/TCZ complex can be used for RA as well as other therapeutic applications	[15]

Table 1. *Cont.*

Photothermal Therapy	Objective of Investigation	Method of Synthesis	Drug Used	In Vitro Model Used	Animal Model Investigated	Outcome of Investigation	Ref
	Improvement of (MTX)-loaded PLGA) gold (Au)/iron (Fe)/gold (Au) half-shell nanoparticles conjugated with arginine–glycine–aspartic acid (RGD) for magnetic targeted chemo-photothermal treatment of RA	MTX-PLGA was prepared by a solvent evaporation method, and RGD was conjugated with MTX-PLGA Au/Fe/Au NPs	Methotrexate	-	CIA Mice	The combined effect of NIR irradiation and external magnetic field enhanced the therapeutic effects of NPs with an MTX dosage of only 0.05% dosage compared to free MTX therapy for the treatment of RA	[20]
	Hydrogels of hyaluronic acid hybridised with triptolide/gold nanoparticles for targeted delivery to rheumatoid arthritis-affected regions combined with photothermal therapy	Hyaluronic acid (HA) hydrogels loaded onto an RGD-attached gold nanoshell containing TP are prepared	Triptolide (TP)	-	CIA mice	Targeted photothermal-chemotherapy using hybrid hydrogels for the treatment of RA is an effective strategy that can maximize the therapeutic effects and reduce dose-related side effects.	[19]
Copper nanomaterial	To investigate combined effect of Au, CuS, and PTT on RA therapy	Seed-mediated and conjugated process	MTX	Murine fibroblast-like synovial cells (FLS) and mouse fibroblast (L929)	CIA rats	VIP-HA-Au NR@CuS-MTX revealed excellent therapeutic effects in vivo and higher NIR absorption than Au NR@CuS. Both HA and VIP played a significant part in controlling or regulating the release of MTX	[9]
Iron oxide nanomaterials	Synthesis of superparamagnetic IONPs to evaluate the efficacy of photothermal effect in arthritis treatment.	Solvothermal method	-	Cytotoxicity study (MTT assay) -RAW 264.7 cells	(CIA) mouse model	The prepared Fe_3O_4 nanoparticles showed promising results, with a size of 220 nm, high photothermal efficiency, and better targeting in the inflamed joint, thereby alleviating the symptoms of RA.	[26]
Black phosphorus nanosheets	Preparation of BPNs/Chitosan/PRP to study their synergistic effect with PTT and PDT therapy for improving osteogenesis in the treatment of RA.	Liquid exfoliation method	Methotrexate	Cytotoxicity assay using CCK-8 standard method (RAW264.7 cells, L929 cells and MSC cells)	(CIA) mouse model	Based on the distinguished concurrent PTT and PDT attributes of BPNs, chitosan/PRP thermos-responsive hydrogel effectively removed proliferating synoviocytes. In addition, BPNs accomplished calcium-extracted biomineralization via the in situ phosphorus-driven activation of BPNs in the target physiological microenvironment, which affected the entire course of treatment and provided improved therapeutic outcome.	[27]

Table 1. *Cont.*

Photothermal Therapy	Objective of Investigation	Method of Synthesis	Drug Used	In Vitro Model Used	Animal Model Investigated	Outcome of Investigation	Ref
Quantum dots	Synthesis of FAGM involved the following steps: 1. Formation of CTAB-coated gold nanorods (GNR) 2. GNR and methotrexate were coated with mesoporous silica shell 3. Synthesis of folic acid-functionalised GM.	1. Formation of CTAB-coated gold nanorods (GNR). The GNRs were prepared via the reduction of $HAuCl_4$ with sodium borohydride. Afterward, CTAB, $HauCl_4$, $AgNO_3$, and H_2SO_4 were added to the above solution in a certain amount. The solution was stirred at 28 °C for 30 min. Then, ascorbic acid was added dropwise, and a colour change was observed from light yellow to a state of colourlessness. The solution was kept overnight and stirred at 28 °C. 2. The prepared CTAB-GNR was again centrifuged and then resuspended. TEOS was added as a silica source and stirred for 30 min. Centrifuged and removed excess CTAB and FAGMs were collected. 3. Methotrexate was mixed with the prepared FAGM, and MTX-FAGM was formed.	Methotrexate	Cytotoxicity assay RAW 246.7 cells	Adjuvant-induced arthritis (AIA) rat model	In conclusion, for the therapy of RA, nanoscale MTX-FAGMs were prepared and their precisely targeted cytotoxicity towards active macrophages was confirmed under NIR laser irradiation. Synergistic action was observed between the photothermal therapy and chemotherapy	[28]

Abbreviations used in above table: PCl-PEG: (poly (ethylene glycol)-block-poly (ε-caprolactone); MTX: Methotrexate; PLGA: poly(DL-lactic-co-glycolic acid); NO: Nitric oxide, Hb: hemoglobin, SiRNA: Small interfering ribonucleic acid, PEG: polyethylene glycol; PTT: Photothermal therapy; CIA: Collagen-induced arthritic mice; FLS: Fibroblast-like synovial cells; CuS: Copper sulphate; TP: Triptolide; HA: Hyaluronic acid; RGD: Arginine–glycine–aspartic acid.

3.5. Iron Oxide Nanomaterial

Due to their outstanding biocompatibility and low toxicity, iron oxide NPs (IONPs) have generated a great deal of interest with respect to biological applications. One intriguing characteristic of IONPs is their special superparamagnetism, which allows them to simultaneously function as heat therapy transducers, carriers for targeted drug delivery, and magnetic resonance imaging (MRI) contrast agents for diagnosis. Iron-based NPs can be used to magnetically transport NPs to a targeted region for the more concentrated, more potent, and longer-lasting retention of NPs in the targeted tissue and, subsequently, improved therapeutic efficacy [29]. IO has received FDA (Food and drug administration) approval as a contrast agent for MRI due to the superior biocompatibility of its iron-based composition [30]. IONPS are coated with different polymer-based coatings that help them stabilize. Examples include various polymers such as alginate, chitosan, dextran, poloxamers, polyvinylpyrrolidone (PVP), polyacrylic acid (PA), polyethylene glycol (PEG), and polyvinyl alcohol (PVA) [28]. As these polymers were found to be biocompatible in nature, they are widely used to stabilize IONPS.

Researchers found that colloidal gold-coated superparamagnetic IONPs, also known as AuSPIONs, had a significant therapeutic effect on RA murine models and also caused a reduction in the circulation of substantial organs [31].

To improve the targeting efficacy of SPIONs, researchers have studied the effects of NPs' size on the efficiency of PTT for arthritis treatment, as shown in Figure 6 [26]. They have studied the size-based influence of Fe_3O_4 NPs in the range of 70–350 nm. However, it was observed that smaller NPs were quickly engulfed by the normal cells, leading to off-targeting. While the NPs with 220 nm size showed a rise in temperature under laser irradiation and gave exact targeting, they could not be engulfed by the normal cells. Still, they easily penetrated the inflamed arthritic tissue and were retained within it with as little difficulty. As a result, the use of NPs of the appropriate size benefits the PTT of RA, which should be appropriately tuned to produce the desired results.

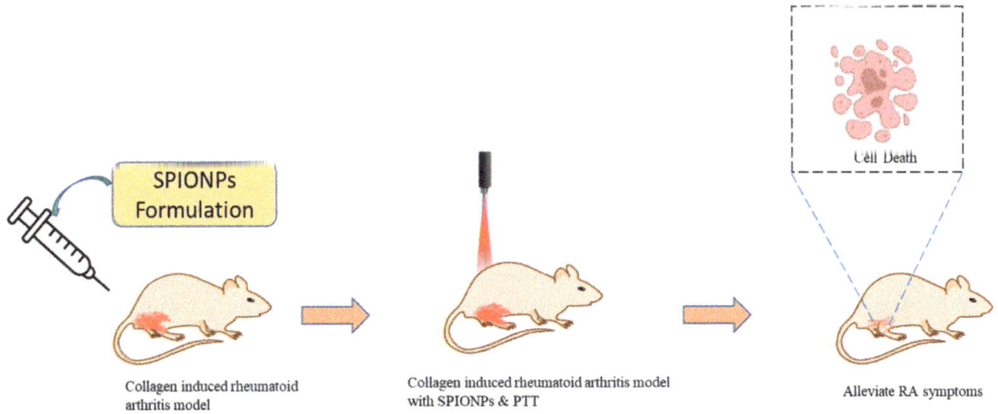

Figure 6. Injection of superparamagnetic IONPs (SPIONPS) combined with photothermal therapy in the collagen-induced arthritic model for the alleviation of arthritis symptoms.

3.6. Black Phosphorus Nanosheets

Black phosphorous nanosheets (BPNs) are superior photo-responsive materials. They are widely used because of their excellent phototherapy effects, strong biocompatibility, and promising osteogenic qualities; they have drawn a great deal of attention [32,33]. During NIR irradiation, black phosphorous nanosheets can convert light energy into heat energy, thus providing a practical photothermal effect. In addition to being necessary for the human body, particularly in the bones, phosphorus is also a component of natural resources [34].

Recently, researchers prepared a black phosphorous nanosheet-based thermos-responsive hydrogel with the help of platelet-rich plasma (PRP)-chitosan as a temperature-responsive polymer [27]. It was found that BPNs can generate reactive oxygen species after irradiation with an NIR laser at 808 nm to remove hyperplastic synoviocytes. The thermos-responsive hydrogel could also regulate the release of the MTX by providing the essential elements for osteogenesis from the degraded products of the BPNs. Further, an in vivo study revealed a reduction in oedema in the developed CIA mouse model [17]. A few other recent studies are discussed in Table 2.

Table 2. Polymer-hybridised nanomaterials in photothermal combined therapy of arthritis.

Photothermal Therapy	Nanomaterial Used	Combination Therapy	In Vitro Model Used	Animal Model Investigated	Outcome of Investigation	Ref
Photothermal Therapy alone	Palladium nanosheets		-	CIA Mouse	Pd-MTX nanosheets reduced the toxicity of MTX and prevented VEGF from proliferating as well as the generation of pro-inflammatory cytokines such as TNF-a and COX-2.	[17]
	Gold nanorods	PTT combined with chemotherapy	-	CIA mouse model	Aggregation of NPs increased in the presence of an external magnetic field, which raised the temperature in the exposed region and expedited the release of MTX.	[17]
	Magnetic iron oxide (Fe_3O_4) NPs	PTT combined with chemotherapy		CIA mouse model	Since NPs' small size allowed them to enter inflammatory cells, more of them accumulated in inflamed joints.	[17]
Chemo-photothermal therapy	Gold half-shelled Nanoparticles	Chemotherapy with photothermal therapy		CIA mouse model	In the inflammatory joints, nanoparticle accumulation was elevated.	[15]
PDT combined Photothermal therapy	Palladium nanosheets	PTT combined with photothermal therapy	Fibroblast cells	CIA Mouse	Pd-MTX nanosheets reduced the toxicity of MTX and prevented VEGF from proliferating as well as the generation of pro-inflammatory cytokines such as TNF-a and COX-2.	[14]
	Copper-based Nanoparticles		-	CIA Mouse	$Cu_{7.2}S_4$ NPs act on joints that showed good bone retention and resembled normal joints	[35]
Gene therapy combined Photothermal therapy	Poly (lactide-co-glycolide) nanoparticles (NPs)	Gene therapy combined with photothermal therapy	-	CIA Mice	RGD functionalised PLGA nanoparticles encapsulating STAT1-targeted siRNAs may be more efficient, presumably by selectively inhibiting macrophage and dendritic cell activation.	[36]
	Dexamethasone-loaded PLGA nanospheres	Gene therapy combined with photothermal therapy	-	CIA Mouse Human TNF Transgene model	The effectiveness of PLGA-PEG NPs for the delivery of a therapeutic medicine in the affected tissues in rheumatoid arthritis was low, perhaps because of the leaky vasculature, angiogenesis, and associated ELVIS effect that occur in the affected joints.	[36]

3.7. Use of Quantum Dots Together with Photothermal Therapy of Arthritis

Semiconducting polymeric quantum dots are a new class of functional nanomaterials being developed for photothermal therapy. Recently, scientists have prepared folate receptors that target semiconductive polymeric quantum dots [28]. To stabilize the quantum dots, they used cetrimonium bromide (CTAB) as a stabilizer. In addition to this stabilizer, the prepared quantum dots were further surface-functionalised to form a structure similar to mesoporous silica. This mesoporous silica nanoparticles acted as a reservoir for the hypoxia-activated prodrug Tirapazamine. These prepared polymeric quantum dots produced a localised photothermal effect in the activated macrophages and generated reactive oxygen species during irradiation with an NIR laser 808 nm, thus reducing the symptoms of arthritis.

4. Biological Applications of Polymer-Based Nanomaterials for PTT of Arthritis

4.1. Photothermal Therapy of Arthritis

PTT has recently received recognition as an effective RA treatment method (Gadeval A, 2021). Dual or multiple stimuli-responsive drug delivery systems, compared to single stimuli-responsive drug systems, offer greater control over medication release since they are more flexible and sensitive in terms of reacting to inflammatory situations. In one study, the investigators provided a method for constructing a dual or multiple-stimuli-responsive drug delivery system to treat RA. In the past, lipid NPs were produced by assembling triglycerol monostearate (TGMS) and 1,2-rdistearoyl-sn-glycero-3-phosphoethanolamine-carboxyl (DSPEPEG). In the current study, the matrix metalloproteinases' cleavable TGMS segment was conjugated with PEG-COOH using 3-amino phenophenyl boronic acid (PBA) as a linker.

Compared to other stimuli-responsive drug delivery systems that have been previously discussed in the literature, the PEG has stealth features. They show that the PEG shell's defence stabilizes Dex-loaded PPT micelles' circulation. In addition, it accurately recognizes scavenger receptors through ligand–receptor interaction, wherein hydrophilic block Dextran Sulfate (DS) was selected as the active target. In one study, Polycaprolactone (PCL) and DS were combined to generate DS-b-PCL, which demonstrated improved biodistribution in the rear feet of CIA mice following each systemic injection compared to little in wild-type animals. Wang et al. [12] demonstrated that Dex-loaded PCL-PEG micelles significantly reduced bone resorption and joint swelling in arthritic rats at a low Dex dose without causing adverse effects.

In order to deliver MTX, researchers have conjugated a sialic acid-dextran-octadecanoic acid to MTX and prepared a MTX-loaded micelle. These micelles were able to enhance bone healing while also obtaining an anti-inflammatory benefit. Through the interaction of sialic acid and vascular endothelial cells, this MTX-loaded micelle greatly increased the drug's accumulation in arthritic joints while showing good therapeutic advantages and few side effects. Effective therapies for RA include PTT and antioxidant combination therapy using Pd and Se NP co-delivery devices [37]. During the simple polymerization process, poly (ethyleneimine) (PEI) was used as a stabilizing agent that produced polypyrrole (PPy) NPs. The produced PPY-PEI NPs showed good photothermal performance and a sizable level of NIR absorption. For the treatment of RA, a strontium ranelate-loaded (SrR) methylcellulose hydrogel and PPY-PEI were used [38].

MTX and gold-based nanomaterials are routinely used for treating RA using PTT and chemotherapy, which induce a significant drop in inflammatory cytokine levels. When combined, gold nanorods and shells produce heat. Therefore, they are treated with NIR light and have improved therapeutic efficacy when combined with MTX. Magnetic iron oxide (Fe_3O_4) NPs have been utilised to treat and diagnose RA. Their biodistribution is ultimately impacted by the differences in their properties concerning size, charge, roughness, and elemental makeup, as depicted in Figure 7. Fe_3O_4 NPs exhibit remarkable biocompatibility and stability upon exposure to NIR light at higher temperatures [17].

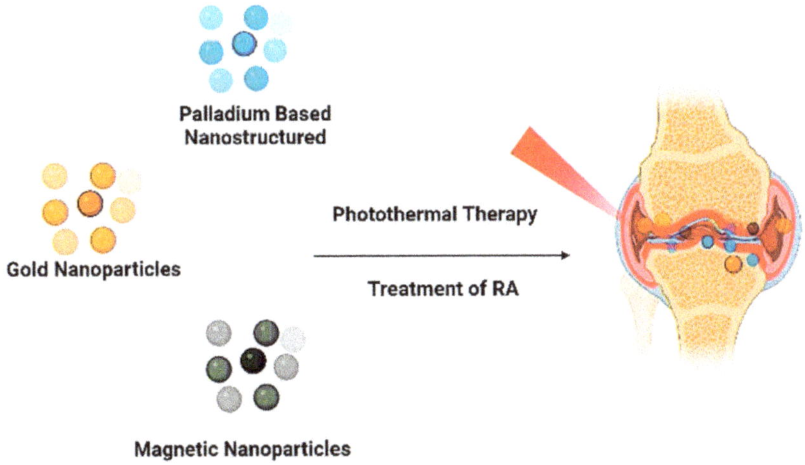

Figure 7. Photothermal Therapy.

Platinum NPs with a monodisperse octahedral crystal structure are uniformly layered on the outer surface of a metal-organic framework (MOF). They depend on Perovskite quantum dots and MOF-loaded polydopamine to enable H_2-based thermal therapy. H_2 can rapidly diffuse across cell membranes and acts as an antioxidant stress agent that reduces harmful reactive oxygen species and protect cells from damage. H_2 is more effectively generated through photocatalysis. Additionally, H_2 can protect healthy cells from the damage caused by overheating [38].

4.2. Chemophotothermal Therapy of Arthritis

Chemotherapy is the primary method of treatment for RA, along with surgery. Most patients can enter remission after employing treatments, biological agents, glucocorticoids, non-steroidal anti-inflammatory medications (NSAIDs), and disease-modifying anti-rheumatic pharmaceuticals. NSAIDs such as aspirin reduce the signs and symptoms of inflammation by inhibiting COX-2 activity [31]. The curative effects of DMARDs such as MTX are also influenced by immunosuppression. However, because of its slow therapeutic effect, DMARD treatment in clinics must be coupled with other drugs. Additionally, this treatment may cause digestive issues, bone marrow suppression, liver, and kidney damage.

One of the most well-supported nanocarriers, liposomes, have been used to successfully distribute Dex, an anti-RA drug, to RA patients. Liposomes incorporating synthetic materials of 100 nm in diameter and surface functionalised with 10% 5 kDa PEG that is slightly negatively charged have been shown to improve their in vivo half-life and capacity to target inflamed joints. In addition, the anti-arthritic activity of Dex is significantly boosted after being enclosed within liposomes according to pharmacodynamic studies using CIA mice [39]. For the treatment of RA, numerous studies have shown the synergistic advantages of combining PTT, chemotherapy, and targeted delivery. Therefore, the use of metallic NPs for specific treatment in combination with targeted drug loading and a hyperthermia agent is conceivable.

In this regard, Kim et al. [20] used PLGA loaded with gold [Au]/iron [Fe]/coupled RGD with MTX to maintain a high MTX concentration in painful joint regions. NIR radiation accelerates the release of MTX from PLGA NPs and induces localised heat generation due to Au half-shells' NIR resonance. The Fe half-shell layer also enables the magnetic transport of the NPs to the target site via the use of a magnetic field, and their retention can be improved in the presence of an external magnetic field. These NPs were injected intravenously into CIA mice. NIR absorbance and MRI revealed an accumulation of NPs in the inflamed paws, and an external magnetic field increased this accumulation [40].

1.3. Image Guided Photothermal Therapy of Arthritis

An innovative method among the phototherapeutic approaches is the image-guided application of light, in which only the target region with a high photosensitizer concentration should be illuminated. Several procedures have involved the provision of WS2 or Au NPs to CT-PTT [41]. Trigger-responsive engineered nanocarriers (ENCs), a recent advancement in the treatment of RA, assist in resolving issues typically associated with conventional medicine-delivery techniques. Due to image guidance, the development of specialised ENCs that improve drug targeting while reducing off-target toxicity is now possible. Numerous imaging methods have been utilised to examine inflammatory joints in RA patients using a range of probes, including computed tomography (CT), positron emission tomography, photoacoustic (PA) imaging, MRI, ultrasound, and X-ray.

A multipurpose imaging tool that can "picture" and "trigger" the targeted, controlled release of medication from ENCs is required. Only a handful of proof-of-concept studies have examined the efficacy of these RA therapeutic options, which are currently in the early stages of development.

Due to the progressive, debilitating nature of the disease; its unknown cause; and its initial similarities to other inflammatory diseases, RA is notoriously difficult to diagnose in its early stages. NIR-II photoacoustic molecular imaging (PMI) is being developed as a promising new strategy for the precise diagnosis and direction of RA treatment due to its exceptional sensitivity and specificity upon deep penetration.

In one study, TCZ, a targeted anti-rheumatic drug, was combined with polymer NPs to provide the first NIR-II-based theragnostic nanoplatform, or TCZ-Polymer NPs, for the PA-imaging-guided therapy of RA. NIR-II PMI data have shown that TCZ-Polymer NPs have better targeting capabilities for the effective non-invasive diagnosis of RA joint tissue, with a high signal-to-noise ratio (SNR) of 35.8 dB in 3D PA tomography images. Additionally, it has been demonstrated that the results of clinical micro-CT and histological analyses are in agreement with respect to the NIR-II PMI treatment evaluation of RA mice.

Six Harlan rats developed arthritis in their knee and ankle joints with the assistance of peptidoglycan polysaccharide polymers. Three rats comprised the untreated controls. Before and up to 24 h after the intravenous administration of 10 mg/kg ICG, optical imaging of the knee and ankle joints was performed using an integrated OI/X-ray system. The fluorescence signal intensities of the joints with and without arthritis were analysed for any discernible differences using generalised estimating equation models. Joint samples from the knee and ankle were further examined and evaluated using histology (Meier, 2010).

Nimesulide and low-dose MTX would be present in therapeutic RGD-mediated polymeric micelles that would be used to actively and passively target RA. As a result, they adhered RGD to NHS-PEG3400-PLA2000, an FDA-approved amphiphilic copolymer that is biocompatible and biodegradable. Additionally, we conducted real-time fluorescence imaging analysis to examine the in vivo distribution of the fluorescence-labeled R-M/N-PMs and in vivo investigations in a rat model of adjuvant-induced arthritis to evaluate the anti-inflammatory activity of R-M/N-PMs.

This problem may be resolved through the creation and application of magnetic NPs, namely, ultrasmall paramagnetic iron oxide (USPIO) (Over the past ten years, many researchers have carefully investigated the application of USPIO to MRI. Highly magnetically sensitive hydrophobic NPs are destroyed by high temperatures. However, the hydrophobicity of the NPs necessitates further modification before they can be transferred to the aqueous phase, which makes the NMs extremely unstable or agglomerated. By using in situ polyol polymerization, an amphiphilic block polymer called polyethylene glycol-tert-butyl polyacrylate (PEG-b-PAA) was created, coated with USPIO, and conjugated with folic acid (FA) with a targeted function towards the PEG-b-PAA/USPIO surface; subsequently, researchers were able to develop a highly stable and focused nanocomposite contrast agent (referred as FA-PEG-b-PA). Herein, the authors also investigated the impact of FA-PEG-b-PAA/USPIO on RA's early diagnosis.

When PTT is used with external magnetic field-targeted therapy, therapeutic efficacy increases, and adverse effects are reduced with fewer doses. Additionally, iron-based parts enabled T2-weighted magnetic resonance imaging (MRI) in vivo, thus supplying the RA treatment's guiding and monitoring capability. Magnetic iron oxide NPs (IONPs) have a unique ability to convert light to heat. The FDA has approved IO as a contrast agent for MRI because of the high biocompatibility of its iron-based composition. Colloidal gold-coated superparamagnetic IONPs, also known as AuSPIONs, were found to have a significant therapeutic effect on RA murine models and provoke a reduction in circulation in the principal organs by [26] investigated the influence of 70–350 nm-sized Fe_3O_4 NPs on the effectiveness of PTT for RA treatment to improve biosafety and its targeting impact.

4.4. PDT Combined Photothermal Therapy of Arthritis

Photodynamic therapy, which uses photosensitizers or non-toxic dyes along with safe, visible light, has been used for more than a century, but is just now gaining popularity [42,43]. In addition to oxygen and light, photosensitizers are among the three essential components of PDT. According to their definition, these dyes are compounds that may absorb light of a specific wavelength and cause photochemical or photophysical processes. The three non-toxic components that make up the molecular basis of photodynamic therapy—a suitable light wavelength, a photosensitizer, and oxygen dissolved in cells—interact to generate the intended effects only when they are present in pathogenic tissues [44].

PDT primarily relies on the presence of oxygen present inside the cells. When a photosensitizer enters the cell, it is exposed to light at a wavelength that coincides with the photosensitizer's absorption spectrum. The absorption of a photon causes the excitation of the photosensitizer from the ground state to the excited state. The energy is released as fluorescence while the residual energy causes the excitation of the photosensitizer towards the excited triplet state, thus generating phosphorescence [45]. Co-delivery for trigger response is the most frequently used technique, particularly when combined with PTT. There are two light-triggered therapies: PTT and PDT. To properly treat RA, a combination of PTT and PDT may provide more significant bone and cartilage preservation, less synovial invasion, and a more substantial anti-inflammatory effect (Figure 8). Important light triggers for both treatments include photothermal agents and photosensitizers [35].

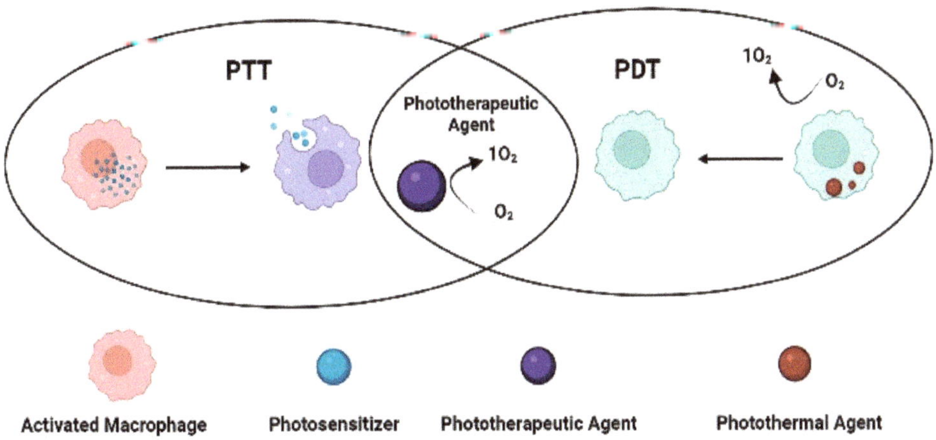

Figure 8. PDT combined photothermal therapy.

As a result, they are frequently co-delivered with conventional agents. Palladium (Pd) blue was the term given to the hexagonal palladium nanosheets developed by Chen et al. [46] with a blue colour. The ultrathin Pd nanosheets display precise and predictable NIR surface plasmon resonance peaks and significant photothermal effects in vitro. They can be utilised as photothermal agents in PTT therapy. However, it is challenging to use inorganic materials alone due to their poor biocompatibility. Considering the photothermal properties of Pd, Xu et al. [21] developed and manufactured nanosheets to treat RA by targeting inflammatory cells and controlling MTX release with Pd particles and RGD peptides. RGD peptides can change Pd NPs, an inorganic photothermal agent, such that their capacity for targeting is enhanced and the disadvantages of inorganic photothermal agents' poor biocompatibility are overcome. The nanosheets showed exceptional PTT efficacy in terms of treating RA-afflicted animals, as it accumulated and significantly reduced inflammatory responses, cartilage degradation, and MTX toxicity in inflamed paws. This indicates that the nanosheets have excellent in vitro anti-inflammatory properties, superior stability, and photothermal conversion efficiency.

Chen et al. [46] reported the design and synthesis of PdNps (Pd-Cys@MTX@RGD) as a nanotherapy agent that can target inflammatory cells and control MTX release. The Pd-Cys@MTX@RGD nanosheets served as an NIR photothermal agent with selective targeting potential towards peptides overexpressed on the surface of inflamed arthritic tissues. Pd-Cys@MTX@RGD nanosheets were observed to mediate the controlled release of MTX upon NIR 808 nm laser irradiation with a power density of 0.3 W cm^2. This was associated with a significant reduction in the toxicity issues associated with MTX. It was advocated that, like MTX, other therapeutic agents can be combined with PTT to amplify its medicinal benefits (Figure 9). The agent in question inhibited the inflammatory response induced by vascular endothelial growth factor (VEGF) and IL-1β. This was correlated with its in vivo performance, which significantly inhibited RA and led to a reduced level of pro-inflammatory cytokines (TNF-α, COX-2).

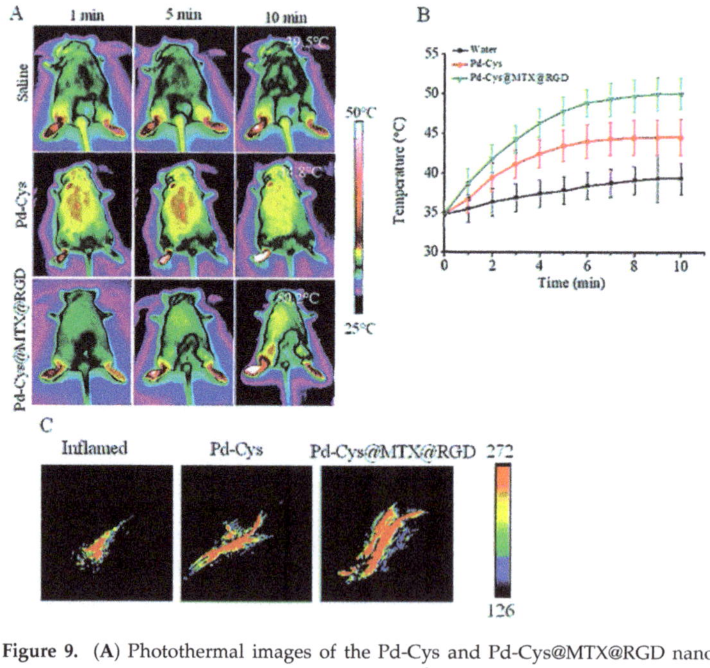

Figure 9. (**A**) Photothermal images of the Pd-Cys and Pd-Cys@MTX@RGD nanosheets in RA mice irradiated at 808 nm for 10 min (0.3 W cm^{-2}). (**B**) Temperature profiles of the Pd-Cys and Pd-Cys@MTX@RGD nanosheets in RA mice irradiated at 808 nm for 10 min. (**C**) Photoacoustic signal images of the Pd-Cys and Pd-Cys@MTX@RGD nanosheets in RA mice excited with 808 nm irradiation. Adapted with permission from reference [46]. Copyright 2022.

PTT and antioxidant combination therapies employing Pd and Se NPs delivery devices are effective treatments for RA. The straightforward polymerization procedure that created polypyrrole (PPy) NPs employed PEI as a stabilizer. The created PPY-PEI NPs showed significant NIR absorption and exhibited good photothermal stability and performance [47]. A methylcellulose hydrogel co-loaded with PPY-PEI and strontium ranelate (SrR) was utilised to treat RA.

A unique method for the treatment of RA involved the use of CuS NPs based on copper NPs and a novel l-cysteine-assisted synthesis. In addition to stimulating chondrogenesis and osteogenesis, copper acts as a photothermal and photosensitizer. In vitro investigations have shown outstanding biocompatibility and excellent photothermal and photodynamic effects.

4.5. Gene Therapy-Combined Photothermal Therapy of Arthritis

RNA interference (RNAi) is a cellular process that silences messenger RNA (mRNA)-based genes post-transcriptionally, and small interference RNA (siRNA)-based silencing is only temporary. Hence, new methods have been devised and described to produce longer-lasting silencing. The use of short hairpin RNA (shRNA), for example, is a vector-encoded technique that can be employed for long-lasting, reliable cell silencing. Small interfering RNA (siRNA), which is highly specific and effective in terms of silencing genes, is a critical component that is mediated by RNA interference [28]. An intravenous dose of Cy5-labeled siRNA/WS was used by Pan et al. [27] to test the therapeutic activity of their wrap some (WS)-containing siRNA (with a core constituted of a cationic lipid bilayer and an siRNA complex enveloped in a neutral lipid bilayer in PEG over the surface). In addition, they used the arthritis scores among CIA mice to demonstrate the effectiveness of siRNA's ability to target tumour necrosis factor (TNF-/WS).

RGD-functionalised, siRNA-loaded poly (lactide-co-glycolide) NPs were created by Scheinman et al. [48] as a nanosystem for STAT1 siRNA delivery to the joint tissues of a CIA mice model. After siRNA was enclosed, the researchers examined the stability and nanoparticle-related properties. According to morphological analysis, RGD-NPs have a size range of 100–200 nm and a net positive charge, which may result from RGD functionalization. The presence of the RGD peptide on the NPs outer surface improves tissue uptake by 10–200 times in arthritis-prone mice, and PLGA NPs shield siRNA from serum breakdown. Compared to the control group, the group of arthritic mice that was treated with RGD functionalization showed improved lung delivery of NPs. For the treatment of arthritis, RGD-functionalised PLGA NPs encapsulating STAT1-targeted siRNAs may be more efficient, presumably by selectively inhibiting macrophage and dendritic cell activation.

For the treatment of RA, Park et al. created PLGA NPs that were loaded with Dex and siRNA. In this study, PLGA NPs were first loaded with Dex before being combined with PEI/siRNA [23]. Then, green fluorescence protein siRNA (GFP siRNA) and medications were transfected into the human chondrocyte cell line (C28/I2) to evaluate the co-delivery of siRNA and Dex. For the treatment of persistent skin inflammations, Desai et al. [49] created cyclic cationic head lipid–polymer hybrid nanocarriers (CyLiPns) packed with the analgesic Capsaicin (Cap) and anti-TNF siRNA (siTNF). A few of the photothermal combined therapies are illustrated in Table 2.

5. Conclusions

Polymer-based NMs and PTT have led to tremendous growth in nanomedicine, which has been proven through various publications and clinical trials over the past years. The combined effect of PTT and polymer NMs has revealed multiple advantages compared to conventional therapies such as radiotherapy, chemotherapy, and immunotherapy for treating diseases such as cancers and RA. Polymer-based NMs are cost-effective, biocompatible, and beneficial materials that are easily functionalised with functional groups. They can convert light to heat energy and thus enhance their own photothermal conversion efficiency. They are used as nanocarriers for controlled drug release that improve the efficacy of RA treatment and reduce the disease's side effects. With the help of NMs, drugs bind to specific target sites, produce signals for imaging determinations, and generate ROS to support

PTT and PDT performance. This review expounds on various polymer-based NMs for the photothermal therapy of RA, including Au, Cu, iron oxide, black phosphorous, quantum dots, and palladium. These NMs have achieved excellent improvement in the treatment of RA via their combination with PTT. However, many issues still need to be discussed, such as their low photothermal conversion efficacy, safety, and low in vivo efficiency, which limit their clinical applications. In addition, their accumulation in non-targeted cells or tissues is another problematic issue. Thus, besides their effective treatment efficiency, the interaction of such polymer NMs with tissues is a matter of concern that requires further investigation to determine their potential applications. In the future, we believe that nanotechnology and advanced polymer-based NMs will play a vital role in treating RA.

Author Contributions: M.T.: Writing—original draft, data curation, review and editing, formal analysis, P.P.: review and editing, R.G.: review, editing, data curation, B.P.: review and editing, validation, R.K.T.: review and editing; and M.C.S.: supervision, project administration. All authors have read and agreed to the published version of the manuscript.

Funding: This research received no external funding.

Institutional Review Board Statement: Not applicable.

Informed Consent Statement: Not applicable.

Data Availability Statement: Not applicable.

Acknowledgments: MT would like to acknowledge the School of Pharmacy, Devi Ahilya Vishwavidyalaya, for providing research and administrative support.

Conflicts of Interest: The authors declare no conflict of interest.

Abbreviations

AuSPIONs	Gold-coated superparamagnetic IONPs
BPNs	Black phosphorous nanosheets
CIA	Collagen-induced mouse model
CTAB	Cetrimonium bromide
IONPs	Iron oxide nanoparticles
MRI	Magnetic resonance imaging
PRP	Platelet-rich plasma
PTT	Photothermal therapy
RA	Rheumatoid arthritis
ROS	Reactive oxygen species
NMs	Nanomaterials
NPs	Nanoparticles
MTX	Methotrexate
DMARDs	Disease-modifying anti-rheumatic drugs
NIR	Near infra-red
PEG	Polyethylene glycol
RGD	Arginine–glycine–aspartic acid
PDT	Photodynamic therapy
HA	Hyaluronic acid
TNF-α	Tumour necrosis factor-α
NSAID	Non-steroidal anti-inflammatory medications
VIP	Vasoactive intestinal peptide
ENCs	Engineered nanocarriers
PMI	Photoacoustic molecular imaging
TCZ	Tocilizumab
Dex	Dexamethasone
PA	Photoacoustic
RNAi	RNA interference
siRNA	Small interference RNA

References

1. Smolen, J.S.; Aletaha, D.; McInnes, I.B. Rheumatoid arthritis. *Lancet* **2016**, *388*, 2023–2038. [CrossRef]
2. Liang, W.; Yu, Y.; Liu, Z.; Ming, W.; Lin, H.; Long, H.; Zhao, J. The Therapeutic Potential of Targeted Nanoparticulate Systems to Treat Rheumatoid Arthritis. *J. Nanomater.* **2022**, *2022*, 8900658. [CrossRef]
3. Dolati, S.; Sadreddini, S.; Rostamzadeh, D.; Ahmadi, M.; Jadidi-Niaragh, F.; Yousefi, M. Utilization of nanoparticle technology in rheumatoid arthritis treatment. *Biomed. Pharmacother.* **2016**, *80*, 30–41. [CrossRef]
4. Shang, H.; Gu, H.; Zhang, N. From traditional to novel treatment of arthritis: A review of recent advances in nanotechnology-based thermal therapy. *Nanomedicine* **2021**, *16*, 2117–2132. [CrossRef]
5. Zhang, X.; Koo, S.; Kim, J.H.; Huang, X.; Kong, N.; Zhang, L.; Zhou, J.; Xue, J.; Harris, M.B.; Tao, W. Nanoscale materials-based platforms for the treatment of bone-related diseases. *Matter* **2021**, *4*, 2727–2764. [CrossRef]
6. Münz, C.; Lünemann, J.D.; Getts, M.T.; Miller, S.D. Antiviral immune responses: Triggers of or triggered by autoimmunity? *Nat. Rev. Immunol.* **2009**, *9*, 246–258. [CrossRef]
7. Beech, J.T.; Andreakos, E.; Ciesielski, C.J.; Green, P.; Foxwell, B.M.; Brennan, F.M. T-cell contact-dependent regulation of CC and CXC chemokine production in monocytes through differential involvement of NFκB: Implications for rheumatoid arthritis. *Arthritis Res. Ther.* **2006**, *8*, 1–10. [CrossRef]
8. Tran, C.N.; Lundy, S.K.; White, P.T.; Endres, J.L.; Motyl, C.D.; Gupta, R.; Wilke, C.M.; Shelden, E.A.; Chung, K.C.; Urquhart, A.G. Molecular interactions between T cells and fibroblast-like synoviocytes: Role of membrane tumor necrosis factor-α on cytokine-activated T cells. *Am. J. Pathol.* **2007**, *171*, 1588–1598. [CrossRef]
9. Huang, Q.-Q.; Pope, R.M. The role of toll-like receptors in rheumatoid arthritis. *Curr. Rheumatol. Rep.* **2009**, *11*, 357–364. [CrossRef]
10. Mavers, M.; Ruderman, E.M.; Perlman, H. Intracellular signal pathways: Potential for therapies. *Curr. Rheumatol. Rep.* **2009**, *11*, 378–385. [CrossRef]
11. Gibofsky, A. Epidemiology, pathophysiology, and diagnosis of rheumatoid arthritis: A Synopsis. *Am. J. Manag. Care* **2014**, *20*, S128–S135.
12. Wang, Q.; Jiang, J.; Chen, W.; Jiang, H.; Zhang, Z.; Sun, X. Targeted delivery of low-dose dexamethasone using PCL–PEG micelles for effective treatment of rheumatoid arthritis. *J. Control. Release* **2016**, *230*, 64–72. [CrossRef]
13. Qindeel, M.; Khan, D.; Ahmed, N.; Khan, S.; Rehman, A.U. Surfactant-free, self-assembled nanomicelles-based transdermal hydrogel for safe and targeted delivery of methotrexate against rheumatoid arthritis. *ACS Nano* **2020**, *14*, 4662–4681. [CrossRef]
14. Chen, X.; Liu, Y.; Wen, Y.; Yu, Q.; Liu, J.; Zhao, Y.; Liu, J.; Ye, G. A photothermal-triggered nitric oxide nanogenerator combined with siRNA for precise therapy of osteoarthritis by suppressing macrophage inflammation. *Nanoscale* **2019**, *11*, 6693–6709. [CrossRef]
15. Lee, S.-M.; Kim, H.J.; Ha, Y.-J.; Park, Y.N.; Lee, S.-K.; Park, Y.-B.; Yoo, K.-H. Targeted chemo-photothermal treatments of rheumatoid arthritis using gold half-shell multifunctional nanoparticles. *ACS Nano* **2013**, *7*, 50–57. [CrossRef]
16. Pandey, P.K.; Maheshwari, R.; Raval, N.; Gondaliya, P.; Kalia, K.; Tekade, R.K. Nanogold-core multifunctional dendrimer for pulsatile chemo-, photothermal-and photodynamic-therapy of rheumatoid arthritis. *J. Colloid Interface Sci.* **2019**, *544*, 61–77. [CrossRef]
17. Gadeval, A.; Chaudhari, S.; Bollampally, S.P.; Polaka, S.; Kalyane, D.; Sengupta, P.; Kalia, K.; Tekade, R.K. Integrated nanomaterials for non-invasive photothermal therapy of rheumatoid arthritis. *Drug Discov. Today* **2021**, *26*, 2315–2328. [CrossRef]
18. Sibuyi, N.R.S.; Moabelo, K.L.; Fadaka, A.O.; Meyer, S.; Onani, M.O.; Madiehe, A.M.; Meyer, M. Multifunctional Gold Nanoparticles for Improved Diagnostic and Therapeutic Applications: A Review. *Nanoscale Res. Lett.* **2021**, *16*, 174. [CrossRef]
19. Li, C.; Liu, R.; Song, Y.; Zhu, D.; Yu, L.; Huang, Q.; Zhang, Z.; Xue, Z.; Hua, Z.; Lu, C. Intra-articular Administrated Hydrogels of Hyaluronic Acid Hybridized with Triptolide/Gold Nanoparticles for Targeted Delivery to Rheumatoid Arthritis Combined with Photothermal-chemo Therapy. *Res. Sq.* **2021**. [CrossRef]
20. Kim, H.J.; Lee, S.-M.; Park, K.-H.; Mun, C.H.; Park, Y.-B.; Yoo, K.-H. Drug-loaded gold/iron/gold plasmonic nanoparticles for magnetic targeted chemo-photothermal treatment of rheumatoid arthritis. *Biomaterials* **2015**, *61*, 95–102. [CrossRef]
21. Xu, C.; Chen, J.; Li, L.; Pu, X.; Chu, X.; Wang, X.; Li, M.; Lu, Y.; Zheng, X. Promotion of chondrogenic differentiation of mesenchymal stem cells by copper: Implications for new cartilage repair biomaterials. *Mater. Sci. Eng. C* **2018**, *93*, 106–114. [CrossRef]
22. Huang, R.; Zhang, C.; Bu, Y.; Li, Z.; Zheng, X.; Qiu, S.; Machuki, J.O.A.; Zhang, L.; Yang, Y.; Guo, K. A multifunctional nano-therapeutic platform based on octahedral yolk-shell Au NR@ CuS: Photothermal/photodynamic and targeted drug delivery tri-combined therapy for rheumatoid arthritis. *Biomaterials* **2021**, *277*, 121088. [CrossRef]
23. Park, J.S.; Yang, H.N.; Jeon, S.Y.; Woo, D.G.; Kim, M.S.; Park, K.-H. The use of anti-COX2 siRNA coated onto PLGA nanoparticles loading dexamethasone in the treatment of rheumatoid arthritis. *Biomaterials* **2012**, *33*, 8600–8612. [CrossRef]
24. Chen, X.; Zhu, X.; Ma, L.; Lin, A.; Gong, Y.; Yuan, G.; Liu, J. A core–shell structure QRu-PLGA-RES-DS NP nanocomposite with photothermal response-induced M2 macrophage polarization for rheumatoid arthritis therapy. *Nanoscale* **2019**, *11*, 18209–18223. [CrossRef]
25. Lee, H.; Lee, M.-Y.; Bhang, S.H.; Kim, B.-S.; Kim, Y.S.; Ju, J.H.; Kim, K.S.; Hahn, S.K. Hyaluronate–gold nanoparticle/tocilizumab complex for the treatment of rheumatoid arthritis. *ACS Nano* **2014**, *8*, 4790–4798. [CrossRef]
26. Zhang, S.; Wu, L.; Cao, J.; Wang, K.; Ge, Y.; Ma, W.; Qi, X.; Shen, S. Effect of magnetic nanoparticles size on rheumatoid arthritis targeting and photothermal therapy. *Colloids Surf. B Biointerfaces* **2018**, *170*, 224–232. [CrossRef]

27. Pan, W.; Dai, C.; Li, Y.; Yin, Y.; Gong, L.; Machuki, J.O.A.; Yang, Y.; Qiu, S.; Guo, K.; Gao, F. PRP-chitosan thermoresponsive hydrogel combined with black phosphorus nanosheets as injectable biomaterial for biotherapy and phototherapy treatment of rheumatoid arthritis. *Biomaterials* **2020**, *239*, 119851. [CrossRef]
28. Li, X.; Zhang, S.; Zhang, X.; Hou, Y.; Meng, X.; Li, G.; Xu, F.; Teng, L.; Qi, Y.; Sun, F.; et al. Folate receptor-targeting semiconducting polymer dots hybrid mesoporous silica nanoparticles against rheumatoid arthritis through synergistic photothermal therapy, photodynamic therapy, and chemotherapy. *Int. J. Pharm.* **2021**, *607*, 120947. [CrossRef]
29. Xiao, L.; Wu, Z.; Zhang, J.; Wang, G.; Ma, Y.; Ding, Y.; He, X.; Zhang, S.; Zhang, Z. Synthesis, photothermal effect and cytotoxicity of Fe_3O_4@Au nanocomposites. *J. Nanosci. Nanotechnol.* **2019**, *19*, 2467–2473. [CrossRef]
30. Chen, C.-L.; Siow, T.Y.; Chou, C.-H.; Lin, C.-H.; Lin, M.-H.; Chen, Y.-C.; Hsieh, W.-Y.; Wang, S.-J.; Chang, C. Targeted superparamagnetic iron oxide nanoparticles for in vivo magnetic resonance imaging of T-cells in rheumatoid arthritis. *Mol. Imaging Biol.* **2017**, *19*, 233–244. [CrossRef]
31. Carneiro, M.F.H.; Machado, A.R.T.; Antunes, L.M.; Souza, T.E.; Freitas, V.A.; Oliveira, L.C.; Rodrigues, J.L.; Pereira, M.C.; Barbosa, F. Gold-coated superparamagnetic iron oxide nanoparticles attenuate collagen-induced arthritis after magnetic targeting. *Biol. Trace Elem. Res.* **2020**, *194*, 502–513. [CrossRef]
32. Liu, J.; Du, P.; Mao, H.; Zhang, L.; Ju, H.; Lei, J. Dual-triggered oxygen self-supply black phosphorus nanosystem for enhanced photodynamic therapy. *Biomaterials* **2018**, *172*, 83–91. [CrossRef]
33. Yang, X.; Wang, D.; Shi, Y.; Zou, J.; Zhao, Q.; Zhang, Q.; Huang, W.; Shao, J.; Xie, X.; Dong, X. Black phosphorus nanosheets immobilizing Ce6 for imaging-guided photothermal/photodynamic cancer therapy. *ACS Appl. Mater. Interfaces* **2018**, *10*, 12431–12440. [CrossRef]
34. Itkonen, S.T.; Rita, H.J.; Saarnio, E.M.; Kemi, V.E.; Karp, H.J.; Kärkkäinen, M.U.; Pekkinen, M.H.; Laitinen, E.K.; Risteli, J.; Koivula, M.-K. Dietary phosphorus intake is negatively associated with bone formation among women and positively associated with some bone traits among men—A cross-sectional study in middle-aged Caucasians. *Nutr. Res.* **2017**, *37*, 58–66. [CrossRef]
35. Lu, Y.; Li, L.; Lin, Z.; Wang, L.; Lin, L.; Li, M.; Zhang, Y.; Yin, Q.; Li, Q.; Xia, H. A new treatment modality for rheumatoid arthritis: Combined photothermal and photodynamic therapy using $Cu_{7.2}S_4$ nanoparticles. *Adv. Healthc. Mater.* **2018**, *7*, 1800013. [CrossRef]
36. Simón-Vázquez, R.; Tsapis, N.; Lorscheider, M.; Rodríguez, A.; Calleja, P.; Mousnier, L.; de Miguel Villegas, E.; González-Fernández, Á.; Fattal, E. Improving dexamethasone drug loading and efficacy in treating arthritis through a lipophilic prodrug entrapped into PLGA-PEG nanoparticles. *Drug Deliv. Transl. Res.* **2022**, *12*, 1270–1284. [CrossRef]
37. Zheng, C.; Wu, A.; Zhai, X.; Ji, H.; Chen, Z.; Chen, X.; Yu, X. The cellular immunotherapy of integrated photothermal anti-oxidation Pd–Se nanoparticles in inhibition of the macrophage inflammatory response in rheumatoid arthritis. *Acta Pharm. Sin. B* **2021**, *11*, 1993–2003. [CrossRef]
38. Zhang, S.; Zhang, M.; Li, X.; Li, G.; Yang, B.; Lu, X.; Gao, Y.; Sun, F. Nano-Based Co-Delivery System for Treatment of Rheumatoid Arthritis. *Molecules* **2022**, *27*, 5973. [CrossRef]
39. Ren, H.; He, Y.; Liang, J.; Cheng, Z.; Zhang, M.; Zhu, Y.; Hong, C.; Qin, J.; Xu, X.; Wang, J. Role of liposome size, surface charge, and PEGylation on rheumatoid arthritis targeting therapy. *ACS Appl. Mater. Interfaces* **2019**, *11*, 20304–20315. [CrossRef]
40. Pirmardvand Chegini, S.; Varshosaz, J.; Taymouri, S. Recent approaches for targeted drug delivery in rheumatoid arthritis diagnosis and treatment. *Artif. Cells Nanomed. Biotechnol.* **2018**, *46* (Suppl. 2), 502–514. [CrossRef]
41. Hashemkhani, M.; Muti, A.; Sennaroğlu, A.; Acar, H.Y. Multimodal image-guided folic acid targeted Ag-based quantum dots for the combination of selective methotrexate delivery and photothermal therapy. *J. Photochem. Photobiol. B Biol.* **2020**, *213*, 112082. [CrossRef]
42. Castano, A.P.; Demidova, T.N.; Hamblin, M.R. Mechanisms in photodynamic therapy: Part one-photosensitizers, photochemistry and cellular localization. *Photodiagnosis Photodyn. Ther.* **2004**, *1*, 279–293. [CrossRef]
43. Hou, Y.-J.; Yang, X.-X.; Liu, R.-Q.; Zhao, D.; Guo, C.-X.; Zhu, A.-C.; Wen, M.-N.; Liu, Z.; Qu, G.-F.; Meng, H.-X. Pathological mechanism of photodynamic therapy and photothermal therapy based on nanoparticles. *Int. J. Nanomed.* **2020**, *15*, 6827. [CrossRef]
44. Kwiatkowski, S.; Knap, B.; Przystupski, D.; Saczko, J.; Kędzierska, E.; Knap-Czop, K.; Kotlińska, J.; Michel, O.; Kotowski, K.; Kulbacka, J. Photodynamic therapy—Mechanisms, photosensitizers and combinations. *Biomed. Pharmacother.* **2018**, *106*, 1098–1107. [CrossRef]
45. Calixto, G.M.F.; Bernegossi, J.; de Freitas, L.M.; Fontana, C.R.; Chorilli, M. Nanotechnology-based drug delivery systems for photodynamic therapy of cancer: A review. *Molecules* **2016**, *21*, 342. [CrossRef]
46. Chen, X.; Zhu, X.; Xu, T.; Xu, M.; Wen, Y.; Liu, Y.; Liu, J.; Qin, X. Targeted hexagonal Pd nanosheet combination therapy for rheumatoid arthritis via the photothermal controlled release of MTX. *J. Mater. Chem. B* **2019**, *7*, 112–122. [CrossRef]
47. Wang, Y.; Meng, H.-M.; Song, G.; Li, Z.; Zhang, X.-B. Conjugated-polymer-based nanomaterials for photothermal therapy. *ACS Appl. Polym. Mater.* **2020**, *2*, 4258–4272. [CrossRef]

48. Scheinman, R.I.; Trivedi, R.; Vermillion, S.; Kompella, U.B. Functionalized STAT1 siRNA nanoparticles regress rheumatoid arthritis in a mouse model. *Nanomedicine* **2011**, *6*, 1669–1682. [CrossRef]
49. Desai, P.R.; Marepally, S.; Patel, A.R.; Voshavar, C.; Chaudhuri, A.; Singh, M. Topical delivery of anti-TNFα siRNA and capsaicin via novel lipid-polymer hybrid nanoparticles efficiently inhibits skin inflammation in vivo. *J. Control. Release* **2013**, *170*, 51–63. [CrossRef]

pharmaceutics

Article

An Efficient, Short Stimulus PANC-1 Cancer Cell Ablation and Electrothermal Therapy Driven by Hydrophobic Interactions

Maria P. Meivita [1], Denise Lee [1], J Shamita Naikar [1,2], Shao-Xiang Go [1], Wey Chyi Teoh [2,3], Yaw Sing Tan [4], Natasa Bajalovic [1,*] and Desmond K. Loke [1,2,*]

1 Department of Science, Mathematics and Technology, Singapore University of Technology and Design, Singapore 487372, Singapore
2 Office of Innovation, Changi General Hospital, Singapore 529889, Singapore
3 Department of Radiology, Changi General Hospital, Singapore 529889, Singapore
4 Bioinformatics Institute, Agency for Science, Technology and Research (A*STAR), Singapore 138671, Singapore
* Correspondence: natasa_bajalovic@sutd.edu.sg (N.B.); desmond_loke@sutd.edu.sg (D.K.L.)

Abstract: Promising results in clinical studies have been demonstrated by the utilization of electrothermal agents (ETAs) in cancer therapy. However, a difficulty arises from the balance between facilitating the degradation of ETAs, and at the same time, increasing the electrothermal performance/stability required for highly efficient treatment. In this study, we controlled the thermal signature of the MoS_2 by harnessing MoS_2 nanostructures with M13 phage (MNM) via the structural assembling (hydrophobic interaction) phenomena and developed a combined PANC-1 cancer cell–MNM alternating current (AC)-stimulus framework for cancer cell ablation and electrothermal therapy. A percentage decrease in the cell viability of ~23% was achieved, as well as a degradation time of 2 weeks; a stimulus length of 100 μs was also achieved. Molecular dynamics (MD) simulations revealed the assembling kinetics in integrated M13 phage–cancer cell protein systems and the structural origin of the hydrophobic interaction-enabled increase in thermal conduction. This study not only introduced an 'ideal' agent that avoided the limitations of ETAs but also provided a proof-of-concept application of MoS_2-based materials in efficacious cancer therapy.

Keywords: ablation; cancer cell; phage; MoS_2; molecular dynamics; agent

Citation: Meivita, M.P.; Lee, D.; Naikar, J.S.; Go, S.-X.; Teoh, W.C.; Tan, Y.S.; Bajalovic, N.; Loke, D.K. An Efficient, Short Stimulus PANC-1 Cancer Cell Ablation and Electrothermal Therapy Driven by Hydrophobic Interactions. *Pharmaceutics* **2023**, *15*, 106. https://doi.org/10.3390/pharmaceutics15010106

Academic Editors: Manivasagan Panchanathan and Eue-Soon Jang

Received: 29 October 2022
Revised: 19 December 2022
Accepted: 20 December 2022
Published: 28 December 2022

1. Introduction

Pancreatic cancer (PC) is the fourth leading cause of death in both men and women worldwide and has the lowest survival rate of all major organ cancers [1–4]. For instance, approximately 56,770 people were diagnosed with PC in 2019, and 45,750 deaths from this disease occurred in the United States [5,6]. Moreover, 80–85% of patients are diagnosed with advanced-stage disease, with a five-year overall survival rate of ~10% [7–9]. In Singapore, PC is the fourth and fifth most typical cause of cancer death in women and men, respectively [10,11]. The archetypal PC type, adenocarcinoma, is inoperable by the time patients are symptomatic [9,12]. This arises from the propensity of pancreatic tumors to spread to adjacent structures and blood vessels in the early stage [13,14]. Thus, in this work, the interest is in the development of alternative technologies that treat locally advanced diseases, which are otherwise not operable.

Cancer thermal-based therapy (TBT) is a promising candidate for achieving a minimally invasive, minimized ablation zone and a highly efficient therapeutic modality [15–19]. TBT operations, based on the utilization of thermal agents (TAs) to generate local hyperpyrexia, enabling the thermal elimination of tumors, have demonstrated excellent success in preclinical and clinical trials [20–22]. Clinical studies have disclosed that the TBT-facilitated ablation of tumors exhibits success in ~90% of patients without noticeable side effects, severe complications, or harmful changes in organ functions [19,23]. TBT is expected to have not only significant but also revolutionary clinical impact due to these promising results, as well as the recent clinical approval of the utility of metal oxide nanoparticles for the TBT in Europe [24,25]. Moreover, experiments have demonstrated enhanced pancreatic tumor ablation and photothermal therapy using polyprrole-based nanoparticles [26]. However, developing TAs with excellent thermal effects (viz., high thermal stability/performance to achieve effective therapeutic outcomes) and rapid degradability (i.e., fast degradation to address safety concerns) is extremely challenging [22,27,28].

TAs that exhibit high thermal performance under stimulations (e.g., inorganic nanomaterials) have been demonstrated in preclinical studies [29,30]. However, traditional high-performance TAs degrade with difficulty or slowly generate potential excretion problems and biosafety concerns [31,32]. Fewer safety and biocompatibility issues result when TAs with a short degradation time (viz., indocyanine green) are harnessed [33,34]. A difficulty arises from the compromise of the thermal performance, which is required for superior therapeutic efficacy, due to the rapid degradation of conventional highly degradable TAs. For instance, upon the application of a short or small number of stimulations, thermal functions disappear [35,36]. Moreover, the degradation of typical TAs further enhances upon the application of stimulations [37,38]. After incubation in aqueous solutions, two dimensional (2D) materials with metal ions that reveal a lower absorption compared to that of pristine 2D materials, indicating that the metal ions facilitate the degradation of the 2D material, have also been demonstrated in recent studies [22]. Furthermore, experiments have shown that, upon the incubation in an aqueous solution, bare nanostructures disclose a greater loss in the absorption than that of nanostructures with a polymer coating, which indicates faster degradation [39]. As a result, an impediment to fulfilling the clinical promise of TBT is represented by the balance between increasing the thermal performance and simultaneously enhancing the degradation of traditional TAs.

The molybdenum disulfide (MoS_2) material system is a leading contender for next-generation TAs with different nanostructure types, since it is a biodegradable and biocompatible 2D material. MoS_2 oxidizes/degrades in air and dissolves in aqueous solutions rapidly, which enhances the safety of TBT [40,41]. However, the thermal performance of traditional MoS_2 systems is limited [42,43]. The M13 is a cylindrical, ~880 nm long, and 5–6 nm in diameter bacteriophage composed of specified proteins [44,45]. We postulate that the integration of the MoS_2 and M13 results in MoS_2 nanostructures with M13 (we call it MNM) that exhibit strong thermal performance, and at the same time, maintains excellent degradability. This is conceived by considering that: (i) MoS_2 systems and M13 phages conjugate to polyethylene glycol (PEG) molecules, which maintains good degradation [46,47] and (ii) the M13 enables high thermal performance, since it carries elements to targets and assembles on tumors/cancer cells well [45].

Herein we disclose that by stimulating and altering structural assembling processes, we were able to control the thermal character of the MoS_2 by utilizing MNM, along with developing a combined alternating-current (AC) stimulus PANC-1 cancer cell MNM framework for cancer cell ablation and electrothermal therapy (Figure 1). A degradation time of 2 weeks was achieved, together with a stimulus length of 100 μs and a percentage decrease in the cell viability of ~23% upon the application of electrical stimuli. This work developed an excellent strategy to avoid the previous impasse between increasing the electrothermal performance/stability and simultaneously enhancing the degradation of electrothermal agents (ETAs).

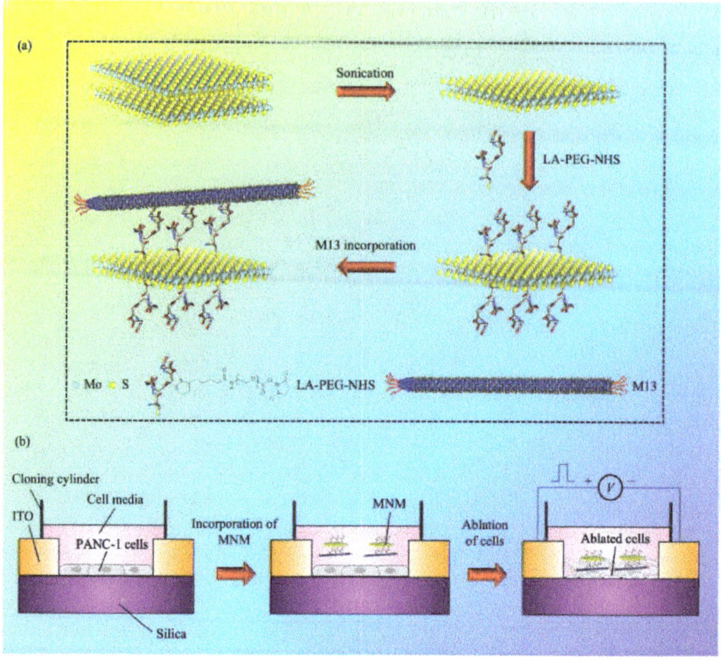

Figure 1. An AC-stimulus cancer cell platform based on MNM. (**a**) The bindings/reactions between the MoS_2/M13 phage and the PEG molecules are illustrated. The schematic illustration was adapted and modified from [48]. (**b**) The MNM is incorporated in the AC-stimulus PANC-1 cancer cell system for cancer cell ablation and electrothermal therapy.

2. Materials and Methods

2.1. Molecular Modelling

The crystal structures of the N1 and N2 domains of the M13 bacteriophage minor coat gene 3 protein (G3P) (PDB ID: 1G3P) and the extracellular domain of the programmed death-ligand 1 (PD-L1) (PDB ID: 3BIK) were retrieved from the Protein Data Bank (PDB) [49]. The oxidized Trp21 in the G3P structure was converted to Trp. Missing loop residues in G3P (residues 66–90) were modeled using the ModLoop web server [50]. Residue 18 was removed from the PD-L1 chain, as it was part of the signal peptide. The two proteins were then docked to each other using the ClusPro web server (accessed on 1 April 2022) [51]. The top five docked models were evaluated, and two were rejected, as they had G3P bound too close to the transmembrane region of the PD-L1, which would result in clashing with the cell membrane. The top-, second-, and fourth-ranked models were selected for further evaluation in molecular dynamics (MD) simulations.

The C-terminus of both protein chains was capped by an N-methyl group. Protonation states were assigned using the PDB2PQR [52] web server and then checked manually. TIP3P water molecules were utilized to solvate the systems with a minimum distance of 10 Å between the proteins and the edge of the periodic truncated octahedron solvent box. The systems were then neutralized by adding sodium ions.

Four independent MD simulation runs were performed for each of the three selected docked models. Energy minimizations and MD simulations were performed with the PMEMD module of AMBER 18 [53] using the ff14SB force field [54]. All bonds involving hydrogen atoms were constrained by the SHAKE algorithm [55], thus allowing for a time step of 2 fs. Nonbonded interactions were truncated at 9 Å. The particle mesh Ewald method [56] was used to treat long-range electrostatic interactions under periodic boundary conditions. Energy minimization was carried out using the steepest descent algorithm for

1000 steps, followed by another 1000 steps with the conjugate gradient algorithm. The systems were then annealed gradually to 300 K over 50 ps at constant volume before equilibration at a constant pressure of 1 atm for another 50 ps. The protein non-hydrogen atoms were kept fixed with a harmonic positional restraint of 2.0 kcal mol^{-1} Å^{-2} during these minimization and equilibration steps. Subsequent unrestrained equilibration (2 ns) and production (300 ns) runs were carried out at 300 K, using a Langevin thermostat [57] with a collision frequency of 2 ps^{-1}, and 1 atm, using a Berendsen barostat [58] with a pressure relaxation time of 2 ps.

2.2. Binding Free Energy Calculations

Binding free energies for the G3P–PD-L1 complex were calculated using the molecular mechanics/Poisson–Boltzmann surface area (MM/PBSA) method [59] implemented in AMBER 18. Two hundred equally spaced snapshot structures were extracted from the last 100 ns of each of the trajectories, and their molecular mechanical energies were calculated with the sander module. The polar contribution to the solvation free energy was calculated by the pbsa [60] program, with the solute dielectric constant set to 2 and the exterior dielectric constant set to 80, while the nonpolar contribution was estimated from the solvent-accessible surface area using the molsurf [61] program, with γ = 0.00542 kcal Å^{-2}, and β = 0.92 [62]. Entropies were estimated by normal mode analysis [63] using the nmode program.

2.3. Electrothermal Simulations

A finite element method (FEM) was utilized in the Ansys software to analyze the thermal distribution of the integrated MNM cancer cell AC-stimulus system. Supplementary Table S1 shows the parameters utilized in the simulation. The heat conduction equation was harnessed to model the heat transfer

$$\nabla \cdot k \nabla T + Q = \rho c \frac{\partial T}{\partial t} \tag{1}$$

where k is the thermal conductivity, T is the temperature, Q is the flow of Joule heat through a unit volume per unit of time, t is the time, ρ is the density, and c is the specific heat. The electrical stimulus with the amplitude of 1–5 V and stimulus length of 100 µs was administered to the system. The initial temperature was fixed at 37 °C (optimal temperature for mammalian cell culture).

2.4. Cell Lines and Cell Culture

The PANC-1 cell line was purchased from American Type Culture Collection (ATCC) and cultured in Dulbecco's Modified Eagle's Medium (DMEM) (Nacalai Tesque Inc., Kyoto, Japan), supplemented with 10% fetal bovine serum (FBS) (Gibco Inc., Carlsbad, CA, USA) and 1% L-glutamine (Gibco Inc., Carlsbad, CA, USA). Cells were incubated at 37 °C in the humidified incubator at an atmosphere of 5% CO_2.

2.5. Escherichia coli (E. coli) and M13 Phage Propagation

The 5-alpha F′Iq competent *E. coli* (high efficiency) was purchased from New England Biolabs (NEB) as the host cell for M13 phage propagation. Overnight culture (O.C.) of *E. coli* was made with tetracycline (TET) and left on the shaker to incubate at 37 °C for 4–6 h at 90 rpm until the mixture became cloudy with an optical absorbance of 0.4 (OD600). New culture (N.C.) was prepared by incubating the O.C. in Lennox L Broth Base (LB Broth Base) at 37 °C for 4–6 h at 90 rpm. First and second precipitations were performed according to the M13 amplification protocol recommended by the manufacturer. The concentration of the M13 phage was measured using a µDrop plate (Thermo Fisher Scientific Inc., Singapore). The bacteriophage M13 15669-B1 (M13 phage) was purchased from ATCC and revived as per the phage recovery and propagation protocol (ATCC).

2.6. MNM Conjugation

The MoS_2 in sterile, deionized water (DI water) was purchased from 2D Semiconductors, Inc. MoS_2 suspension was sonicated prior to MNM conjugation to establish a homogenous mixture. The lipoic acid (LA), PEG, and N-hydroxysuccinimide ester branches (NHS) (LA-PEG-NHS) were purchased from Nanocs Inc. and reconstituted in DI water. The mixture of the MoS_2 and LA-PEG-NHS was left to incubate on the shaker at 25 °C for 48 h under gentle shaking. The M13 phage was filtered to ensure sterility. The filtered M13 phage was then added to the mixture and left to incubate at 25 °C for 48 h under gentle shaking. Conjugated MNM was resuspended in Dulbecco's phosphate-buffered saline (DPBS).

2.7. Material Characterization

Atomic force microscopy (AFM) was performed using the Bruker Dimension Icon system (Bruker Cooperation, Billerica, MA, USA) with a 1.5 μm × 1.5 μm scanning area. Samples were drop-cased on the Si substrate before AFM testing. Raw AFM data analysis was performed with the NanoScope analysis software. Raman analysis was performed using the inVia Raman microscope (Renishaw, Hoffman Estates, IL, USA) with a 532 nm excitation laser via a measured wave number in the 100–3000 cm^{-1} range. Transmission electron microscopy (TEM) imaging was performed with the field emission TEM via the FEI Talos F200 system (Thermo Scientific, Hillsboro, OR, USA), operated at a 120 kV acceleration voltage. Samples were negatively stained and drop-casted on the carbon film-coated Cu grids prior to imaging. Fourier-transform infrared (FTIR) was performed using an infrared spectroscopy (PerkinElmer, Shelton, CT, USA) after drop-casting the samples on a silicon substrate. The stability studies were performed using the MNM with 30% MN. The MNM was dispersed in the DMEM solution and maintained at 37 °C in the humidified incubator at 5% CO_2. At predetermined time intervals, the absorbance of the MNM was measured at λ = 500–600 nm. Thermal analysis was carried out using the lock-in infrared (IR) thermography with the ELITE system (Thermo Fisher Scientific, Waltham, MA, USA) to identify hotspot locations upon the application of direct current (DC) electrical stimulus.

2.8. Cell Viability Studies

PANC-1 cells were plated in a 96-well plate at a seeding density of 3 × 10^3 cells per well and left to incubate for 24 h at 37 °C in a humidified incubator at an atmosphere of 5% CO_2. Different concentrations of the MN were added to the cells, and the cells with the MNM were allowed to incubate for 24 h at 37 °C in the humidified incubator at an atmosphere of 5% CO_2. Cytotoxicity of the MNM was determined by the WST-1 assay performed 24 h after adding the MNM to the cells. The cells were washed once with DPBS prior to the addition of WST-1 assay.

2.9. Electrothermal Ablation Studies

PANC-1 cells with a density of 3 × 10^3 cells per well were seeded in the hybrid MNM cancer cell AC-stimulus system. The system comprised two 650 nm-thick left and right ITO electrodes on the glass substrate (Latech), with a cloning cylinder secured using silicone adhesive (Sigma-Aldrich). The size of the gap between the electrodes was chosen to be 0.1 mm. To allow attachment to the glass surface, cells were cultured for 24 h. The cells were then incubated with MNM. After 24 h, electrical stimuli were applied to the system (amplitude = 1 V, 2 V, and 5 V; stimulus length = 10 μs and 100 μs; and number of stimuli = 10,000). Subsequently, the WST-1 assay was utilized 24 h after applying the stimuli to measure cell viability.

3. Results

3.1. MD Simulations

Experiments have shown that overexpression of the transmembrane protein PD-L1 occurs in PC cells [64,65]. Therefore, we hypothesize that the interaction between the

M13 coat protein, G3P, and the extracellular domain of the PD-L1 could account for the selectivity targeting of PC cells by MNMs. We docked the M13 minor coat protein to the PD-L1 using the ClusPro web server, and out of the top five docked models, three were selected for further evaluation in MD simulations. Based on the computed binding free energies obtained from the MD simulation, the fourth-ranked model attained from the ClusPro web server was the most stable. Its average binding free energy of 28.4 kcal/mol was lower than/comparable to that of other peptide–cancer cell protein models (Supplementary Figure S1). The time evolution of the root-mean-square displacement (RMSD) of the protein backbone atoms from the starting structure showed a gradual increase in values over the initial period (from 0 to ~120 ns) and remained stable afterwards (up to 300 ns), indicating that the system reached equilibrium (Figure 2a).

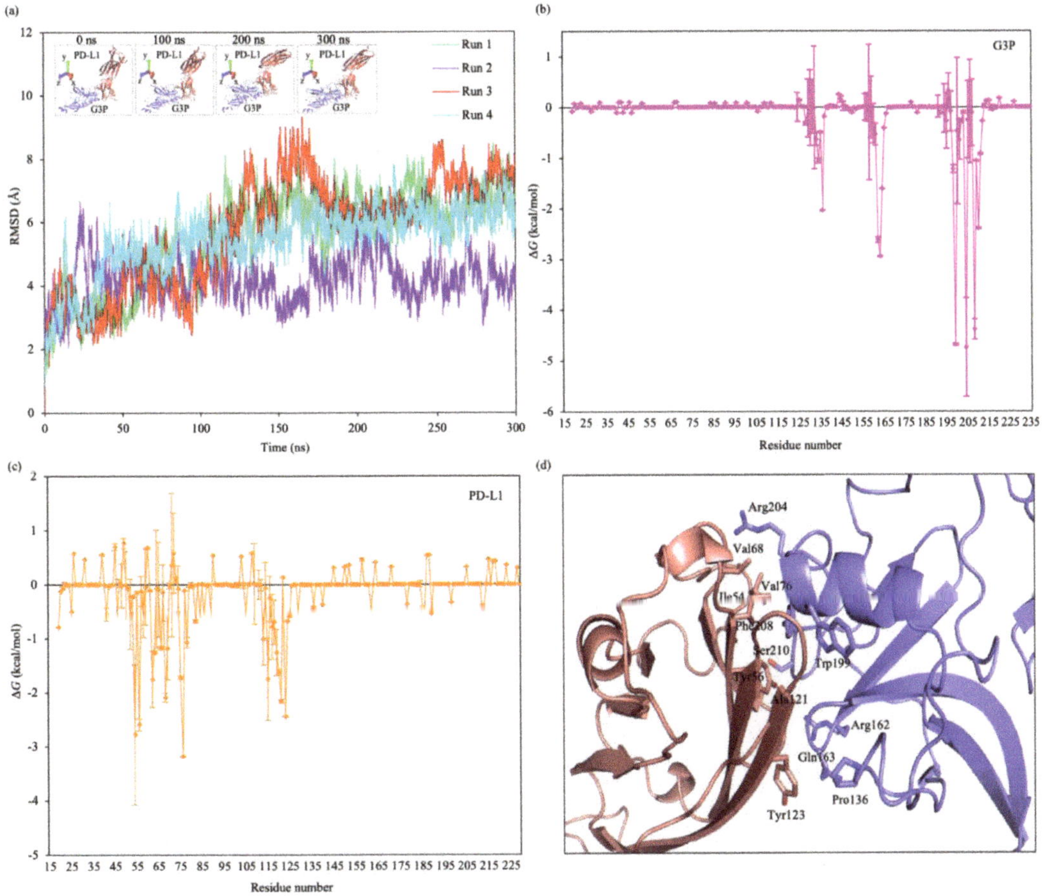

Figure 2. MD simulations. (**a**) Time evolution of the RMSD of backbone atoms of the M13 G3P–PD-L1 protein–protein complex. Inset, snapshots of the M13 phage protein G3P and cancer cell protein PD-L1 for different periods of run 1. (**b**,**c**) Binding free energy contributions of the (**b**) G3P and (**c**) PD-L1 residues. (**d**) Binding interface of the G3P–PD-L1 complex (purple, G3P; pink, PD-L1). The major interacting residues are shown in sticks.

The binding interface of the complex in this model was located at the N2 domain of the G3P and the N-terminal V domain of PD-L1. To determine the residue types that are important for the stability of the interaction between the G3P and PD-L1, we performed the binding free energy decomposition based on the structures of the G3P–PD-L1 complex obtained from MD simulations. The G3P residues Pro136, Arg162, Gln163, Trp199, Arg204, Phe208, and Ser210 and PD-L1 residues Ile54, Tyr56, Val68, Val76, Ala121, and Tyr123 contributed to most of the binding free energy (Figure 2b,c). The hydrophobic nature of the binding interface between nonpolar residues on the G3P and PD-L1 caused the residues to be excluded from the polar aqueous environment and be associated with one another, allowing the proteins to interact in an end-to-end fashion. Thus, the G3P–PD-L1 interaction was stabilized by the hydrophobic core and augmented by peripheral polar interactions (Figure 2d). The breakdown of the components of the computed binding free energy (Supplementary Table S2) also indicates that the complex formation was dominated by hydrophobic interactions.

The M13 phage is also able to carry the nanomaterial to the cancer cell if the phage carries nanomaterials [66,67]. When an electrical stimulus is applied to the system, strong Joule heating occurs because the nanomaterial, e.g., MoS_2, exhibits large electrical conductance [68,69]. As a result, the peak temperature in the cell layer (~319 K) (Supplementary Figures S2 and S3) reaches a value above the temperature required to induce cell death (315 K) [70,71], leading to excellent ablation of cancer cells.

3.2. Synthesis and Characterization of MNM

MoS_2 nanostructure (MN) results from the ultrasonication of bulk MoS_2 samples. The AFM image/cross-sectional plot discloses that the MN exhibits an average thickness of ~6.7 nm (Figure 3a,b), suggesting a stack of two triple-decker layers of MoS_2. Moreover, Raman spectroscopy of the MN is shown in Figure 3c. The Raman spectra of the MoS_2 is dominated by two peaks: (i) the A_{1g} peak, which corresponds to the out-of-plane vibration of S atoms in opposite directions and (ii) the E^1_{2g} peak, which is due to in-plane vibrations of two S atoms, with respect to the Mo atom [72,73]. The MN exhibits the Raman peaks E^1_{2g} and A_{1g} at ~382 and 406 cm^{-1}, respectively, indicating the excellent crystal quality and structure of MN.

To enhance the stability and biocompatibility, we further modified the MN with PEG. Moreover, the M13 phage that assembles on PD-L1 cancer cell proteins was conjugated to PEG molecules to confer the cell-targeting ability to MN. The PD-L1 is a subtype of the integrin protein family that regulates angiogenesis and cancer metastasis, making it an attractive tumor cell and angiogenesis therapeutic target [74,75]. The M13/MN was conjugated to PEG molecules through amine reaction/disulfide binding (Figure 3d) [68,76,77]. The conjugation was performed by utilizing the mixture of the LA-PEG-NHS as a linker between the M13 and PEG molecules. The NHS of the LA-PEG-NHS reacted with the amine group on the M13 phage, while the LA bound to the MN via disulfide binding. Finally, the MNM was prepared, as illustrated in Figure 3d. The Fourier transform infrared (FTIR) spectroscopy was also utilized to investigate the grafting of the LA-PEG-NHS on MoS_2 (Supplementary Figure S4). The prototypical stretching vibration of the carbonyl group in the PEG at ~1090 cm^{-1} was disclosed by the FTIR spectrum of the MNM, indicating the surface presence of PEG. The TEM image revealed that the MNM exhibited a flower-like morphology, as well as a head/sheet-type lateral size of ~200 nm (Figure 3e). For the MN, experiments have demonstrated that the diameter of the MoS_2 nanosheet is large (~500 nm). However, for the MNM, when the PEGylation is carried out, the sonication process can partially break down these nanosheets, leading to a decrease in the diameter of MoS_2 nanosheets. The average thickness of the MNM was ~21 nm, as shown in the AFM image/cross-sectional plot (Figure 3f,g), with an increase in the sample thickness induced by the M13 phage and PEG/polymer coating. Recent studies have shown that the M13 phage revealed a thickness of ~10 nm [78–80], whereas a thickness of a few nm was disclosed for the case of PEG [81–83].

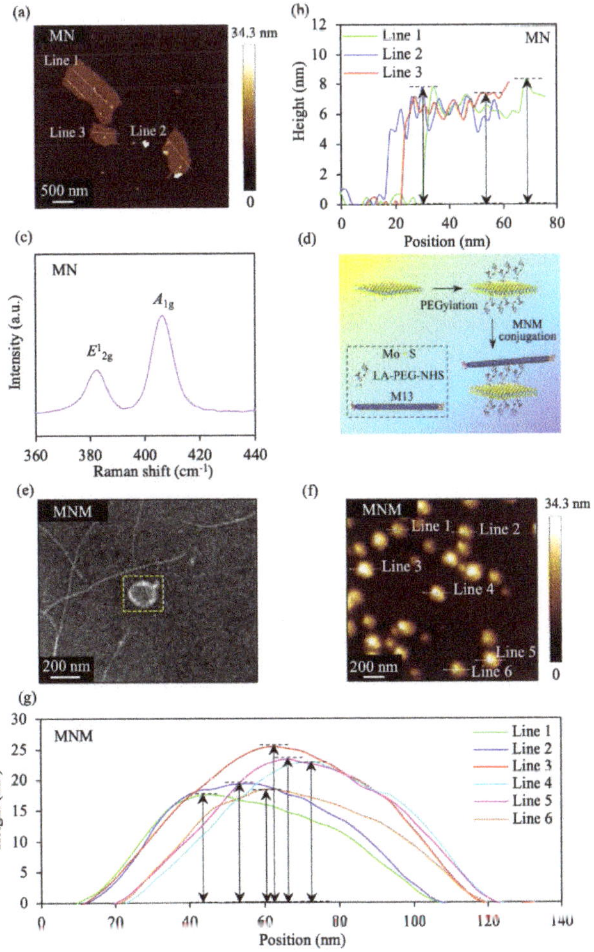

Figure 3. Characterization of the MN and MNM. (**a**) AFM image of MN. (**b**) Height profiles of the MN, along the white lines in (**a**). (**c**) Raman spectra of MN. (**d**) Schematic diagram of the composition and process utilized to construct MNM. (**e**) TEM image of MNM. The yellow dashed boxed area contains the head/ sheet-type structure. TEM images of the M13 phage with a diameter of ~10 nm and also a length of ~1 µm have been shown in M13-type samples harnessed by other research groups [78–80,84]. Our results are similar, since a similar TEM image was achieved for the sample utilized in this work. (**f**) AFM image of MNM. (**g**) Height profiles of the MNM, along the white lines in (**f**).

3.3. MNM Thermal/Stability Signatures

The MNM in the DMEM solution at different times was examined. To measure the absorbance of samples, the ultraviolet visible (UV-Vis) spectrometer was utilized. For materials with a high degree of degradation, a low absorbance/normalized absorbance resulted, whereas a high absorbance/normalized absorbance value occurred for the case of the materials with a low degree of degradation [22,39]. The normalized absorbance (A_N) is given by

$$A_N = \frac{A_x - A_0}{A_0} \tag{2}$$

where A_λ is the absorbance of the MNM in the DMEM at the specified week, and A_0 is the absorbance of the MNM in the DMEM at week 0. At week 0, the MNM in the DMEM revealed a high normalized absorbance value (1.0), while the MNM in the DMEM solution disclosed a low normalized absorbance at week 2 (~0.56) (Figure 4a,b). Thus, a low degree of degradation was demonstrated by the MNM at week 0, and in the week-2 case, a high degree of degradation was revealed by the MNM. Experiments have also shown that most of the Mo element in MoS_2 nanosheets were oxidized to the high valence state (Mo^{VI}), suggesting the oxidation of the $Mo^{IV}S_2$ with a dark brown color into a colorless, water-soluble Mo^{VI}-oxide species, e.g., MoO_4^{2-} [85]. Furthermore, the stimulus current-dependent change in the peak temperature (ΔT) of the MoS_2 sample was investigated (ΔT = temperature achieved at a specified current T_x − reference temperature T_0 (300 K)). The ΔT increased with increasing current (Figure 4c,d), indicating that the temperature in the system can be modified with different stimulus conditions.

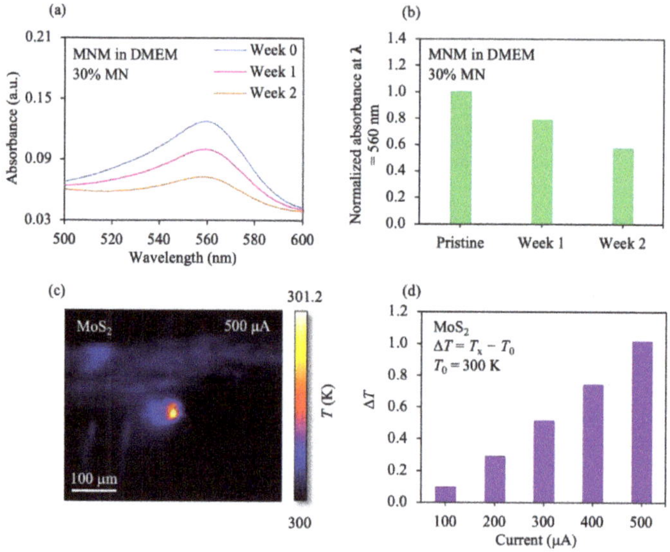

Figure 4. Stability and electrothermal effects of MNM. (**a**) UV-vis absorbance spectra of the MNM in the DMEM solution, with 30% MN for different weeks. The 2D material-based samples utilized by other research groups have demonstrated absorbance spectra with a decrease in the absorbance due to degradation [22,39]. A similar set of spectra were obtained for the samples used in this work, indicating that our results are similar. (**b**) Variation of the normalized absorbance at λ = 560 nm in (**a**) for the MNM in the DMEM, with 30% MN for different periods. (**c**) Thermographic map of the MoS_2 sample upon the application of 500 μA stimulus. (**d**) Change in temperature (ΔT) of the sample for different stimulus currents. T_x is the temperature obtained at the targeted current.

3.4. *Influence of MNM on Cancer Cell Ablation*

We evaluated the influence of MN concentration on the cellular response to MNM. Experiments have demonstrated that nanosheets in cancer cells generate strong Joule heating that leads to cell death [86,87]. As the MN concentration modulates the Joule heating in the MNM, we conceived that the therapeutic influence of the MNM could be controlled by the variation in MN concentrations [28,69]. The PANC-1 cells were incubated with different MN concentrations, and the cell viability was measured to test this hypothesis. The relative cell viability is represented by

$$Relative\ cell\ viability = \frac{A_{pc,\,x} - A_{pc}}{A_{pc}} \times 100\% \quad (3)$$

where $A_{pc,x}$ is the absorbance of PANC-1 cells with a targeted MN concentration, and A_{pc} is the absorbance of PANC-1 cells only. PANC-1 cells with 10–30% MN disclosed a high relative cell viability (~100%) (Figure 5a). On the other hand, a low relative viability was observed for the cells incubated with 90% MN (~61%). Based on these findings, and to achieve a high initial cell viability, as well as a large conductance for strong Joule heating, samples with 10–30% MN were chosen. Upon the application of electrical stimuli, the relative viability of the cells incubated with 30% MN decreased from 100% to 77% (the cell viability decreased by 23%) (Figure 5b). Moreover, with an increase in the MN concentration from 10% to 30%, the relative cell viability after applying the stimuli decreased from 93% to 77%. Additionally, the change in stimulus conditions affected cell viability. When stimuli with an increased length were administered to the cells with 30% MN (from 10 μs to 100 μs), the relative cell viability decreased from 87% to 77%, as shown in Figure 5c. The relative viability of the cells incubated with 30% MN further decreased from 100% to 77% with increasing stimulus voltage (from 1 V to 5 V) (Figure 5d). An important part of cancer cell studies is cell morphology. The morphology of the cells with 2D materials after different treatments has been demonstrated by experiments [22,88,89]. As the samples were modified from the cells with only 2D materials to the cells treated with 2D materials and metal ions/the cells treated with 2D materials and metal ions and conditioned with the targeting agent, the cell morphology changed. The cell morphology also changed when the cells were adjusted from the cells with only 2D materials to the cells with 2D materials and metal ions and conditioned with the targeting agent, including being exposed to optical stimulation. Notably, this work demonstrated the control of thermal signatures using MNM structures, which has not been performed before, as well as developed a previously unreported combined AC-stimulus MNM cancer cell platform for cancer cell ablation and electrothermal therapy. This study also disclosed MD simulations of integrated cancer cell–M13 phage protein systems, which has not been previously carried out. These approaches allow ETAs to maintain excellent degradation, and at the same time, enhance electrothermal performance.

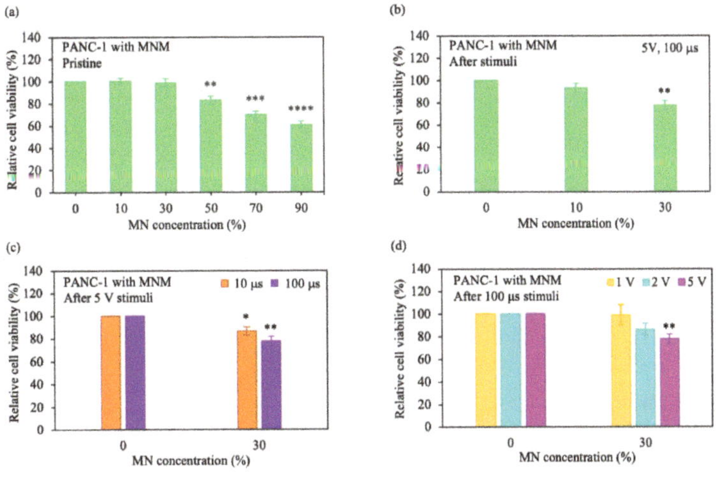

Figure 5. Cytotoxicity and electrothermal performance of MNM on ablating cancer cells. (**a**,**b**) Variation of relative cell viabilities of the (**a**) PANC-1 cells with MNM only and (**b**) PANC-1 cells with MNM after electrical stimuli for different MN concentrations. The cell viability was measured by the WST-1 assay, and the cells were subjected to different concentrations of MN. (**c**,**d**) Relative cell viability variations for different (**c**) stimulus lengths and (**d**) pulse amplitudes. The error bars indicate the standard error of the mean (SEM) from three independent experiments (n = 6). The significance values were calculated using Student's t-test and are indicated as follows: * ($p < 0.05$), ** ($p < 0.01$), *** ($p < 0.001$), and **** ($p < 0.0001$). The non-significance values were unmarked.

4. Discussion

Due to several requirements, applications such as cancer cell ablation and electrothermal therapy are challenging. The requirements are: (1) understanding assembly processes between the phage protein and cancer cell proteins, (2) excellent electrothermal performance, (3) good degradability, and (4) short stimulus time. Currently, the number of traditional ETAs that fulfill the requirements listed above are limited. The current state of the MNM is able to achieve most of these requirements with reasonable performance in relation to conventional ETAs, as indicated by examples disclosed in this work. The vital importance of the MNM to enable the applications is the deeper levels of insights into the hydrophobic interaction-facilitated increase in thermal conduction, which has not been demonstrated before. These results opened the door for utilizing M13 phage cancer cell interactions in corresponding systems specifically for applications in materials science and medicine. Moreover, a percentage decrease in the cell viability of 23% in the integrated MNM cancer cell system under AC stimulation was achieved, which is ~2 times higher than the average of ~10% in current thermal-based therapy systems (Supplementary Figure S6). These results indicate that cell death in a larger population was induced for facilitating effective treatments. Additionally, the MNM exhibited a degradation time of ~2 weeks, which is 73.3% faster than the average of 7.5 weeks in existing MoS_2-based systems in physiological media (Supplementary Figure S7), indicating that the elimination of ETAs from the systems utilized in this work occurred rapidly for enabling safe therapeutics. Furthermore, a stimulus length of 100 µs was achieved in the combined AC-stimulus MNM cancer cell platform, which is ~99.7% shorter than the average of 45 ms in state-of-the-art thermal-type therapy systems (Supplementary Figure S8). This indicates that the thermal generation is rapid for enhancing the treatment quality of patients.

5. Conclusions

These strong decreases in the cell viability and short stimulus length, together with good degradation, are achieved through hydrophobic interactions in integrated AC-stimulus PANC-1 cancer cell MNM systems that alter the thermal signature of MoS_2. A different ETA concept with the potential as a unique resolution for the impasse between increasing the degradability of ETAs, and simultaneously enhancing the electrothermal performance/stability, is represented by the utilization of MNM. This study paves the way for the potential application of the MoS_2 in combined cancer therapies.

Supplementary Materials: The following supporting information can be downloaded at: https://www.mdpi.com/article/10.3390/pharmaceutics15010106/s1, Table S1: Thermoelectric properties of the cell layer/nanostructure model utilized in electrothermal simulations; Table S2: Components of the computed binding free energy (kcal/mol) for the interaction of the G3P with PD-L1; Table S3: References for Figure S1; Table S4: References for Figure S6; Table S5: References for Figure S7; Table S6: References for Figure S8; Figure S1: Comparison of the averaged computed binding-free energy of the G3P–PD-L1 complex with that of current peptide-cancer cell protein models. The information of the references can be found in Table S3; Figure S2: (a) Thermal distribution of the cell-layer/nanostructure model. The MoS_2 and PEG/M13 was inserted in the middle of the cell layer, and a square-based electrical stimulus was applied. (b) Variation of the peak temperature in the cell layer for different stimulus amplitudes; Figure S3: Thermal profile of the cell layer upon the application of 5-V stimulus; Figure S4: Fourier transform infrared (FTIR) spectra of the MNM, LA-PEG-NHS, M13 and MN; Figure S5: Variation of the relative cell viability of PANC-1 cells with MN for different MN concentrations; Figure S6: Comparison of the percentage decrease in the cell viability of integrated MNM cancer cell AC-stimulus systems with that of existing thermal-based therapy platforms. The information of the references can be found in Table S4; Figure S7: Comparison of the degradation time of the MNM with that of state-of-the-art MoS_2-based systems in physiological media. The information of the references can be found in Table S5; Figure S8: Comparison of the stimulus length of the combined MNM AC-stimulus cancer cell platform with that of current thermal-type therapy systems. The information of the references can be found in Table S6.

Author Contributions: Investigation, M.P.M., D.L., J.S.N., S.-X.G., W.C.T. and Y.S.T.; Supervision, N.B. and D.K.L. All authors have read and agreed to the published version of the manuscript.

Funding: This work was financially supported by the Ministry of Education (Singapore) (MOE-T2EP50220-0022), SUTD-MIT International Design Center (Singapore), SUTD Kickstarter Initiative (SKI 2021_02_03, SKI 2021_02_17, SKI 2021_01_04), SUTD-ZJU IDEA Grant Program (SUTD-ZJU (VP) 201903), Changi General Hospital (Singapore) (CGH-SUTD-HTIF2019-001), Agency of Science, Technology, and Research (Singapore) (A20G9b0135), and National Supercomputing Centre (Singapore) (15001618). M.P.M. thanks the Singapore University of Technology and Design for their scholarship support.

Data Availability Statement: The authors declare that all data supporting the findings of this study are available within the article and the Supplementary Information. Other data are available from the corresponding authors upon reasonable request.

Acknowledgments: We thank C. Cheung, X. Liu, N. Thiruchelvam, and J. J.-X. Lee for their valuable discussions.

Conflicts of Interest: The authors declare no competing financial interests.

References

1. Cheng, H.; Liu, C.; Jiang, J.; Luo, G.; Lu, Y.; Jin, K.; Guo, M.; Zhang, Z.; Xu, J.; Liu, L.; et al. Analysis of CtDNA to Predict Prognosis and Monitor Treatment Responses in Metastatic Pancreatic Cancer Patients. *Int. J. Cancer* **2017**, *140*, 2344–2350. [CrossRef] [PubMed]
2. Maisonneuve, P.; Amar, S.; Lowenfels, A.B. Periodontal Disease, Edentulism, and Pancreatic Cancer: A Meta-Analysis. *Ann. Oncol.* **2017**, *28*, 985–995. [CrossRef] [PubMed]
3. Neoptolemos, J.P.; Kleeff, J.; Michl, P.; Costello, E.; Greenhalf, W.; Palmer, D.H. Therapeutic Developments in Pancreatic Cancer: Current and Future Perspectives. *Nat. Rev. Gastroenterol. Hepatol.* **2018**, *15*, 333–348. [CrossRef]
4. Pourshams, A.; Sepanlou, S.G.; Ikuta, K.S.; Bisignano, C.; Safiri, S.; Roshandel, G.; Sharif, M.; Khatibian, M.; Fitzmaurice, C.; Nixon, M.R.; et al. The Global, Regional, and National Burden of Pancreatic Cancer and Its Attributable Risk Factors in 195 Countries and Territories, 1990–2017: A Systematic Analysis for the Global Burden of Disease Study 2017. *Lancet Gastroenterol. Hepatol.* **2019**, *4*, 934–947. [CrossRef] [PubMed]
5. Siegel, R.L.; Miller, K.D.; Jemal, A. Cancer Statistics, 2019. *CA A Cancer J. Clin.* **2019**, *69*, 7–34. [CrossRef] [PubMed]
6. Khalaf, N.; El-Serag, H.B.; Abrams, H.R.; Thrift, A.P. Burden of Pancreatic Cancer—From Epidemiology to Practice. *Clin. Gastroenterol. Hepatol.* **2021**, *19*, 876–884. [CrossRef] [PubMed]
7. ter Veer, E.; van Rijssen, L.B.; Besselink, M.G.; Mali, R.M.A.; Berlin, J.D.; Boeck, S.; Bonnetain, F.; Chau, I.; Conroy, T.; Van Cutsem, E.; et al. Consensus Statement on Mandatory Measurements in Pancreatic Cancer Trials (COMM-PACT) for Systemic Treatment of Unresectable Disease. *Lancet Oncol.* **2018**, *19*, e151–e160. [CrossRef]
8. Mizrahi, J.D.; Surana, R.; Valle, J.W.; Shroff, R.T. Pancreatic Cancer. *Lancet* **2020**, *395*, 2008–2020. [CrossRef]
9. Park, W.; Chawla, A.; O'Reilly, E.M. Pancreatic Cancer: A Review. *JAMA* **2021**, *326*, 851–862. [CrossRef]
10. What Is Cancer | Cancer Facts and Statistics in Singapore. Available online: https://www.gleneagles.com.sg/facilities-services/centre-excellence/cancer-care/what-is-cancer (accessed on 12 September 2022).
11. Cancer Registry—National Registry Of Diseases Office. Available online: https://nrdo.gov.sg/publications/cancer (accessed on 12 September 2022).
12. Singhi, A.D.; Koay, E.J.; Chari, S.T.; Maitra, A. Early Detection of Pancreatic Cancer: Opportunities and Challenges. *Gastroenterology* **2019**, *156*, 2024–2040. [CrossRef]
13. Nguyen, D.-H.T.; Lee, E.; Alimperti, S.; Norgard, R.J.; Wong, A.; Lee, J.J.-K.; Eyckmans, J.; Stanger, B.Z.; Chen, C.S. A Biomimetic Pancreatic Cancer On-Chip Reveals Endothelial Ablation via ALK7 Signaling. *Sci. Adv.* **2019**, *5*, eaav6789. [CrossRef] [PubMed]
14. Bulle, A.; Lim, K.-H. Beyond Just a Tight Fortress: Contribution of Stroma to Epithelial-Mesenchymal Transition in Pancreatic Cancer. *Sig. Transduct. Target.* **2020**, *5*, 1–12. [CrossRef] [PubMed]
15. Abbas, M.; Zou, Q.; Li, S.; Yan, X. Self-Assembled Peptide- and Protein-Based Nanomaterials for Antitumor Photodynamic and Photothermal Therapy. *Adv. Mater.* **2017**, *29*, 1605021. [CrossRef] [PubMed]
16. Zhao, X.; Yang, C.-X.; Chen, L.-G.; Yan, X.-P. Dual-Stimuli Responsive and Reversibly Activatable Theranostic Nanoprobe for Precision Tumor-Targeting and Fluorescence-Guided Photothermal Therapy. *Nat. Commun.* **2017**, *8*, 14998. [CrossRef] [PubMed]
17. Aioub, M.; Panikkanvalappil, S.R.; El-Sayed, M.A. Platinum-Coated Gold Nanorods: Efficient Reactive Oxygen Scavengers That Prevent Oxidative Damage toward Healthy, Untreated Cells during Plasmonic Photothermal Therapy. *ACS Nano* **2017**, *11*, 579–586. [CrossRef]
18. Yang, Y.; Fan, X.; Li, L.; Yang, Y.; Nuernisha, A.; Xue, D.; He, C.; Qian, J.; Hu, Q.; Chen, H.; et al. Semiconducting Polymer Nanoparticles as Theranostic System for Near-Infrared-II Fluorescence Imaging and Photothermal Therapy under Safe Laser Fluence. *ACS Nano* **2020**, *14*, 2509–2521. [CrossRef] [PubMed]

19. Li, X.; Lovell, J.F.; Yoon, J.; Chen, X. Clinical Development and Potential of Photothermal and Photodynamic Therapies for Cancer. *Nat. Rev. Clin. Oncol.* **2020**, *17*, 657–674. [CrossRef]
20. Tao, W.; Kong, N.; Ji, X.; Zhang, Y.; Sharma, A.; Ouyang, J.; Qi, B.; Wang, J.; Xie, N.; Kang, C.; et al. Emerging Two-Dimensional Monoelemental Materials (Xenes) for Biomedical Applications. *Chem. Soc. Rev.* **2019**, *48*, 2891–2912. [CrossRef]
21. Lv, K.; Lin, H.; Qu, F. Biodegradable Hollow Co_3S_4@N-Doped Carbon as Enhanced PTT/PDT Agent for Multimodal MR/Thermal Imaging and Synergistic Antitumor Therapy. *Chem. Eng. J.* **2020**, *392*, 124555. [CrossRef]
22. Hu, K.; Xie, L.; Zhang, Y.; Hanyu, M.; Yang, Z.; Nagatsu, K.; Suzuki, H.; Ouyang, J.; Ji, X.; Wei, J.; et al. Marriage of Black Phosphorus and Cu2+ as Effective Photothermal Agents for PET-Guided Combination Cancer Therapy. phNat. Commun. **2020**, *11*, 2778. [CrossRef]
23. Rastinehad, A.R.; Anastos, H.; Wajswol, E.; Winoker, J.S.; Sfakianos, J.P.; Doppalapudi, S.K.; Carrick, M.R.; Knauer, C.J.; Taouli, B.; Lewis, S.C.; et al. Gold Nanoshell-Localized Photothermal Ablation of Prostate Tumors in a Clinical Pilot Device Study. *Proc. Natl. Acad. Sci. USA* **2019**, *116*, 18590–18596. [CrossRef]
24. Mahmoudi, K.; Bouras, A.; Bozec, D.; Ivkov, R.; Hadjipanayis, C. Magnetic Hyperthermia Therapy for the Treatment of Glioblastoma: A Review of the Therapy's History, Efficacy and Application in Humans. *Int. J. Hyperth.* **2018**, *34*, 1316–1328. [CrossRef] [PubMed]
25. Soetaert, F.; Korangath, P.; Serantes, D.; Fiering, S.; Ivkov, R. Cancer Therapy with Iron Oxide Nanoparticles: Agents of Thermal and Immune Therapies. *Adv. Drug Deliv. Rev.* **2020**, *163–164*, 65–83. [CrossRef] [PubMed]
26. Cheng, Y.; Zheng, X.; Zhang, L.; Zhao, J.; Hu, L.; Wang, S. Enhanced Photothermal and Chemotherapy of Pancreatic Tumors by Degrading the Extracellular Matrix. *Colloids Surf. B Biointerfaces* **2023**, *221*, 113010. [CrossRef] [PubMed]
27. Chen, L.; Feng, Y.; Zhou, X.; Zhang, Q.; Nie, W.; Wang, W.; Zhang, Y.; He, C. One-Pot Synthesis of MoS_2 Nanoflakes with Desirable Degradability for Photothermal Cancer Therapy. *ACS Appl. Mater. Interfaces* **2017**, *9*, 17347–17358. [CrossRef] [PubMed]
28. Li, J.; Zhang, W.; Ji, W.; Wang, J.; Wang, N.; Wu, W.; Wu, Q.; Hou, X.; Hu, W.; Li, L. Near Infrared Photothermal Conversion Materials: Mechanism, Preparation, and Photothermal Cancer Therapy Applications. *J. Mater. Chem. B* **2021**, *9*, 7909–7926. [CrossRef]
29. Fernandes, N.; Rodrigues, C.F.; Moreira, A.F.; Correia, I.J. Overview of the Application of Inorganic Nanomaterials in Cancer Photothermal Therapy. *Biomater. Sci.* **2020**, *8*, 2990–3020. [CrossRef] [PubMed]
30. Wang, Y.; Meng, H.-M.; Li, Z. Near-Infrared Inorganic Nanomaterial-Based Nanosystems for Photothermal Therapy. *Nanoscale* **2021**, *13*, 8751–8772. [CrossRef]
31. Zou, Q.; Abbas, M.; Zhao, L.; Li, S.; Shen, G.; Yan, X. Biological Photothermal Nanodots Based on Self-Assembly of Peptide–Porphyrin Conjugates for Antitumor Therapy. *J. Am. Chem. Soc.* **2017**, *139*, 1921–1927. [CrossRef]
32. An, D.; Fu, J.; Zhang, B.; Xie, N.; Nie, G.; Ågren, H.; Qiu, M.; Zhang, H. NIR-II Responsive Inorganic 2D Nanomaterials for Cancer Photothermal Therapy: Recent Advances and Future Challenges. *Adv. Funct. Mater.* **2021**, *31*, 2101625. [CrossRef]
33. Jiang, K.; Zhang, L.; Hu, Q.; Yue, D.; Zhang, J.; Zhang, X.; Li, B.; Cui, Y.; Yang, Y.; Qian, G. Indocyanine Green–Encapsulated Nanoscale Metal–Organic Frameworks for Highly Effective Chemo-Photothermal Combination Cancer Therapy. *Mater. Today Nano* **2018**, *2*, 50–57. [CrossRef]
34. Wu, J.; Williams, G.R.; Niu, S.; Gao, F.; Tang, R.; Zhu, L.-M. A Multifunctional Biodegradable Nanocomposite for Cancer Theranostics. *Adv. Sci.* **2019**, *6*, 1802001. [CrossRef]
35. Hussein, E.A.; Zagho, M.M.; Nasrallah, G.K.; Elzatahry, A.A. Recent Advances in Functional Nanostructures as Cancer Photothermal Therapy. *Int. J. Nanomed.* **2018**, *13*, 2897–2906. [CrossRef]
36. Yin, C.; Li, X.; Wang, Y.; Liang, Y.; Zhou, S.; Zhao, P.; Lee, C.-S.; Fan, Q.; Huang, W. Organic Semiconducting Macromolecular Dyes for NIR-II Photoacoustic Imaging and Photothermal Therapy. *Adv. Funct. Mater.* **2021**, *31*, 2104650. [CrossRef]
37. Sheng, Y.; Wang, Z.; Neubi, G.M.N.; Cheng, H.; Zhang, C.; Zhang, H.; Wang, R.; Zhou, J.; Ding, Y. Lipoprotein-Inspired Penetrating Nanoparticles for Deep Tumor-Targeted Shuttling of Indocyanine Green and Enhanced Photo-Theranostics. *Biomater. Sci.* **2019**, *7*, 3425–3437. [CrossRef] [PubMed]
38. Shen, J.; Chen, D.; Liu, Y.; Gao, G.; Liu, Z.; Wang, G.; Wu, C.; Fang, X. A Biodegradable Nano-Photosensitizer with Photoactivatable Singlet Oxygen Generation for Synergistic Phototherapy. *J. Mater. Chem. B* **2021**, *9*, 4826–4831. [CrossRef]
39. Shao, J.; Xie, H.; Huang, H.; Li, Z.; Sun, Z.; Xu, Y.; Xiao, Q.; Yu, X.-F.; Zhao, Y.; Wang, H.; et al. Biodegradable Black Phosphorus-Based Nanospheres for in Vivo Photothermal Cancer Therapy. *Nat. Commun.* **2016**, *7*, 12967. [CrossRef]
40. Wang, Z.; von dem Bussche, A.; Qiu, Y.; Valentin, T.M.; Gion, K.; Kane, A.B.; Hurt, R.H. Chemical Dissolution Pathways of MoS_2 Nanosheets in Biological and Environmental Media. *Environ. Sci. Technol.* **2016**, *50*, 7208–7217. [CrossRef]
41. Kurapati, R.; Muzi, L.; de Garibay, A.P.R.; Russier, J.; Voiry, D.; Vacchi, I.A.; Chhowalla, M.; Bianco, A. Enzymatic Biodegradability of Pristine and Functionalized Transition Metal Dichalcogenide MoS_2 Nanosheets. *Adv. Funct. Mater.* **2017**, *27*, 1605176. [CrossRef]
42. Wang, S.; Li, K.; Chen, Y.; Chen, H.; Ma, M.; Feng, J.; Zhao, Q.; Shi, J. Biocompatible PEGylated MoS_2 Nanosheets: Controllable Bottom-up Synthesis and Highly Efficient Photothermal Regression of Tumor. *Biomaterials* **2015**, *39*, 206–217. [CrossRef]
43. Chan, S.S.; Lee, D.; Meivita, M.P.; Li, L.; Tan, Y.S.; Bajalovic, N.; Loke, D.K. Ultrasensitive two-dimensional material-based MCF-7 cancer cell sensor driven by perturbation processes. *Nanoscale Adv.* **2021**, *3*, 6974–6983. [CrossRef] [PubMed]
44. Li, L.; Belcher, A.M.; Loke, D.K. Simulating selective binding of a biological template to a nanoscale architecture: A core con-cept of a clamp-based binding-pocket-favored N-terminal-domain assembly. *Nanoscale* **2020**, *12*, 24214–24227. [CrossRef] [PubMed]

45. Bortot, B.; Apollonio, M.; Baj, G.; Andolfi, L.; Zupin, L.; Crovella, S.; di Giosia, M.; Cantelli, A.; Saporetti, R.; Ulfo, L.; et al. Advanced Photodynamic Therapy with an Engineered M13 Phage Targeting EGFR: Mitochondrial Localization and Autophagy Induction in Ovarian Cancer Cell Lines. *Free Radic. Biol. Med.* **2022**, *179*, 242–251. [CrossRef] [PubMed]
46. Emami, F.; Banstola, A.; Vatanara, A.; Lee, S.; Kim, J.O.; Jeong, J.-H.; Yook, S. Doxorubicin and Anti-PD-L1 Antibody Conjugated Gold Nanoparticles for Colorectal Cancer Photochemotherapy. *Mol. Pharm.* **2019**, *16*, 1184–1199. [CrossRef] [PubMed]
47. Kadkhoda, J.; Akrami-Hasan-Kohal, M.; Tohidkia, M.R.; Khaledi, S.; Davaran, S.; Aghanejad, A. Advances in Antibody Nanoconjugates for Diagnosis and Therapy: A Review of Recent Studies and Trends. *Int. J. Biol. Macromol.* **2021**, *185*, 664–678. [CrossRef]
48. Yoo, S.Y.; Merzlyak, A.; Lee, S.-W. Synthetic Phage for Tissue Regeneration. *Mediat. Inflamm.* **2014**, *2014*, 1–11. [CrossRef]
49. Berman, H.M.; Westbrook, J.; Feng, Z.; Gilliland, G.; Bhat, T.N.; Weissig, H.; Shindyalov, I.N.; Bourne, P.E. The Protein Data Bank. *Nucleic Acids Res.* **2000**, *28*, 235–242. [CrossRef]
50. Fiser, A.; Sali, A. ModLoop: Automated Modeling of Loops in Protein Structures. *Bioinformatics* **2003**, *19*, 2500–2501. [CrossRef]
51. Kozakov, D.; Hall, D.R.; Xia, B.; Porter, K.A.; Padhorny, D.; Yueh, C.; Beglov, D.; Vajda, S. The ClusPro Web Server for Protein-Protein Docking. *Nat. Protoc.* **2017**, *12*, 255–278. [CrossRef]
52. Dolinsky, T.J.; Nielsen, J.E.; McCammon, J.A.; Baker, N.A. PDB2PQR: An Automated Pipeline for the Setup of Poisson-Boltzmann Electrostatics Calculations. *Nucleic Acids Res.* **2004**, *32*, W665–W667. [CrossRef]
53. Case, D.A.; Ben-Shalom, I.Y.; Brozell, S.R.; Cerutti, D.S.; Cheatham, T.E., III; Cruzeiro, V.W.D.; Darden, T.A.; Duke, R.E.; Ghoreishi, D.; Gilson, M.K.; et al. *AMBER 2018*; University of California: San Francisco, CA, USA, 2018.
54. Maier, J.A.; Martinez, C.; Kasavajhala, K.; Wickstrom, L.; Hauser, K.E.; Simmerling, C. Ff14SB: Improving the Accuracy of Protein Side Chain and Backbone Parameters from Ff99SB. *J. Chem. Theory Comput.* **2015**, *11*, 3696–3713. [CrossRef] [PubMed]
55. Ryckaert, J.-P.; Ciccotti, G.; Berendsen, H.J.C. Numerical Integration of the Cartesian Equations of Motion of a System with Constraints: Molecular Dynamics of n-Alkanes. *J. Comput. Phys.* **1977**, *23*, 327–341. [CrossRef]
56. Darden, T.; York, D.; Pedersen, L. Particle Mesh Ewald: An N·log(N) Method for Ewald Sums in Large Systems. *J. Chem. Phys.* **1993**, *98*, 10089–10092. [CrossRef]
57. Izaguirre, J.A.; Catarello, D.P.; Wozniak, J.M.; Skeel, R.D. Langevin Stabilization of Molecular Dynamics. *J. Chem. Phys.* **2001**, *114*, 2090–2098. [CrossRef]
58. Berendsen, H.J.C.; Postma, J.P.M.; van Gunsteren, W.F.; DiNola, A.; Haak, J.R. Molecular Dynamics with Coupling to an External Bath. *J. Chem. Phys.* **1984**, *81*, 3684–3690. [CrossRef]
59. Srinivasan, J.; Cheatham, T.E.; Cieplak, P.; Kollman, P.A.; Case, D.A. Continuum Solvent Studies of the Stability of DNA, RNA, and Phosphoramidate–DNA Helices. *J. Am. Chem. Soc.* **1998**, *120*, 9401–9409. [CrossRef]
60. Accelerated Poisson-Boltzmann Calculations for Static and Dynamic Systems—PubMed. Available online: https://pubmed.ncbi.nlm.nih.gov/12210150/ (accessed on 29 October 2022).
61. Connolly, M.L. Analytical Molecular Surface Calculation. *J. Appl. Crystallogr.* **1983**, *16*, 548–558. [CrossRef]
62. Sitkoff, D.; Sharp, K.A.; Honig, B. Accurate Calculation of Hydration Free Energies Using Macroscopic Solvent Models. Available online: https://pubs.acs.org/doi/pdf/10.1021/j100058a043 (accessed on 29 October 2022).
63. Brooks, B.R.; Janežič, D.; Karplus, M. Harmonic Analysis of Large Systems. I. Methodology. *J. Comput. Chem.* **1995**, *16*, 1522–1542. [CrossRef]
64. Zhuan-Sun, Y.; Huang, F.; Feng, M.; Zhao, X.; Chen, W.; Zhu, Z.; Zhang, S. Prognostic Value of PD-L1 Overexpression for Pancreatic Cancer: Evidence from a Meta-Analysis. *Onco Targets* **2017**, *10*, 5005–5012. [CrossRef]
65. Wang, X.; Li, X.; Wei, X.; Jiang, H.; Lan, C.; Yang, S.; Wang, H.; Yang, Y.; Tian, C.; Xu, Z.; et al. PD-L1 Is a Direct Target of Cancer-FOXP3 in Pancreatic Ductal Adenocarcinoma (PDAC), and Combined Immunotherapy with Antibodies against PD-L1 and CCL5 Is Effective in the Treatment of PDAC. *Sig. Transduct. Target.* **2020**, *5*, 1–12. [CrossRef]
66. Suthiwangcharoen, N.; Li, T.; Li, K.; Thompson, P.; You, S.; Wang, Q. M13 Bacteriophage-Polymer Nanoassemblies as Drug Delivery Vehicles. *Nano Res.* **2011**, *4*, 483–493. [CrossRef]
67. Loke, D.K.; Clausen, G.J.; Ohmura, J.F.; Chong, T.C.; Belcher, A.M. Biological-templating of a segregating binary alloy for nanowire-like phase-change materials and memory. *ACS Appl. Nano Mater.* **2018**, *1*, 6556–6562. [CrossRef]
68. Feng, W.; Chen, L.; Qin, M.; Zhou, X.; Zhang, Q.; Miao, Y.; Qiu, K.; Zhang, Y.; He, C. Flower-like PEGylated MoS_2 Nanoflakes for near-Infrared Photothermal Cancer Therapy. *Sci. Rep.* **2015**, *5*, 17422. [CrossRef]
69. Chan, S.S.; Tan, Y.S.; Wu, K.X.; Cheung, C.; Loke, D.K. Ultra-High Signal Detection of Human Embryonic Stem Cells Driven by Two-Dimensional Materials. *ACS Appl. Bio Mater.* **2018**, *1*, 210–215. [CrossRef] [PubMed]
70. Han, H.S.; Choi, K.Y. Advances in Nanomaterial-Mediated Photothermal Cancer Therapies: Toward Clinical Applications. *Biomedicines* **2021**, *9*, 305. [CrossRef] [PubMed]
71. Luo, L.; Qin, B.; Jiang, M.; Xie, L.; Luo, Z.; Guo, X.; Zhang, J.; Li, X.; Zhu, C.; Du, Y.; et al. Regulating Immune Memory and Reversing Tumor Thermotolerance through a Step-by-Step Starving-Photothermal Therapy. *J. Nanobiotechnol.* **2021**, *19*, 297. [CrossRef]
72. Fu, S.; Kang, K.; Shayan, K.; Yoshimura, A.; Dadras, S.; Wang, X.; Zhang, L.; Chen, S.; Liu, N.; Jindal, A.; et al. Enabling Room Temperature Ferromagnetism in Monolayer MoS_2 via in Situ Iron-Doping. *Nat. Commun.* **2020**, *11*, 2034. [CrossRef]
73. Quan, J.; Linhart, L.; Lin, M.-L.; Lee, D.; Zhu, J.; Wang, C.-Y.; Hsu, W.-T.; Choi, J.; Embley, J.; Young, C.; et al. Phonon Renormalization in Reconstructed MoS_2 Moiré Superlattices. *Nat. Mater.* **2021**, *20*, 1100–1105. [CrossRef]

74. Sun, Z.; Zhang, Y.; Cao, D.; Wang, X.; Yan, X.; Li, H.; Huang, L.; Qu, X.; Kong, C.; Qin, H.; et al. PD-1/PD-L1 Pathway and Angiogenesis Dual Recognizable Nanoparticles for Enhancing Chemotherapy of Malignant Cancer. *Drug Deliv.* **2018**, *25*, 1746–1755. [CrossRef]
75. Yang, Y.; Xia, L.; Wu, Y.; Zhou, H.; Chen, X.; Li, H.; Xu, M.; Qi, Z.; Wang, Z.; Sun, H.; et al. Programmed Death Ligand-1 Regulates Angiogenesis and Metastasis by Participating in the c-JUN/VEGFR2 Signaling Axis in Ovarian Cancer. *Cancer Commun.* **2021**, *41*, 511–527. [CrossRef]
76. Chung, W.-J.; Lee, D.-Y.; Yoo, S.Y. Chemical Modulation of M13 Bacteriophage and Its Functional Opportunities for Nanomedicine. *Int. J. Nanomed.* **2014**, *9*, 5825–5836. [CrossRef]
77. Yin, F.; Anderson, T.; Panwar, N.; Zhang, K.; Tjin, S.C.; Ng, B.K.; Yoon, H.S.; Qu, J.; Yong, K.-T. Functionalized MoS_2 Nanosheets as Multi-Gene Delivery Vehicles for In Vivo Pancreatic Cancer Therapy. *Nanotheranostics* **2018**, *2*, 371–386. [CrossRef] [PubMed]
78. Whaley, S.R.; English, D.S.; Hu, E.L.; Barbara, P.F.; Belcher, A.M. Selection of Peptides with Semiconductor Binding Specificity for Directed Nanocrystal Assembly. *Nature* **2000**, *405*, 665–668. [CrossRef]
79. Zhu, H.; White, I.M.; Suter, J.D.; Zourob, M.; Fan, X. Opto-Fluidic Micro-Ring Resonator for Sensitive Label-Free Viral Detection. *Analyst* **2008**, *133*, 356–360. [CrossRef] [PubMed]
80. Torrisi, L.; Silipigni, L.; Kovacik, L.; Lavrentiev, V.; Cutroneo, M.; Torrisi, A.; De Plano, L.M.; Franco, D.; Guglielmino, S. M13 Phages Uptake of Gold Nanoparticles for Radio- and Thermal-Therapy and Contrast Imaging Improvement. *Appl. Sci.* **2021**, *11*, 11391. [CrossRef]
81. Cheng, L.; Liu, J.; Gu, X.; Gong, H.; Shi, X.; Liu, T.; Wang, C.; Wang, X.; Liu, G.; Xing, H.; et al. PEGylated WS_2 Nanosheets as a Multifunctional Theranostic Agent for in Vivo Dual-Modal CT/Photoacoustic Imaging Guided Photothermal Therapy. *Adv. Mater.* **2014**, *26*, 1886–1893. [CrossRef]
82. Liu, T.; Wang, C.; Gu, X.; Gong, H.; Cheng, L.; Shi, X.; Feng, L.; Sun, B.; Liu, Z. Drug Delivery with PEGylated MoS_2 Nano-Sheets for Combined Photothermal and Chemotherapy of Cancer. *Adv. Mater.* **2014**, *26*, 3433–3440. [CrossRef]
83. Pan, W.; Liu, C.; Li, Y.; Yang, Y.; Li, W.; Feng, C.; Li, L. Ultrathin Tellurium Nanosheets for Simultaneous Cancer Thermo-Chemotherapy. *Bioact. Mater.* **2022**, *13*, 96–104. [CrossRef]
84. Li, W.; Suez, I.; Szoka, F.C. Reconstitution of the M13 Major Coat Protein and Its Transmembrane Peptide Segment on a DNA Template. *Biochemistry* **2007**, *46*, 8579–8591. [CrossRef]
85. Hao, J.; Song, G.; Liu, T.; Yi, X.; Yang, K.; Cheng, L.; Liu, Z. In Vivo Long-Term Biodistribution, Excretion, and Toxicology of PEGylated Transition-Metal Dichalcogenides MS_2 (M = Mo, W, Ti) Nanosheets. *Adv. Sci. (Weinh)* **2016**, *4*, 1600160. [CrossRef]
86. Chan, S.S.Y.; Go, S.X.; Meivita, M.P.; Lee, D.; Bajalovic, N.; Loke, D.K. Ultra-Efficient Highly-Selective MFC-7 Cancer Cell Therapy Enabled by Combined Electric-Pulse Carbon 1D-Nanomaterials Platforms. *Mater. Adv.* **2022**, *3*, 3915–3924. [CrossRef]
87. Meivita, M.P.; Chan, S.S.Y.; Go, S.X.; Lee, D.; Bajalovic, N.; Loke, D.K. WS_2/Polyethylene Glycol Nanostructures for Ultra-Efficient MCF-7 Cancer Cell Ablation and Electrothermal Therapy. *ACS Omega* **2022**, *7*, 23075–23082. [CrossRef] [PubMed]
88. Lu, H.; Xu, Y.; Qiao, R.; Lu, Z.; Wang, P.; Zhang, X.; Chen, A.; Zou, L.; Wang, Z. A Novel Clustered SPIO Nanoplatform with Enhanced Magnetic Resonance T2 Relaxation Rate for Micro-Tumor Detection and Photothermal Synergistic Therapy. *Nano Res.* **2020**, *13*, 2216–2225. [CrossRef]
89. Wang, Y.; Li, M.; Luo, T.; Jiao, M.; Jin, S.; Dou, P.; Zuo, F.; Wu, C.; Han, C.; Li, J.; et al. Development of FL/MR Dual-Modal Au Nanobipyramids for Targeted Cancer Imaging and Photothermal Therapy. *Mater. Sci. Eng. C* **2021**, *127*, 112190. [CrossRef]

pharmaceutics

MDPI

Review

Mitophagy Induced by Metal Nanoparticles for Cancer Treatment

Deepa Mundekkad [1] and William C. Cho [2,*]

[1] Centre for NanoBioTechnology (CNBT), Vellore Institute of Technology, Vellore 632014, India
[2] Department of Clinical Oncology, Queen Elizabeth Hospital, Kowloon, Hong Kong SAR, China
* Correspondence: chocs@ha.org.hk or williamcscho@gmail.com

Abstract: Research on nanoparticles, especially metal nanoparticles, in cancer therapy is gaining momentum. The versatility and biocompatibility of metal nanoparticles make them good for various applications in cancer therapy. They can bring about apoptotic cell death in cancer cells. In addition to apoptosis, nanoparticles mediate a special type of autophagy facilitated through mitochondria called mitophagy. Interestingly, nanoparticles with antioxidant properties are capable of inducing mitophagy by altering the levels of reactive oxygen species and by influencing signaling pathways like PINK/Parkin pathway and P13K/Akt/mTOR pathway. The current review presents various roles of metal nanoparticles in inducing mitophagy in cancer cells. We envision this review sheds some light on the blind spots in the research related to mitophagy induced by nanoparticles for cancer treatment.

Keywords: metal oxide nanoparticles; mitophagy; dysfunctional mitochondria; cancer; oxidative stress-related pathway

1. Introduction

Nanoparticles have wide applications in the medical field with important roles to play in cancer management. They help in imaging as contrast agents, drug carriers in gene delivery, etc. [1]. They are explicitly used in cancer therapy due to their capacity to deliver drugs to remote regions of the body that are normally inaccessible [2]. Among the different types of nanoparticles, metal nanoparticles are exceedingly used for medical applications due to their thermal, chemical, and optical features [3]. Compared to their counterparts, the metal nanoparticles possess many highly active uncoordinated sites with a large surface-to-volume ratio which makes them attractive. They are found to have a catalytic effect due to their special structural and physical features like the active surface atoms that can change size and shape rendering structural flexibility. The optical polarizability, electrical conductivity, ease with which they can complex with biopolymers, etc. make them ideal for biomedical applications [4] including cancer therapy.

Citation: Mundekkad, D.; Cho, W.C. Mitophagy Induced by Metal Nanoparticles for Cancer Treatment. *Pharmaceutics* **2022**, *14*, 2275. https://doi.org/10.3390/pharmaceutics14112275

Academic Editors: Manivasagan Panchanathan and Eue-Soon Jang

Received: 10 September 2022
Accepted: 19 October 2022
Published: 24 October 2022

Research reveals that the antitumor potency increases when conventional drugs are conjugated with nanoparticles. This is mainly due to the subcellular performance of nanoparticles to penetrate and reach organelles such as endosomes, nucleus, mitochondria, endoplasmic reticulum, Golgi apparatus, etc. Metal nanoparticles have a specific affinity towards mitochondria due to the membrane potential difference and can bring about the inhibition of mitochondrial respiration and can induce cell death in tumor cells by various mechanisms such as apoptosis and autophagy [5]. The role of nanoparticles in mitochondria-mediated cell death (mitophagy) is a relatively less exploited area of research. This review focuses on the role of metal-based nanoparticles in inducing mitophagy in tumor cells.

Even though mitochondria are best known as 'the powerhouse of the cell', it possesses other important roles in cellular metabolism. Mitochondria are recognized as the center of oxidative phosphorylation producing highly reactive oxygen species (ROS). It is the converging point of many metabolic processes and is also actively involved in the cell cycle process, cell differentiation, and cell death. That is why mitochondrial dysfunction is involved in the proliferation and progression of cancer cells. Any dysfunction in the mitochondria can lead to the disruption of oxidative phosphorylation, resulting in reduced energy metabolism, ROS accumulation, inflammation, etc. which render cancer progression. Therefore, targeting mitochondria is an emerging trend in cancer therapy to induce apoptosis [6].

Molecular mechanisms such as autophagy and apoptosis take up the housekeeping role to help the cells to eliminate faulty organelles. Nanoparticles can act as autophagy modulators due to their effect on signaling pathways, thus creating an overstimulating signaling cascade in cancer cells than in non-cancerous cells. ROS-induced autophagic cell death brought about by silver nanoparticles in cancer cells was studied as a selective mechanism of autophagic cell death [7]. The unique property of nanoparticles to induce or inhibit ROS to induce toxicity in cancer cells has enabled their usage in the medical field. In normal skin cells treated with zinc oxide nanoparticles, abnormal accumulation of autophagosome was observed resulting from ROS accumulation in a concentration dependent manner. It was also found to activate autophagy through the inhibition of PI3K/Akt/mTOR signaling pathway [8]. The differences and similarities between autophagy and mitophagy are summarized in Table 1 and the various effects brought about by nanoparticles while inducing autophagy are listed in Table 2.

Table 1. Differences and similarities between autophagy and mitophagy.

	Autophagy	Mitophagy
Type	General form of degradation of cellular components including organelles	Specific degradation of mitochondria
Regulation	Dependent on the nutrient/energy/stress signals	Independent of the nutrient/energy/stress signals
Stimulus	Nutrient and energy stress, ER stress, pathogen-associated molecular patterns (PAMPs), danger-associated molecular patterns (DAMPs), hypoxia, redox stress, mitochondrial damage	Mitochondrial membrane depolarization, changes in cytosolic pH
Malleability	Malleable as it can degrade specific targets, entire organelles, and large portions of cytoplasm	Not malleable at all as it degrades only mitochondria
Substrate	Sequestosome 1/p62 (SQSTM/p62)	Mitophagy involves unique and additional substrate identification mechanisms, notably PTEN-induced kinase 1 (PINK1) and parkin RBR E3 ubiquitin protein ligase (PRKN)
Types	Three types— 1. Chaperone-mediated autophagy (initiated by chaperone Hsc70 and recognizes one protein at a time) 2. Microautophagy (initiated by invagination of lysosomal membranes. Lipid, protein or organelles can be degraded) and 3. Macroautophagy (double membraned organelles are degraded)	Only one type where 1. Only mitochondria are degraded
Methods of detection	1. Conventional electron microscopy to detect the autophagosome number and autophagic flux 2. Fluorescence microscopy to count the average number of punctate structures per cell (puncta formation assay) 3. Immunoblotting to detect the conversion of the cytosolic form of LC3 (LC3-1) to its membrane-bound lipidated form (LC3-II) 4. Flow cytometry to detect the degradation of autophagy-selective substrates	1. Electron microscopy 2. Fluorescence microscopy to detect co-localization of mitochondria with autophagosomes or lysosomes 3. Western blotting to measure the degradation of mitochondrial proteins (like LC3) 4. Fluorescent protein-tagged assays like MitoTimer, mt-Keima, and mito-QC

Table 1. *Cont.*

Similarities
Activation of similar molecular machinery (i.e., BECN1, ULK1/2, LC3)
Generation of ROS molecules can act as a trigger for both processes
Both processes result in cellular damage and apoptosis
Essential for quality control of the cells' defense mechanism
Both processes are forms of degradation mechanisms rather than cell death mechanisms (unlike apoptosis).

SQSTM1—Sequestosome 1, BECN1—the gene encoding the protein Beclin 1, ULK 1/2—Unc-51 such as autophagy activating kinase (1/2), LC3—Microtubule-associated protein 1A/1B-light chain 3.

Table 2. Autophagy induced by different nanoparticles.

Nanoparticle (Size)	Mode of Cellular Recycling	Mechanism/Outcome	Conjugation	In Vitro/In Vivo Model	Ref.
Polydopamine nanoparticle (101.96 ± 6.70 nm)	Autophagy	Photothermal cell killing	Beclin 1- derived peptide (Beclin 1), polyethylene glycol (PEG) and cyclic Arg-Gly-Asp (RGD) peptides (PPBR)	NIH3T3 cells, HeLa cells	[9]
Au NPs (15 nm)	Autophagy	ROS generation by cellular uptake	Poly (acryloyl-L,D) and racemic valine (PAV)	MDA-MB-231 cells	[10]
Autophagy Cascade Amplification NPs (Self-assembled peptide-cholesterol monomers) (150 nm)	Autophagy	Overactivated autophagy and enhanced tumor antigen processing	Oxaliplatin prodrug (HA-OXA)	CT26 tumor-bearing mice	[11]
ZnO NP (300 nm)	Autophagy	ROS-mediated enhanced tumor chemotherapy by overstimulated autophagy	Bare	P-gp-mediated multi-drug resistant human breast cancer cells (MCF-7/ADR cells), BALB/c mice	[12]
Nickel oxide NPs (24.05 ± 2.9 nm)	Autophagy	Oxidative stress, JNK activation	Bare	HeLa	[13]
Silver NPs (78 nm)	Autophagy	Lysosome injury and cell hypoxia	Bare	Prostate cancer cell line (PC-3)	[14]
Silver NPs (11–23 nm)	Enhanced autophagy	Inhibition of NLRP3 inflammasome	Bare	THP-1 cells, AMJ-13 cells, HBL cells	[15]
Gold NPs (60.00 ± 4.24 nm)	Autophagy	ROS-mediated cell death	Bare	Human ovarian adenocarcinoma cells (SKOV-3)	[16]
Hollow mesoporous titanium dioxide nanoparticles (HMTNPs) (~100 nm)	Escape from macrophage phagocytosis	Sonodynamic Therapy	Hydroxychloroquine sulphate (HCQ)	MCF-7, MDA-MB-231, HepG2, NIH3T3,	[17]
Copper oxide NPs (5.4 ± 1.2 nm)	Destructive autophagy	Enhanced radio-sensitizing effect	Bare	MCF-7	[18]
ZIF-82-PVP nanocrystals (~240 nm)	Autophagy	Apoptosis promoted by X-ray-induced nitrosative stress	Conjugation with PVP	MDA-MB-231, 4T1 and Panc-1 cells	[19]

Table 2. Cont.

Nanoparticle (Size)	Mode of Cellular Recycling	Mechanism/Outcome	Conjugation	In Vitro/In Vivo Model	Ref.
Fe_3O_4 NP (26.3 ± 4.42 nm)	Autophagy	ROS-mediated NF-κB and TGF-β signaling pathway activation	Polyethyleneimine	HeLa	[20]
Branched Au-Ag NPs (~200 nm)	Autophagy	Photothermal toxicity	Polydopamine-coated	Human bladder cancer cells (T24 cells)	[21]
TiO_2 NPs (20–30 nm)	Autophagy blockage	Cytotoxicity and apoptosis induction with enhanced chemotherapeutic effect	5-fluorouracil	Human AGS gastric cancer cells	[22]
Au NPs (42.6 ± 5.3 nm)	Autophagy	Immunogenic cancer cell death	Polydopamine	MCF-7 and MDA-MB231	[23]
Copper-palladium alloy tetrapod nanoparticle (~50 nm)	Pro-survival autophagy	TNP-1-mediated photothermal therapy	-	Triple-negative (4T1), drug-resistant (MCF-7/MDR) and patient-derived breast cancer models	[24]
Spheroid fluorescent polystyrene nanoparticles (PS-NPs) (30 nm)	Selective autophagy	Inhibition of autophagosome formation and rescued ATG4-mediated autophagy	Functionalized with amino groups	OAW42 cells	[25]
Chitosan nanoparticles (100.0 ± 6.7 nm)	Cytoprotective autophagy	ROS generation	-	Hela cells and SMMC-7721 cells	[26]
Selenium nanoparticles (60 nm)	Activation of early autophagy but inhibition of late autophagy	Promoting apoptosis	Laminarin polysaccharides	HepG2 cells	[27]
Cuprous oxide NPs (200 nm)	ERK-dependent autophagy	ROS-mediated apoptosis	-	T24, J82, 5637, and UMUC3	[28]

Mitophagy is the molecular event where dysfunctional or damaged mitochondria are effectively degraded and eliminated (Figure 1). Programmed mitophagy generally happens during the developmental process as a specialized form of autophagy. Persistent damage to mitochondria during stress and other pathophysiological conditions can lead to mitophagy. Starvation or hypoxia can also lead to mitophagy. Impaired mitophagy is common among autoimmune diseases, cardiovascular diseases, neurodegenerative diseases, metabolic disorders, and various types of cancers [29]. As metabolic reprogramming is common in cancer, the mitophagic processes are dysfunctional; mitophagy behaves as a tumor promotor or suppressor depending on the components, the type of cancer, and the microenvironment of the cancer cells. For example, BCl_2 Interacting Protein 3 (BNIP3) is a pro-mitophagic receptor that is supposed to induce mitophagy. However, it functions as a tumor suppressor in breast cancer but acts as a promoter of tumor activities in the case of melanoma, renal cell carcinoma, and pancreatic cancer [30]. Other modulators or regulators of mitophagy such as PTEN-induced putative protein kinase 1 (PINK1) and Parkin, also have distinct roles in mitophagy. PINK1/Parkin has tumor suppressive activity and is frequently deleted in several types of cancers such as breast, ovarian, bladder, etc. [31] confirming that the loss of function of PINK1/Parkin can lead to the inhibition of mitophagy and in turn promote tumorigenesis in various types of tumors [32]. Overall, this indicates that mitophagy may have dual roles of inducing cell death as well as promoting cell survival [33].

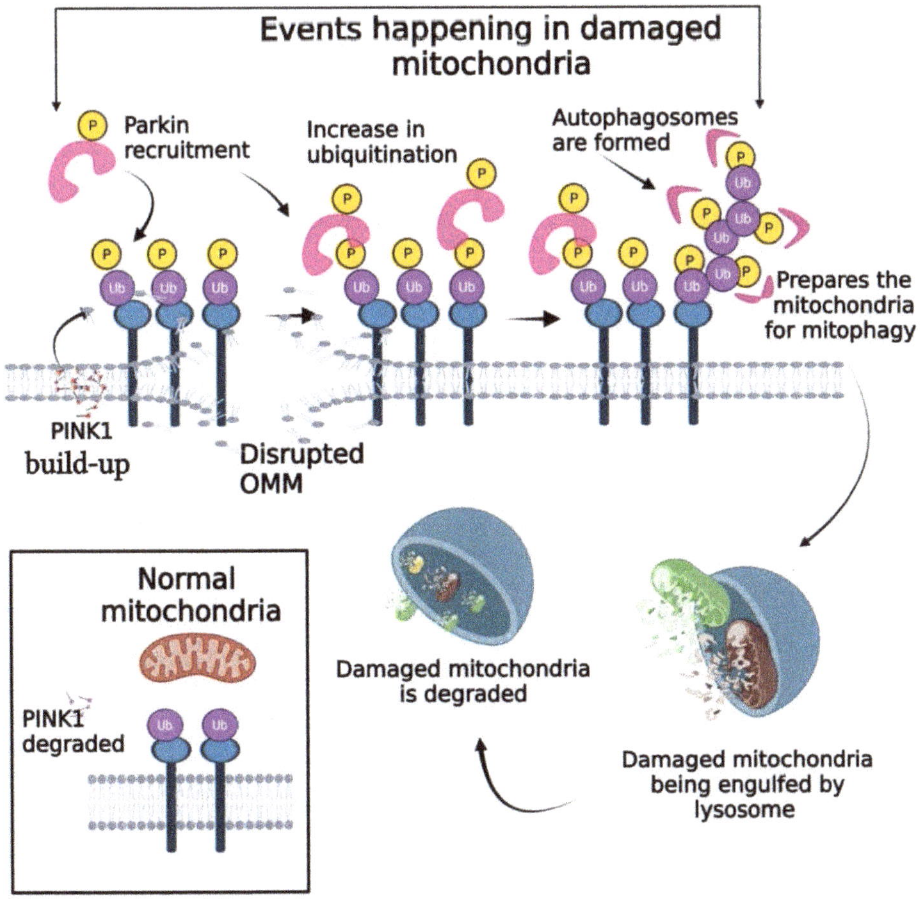

Figure 1. The process of mitophagy. When the outer mitochondrial membrane is disrupted, PINK1 is accumulated which in turn helps in the recruitment of Parkin and other receptor proteins. Autophagosomes formed prepare the mitochondria for mitophagy where the damaged or dysfunctional mitochondria are engulfed by lysosomes and degraded.

Understanding the role of various components such as Pink1/Parkin, BNIP3, FUN14 domain-containing protein 1 (FUNDC1), optineurin (OPTN), microtubule-associated protein 1A/1B-light chain 3 (LC3), etc. in mitophagy will help in managing tumor progression. Chemotherapeutic drugs can induce cytotoxic effects through the induction of mitochondrial dysfunction as a result of oxidative stress that has an inhibitory effect on these components [33]. Many natural compounds are found to have varying effects in the induction of mitophagy [29]. However, at times, the complex nature of mitochondria makes it a challenge to target them for mitophagy. Consequently, a series of research have revealed that conjugating the drugs with ligands that target mitochondria can selectively perturb mitochondrial functions. Some of these studies used nanoparticles as effective drug carriers as they are very responsive to photosensitizers, radiosensitizers, and theranostic agents. They can target the energy machinery of tumor cells and effectively manipulate the underlying functional mechanisms in tumor cells. Nanoparticle-induced toxicity is an emerging area of research with special emphasis given to mitophagy as the means of cell death induced by nanoparticles [34] (Figure 2).

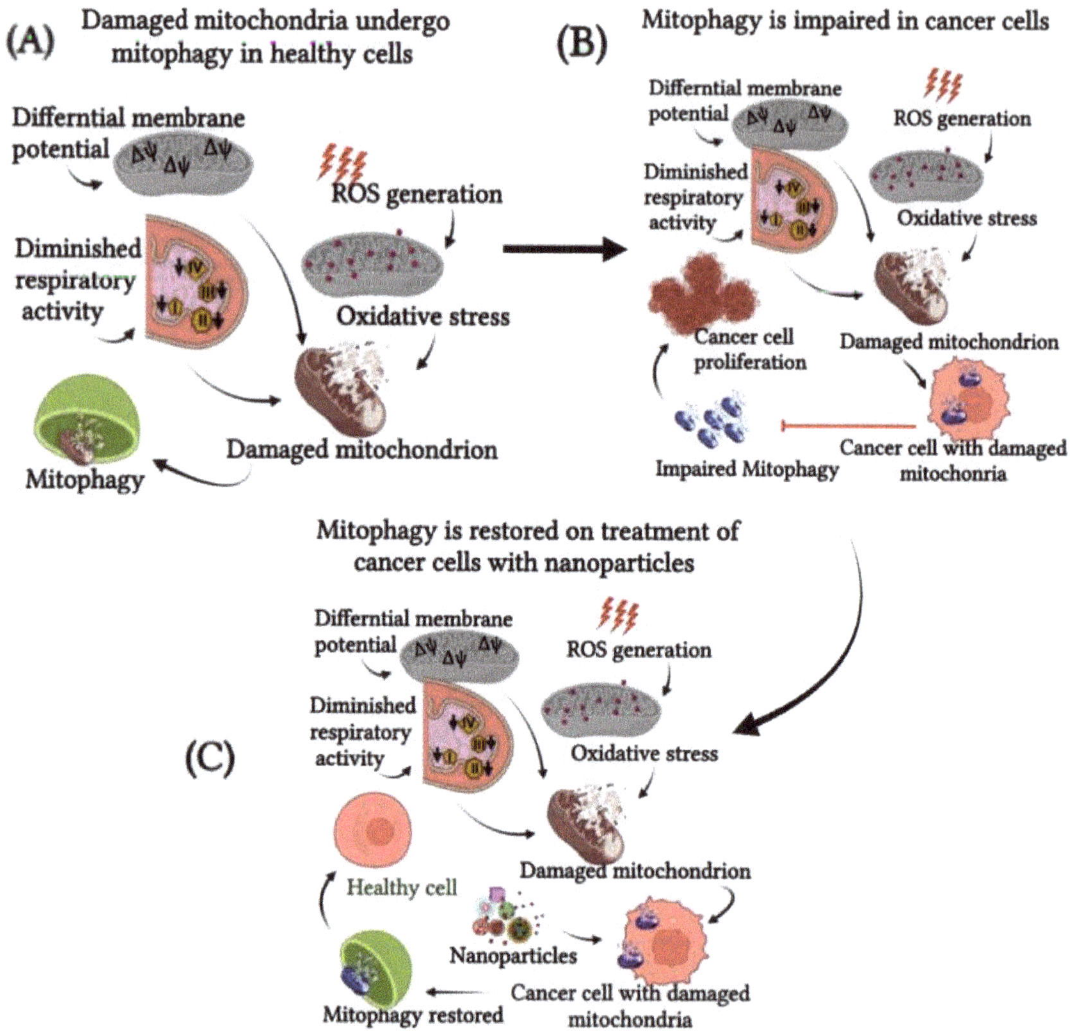

Figure 2. Effect of nanoparticles on mitophagy in cancer cells. Healthy cells undergo mitophagy (**A**) whereas cancer cells accumulate damaged mitochondria and mitophagy is impaired (**B**). Nanoparticles can restore normal homeostasis by restoring mitophagy and thus protecting the cells (**C**).

2. Different Metal-Based Nanoparticles and Their Mode of Action

Metal-based nanoparticles are used extensively in the medical field as drugs and imaging agents [35]. One way by which the nanoparticles exert their therapeutic effect is the covalent bonding with biomolecules. The covalent binding of metal-based nanoparticles with other biomolecules is responsible for the extensive ligand exchange chemistry of the drug. The ligand exchange property of nanoparticles helps to understand and interpret the molecular events at atomic levels so as to study the variations in the property of the drug. The formation of a covalent amide bond with carboxylic acid on the surface of a mesoporous silica nanoparticle, in synergy with zinc oxide quantum dots, was responsible for the inherent anticancer properties. Induction of mitophagy was found to be associated with the anticancer property triggered by excessive ROS formation [36]. The redox balance is important for cellular homeostasis that regulates a plethora of biological processes deciding the physiological well-being of the cell. Zinc oxide nanoparticle was found to induce ROS-mediated autophagy in CAL 27 oral cancer cell line [37]. PINK1/Parkin-mediated mitophagy was reported as the basis for the anticancer activity of the metal oxide nanoparticle. The high expression of ROS in cancer cells results in the oxidation of cell components resulting in the loss of cell function and triggers autophagy. HC11 cells, treated with silver nanoparticle increased the expression of proteins associated with oxidative stress [38]. Hemeoxygnase-1 (HO-1), Kelch-like ECH-associated protein 1 (Kaep1), BTB and CNC homology proteins 1 (Bach1) and nuclear factor erythroid related factor 2 (Nrf2) were found to increase in the cells treated with the nanoparticles.

Another mode of action of nanoparticles is the activation of the immune system by phototherapy. Metal-based nanoparticles are excellent photosensitizers; they can be easily sensitized even by low-intensity light sources as in photodynamic therapy (PDT) and the irradiation creates ROS in cancer cells that can affect the mitochondrial membrane potential to induce mitophagy. In photothermal therapy (PTT), the nanoparticles sensitized with intense light and electromagnetic waves can instigate the expression of heat shock proteins (HSP) on the cell surface which simultaneously leads to the release of cytokines and other inflammatory regulators [2]. The immune system of the cells' defense mechanism is activated upon the release of antigens in photo-immune therapy (PIT). The interventional PDT and PTT can trigger signaling molecules that act on the mitochondrial proteins directing the mitochondria for destruction by mitophagy. Iron oxide nanoparticle was directed to induce mitophagy in MCF-7 cell line through a photothermal effect. When the photosensitized nanoparticles enter the cancer cells, they form aggregates in lysosomes and exocytosis is inhibited [39].

3. Selectivity and Targeting of Mitophagy by Nanoparticles

Early reports on the depolarized mitochondria being deported for damage control indicated mitochondrial autophagy through selective targeting. The involvement of specific proteins such as the key component ATG32 and BNIP3L/NIX, Drp1, etc. were identified and found to have a decisive role in selective targeting [40]. As a sign of cell survival, this mechanism is responsible for clearing the damaged, superfluous, or aged mitochondria with a definite advantage of complete turnover of all the components including the membrane and its associated proteins. Mitochondrial degradation needs to be highly specific and selective as even in extreme conditions such as natural or induced starvation, the mitochondria should be preserved as a source of energy. In such cases, restricted mitochondrial fission followed by a fused mitochondrial network will prevent mitophagy. The bite-sized fragments formed because of this selective induction of mitophagy are degraded. Thus, selective mitophagy throughout the prolonged starvation period is an adaptive response by the cell's defense mechanism to augment and optimize the mitochondrial population. Activation of mitophagy and blockage of mass autophagy was observed in human hepatocellular carcinoma (HepG2) cells treated with silver nanoparticles. Mitochondrial fission induced by Drp1 and oxidative stress promoted mitochondrial degradation but blocked autophagic flux [41]. Similar results were observed in NSCLC cell lines when gold nanoparticle promoted Drp1-dependent mitophagy activation [42]. In addition to the mechanism of action, the size, shape, and conjugation of the nanoparticles with other biomolecules also decide the selectivity of mitochondria in nanoparticle-induced mitophagy. The effective strategy is to conjugate nanoparticles with peptides and amino acids to effectively deliver them to the receptors on the mitochondrial membrane and make them ready for mitophagy rather than any other type of programmed cell death (such as apoptosis or necrosis). However, understanding the relationship between the size, shape and conjugation of nanoparticles will help to target their interactions as it acts as a feedback mechanism for inducing mitophagy. Furthermore, these interactions are believed to have a decisive role in the distribution and intracellular trafficking behaviors of nanoparticles in terms of mitophagy [43]. The feedback mechanisms for the initiation of mitophagy demand autophagosome sequestration and nanoparticles can be selectively manipulated to induce or inhibit mitophagy depending on how this differential role is required to affect the treatment outcome of various cancers [7]. The range of structural diversity of nanoparticles provides them with unlimited combinations that can be employed in the targeted delivery of cancer-specific drugs to mitochondria to effect mitophagy. Surface coating of the nanoparticles such as iron oxide with dimercaptosuccinic acid (DMSA), or 3-aminopropyl-trietoxysilane (APS) is found to affect the intracellular trafficking of drugs [44]. These mechanisms will help in directing the nanoparticles to induce mitophagy.

Mitochondrial targeting can be active or passive. Active targeting is achieved by surface functionalization of the nanoparticles with mitochondria-specific ligands. These ligands could be special moieties that can either be loaded onto the drug or can be standalone. Ligands such as natural products (glycyrrhetinic acid), mitochondrial peptides, α-tocopherol, etc. can be used for active targeting of nanoparticles to the mitochondrial membranes. However, the limitations such as immunogenicity, high production costs, complexities during synthesis, off-target toxicity, prolonged blood circulation period, and delay in clearance from the biological system must be taken into account when designing the proper nanoparticle for active targeting. Passive targeting depends on the physiological and chemical microenvironment of the mitochondria. pH, surface charge, the potential difference between both membranes, surface functionalization, etc. are factors affecting the passive targeting of nanoparticles to mitochondria. This offers an advantage compared to active targeting as the flexibility of nanoparticles allows the manipulation of their physical properties according to the requirements. However, the disadvantage is the possibility of the formation of nanoparticle aggregates resulting in rapid clearance from the host system.

4. Metal Nanoparticles and Their Roles in Mitophagy

In contrast with bulk autophagy, selective autophagy (like mitophagy) identifies specific organelles for degradation depending on cargo-specific receptor proteins. These receptor proteins act as chaperons for translocators to induce the effect and nanoparticles are found to be good candidates for inducing mitophagy due to their physiochemical and biochemical aspects [34]. Even though toxicity is a limiting factor in the extensive use of nanoparticles in medicine, there is research describing nanoparticle-induced toxicity being redirected for inducing apoptosis, oxidative stress, autophagy and even mitophagy [45]. Most of these nanoparticles can effortlessly cross biological barriers and therefore can promote mitophagy. It is essentially due to the physiochemical and biological advantages of nanoparticles that they are capable of crossing biological barriers. Nanoparticles are known to have enhanced permeability and retention effect (EPR) that help them to easily accumulate in the permeable vasculature. The EPR effect along with the ability of nanoparticles to reach specific locations to release drugs in a controlled mode can enhance their therapeutic index.

Pink/Parkin and BNIP3 are the two major pathways involved in mitophagy. Several nanoparticles such as gold, iron, silver, zinc, etc., and their oxides can be found to induce mitophagy in cancer cells through various mechanisms involving these pathways (Figure 3). Some of the nanoparticles or their oxide forms are capable of inducing mitophagy by disturbing the membrane potential of the mitochondria, or increasing the ROS content in the cell, or influencing the signaling pathway [46]. They are even capable of acting as tracking molecules that can trace the routes of various components of mitophagy.

4.1. Gold-Based Nanoparticles

Gold nanoparticles (GNPs/Au NPs) are one of the most studied nanoparticles in cancer research. GNPs can be used to sensitize the tumor cells so that mitochondrial function can be altered accordingly. Human breast cancer cell lines (MDA-MB-231) incubated with GNPs were irradiated with 225 kVp X-rays and were found to influence mitochondrial function resulting in decreased cell survival. The GNPs induced oxidation in the mitochondrial membrane, and mitochondrial polarization was observed [47]. Fluorescently labeled Au NPs (Cy5@ Au NPs) were found to have a high tolerance to lysosomal proteins whereby they could tolerate photobleaching and thus can be used for tracking lysosomes to image mitophagy [48]. Induction of mitophagy, concomitant with apoptosis, was observed in THP-1 cells exposed to gold nanoparticles. These functionalized nanoparticles affected oxidative phosphorylation and protein ubiquitination also. In another study, it was observed that the gold nanoparticle-peptide conjugate can induce mitophagy with a change in the mitochondrial membrane potential. This type of association with nanoparticles and peptides was found to assist cell metabolism as well, even as intracellular trafficking was activated [43]. Autophagic mitochondrial fission was observed in NSCLC cell line treated with Au NP. The Au NP was found to cause excessive mitochondrial fragmentation in the cells under study. This was accompanied by a drastic increase in the recruitment of dynamin-related protein 1 (Drp1), mitochondrial dysfunctions, and enhanced induction of autophagy [42].

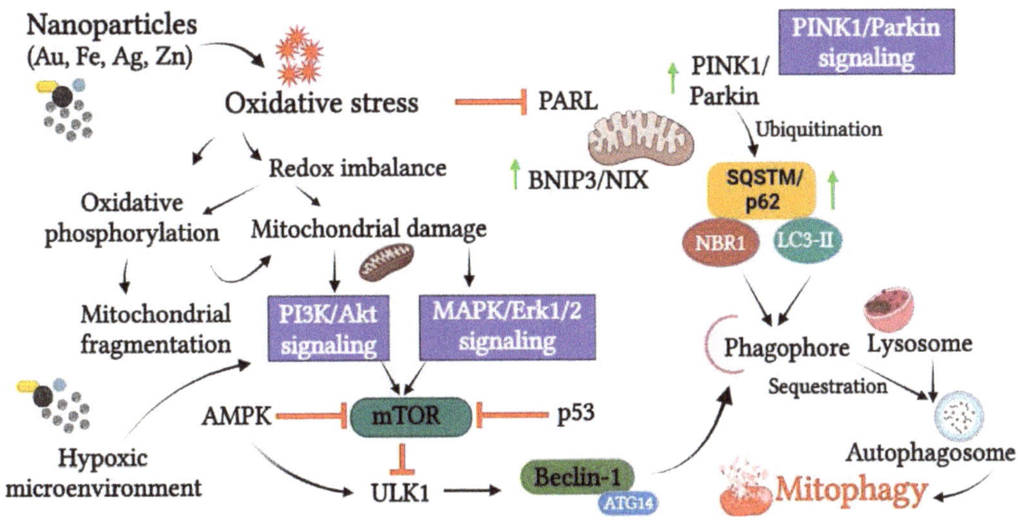

Figure 3. Major molecular pathways implicated in mitophagy that is induced by nanoparticles. Au—gold, Fe—iron, Ag—silver, Zn—zinc, PARL—presenilin-associated rhomboid-like protein, BNIP3—BCL2/adenovirus E1B 19 kDa protein-interacting protein 3, NBR1—neighbor of BRCA1, LC3-II—microtubule-associated protein 1A/1B-light chain 3, AMPK—AMP-activated protein kinase, PI3K—phosphatidylinositol-3-kinase, Akt—protein kinase B, MAPK—mitogen activated protein kinase, Erk 1/2—extracellular signal-regulated kinase 1/2, AMPK—AMP-activated protein kinase, mTOR—mammalian target of rapamycin, ULK1—Unc-51-like kinase 1, ATG14—autophagy-related 14.

4.2. Iron-Based Nanoparticles

Iron nanoparticles (Fe NPs), especially superparamagnetic iron oxide nanoparticles (SPIONs), are found to induce mitophagy in cancer cells. Iron oxide nanoparticles were traditionally made use of in magnetic resonance imaging where they act as contrast agents. SPIONs were found to recruit PARKIN from the cytoplasm to mitochondria, mediated by the protein PINK-1 located on the outer mitochondrial membrane [45]. The ubiquitination of PARKIN makes them susceptible to degradation by lysosomes and thus, mitophagy is induced in the cells under study. The increased involvement of mitochondrial proteins LC3-II and p62 in cells treated with iron oxide nanoparticle are further proof of the execution of mitophagy. Furthermore, iron oxide nanoparticles exhibit enzyme mimicking properties [49] that could be translated to antitumor properties. The inherent enzyme-like activity of iron oxide nanoparticles could initiate mitophagy and protect the cell from oxidative damage mediated by ROS molecules [50]. The ultra-small size of the iron nanoparticles is favorable for their easy transport to mitochondria where they can induce mitophagy after compromising the integrity of the mitochondrial membrane [51]. Cellular internalization was maximized when spindle-shaped iron oxide nanoparticles were used as nano transducers in mitochondria [52].

4.3. Silver-Based Nanoparticles

Silver nanoparticles (Ag NPs) are known to cause damage to DNA mediated by oxidative stress and mitochondrial dysfunction leading to cell death [53]. Silver ions can attach to protein receptors on the cell surface and bring about the denaturation of proteins resulting in pores in the cell membrane. This can lead to the disparity of membrane potential in mitochondria. In the A549 cell line, excessive ROS production and oxidative imbalance due to Ag NPs were found to induce autophagy. This led to mitophagy mediated by PINK1/Parkin pathway as evidenced by upregulation of LC3 II/I, p62 expressions. Mitochondrial membrane potential was reduced accompanied by the upregulation of caspases 3 and 9, cytochrome c and BAX/BCl_2. Human hepatocellular carcinoma (HepG2) cells, when treated with Ag NPs, were found to stimulate mitochondrial fission and oxidative stress. There was a crosstalk between dynamin-related protein 1 (Drp1)-dependent fission and oxidative stress that triggered the Ag NP-mediated mitophagy [41,54]. Ag NPs can also decrease the membrane potential of mitochondria to stimulate mitophagy and apoptosis as observed in glioma cells [55]. As potent ROS inducers, Ag NPs contribute to autophagic flux through redox signaling that involves hypoxia-inducing factors such as HIF-α, thus triggering mitophagy. The combined effect of ionizing radiation and Ag NP on a panel of lung cancer cell lines revealed a dose- and time-dependent increase in protein oxidation releasing mitochondrial ROS. The exposure was found to result in decreased cell proliferation and caused cell cycle arrest in the cells under study [56]. (3-(2,4-dioxocyclohexyl)propyl-Net_2-Coumarin (DCP-Net_2C)) is a probe developed to analyze the sulfenylated proteins in mitochondria that are affected by treatment with Ag NPs [57]. Rather than acting on the mitochondrial membrane permeability transition pore proteins, the silver nanocrystals coated with bovine serum albumin are found to interact with the phospholipid bilayer to induce mitochondrial membrane permeability transition (MPT) resulting in rupture of the mitochondrial membrane [58].

4.4. Zinc-Based Nanoparticles

The treatment of human tongue cancer cells (CAL 27 cell line) with zinc oxide nanoparticles (ZnO NPs) resulted in an increase in the non-functional swelling of mitochondria implying cellular damage to mitochondria resulting in mitophagy and cell death. Further to this, an increase in the intracellular levels of reactive oxygen species along with a decrease in mitochondrial potential was also found in the cells treated with ZnO NPs [37]. The transport of Parkin from the cytosol to the mitochondrial membrane of the treated cells implies the execution of mitophagy by ZnO NPs [59]. The upregulation of hypoxia-inducible factor-1α (HIF-1α) endorsed by the inhibition of prolyl hydrolase and ROS was explained as due to mitophagy induced by ZnO NPs [60]. The synergistic effect of ROS and Zn ions was specially assessed in the upregulation of HIF-1α. The treatment of osteosarcoma cells with ZnO NPs also was found to result in mitophagy. As a result of mitophagy, the cell adhesion protein β-catenin was degraded, and tumor metastasis was impaired in the cells under study [61].

5. Contradiction Where Nanoparticle Treatment Inhibits Mitophagy Instead of Promoting Mitophagy

An interesting research reported a contradictory result where ZnO NP treatment caused aberrant expression of LAMP-2 that resulted in impaired autophagic flux and sequential dysfunctional mitophagy [62]. This resulted in the accumulation of damaged mitochondria and accumulated ROS that was in total disagreement with previous research where ZnO NP induced mitophagy. Furthermore, the intracellular ROS levels were found to be efficient in chemodynamic therapy where the iron-based nanocatalyst effectively inhibited the PINK1/Parkin-mediated mitophagy in endometrial cells [50].

Probable Mechanism of Action of Nanoparticles That Is the Basis for the Contradictory Action

Instead of promoting mitophagy, nanoparticles are found to inhibit mitophagy in some cases. The contradictory action may be because of the variation in biological and physicochemical properties exhibited by the nanoparticles. Though there are limitations to the toxicological assessment of nanoparticles (for example, the inability to quantify the correct dose to get an optimum, quantifiable in vivo effect), the in vitro tests have helped to successfully evaluate the interaction of the nanoparticle with the cellular environment and come up with plausible explanations for the contradictory effect. ROS generation is one of the major reasons for mitophagy induction where membrane potential is affected. Nanoparticles can inhibit intracellular ROS generation thus leading to the inhibition of mitophagy. Platinum nanoparticles retained mitochondrial membrane potential thus inhibiting intracellular ROS in the human brain glioblastoma cancer cell line [63].

6. Pathways Involved in Nanoparticle-Mediated Mitophagy

Mitophagy being an evolutionarily conserved mechanism, the metabolic processes and the cell's defense mechanisms make sure to accurately execute the process of mitophagy to eliminate the damaged mitochondria at the earliest through an interplay of different signaling pathways. It is therefore imperative to understand the components of the pathway that are involved in the process of mitophagy mediated by nanoparticles. The metal nanoparticles are known to have an intrinsic selectivity in activating the pathways leading to mitophagy in cancer, especially as compared to their normal counterparts [64]. Multiple pathways drives mitophagy mediated by nanoparticles, and they may be dependent on regulatory, and signaling pathways, with regular crosstalk between them [29] (Figure 4).

6.1. PINK1/Parkin Pathway

One of the most significant pathways of mitophagy is the PINK1/Parkin pathway which is based on ubiquitin which proceeds to degrade the damaged mitochondria. PINK1 is the first protein to respond to the damage in mitochondria as it can easily sense mitochondrial transmembrane potential loss [65]. PINK1 belongs to the serine/threonine kinase family that is activated under mitochondrial damage. PINK1 is generally very stable in a normal state as they are cleaved by matrix processing peptidase (MPP), and Presenilins-associated rhomboid-like protein (PARL). The accumulation of cleaved PINK1 is prevented by translocating them back to the cytosol to be degraded by proteasomal enzymes [66,67]. However, the cleavage of PINK1 and its further translocation back to the cytosol is impaired upon depolarization due to the loss of transmembrane potential in damaged mitochondria. There is an upsurge in PINK1 followed by the phosphorylation of ubiquitin molecules at serine 65 on the outer mitochondrial membrane. PINK1 along with the phosphorylated ubiquitin then recruits Parkin from the cytosol to the outer mitochondrial membrane where it conjugates with the phosphorylated ubiquitin. Parkin is a cytosolic E3 ubiquitin ligase that promotes the degradation of the ubiquitinated protein. With the help of LC3-II, the ubiquitinated proteins on the outer mitochondrial membrane lead the damaged mitochondria towards the lysosome for destruction by proteasomal enzymes [68]. Other proteins such as NDP52 and optineurin are also involved in the stimulation of mitophagy but they are found to act independent of Parkin [69].

Advanced studies show that when hepatic cells were treated with SPIONs, the immunofluorescent signals given out by PINK1 were increased along with high proportions of mitochondrial LC3-II and p62 [46]. This is indicative of the role of SPIONs in PINK1/Parkin-dependent ubiquitin-mediated mitophagy. Further, biogenically synthesized selenium nanoparticles (Se NPs) tested on IPEC-J2 cells were found to abate the fusion of mitochondria and lysosome, reduce the overproduction of ROS and decrease the mitochondrial membrane potential (MMP). The PINK1 and Parkin expression in the cells were found to be down-regulated confirming mitophagy induction by Se NPs [70] and Au NPs [71].

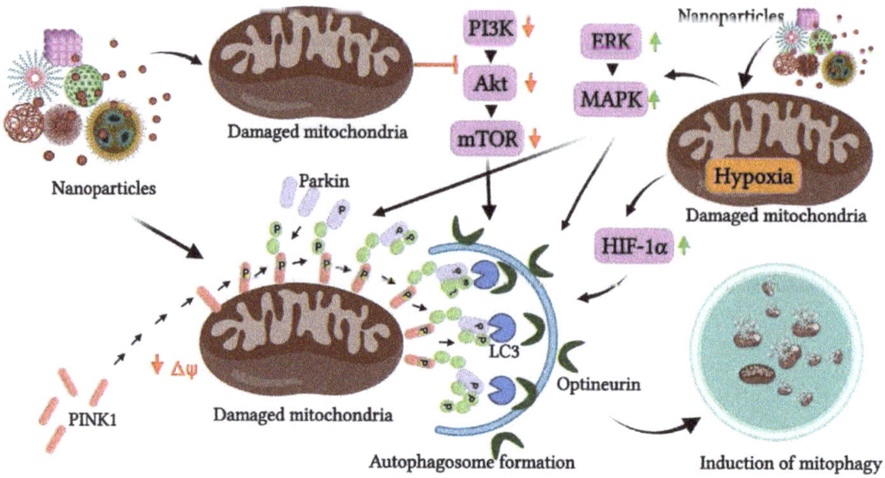

Figure 4. Crosstalk between different signaling pathways in nanoparticle-induced mitophagy. The involvement of nanoparticles in the induction of mitophagy is regulated mainly by PINK1/Parkin pathway with the involvement of additional pathways such as PI3/Akt/mTOR, ERK/MAPK and HIF-1α. (PI3K—phosphatidylinositol 3-kinase, Akt—protein kinase B, mTOR—mammalian target of rapamycin, ERK—extracellular signal-regulated kinase, MAPK—mitogen-activated protein kinase, HIF-1α—hypoxia-inducible factor-1alpha, LC3—microtubule-associated protein 1A/1B-light chain 3).

6.2. PI3K/Akt/mTOR Pathway

PI3K/Akt/mTOR is another major signaling pathway involved in mitophagy. Phosphatidylinositol 3-kinases (PI3Ks) are a group of enzyme transducers involved in a variety of cellular activities including the growth and proliferation of cancer cells. Akt (Protein kinase B) is a serine/threonine kinase involved in supervising the movement of parkin to the damaged mitochondria [72]. The mammalian target of rapamycin (mTOR) is another highly conserved serine/threonine protein kinase that is phosphorylated at Serine 473. While mTOR is important for the formation of autophagosomes, its inactivation is essential for autophagy because hyperactivity of mTOR is found to repress PINK1 expression and this, in turn, will decrease the translocation of Parkin to mitochondria [73]. The proliferation of A549 cell lines was found to be affected by blocking the PI3K/Akt/mTOR pathway heralded by autophagy [74] as this pathway is classified as a negative regulator of autophagosome formation [38]. The decreased levels of mitophagy markers were observed accompanied by the inhibition of the PI3K/Akt/mTOR pathway in the cultured glioblastoma multiforme (GBM) cells treated with solid lipid curcumin particles (SLCP). Low expression levels of mitophagy markers were found after treatment confirming the signaling pathway inhibition. MCF-7 breast cancer cell line treated with gold nanocomplex was found to have differential expression patterns of the genes belonging to the PI3K/Akt pathway [75]. Forkhead Box O1 (FOXO1), the transcriptional factor that is a downstream target of the Akt signaling pathway, was found to be activated by the treatment with nanocomplex. Additional data suggests the suppression of TSC2, a potent inhibitor of mTOR. The study concluded that the crosstalk between PI3K/Akt/mTOR pathways was essential to mediate the various mechanisms involved in the multiple pathways to induce the inhibitory effect on the cancer cell lines under study. mTOR also acts as a mediator in the crosstalk between PI3K/Akt and AMPK pathways and in PC-3 prostate cancer cells, Ag NPs were found to activate autophagy through the AMPK-mTOR pathway [14].

6.3. MAPK/ERK Pathway

MAPK/ERK pathway is a prominent signaling pathway involving mitophagy. Mitogen-activated protein kinase (MAPK) belongs to the line of ubiquitous proline-directed, protein-serine/threonine kinases. They are actively involved in the three-tiered protein kinase cascade controlling some of the major cellular activities leading to the functional and developmental organization of cells in an organism [76]. The extracellular signal-regulated kinase 1/2 (ERK 1/2), c-Jun N-terminal kinase (JNK) and p38 are the major responders to this pathway. While JNK and p38 MAPK pathways are related to stress-mediated apoptosis in cells, the MAPK/ERK pathway is fundamental in the signal transduction network [77]. The anti-apoptotic effect of the signaling pathway has a major role in cancer cell proliferation. Additionally, the MAPK/ERK pathway is found activated in mitophagy in response to oxidative stress [78]. Reports confirm that the activity-based localization of ERK2 to mitochondria is sufficient to induce mitophagy. The constitutive overexpression of active ERK2 further increased fusion proteins such as GLP LC3, a mitochondrial marker for autophagic vesicles, on the mitochondrial membrane emphasizing the overall role of this signaling pathway in mitophagy. The changes in MAPK/ERK pathway are highly dysregulated in malignant tumors.

Sonodynamic therapy based on nano-sensitized liposomes showed a visible increase in MAPK/p38 phosphorylation regulated by ROS formation. An increase in phosphorylation suggested aggravated oxidative stress, and reduced mitophagic vacuolization with impaired Parkin translocation [79]. Absorption of iron oxide nanoparticles led to the failure of respiration and mitophagy of the cells. Gold nanoparticles, on the other hand, were found to induce autophagic flux resulting in impaired lysosomal function.

6.4. Hypoxia-Inducible Factor-1α (HIF-1α) Pathway

There is a high demand for oxygen and nutrients in cancer cells; the failure of balancing the demands will lead to a hypoxic microenvironment in the tumor cells. Tumor cells counter this abnormality by modifying the expression pattern of transcriptional factors that respond to hypoxia. HIF-1 is the major component of the transcriptionally regulated pathway where under normal conditions, the prolines of the HIF-1α subunit become activated and interact with The von Hippel-Lindau (VHL) protein and are degraded. Under conditions of hypoxia, the hydroxylases of proline remain inactive and fail to form the complex with VHL. The unhydroxylated 1α subunit moves freely to the nucleus to associate with the β subunit (HIF-1β) and CBP/p300. This complex then combines with the hypoxia response elements (HRE). The association of the HIF-1β/CBP/p300 complex with HRE elicits more transcriptional activities resulting in the transcription of more genes downstream of the pathway. More than 60 such genes are known to be transcribed during the process. Many cellular metabolic activities are affected by the regulation of these processes [80]. Nanoparticles are known to be hypoxia-responsive once they extravasate into the tumor microenvironment. This quality promotes the efficiency of nanoparticles in tumor therapy [81]. Iron oxide nanoparticle-doxorubicin complexed with hypoxic cell radiosensitizer SAN (sanazole) induced downregulation of HIF-1α [82]. The associated genes, vascular endothelial growth factor (VEGF), and Akt are also downregulated. On the other hand, the treatment of A549 cells with Ag NPs showed an increase in the expression of HIF-1α where exposure of the cells to hypoxia blocked oxidative stress induced by the nanoparticles. It was noted that the autophagic flux was restricted through the regulation of LC3-II and p62 [83].

6.5. Oxidative Stress-Related Pathways

It is well known that levels of ROS can decide the fate of cells; high ROS can lead the cells to apoptotic cell death whereas a low concentration of ROS in cells can lead to mitochondrial dysfunction. This can be a trigger for mitophagy where the damaged mitochondria are removed for cell survival and maintenance of cell homeostasis [84]. The treatment of cells with copper oxide (CuO) NPs aggravates the build-up of excessive ROS in the form of superoxide anions that may result in impaired mitophagic flux. Dysfunctional mitochondria are believed to be the source of ROS accumulation here. Ag NPs were capable of regulating autophagy mediated by injury to mitochondria and lysosomes in A549 cells. Excessive ROS production with an imbalance between the oxidant/antioxidant systems was evident in the cells under study [53].

7. Blind Spots in Research Involving Mitophagy, Nanoparticles, and Cancer

Much research and data are found on the cause and effect of mitophagy in research related to neurodegenerative diseases such as Parkinson's disease. The mitophagic pathway is extensively studied in these disease conditions. It is understandable as there is a direct involvement of PINK1/Parkin in neurodegeneration. However, the fact that there is a crosstalk between the proteins involved in the process of mitophagy in neurodegeneration, inflammation, immunomodulation, and cancer seems to be overlooked. Different cancers such as colon, liver and pancreatic cancers have a serious overlap between inflammation and immunomodulation. Receptor-mediated mitophagy can reveal a lot about the mechanism of action of nanoparticles that can help in developing the nanoparticles as an effective strategy to combat various types of cancer. Further, new models such as 3D cell culture that can mimic human physiology could be developed. The 3D models can emulate the in vivo model in a much better way than the pretend in vitro environment. Experimentation with the 3D models will largely enhance the possibilities for trying multiple target proteins to be studied at the same time. In addition, 3D models will help to minimize research with animal models thus avoiding ethical issues to a larger extent. There are many challenges in studies involving nanoparticle-induced mitophagy, some of which are listed in Figure 5.

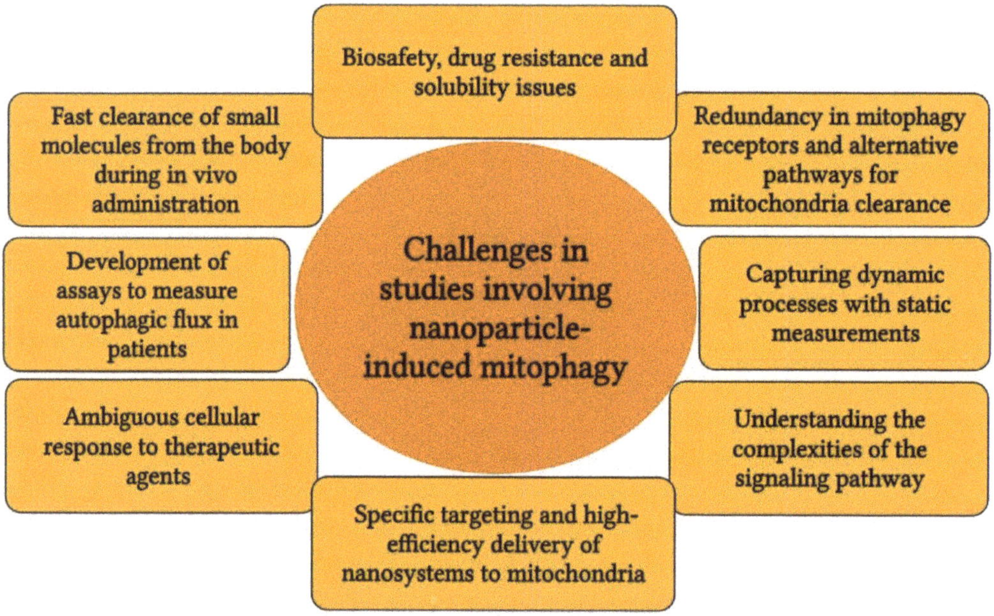

Figure 5. Challenges in studies involving nanoparticle-induced mitophagy.

When the mitochondrial metabolism is targeted by the molecules, it can result in the inhibition of the glycolysis/TCA cycle, redox signaling, or one-carbon metabolism responsible for the production of ROS and antioxidants. Proteins such as SOD, NADH, αTOS, etc. can also be targeted by these molecules resulting in the up- or down-regulation of ROS and other antioxidant enzymes such as GPX (Figure 6).

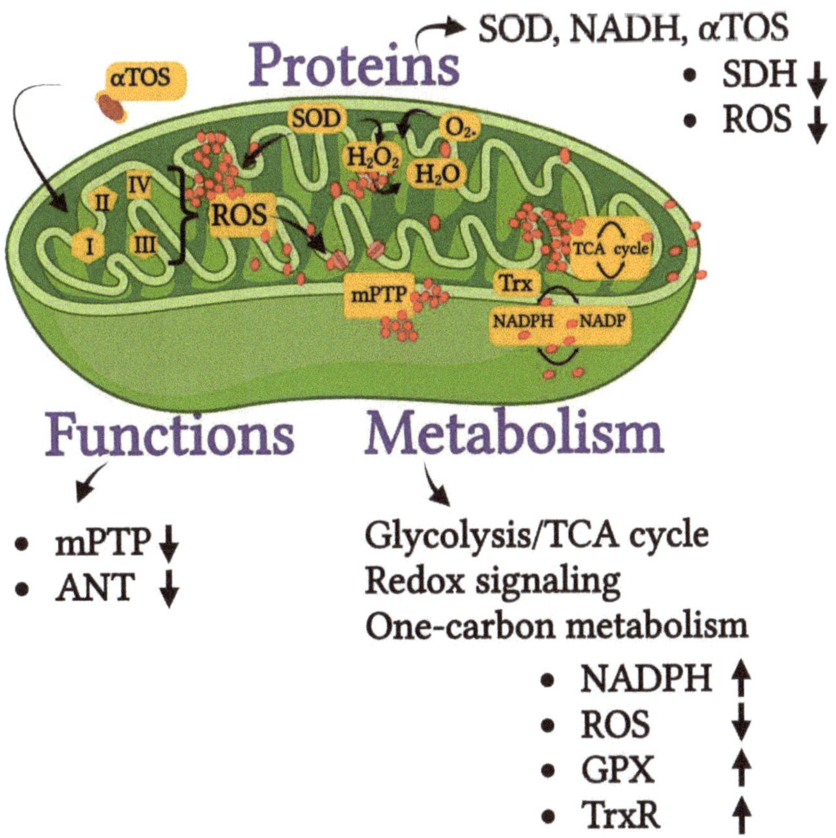

Figure 6. Possible targets to induce mitophagy in cancer. In addition to mitochondrial proteins, the function and metabolism of the mitochondria may be targeted with the metal nanoparticles resulting in mitophagy. The increase in the expression of the reaction product is indicated by the up arrow and the decrease is indicated by the down arrow. SOD—superoxide dismutase, NADH—nicotinamide adenine dinucleotide, αTOS—alpha-tocopherol succinate, SDH—succinate dehydrogenase, ROS—reactive oxygen species, mPTP—mitochondrial permeability transition pore, ANT—adenine nucleotide translocase, TCA—tricarboxylic acid, NAPDH—nicotinamide adenine dinucleotide phosphate, GPX—glutathione peroxidase, TrxR—thioredoxin reductase.

8. Recent Developments in Nanoparticle-Mediated Mitophagy

Through the advancement of the nanobiotechnology, scientists can fabricate new, sophisticated biomaterials that incorporate multiple functions and activities, and can offer a versatile platform for the newly fabricated materials [85]. Enzyme-instructed self-assembly (EISA) and aggregation-induced emission (AIE) are two advanced models employed in cancer therapy.

8.1. Enzyme-Instructed Self-Assembly or EISA

EISA is a process where the self-assembly of cellular components such as protein is mediated by enzymatic processes. Molecular assemblies can be manipulated and modified to stimulate various enzymatic reactions leading to the self-assembly of peptides to form nanostructures. EISA substrates can form supramolecular assemblies that can selectively target cancer cells because of their high penetrating power. They can accumulate in the mitochondrial matrix and induce mitochondrial dysfunction leading to the initiation of mitophagy [86]. Nanofibers, formed by EISA, can be transported to mitochondria where they induce mitochondrial dysfunction even at a relatively lesser concentration. This could lead to cell death activated through multiple signaling pathways [87]. Self-assembled nano peptides could be directed towards PD-L1 (programmed cell death ligand 1) on the cell surface to selectively degrade and thus manipulate their levels so as to avoid immune escape. This high efficiency of nano peptides to bind to immune cells and manipulate them was probably due to the multivalent binding sites found on the surface of the self-assembled nano peptides [88]. Mitochondria-targeted EISA can serve as an alternate strategy to target cancer cells through the production of highly selective, multitargeted nano peptides with minimal drug resistance [89].

8.2. Aggregation-Induced Emission (AIE)

Certain fluorescent molecules can emit higher fluorescence upon entering a crystalline state. Organic compounds with luminescent properties show higher efficiency when aggregated as compared to the solution. This phenomenon called aggregation-induced emission is found to be very high in metal nanoclusters and helps them to locate proteins. Aggregation-induced emission probes are developed that are sensitive to viscosity and regardless of the intensity of the mitochondrial membrane potential, this probe can be discreetly directed to the mitochondria. This is a unique but accurate way of detecting mitophagy [90]. The real-time monitoring of mitochondrial viscosity proved to be better as it is closely associated with the mitochondrial respiratory state reflecting the state of physical wellness or disease of the mitochondria. A near-infrared fluorophore with lipocationic property was developed to selectively accumulate in the mitochondria of cancer cells [91]. This competent multimodal theranostic agent could evaluate mitophagy activities through theranostic approaches. These nanoaggregates could induce mitophagy and block mitophagic flux to accelerate apoptosis in cancer. They were found to be highly beneficial in tracing mitophagy in apoptotic cells in PDT [92,93]. Combining the optoelectronic and sensory properties of these aggregates, they have wide applications in imaging and theranostics. The strong, dynamic intermolecular interactions, polarity, and the power of emitting light confer many more possible roles to the AIE system for cancer therapy.

9. Mechanistic Role of Anti-Tumor Nanoparticles in Inducing Mitophagy

Autophagy is a survival mechanism adopted by cells. In cancer cells, it can be activated to mediate resistance to chemotherapeutic agents. This spontaneous resistance can interfere with the efficacy of the therapeutic agent. However, there are very less interventions to mechanistically regulate autophagy. This is because the drugs that can inhibit or activate autophagy (such as rapamycin and hydroxychloroquine) have not been developed properly for this function [39]. It was also observed that the intrinsic pharmacological specificity of these drugs is very low to explicitly target the components of autophagy. This problem related to specificity arises from the structural complexity of tissues, the wide range of homologous and heterologous interactions that they are subjected to, and due to their failure to specifically target a single cell type. However, these challenges may be overcome by the nanoparticles, at which nanoparticles can explicitly target the decisive molecule in the pathway. Additionally, enhanced efficacy, stability, solubility, and adaptability make them pharmacologically effective in acting as anti-tumor agents.

Increased expression of Beclin-1, XBP1, CHOP, and LC3II in cancer cells suggested the Ag NPs-induced mitophagy in cancer cells. The downregulation of ATG3 and ATG12 was also observed in Ag NP-treated cells [94]. The drug-resistant gene, NPRL2, showed increased expression in cancer cells and upon upregulation, it repressed the mTOR signaling pathway to activate the process of autophagy and suppressed apoptosis. Table 3 explains the role of some of these proteins in the induction of mitophagy. The ROS-scavenging properties of nanoparticles such as iron oxide (Fe_3O_4) [95], ZnO, and silica (Si) [96] nanoparticles can disrupt the antioxidant mechanism of the cell resulting in oxidative imbalance, finally inducing stress-related mitophagy [96–98]. The exposure to graphene oxide (GO) leads to the disruption of autophagic flux and weakening of lysosomal function resulting in the accumulation of the substrate for autophagy such as the autophagosome cargo protein p62 also called sequestosome-1 (p62/SQSTM) [99]. This can lead to mitochondria-mediated apoptosis in the cancer cells. Fe_3O_4 NPs are also found to induce excessive autophagy leading to endothelial dysfunction and inflammation that is assumed to be associated with the Beclin-1/VPS34 complex [100].

Table 3. Some of the main proteins involved in mitophagy other than PINK1 and Parkin.

Protein	Role in Mitophagy	References
Beclin-1	Tumor suppressor protein actively involved in autophagy	[53]
X-Box Binding Protein 1 (XBP1)	Protein released upon oxidative stress that can induce autophagy in cancer cells via JNK activation and eIF2α phosphorylation	[101]
C/EBP homologous protein (CHOP)	Transcription factor required for the initiation of autophagy	[102]
Microtubule-associated protein 1A/1B-light chain 3 (LC3)	Conjugates with phosphatidylethanolamine to form LC3-phosphatidylethanolamine conjugate (LC3-II). It is recruited to the autophagosome membrane to assist degradation by lysosomes	[103]
Autophagy-related 3 (ATG3), Autophagy-related 12 (ATG12)	Proteins that are necessary to induce autophagy	[104]
Nitrogen permease regulator-like 2 (NPRL2)	Repress the mTOR signaling pathway to activate the process of autophagy and suppress apoptosis	[105]
BCL2 and adenovirus E1B19-kDa-interacting protein3 (BNIP3)	Mediate elimination of mitochondria by initiating LC3-dependent mitophagy, promote mitophagy by suppressing PINK1 cleavage	[106]
FUN14domain-containing protein1 (FUNDC1)	Essential role in mitochondrial quality control by mediating mitochondrial clearance by transducing hypoxia signals	[107]
Presenilin-associated rhomboid-like protein (PARL)	Regulator of PINK1 and Parkin	[108]

10. Conclusions

Beyond protein and metal ion corona that can act as probable targets, high-throughput data can be generated to study the behavior of the nanoparticle when exposed to different biological conditions. This will give clarity to the dose-time effect in relation to nanoparticle toxicity. The integration of automated platforms (such as high-throughput screening) and '-omics' technologies could be an effective strategy to fill the existing knowledge gap in nanoparticle-induced mitophagy in cancer. It is expected that the technical advancements in the field of nanotechnology will inspire the scientific community to integrate these techniques (such as EISA) and develop a multi-disciplinary approach to cancer therapy.

Author Contributions: D.M.: conceptualization, investigation, methodology, resources, visualization, writing—original draft. W.C.C.: conceptualization and design, supervision, methodology, visualization, writing—review. All authors have read and agreed to the published version of the manuscript.

Funding: This research received no external funding.

Institutional Review Board Statement: Not applicable.

Informed Consent Statement: Not applicable.

Data Availability Statement: Not applicable.

Conflicts of Interest: The authors declare no conflict of interest.

References

1. Murthy, S.K. Nanoparticles in modern medicine: State of the art and future challenges. *Int. J. Nanomed.* **2007**, *2*, 129.
2. Mundekkad, D.; Cho, W.C. Nanoparticles in clinical translation for cancer therapy. *Int. J. Mol. Sci.* **2022**, *23*, 1685. [CrossRef] [PubMed]
3. Yaqoob, A.A.; Ahmad, H.; Parveen, T.; Ahmad, A.; Oves, M.; Ismail, I.M.I.; Qari, H.A.; Umar, K.; Mohamad Ibrahim, M.N. Recent advances in metal decorated nanomaterials and their various biological applications: A review. *Front. Chem.* **2020**, *8*, 341. [CrossRef]
4. Wijesinghe, W.; Mantilaka, M.; Ruparathna, K.A.A.; Rajapakshe, R.; Sameera, S.A.L.; Thilakarathna, M. Filler matrix interfaces of inorganic/biopolymer composites and their applications. In *Interfaces in Particle and Fibre Reinforced Composites*; Elsevier: Amsterdam, The Netherlands, 2020; pp. 95–112.
5. Liu, C.-G.; Han, Y.-H.; Kankala, R.K.; Wang, S.-B.; Chen, A.-Z. Subcellular performance of nanoparticles in cancer therapy. *Int. J. Nanomed.* **2020**, *15*, 675. [CrossRef]
6. Tabish, T.A.; Hamblin, M.R. Mitochondria-targeted nanoparticles (mitoNANO): An emerging therapeutic shortcut for cancer. *Biomater. Biosyst.* **2021**, *3*, 100023. [CrossRef]
7. Negi, S.; Chaudhuri, A.; Kumar, D.N.; Dehari, D.; Singh, S.; Agrawal, A.K. Nanotherapeutics in autophagy: A paradigm shift in cancer treatment. *Drug Deliv. Transl. Res.* **2022**, *12*, 2589–2612. [CrossRef] [PubMed]
8. Gautam, A.; Rakshit, M.; Nguyen, K.T.; Kathawala, M.H.; Nguyen, L.T.H.; Tay, C.Y.; Wong, E.; Ng, K.W. Understanding the implications of engineered nanoparticle induced autophagy in human epidermal keratinocytes in vitro. *NanoImpact* **2019**, *15*, 100177. [CrossRef]
9. Zhou, Z.; Yan, Y.; Wang, L.; Zhang, Q.; Cheng, Y. Melanin-like nanoparticles decorated with an autophagy-inducing peptide for efficient targeted photothermal therapy. *Biomaterials* **2019**, *203*, 63–72. [CrossRef] [PubMed]
10. Yuan, L.; Zhang, F.; Qi, X.; Yang, Y.; Yan, C.; Jiang, J.; Deng, J. Chiral polymer modified nanoparticles selectively induce autophagy of cancer cells for tumor ablation. *J. Nanobiotechnol.* **2018**, *16*, 55. [CrossRef] [PubMed]
11. Wang, X.; Li, M.; Ren, K.; Xia, C.; Li, J.; Yu, Q.; Qiu, Y.; Lu, Z.; Long, Y.; Zhang, Z. On-Demand Autophagy Cascade Amplification Nanoparticles Precisely Enhanced Oxaliplatin-Induced Cancer Immunotherapy. *Adv. Mater.* **2020**, *32*, 2002160. [CrossRef] [PubMed]
12. Hu, Y.; Zhang, H.-R.; Dong, L.; Xu, M.-R.; Zhang, L.; Ding, W.-P.; Zhang, J.-Q.; Lin, J.; Zhang, Y.-J.; Qiu, B.-S. Enhancing tumor chemotherapy and overcoming drug resistance through autophagy-mediated intracellular dissolution of zinc oxide nanoparticles. *Nanoscale* **2019**, *11*, 11789–11807. [CrossRef] [PubMed]
13. Cho, Y.-L.; Tan, H.W.S.; Saquib, Q.; Ren, Y.; Ahmad, J.; Wahab, R.; He, W.; Bay, B.-H.; Shen, H.-M. Dual role of oxidative stress-JNK activation in autophagy and apoptosis induced by nickel oxide nanoparticles in human cancer cells. *Free Radic. Biol. Med.* **2020**, *153*, 173–186. [CrossRef] [PubMed]
14. Chen, Y.; Yang, T.; Chen, S.; Qi, S.; Zhang, Z.; Xu, Y. Silver nanoparticles regulate autophagy through lysosome injury and cell hypoxia in prostate cancer cells. *J. Biochem. Mol. Toxicol.* **2020**, *34*, e22474. [CrossRef] [PubMed]
15. Jabir, M.S.; Saleh, Y.M.; Sulaiman, G.M.; Yaseen, N.Y.; Sahib, U.I.; Dewir, Y.H.; Alwahibi, M.S.; Soliman, D.A. Green synthesis of silver nanoparticles using Annona muricata extract as an inducer of apoptosis in cancer cells and inhibitor for NLRP3 inflammasome via enhanced autophagy. *Nanomaterials* **2021**, *11*, 384. [CrossRef]
16. Piktel, E.; Ościłowska, I.; Suprewicz, Ł.; Depciuch, J.; Marcińczyk, N.; Chabielska, E.; Wolak, P.; Wollny, T.; Janion, M.; Parlinska-Wojtan, M. ROS-mediated apoptosis and autophagy in ovarian cancer cells treated with peanut-shaped gold nanoparticles. *Int. J. Nanomed.* **2021**, *16*, 1993. [CrossRef]
17. Feng, Q.; Yang, X.; Hao, Y.; Wang, N.; Feng, X.; Hou, L.; Zhang, Z. Cancer cell membrane-biomimetic nanoplatform for enhanced sonodynamic therapy on breast cancer via autophagy regulation strategy. *ACS Appl. Mater. Interfaces* **2019**, *11*, 32729–32738. [CrossRef]
18. Jiang, Y.-W.; Gao, G.; Jia, H.-R.; Zhang, X.; Zhao, J.; Ma, N.; Liu, J.-B.; Liu, P.; Wu, F.-G. Copper oxide nanoparticles induce enhanced radiosensitizing effect via destructive autophagy. *ACS Biomater. Sci. Eng.* **2019**, *5*, 1569–1579. [CrossRef]
19. Li, Y.; Gong, T.; Gao, H.; Chen, Y.; Li, H.; Zhao, P.; Jiang, Y.; Wang, K.; Wu, Y.; Zheng, X. ZIF-Based Nanoparticles Combine X-Ray-Induced Nitrosative Stress with Autophagy Management for Hypoxic Prostate Cancer Therapy. *Angew. Chem.* **2021**, *133*, 15600–15609. [CrossRef]
20. Man, S.; Li, M.; Zhou, J.; Wang, H.; Zhang, J.; Ma, L. Polyethyleneimine coated Fe_3O_4 magnetic nanoparticles induce autophagy, NF-κB and TGF-β signaling pathway activation in HeLa cervical carcinoma cells via reactive oxygen species generation. *Biomater. Sci.* **2020**, *8*, 201–211. [CrossRef]
21. Zhao, X.; Qi, T.; Kong, C.; Hao, M.; Wang, Y.; Li, J.; Liu, B.; Gao, Y.; Jiang, J. Photothermal exposure of polydopamine-coated branched Au–Ag nanoparticles induces cell cycle arrest, apoptosis, and autophagy in human bladder cancer cells. *Int. J. Nanomed.* **2018**, *13*, 6413. [CrossRef]
22. Azimee, S.; Rahmati, M.; Fahimi, H.; Moosavi, M.A. TiO_2 nanoparticles enhance the chemotherapeutic effects of 5-fluorouracil in human AGS gastric cancer cells via autophagy blockade. *Life Sci.* **2020**, *248*, 117466. [CrossRef] [PubMed]
23. Nguyen, L.N.; Kaushik, N.; Bhartiya, P.; Gurmessa, S.K.; Kim, H.-J.; Nguyen, L.Q.; Kaushik, N.K.; Choi, E.H. Plasma-synthesized mussel-inspired gold nanoparticles promote autophagy-dependent damage-associated molecular pattern release to potentiate immunogenic cancer cell death. *J. Ind. Eng. Chem.* **2021**, *100*, 99–111. [CrossRef]

24. Zhang, Y.; Sha, R.; Zhang, L.; Zhang, W.; Jin, P.; Xu, W.; Ding, J.; Lin, J.; Qian, J.; Yao, G. Harnessing copper-palladium alloy tetrapod nanoparticle-induced pro-survival autophagy for optimized photothermal therapy of drug-resistant cancer. *Nat. Commun.* **2018**, *9*, 4236. [CrossRef] [PubMed]
25. Seca, C.; Ferraresi, A.; Phadngam, S.; Vidoni, C.; Isidoro, C. Autophagy-dependent toxicity of amino-functionalized nanoparticles in ovarian cancer cells. *J. Mater. Chem. B* **2019**, *7*, 5376–5391. [CrossRef]
26. Wang, H.; Yu, X.; Su, C.; Shi, Y.; Zhao, L. Chitosan nanoparticles triggered the induction of ROS-mediated cytoprotective autophagy in cancer cells. *Artif. Cells Nanomed. Biotechnol.* **2018**, *46*, 293–301. [CrossRef] [PubMed]
27. Cui, D.; Ma, J.; Liang, T.; Sun, L.; Meng, L.; Liang, T.; Li, Q. Selenium nanoparticles fabricated in laminarin polysaccharides solutions exert their cytotoxicities in HepG2 cells by inhibiting autophagy and promoting apoptosis. *Int. J. Biol. Macromol.* **2019**, *137*, 829–835. [CrossRef] [PubMed]
28. Xiong, Q.; Liu, A.; Ren, Q.; Xue, Y.; Yu, X.; Ying, Y.; Gao, H.; Tan, H.; Zhang, Z.; Li, W. Cuprous oxide nanoparticles trigger reactive oxygen species-induced apoptosis through activation of erk-dependent autophagy in bladder cancer. *Cell Death Dis.* **2020**, *11*, 366. [CrossRef]
29. Palikaras, K.; Lionaki, E.; Tavernarakis, N. Mechanisms of mitophagy in cellular homeostasis, physiology and pathology. *Nat. Cell Biol.* **2018**, *20*, 1013–1022. [CrossRef]
30. Vara-Perez, M.; Felipe-Abrio, B.; Agostinis, P. Mitophagy in cancer: A tale of adaptation. *Cells* **2019**, *8*, 493. [CrossRef]
31. Chourasia, A.H.; Boland, M.L.; Macleod, K.F. Mitophagy and cancer. *Cancer Metab.* **2015**, *3*, 4. [CrossRef]
32. Wang, Y.; Liu, H.-H.; Cao, Y.-T.; Zhang, L.-L.; Huang, F.; Yi, C. The role of mitochondrial dynamics and mitophagy in carcinogenesis, metastasis and therapy. *Front. Cell Dev. Biol.* **2020**, *8*, 413. [CrossRef] [PubMed]
33. Panigrahi, D.P.; Praharaj, P.P.; Bhol, C.S.; Mahapatra, K.K.; Patra, S.; Behera, B.P.; Mishra, S.R.; Bhutia, S.K. The emerging, multifaceted role of mitophagy in cancer and cancer therapeutics. *Semin. Cancer Biol.* **2020**, *66*, 45–58. [CrossRef] [PubMed]
34. Li, Y.; Ju, D. The role of autophagy in nanoparticles-induced toxicity and its related cellular and molecular mechanisms. In *Cellular and Molecular Toxicology of Nanoparticles*; Springer: Berlin/Heidelberg, Germany, 2018; pp. 71–84.
35. Boros, E.; Dyson, P.J.; Gasser, G. Classification of metal-based drugs according to their mechanisms of action. *Chem* **2020**, *6*, 41–60. [CrossRef] [PubMed]
36. Singh, T.A.; Das, J.; Sil, P.C. Zinc oxide nanoparticles: A comprehensive review on its synthesis, anticancer and drug delivery applications as well as health risks. *Adv. Colloid Interface Sci.* **2020**, *286*, 102317. [CrossRef] [PubMed]
37. Wang, J.; Gao, S.; Wang, S.; Xu, Z.; Wei, L. Zinc oxide nanoparticles induce toxicity in CAL 27 oral cancer cell lines by activating PINK1/Parkin-mediated mitophagy. *Int. J. Nanomed.* **2018**, *13*, 3441. [CrossRef]
38. Hou, J.; Zhao, L.; Tang, H.; He, X.; Ye, G.; Shi, F.; Kang, M.; Chen, H.; Li, Y. Silver nanoparticles induced oxidative stress and mitochondrial injuries mediated autophagy in HC11 cells through Akt/AMPK/mTOR pathway. *Biol. Trace Elem. Res.* **2021**, *199*, 1062–1073. [CrossRef]
39. Long, X.; Yan, J.; Zhang, Z.; Chang, J.; He, B.; Sun, Y.; Liang, Y. Autophagy-targeted nanoparticles for effective cancer treatment: Advances and outlook. *NPG Asia Mater.* **2022**, *14*, 71. [CrossRef]
40. Montava-Garriga, L.; Ganley, I.G. Outstanding questions in mitophagy: What we do and do not know. *J. Mol. Biol.* **2020**, *432*, 206–230. [CrossRef]
41. Li, J.; Chang, X.; Shang, M.; Niu, S.; Zhang, W.; Li, Y.; Sun, Z.; Wu, T.; Kong, L.; Zhang, T. The crosstalk between DRP1-dependent mitochondrial fission and oxidative stress triggers hepatocyte apoptosis induced by silver nanoparticles. *Nanoscale* **2021**, *13*, 12356–12369. [CrossRef]
42. Ke, S.; Zhou, T.; Yang, P.; Wang, Y.; Zhang, P.; Chen, K.; Ren, L.; Ye, S. Gold nanoparticles enhance TRAIL sensitivity through Drp1-mediated apoptotic and autophagic mitochondrial fission in NSCLC cells. *Int. J. Nanomed.* **2017**, *12*, 2531. [CrossRef]
43. Zhang, Z.; Zhou, L.; Zhou, Y.; Liu, J.; Xing, X.; Zhong, J.; Xu, G.; Kang, Z.; Liu, J. Mitophagy induced by nanoparticle–peptide conjugates enabling an alternative intracellular trafficking route. *Biomaterials* **2015**, *65*, 56–65. [CrossRef] [PubMed]
44. Portilla, Y.; Mulens-Arias, V.; Paradela, A.; Ramos-Fernández, A.; Pérez-Yagüe, S.; Morales, M.P.; Barber, D.F. The surface coating of iron oxide nanoparticles drives their intracellular trafficking and degradation in endolysosomes differently depending on the cell type. *Biomaterials* **2022**, *281*, 121365. [CrossRef] [PubMed]
45. Vernucci, E.; Tomino, C.; Molinari, F.; Limongi, D.; Aventaggiato, M.; Sansone, L.; Tafani, M.; Russo, M.A. Mitophagy and oxidative stress in cancer and aging: Focus on sirtuins and nanomaterials. *Oxid. Med. Cell. Longev.* **2019**, *2019*, 6387357. [CrossRef] [PubMed]
46. Taggart, L.E.; McMahon, S.J.; Currell, F.J.; Prise, K.M.; Butterworth, K.T. The role of mitochondrial function in gold nanoparticle mediated radiosensitisation. *Cancer Nanotechnol.* **2014**, *5*, 5. [CrossRef]
47. Qiu, K.; Du, Y.; Liu, J.; Guan, J.-L.; Chao, H.; Diao, J. Super-resolution observation of lysosomal dynamics with fluorescent gold nanoparticles. *Theranostics* **2020**, *10*, 6072. [CrossRef]
48. Gallud, A.; Klöditz, K.; Ytterberg, J.; Östberg, N.; Katayama, S.; Skoog, T.; Gogvadze, V.; Chen, Y.-Z.; Xue, D.; Moya, S. Cationic gold nanoparticles elicit mitochondrial dysfunction: A multi-omics study. *Sci. Rep.* **2019**, *9*, 4366. [CrossRef]
49. Mundekkad, D.; Alex, A.V. Analysis of structural and biomimetic characteristics of the green-synthesized Fe_3O_4 nanozyme from the fruit peel extract of Punica granatum. *Chem. Pap.* **2022**, *76*, 3863–3878. [CrossRef]
50. Gong, X.; Pu, X.; Wang, J.; Yang, L.; Cui, Y.; Li, L.; Sun, X.; Liu, J.; Bai, J.; Wang, Y. Enhancing of Nanocatalyst-Driven Chemodynaminc Therapy for Endometrial Cancer Cells Through Inhibition of PINK1/Parkin-Mediated Mitophagy. *Int. J. Nanomed.* **2021**, *16*, 6661. [CrossRef]

51. Rivas-García, L.; Quiles, J.L.; Varela-López, A.; Giampieri, F.; Battino, M.; Bettmer, J.; Montes-Bayón, M.; Llopis, J.; Sánchez-González, C. Ultra-small iron nanoparticles target mitochondria inducing autophagy, acting on mitochondrial dna and reducing respiration. *Pharmaceutics* **2021**, *13*, 90. [CrossRef]
52. Park, W.; Kim, S.-J.; Cheresh, P.; Yun, J.; Lee, B.; Kamp, D.W.; Kim, D.-H. Magneto mitochondrial dysfunction mediated cancer cell death using intracellular magnetic nano-transducers. *Biomater. Sci.* **2021**, *9*, 5497–5507. [CrossRef]
53. Li, J.; Chang, X.; Shang, M.; Niu, S.; Zhang, W.; Zhang, B.; Huang, W.; Wu, T.; Zhang, T.; Tang, M. Mitophagy–lysosomal pathway is involved in silver nanoparticle-induced apoptosis in A549 cells. *Ecotoxicol. Environ. Saf.* **2021**, *208*, 111463. [CrossRef]
54. Holmila, R.; Wu, H.; Lee, J.; Tsang, A.W.; Singh, R.; Furdui, C.M. Integrated redox proteomic analysis highlights new mechanisms of sensitivity to silver nanoparticles. *Mol. Cell. Proteom.* **2021**, *20*, 100073. [CrossRef] [PubMed]
55. Kovács, D.; Igaz, N.; Gopisetty, M.K.; Kiricsi, M. Cancer therapy by silver nanoparticles: Fiction or reality? *Int. J. Mol. Sci.* **2022**, *23*, 839. [CrossRef] [PubMed]
56. Holmila, R.J.; Vance, S.A.; King, S.B.; Tsang, A.W.; Singh, R.; Furdui, C.M. Silver nanoparticles induce mitochondrial protein oxidation in lung cells impacting cell cycle and proliferation. *Antioxidants* **2019**, *8*, 552. [CrossRef] [PubMed]
57. Holmila, R.J.; Vance, S.A.; Chen, X.; Wu, H.; Shukla, K.; Bharadwaj, M.S.; Mims, J.; Wary, Z.; Marrs, G.; Singh, R. Mitochondria-targeted probes for imaging protein sulfenylation. *Sci. Rep.* **2018**, *8*, 6635. [CrossRef] [PubMed]
58. Dong, P.; Li, J.-H.; Xu, S.-P.; Wu, X.-J.; Xiang, X.; Yang, Q.-Q.; Jin, J.-C.; Liu, Y.; Jiang, F.-L. Mitochondrial dysfunction induced by ultra-small silver nanoclusters with a distinct toxic mechanism. *J. Hazard. Mater.* **2016**, *308*, 139–148. [CrossRef]
59. Wei, L.; Wang, J.; Chen, A.; Liu, J.; Feng, X.; Shao, L. Involvement of PINK1/parkin-mediated mitophagy in ZnO nanoparticle-induced toxicity in BV-2 cells. *Int. J. Nanomed.* **2017**, *12*, 1891. [CrossRef]
60. He, G.; Pan, X.; Liu, X.; Zhu, Y.; Ma, Y.; Du, C.; Liu, X.; Mao, C. HIF-1α-Mediated mitophagy determines ZnO nanoparticle-induced human osteosarcoma cell death both in vitro and in vivo. *ACS Appl. Mater. Interfaces* **2020**, *12*, 48296–48309. [CrossRef] [PubMed]
61. He, G.; Nie, J.-J.; Liu, X.; Ding, Z.; Luo, P.; Liu, Y.; Zhang, B.-W.; Wang, R.; Liu, X.; Hai, Y. Zinc oxide nanoparticles inhibit osteosarcoma metastasis by downregulating β-catenin via HIF-1α/BNIP3/LC3B-mediated mitophagy pathway. *Bioact. Mater.* **2023**, *19*, 690–702. [CrossRef]
62. Zhang, J.; Qin, X.; Wang, B.; Xu, G.; Qin, Z.; Wang, J.; Wu, L.; Ju, X.; Bose, D.D.; Qiu, F. Zinc oxide nanoparticles harness autophagy to induce cell death in lung epithelial cells. *Cell Death Dis.* **2017**, *8*, e2954. [CrossRef]
63. Gunes, S.; He, Z.; van Acken, D.; Malone, R.; Cullen, P.J.; Curtin, J.F. Platinum nanoparticles inhibit intracellular ROS generation and protect against cold atmospheric plasma-induced cytotoxicity. *Nanomed. Nanotechnol. Biol. Med.* **2021**, *36*, 102436. [CrossRef] [PubMed]
64. Badawy, M.M.M.; Abdel-Hamid, G.R.; Mohamed, H.E. Antitumor Activity of Chitosan-Coated Iron Oxide Nanocomposite Against Hepatocellular Carcinoma in Animal Models. *Biol. Trace Elem. Res.* **2022**, 1–12. Available online: https://link.springer.com/content/pdf/10.1007/s12011-022-03221-7.pdf (accessed on 9 September 2022). [CrossRef] [PubMed]
65. Chen, G.; Kroemer, G.; Kepp, O. Mitophagy: An emerging role in aging and age-associated diseases. *Front. cell Dev. Biol.* **2020**, *8*, 200. [CrossRef] [PubMed]
66. Sekine, S. PINK1 import regulation at a crossroad of mitochondrial fate: The molecular mechanisms of PINK1 import. *J. Biochem.* **2020**, *167*, 217–224. [CrossRef]
67. Sekine, S.; Youle, R.J. PINK1 import regulation; a fine system to convey mitochondrial stress to the cytosol. *BMC Biol.* **2018**, *16*, 2. [CrossRef]
68. Zimmermann, M.; Reichert, A.S. How to get rid of mitochondria: Crosstalk and regulation of multiple mitophagy pathways. *Biol. Chem.* **2018**, *399*, 29–45. [CrossRef]
69. Padman, B.S.; Nguyen, T.N.; Uoselis, L.; Skulsuppaisarn, M.; Nguyen, L.K.; Lazarou, M. LC3/GABARAPs drive ubiquitin-independent recruitment of Optineurin and NDP52 to amplify mitophagy. *Nat. Commun.* **2019**, *10*, 408. [CrossRef]
70. Yan, S.; Qiao, L.; Dou, X.; Song, X.; Chen, Y.; Zhang, B.; Xu, C. Biogenic selenium nanoparticles by Lactobacillus casei ATCC 393 alleviate the intestinal permeability, mitochondrial dysfunction and mitophagy induced by oxidative stress. *Food Funct.* **2021**, *12*, 7068–7080. [CrossRef]
71. Xu, X.Y.; Tran, T.H.M.; Perumalsamy, H.; Sanjeevram, D.; Kim, Y.-J. Biosynthetic gold nanoparticles of Hibiscus syriacus L. callus potentiates anti-inflammation efficacy via an autophagy-dependent mechanism. *Mater. Sci. Eng. C* **2021**, *124*, 112035. [CrossRef]
72. Soutar, M.P.M.; Kempthorne, L.; Miyakawa, S.; Annuario, E.; Melandri, D.; Harley, J.; O'Sullivan, G.A.; Wray, S.; Hancock, D.C.; Cookson, M.R. AKT signalling selectively regulates PINK1 mitophagy in SHSY5Y cells and human iPSC-derived neurons. *Sci. Rep.* **2018**, *8*, 8855. [CrossRef]
73. Bartolomé, A.; García-Aguilar, A.; Asahara, S.-I.; Kido, Y.; Guillén, C.; Pajvani, U.B.; Benito, M. MTORC1 regulates both general autophagy and mitophagy induction after oxidative phosphorylation uncoupling. *Mol. Cell. Biol.* **2017**, *37*, e00441-17. [CrossRef] [PubMed]
74. Xu, Z.; Han, X.; Ou, D.; Liu, T.; Li, Z.; Jiang, G.; Liu, J.; Zhang, J. Targeting PI3K/AKT/mTOR-mediated autophagy for tumor therapy. *Appl. Microbiol. Biotechnol.* **2020**, *104*, 575–587. [CrossRef] [PubMed]
75. Mahmoud, N.N.; Abuarqoub, D.; Zaza, R.; Sabbah, D.A.; Khalil, E.A.; Abu-Dahab, R. Gold Nanocomplex strongly modulates the PI3K/Akt pathway and other pathways in MCF-7 breast cancer cell line. *Int. J. Mol. Sci.* **2020**, *21*, 3320. [CrossRef] [PubMed]

76. Pearson, G.; Robinson, F.; Beers Gibson, T.; Xu, B.; Karandikar, M.; Berman, K.; Cobb, M.H. Mitogen-activated protein (MAP) kinase pathways: Regulation and physiological functions. *Endocr. Rev.* **2001**, *22*, 153–183.
77. Guo, Y.; Pan, W.; Liu, S.; Shen, Z.; Xu, Y.; Hu, L. ERK/MAPK signalling pathway and tumorigenesis. *Exp. Ther. Med.* **2020**, *19*, 1997–2007. [CrossRef]
78. Park, J.H.; Ko, J.; Park, Y.S.; Park, J.; Hwang, J.; Koh, H.C. Clearance of damaged mitochondria through PINK1 stabilization by JNK and ERK MAPK signaling in chlorpyrifos-treated neuroblastoma cells. *Mol. Neurobiol.* **2017**, *54*, 1844–1857. [CrossRef]
79. Qu, F.; Wang, P.; Zhang, K.; Shi, Y.; Li, Y.; Li, C.; Lu, J.; Liu, Q.; Wang, X. Manipulation of Mitophagy by "All-in-One" nanosensitizer augments sonodynamic glioma therapy. *Autophagy* **2020**, *16*, 1413–1435. [CrossRef]
80. Chen, Z.; Wu, H.; Huang, S.; Li, W.; Zhang, S.; Zheng, P.; Zhou, X.; Liu, W.; Zhang, D. Expression of BNIP3 and its correlations to hypoxia-induced autophagy and clinicopathological features in salivary adenoid cystic carcinoma. *Cancer Biomark.* **2015**, *15*, 467–475. [CrossRef]
81. Wang, Y.; Shang, W.; Niu, M.; Tian, J.; Xu, K. Hypoxia-active nanoparticles used in tumor theranostic. *Int. J. Nanomed.* **2019**, *14*, 3705. [CrossRef]
82. Sreeja, S.; Nair, C.K.K. Chemo-directed specific targeting of nanoparticle-doxorubicin complexes to tumor in animal model. *J. Drug Deliv. Sci. Technol.* **2016**, *31*, 167–175. [CrossRef]
83. Jeong, J.-K.; Gurunathan, S.; Kang, M.-H.; Han, J.W.; Das, J.; Choi, Y.-J.; Kwon, D.-N.; Cho, S.-G.; Park, C.; Seo, H.G. Hypoxia-mediated autophagic flux inhibits silver nanoparticle-triggered apoptosis in human lung cancer cells. *Sci. Rep.* **2016**, *6*, 21688. [CrossRef] [PubMed]
84. Zhang, S.; Pelligra, C.I.; Feng, X.; Osuji, C.O. Directed assembly of hybrid nanomaterials and nanocomposites. *Adv. Mater.* **2018**, *30*, 1705794. [CrossRef] [PubMed]
85. Kim, B.J.; Xu, B. Enzyme-instructed self-assembly for cancer therapy and imaging. *Bioconjug. Chem.* **2020**, *31*, 492–500. [CrossRef] [PubMed]
86. Gao, J.; Zhan, J.; Yang, Z. Enzyme-instructed self-assembly (EISA) and hydrogelation of peptides. *Adv. Mater.* **2020**, *32*, 1805798. [CrossRef]
87. Wang, Y.; Li, X.; Zheng, D.; Chen, Y.; Zhang, Z.; Yang, Z. Selective Degradation of PD-L1 in Cancer Cells by Enzyme-Instructed Self-Assembly. *Adv. Funct. Mater.* **2021**, *31*, 2102505. [CrossRef]
88. Jeena, M.T.; Kim, S.; Jin, S.; Ryu, J.-H. Recent progress in mitochondria-targeted drug and drug-free agents for cancer therapy. *Cancers* **2019**, *12*, 4. [CrossRef]
89. Wang, X.; Fan, L.; Wang, S.; Zhang, Y.; Li, F.; Zan, Q.; Lu, W.; Shuang, S.; Dong, C. Real-time monitoring mitochondrial viscosity during mitophagy using a mitochondria-immobilized near-infrared aggregation-induced emission probe. *Anal. Chem.* **2021**, *93*, 3241–3249. [CrossRef]
90. Li, Y.; Zhuang, J.; Lu, Y.; Li, N.; Gu, M.; Xia, J.; Zhao, N.; Tang, B.Z. High-performance near-infrared aggregation-induced emission luminogen with mitophagy regulating capability for multimodal cancer theranostics. *ACS Nano* **2021**, *15*, 20453–20465. [CrossRef]
91. Wang, J.; Zhu, X.; Zhang, J.; Wang, H.; Liu, G.; Bu, Y.; Yu, J.; Tian, Y.; Zhou, H. AIE-based theranostic agent: In situ tracking mitophagy prior to late apoptosis to guide the photodynamic therapy. *ACS Appl. Mater. Interfaces* **2019**, *12*, 1988–1996. [CrossRef]
92. Zhang, Y.; Wang, L.; Rao, Q.; Bu, Y.; Xu, T.; Zhu, X.; Zhang, J.; Tian, Y.; Zhou, H. Tuning the hydrophobicity of pyridinium-based probes to realize the mitochondria-targeted photodynamic therapy and mitophagy tracking. *Sens. Actuators B Chem.* **2020**, *321*, 128460. [CrossRef]
93. Bao, J.; Jiang, Z.; Ding, W.; Cao, Y.; Yang, L.; Liu, J. Silver nanoparticles induce mitochondria-dependent apoptosis and late non-canonical autophagy in HT-29 colon cancer cells. *Nanotechnol. Rev.* **2022**, *11*, 1911–1926. [CrossRef]
94. Ashrafizadeh, M.; Paskeh, M.D.A.; Mirzaei, S.; Gholami, M.H.; Zarrabi, A.; Hashemi, F.; Hushmandi, K.; Hashemi, M.; Nabavi, N.; Crea, F. Targeting autophagy in prostate cancer: Preclinical and clinical evidence for therapeutic response. *J. Exp. Clin. Cancer Res.* **2022**, *41*, 105. [CrossRef]
95. Mundekkad, D.; Kameshwari, G.V.; Karchalkar, P.; Koti, R. The catalytic and ROS-scavenging activities of green synthesized, antiferromagnetic α-Fe_2O_3 nanoparticle with a prismatic octahedron morphology from pomegranate rind extract. *Nanotechnology* **2021**, *33*, 4. [CrossRef] [PubMed]
96. Guo, C.; Yang, M.; Jing, L.I.; Wang, J.; Yu, Y.; Li, Y.; Duan, J.; Zhou, X.; Li, Y.; Sun, Z. Amorphous silica nanoparticles trigger vascular endothelial cell injury through apoptosis and autophagy via reactive oxygen species-mediated MAPK/Bcl-2 and PI3K/Akt/mTOR signaling. *Int. J. Nanomed.* **2016**, *11*, 5257. [CrossRef] [PubMed]
97. Feng, X.; Zhang, Y.; Zhang, C.; Lai, X.; Zhang, Y.; Wu, J.; Hu, C.; Shao, L. Nanomaterial-mediated autophagy: Coexisting hazard and health benefits in biomedicine. *Part. Fibre Toxicol.* **2020**, *17*, 53. [CrossRef] [PubMed]
98. Roy, R.; Singh, S.K.; Chauhan, L.K.S.; Das, M.; Tripathi, A.; Dwivedi, P.D. Zinc oxide nanoparticles induce apoptosis by enhancement of autophagy via PI3K/Akt/mTOR inhibition. *Toxicol. Lett.* **2014**, *227*, 29–40. [CrossRef] [PubMed]
99. Feng, X.; Chen, L.; Guo, W.; Zhang, Y.; Lai, X.; Shao, L.; Li, Y. Graphene oxide induces p62/SQSTM-dependent apoptosis through the impairment of autophagic flux and lysosomal dysfunction in PC12 cells. *Acta Biomater.* **2018**, *81*, 278–292. [CrossRef] [PubMed]
100. Zhang, L.; Wang, X.; Miao, Y.; Chen, Z.; Qiang, P.; Cui, L.; Jing, H.; Guo, Y. Magnetic ferroferric oxide nanoparticles induce vascular endothelial cell dysfunction and inflammation by disturbing autophagy. *J. Hazard. Mater.* **2016**, *304*, 186–195. [CrossRef]

101. Mehrbod, P.; Ande, S.R.; Alizadeh, J.; Rahimizadeh, S.; Shariati, A.; Malek, H.; Hashemi, M.; Glover, K.K.M.; Sher, A.A.; Coombs, K.M. The roles of apoptosis, autophagy and unfolded protein response in arbovirus, influenza virus, and HIV infections. *Virulence* **2019**, *10*, 376–413. [CrossRef]
102. Hu, H.; Tian, M.; Ding, C.; Yu, S. The C/EBP homologous protein (CHOP) transcription factor functions in endoplasmic reticulum stress-induced apoptosis and microbial infection. *Front. Immunol.* **2019**, *9*, 3083. [CrossRef]
103. Sipos, A.; Kim, K.-J.; Sioutas, C.; Crandall, E.D. Evidence for nanoparticle-induced lysosomal dysfunction in lung adenocarcinoma (A549) cells. *Int. J. Mol. Sci.* **2019**, *20*, 5253. [CrossRef] [PubMed]
104. Shpilka, T.; Weidberg, H.; Pietrokovski, S.; Elazar, Z. Atg8: An autophagy-related ubiquitin-like protein family. *Genome Biol.* **2011**, *12*, 226. [CrossRef] [PubMed]
105. Akbay, B. Regulation of the Akt/mTORC1 Pathway by HIV Transcriptional Activator Tat in B Cells. Ph.D. Thesis, Nazarbayev University, Astana, Kazakhstan, 2021.
106. Walls, K.C.; Ghosh, A.P.; Ballestas, M.E.; Klocke, B.J.; Roth, K.A. bcl-2/Adenovirus E1B 19-kd interacting protein 3 (BNIP3) regulates hypoxia-induced neural precursor cell death. *J. Neuropathol. Exp. Neurol.* **2009**, *68*, 1326–1338. [CrossRef] [PubMed]
107. Liu, H.; Zang, C.; Yuan, F.; Ju, C.; Shang, M.; Ning, J.; Yang, Y.; Ma, J.; Li, G.; Bao, X. The role of FUNDC1 in mitophagy, mitochondrial dynamics and human diseases. *Biochem. Pharmacol.* **2021**, 114891. [CrossRef] [PubMed]
108. Qin, C.; Wang, Y.; Zhao, B.; Li, Z.; Li, T.; Yang, X.; Zhao, Y.; Wang, W. STOML2 Restricts Mitophagy and Increases Chemosensitivity in Pancreatic Cancer through Stabilizing PARL-induced PINK1 degradation. *Res. Sq.* **2022**. [CrossRef]

pharmaceutics MDPI

Article

Methylene Blue-Loaded Mesoporous Silica-Coated Gold Nanorods on Graphene Oxide for Synergistic Photothermal and Photodynamic Therapy

Sun-Hwa Seo, Ara Joe, Hyo-Won Han, Panchanathan Manivasagan and Eue-Soon Jang *

Department of Applied Chemistry, Kumoh National Institute of Technology, Gumi 730-701, Gyeongbuk, Korea
* Correspondence: euesoon@kumoh.ac.kr

Abstract: Photo-nanotheranostics integrates near-infrared (NIR) light-triggered diagnostics and therapeutics, which are combined into a novel all-in-one phototheranostic nanomaterial that holds great promise for the early detection and precise treatment of cancer. In this study, we developed methylene blue-loaded mesoporous silica-coated gold nanorods on graphene oxide (MB-GNR@mSiO_2-GO) as an all-in-one photo-nanotheranostic agent for intracellular surface-enhanced Raman scattering (SERS) imaging-guided photothermal therapy (PTT)/photodynamic therapy (PDT) for cancer. Amine functionalization of the MB-GNR@mSiO_2 surfaces was performed using 3-aminopropyltriethoxysilane (APTES), which was well anchored on the carboxyl groups of graphene oxide (GO) nanosheets uniformly, and showed a remarkably higher photothermal conversion efficiency (48.93%), resulting in outstanding PTT/PDT for cancer. The in vitro photothermal/photodynamic effect of MB-GNR@mSiO_2-GO with laser irradiation showed significantly reduced cell viability (6.32%), indicating that MB-GNR@mSiO_2-GO with laser irradiation induced significantly more cell deaths. Under laser irradiation, MB-GNR@mSiO_2-GO showed a strong SERS effect, which permits accurate cancer cell detection by SERS imaging. Subsequently, the same Raman laser can focus on highly detected MDA-MB-231 cells for a prolonged time to perform PTT/PDT. Therefore, MB-GNR@mSiO_2-GO has great potential for precise SERS imaging-guided synergistic PTT/PDT for cancer.

Keywords: methylene blue; gold nanorods; graphene oxide; diagnosis; phototherapy

Citation: Seo, S.-H.; Joe, A.; Han, H.-W.; Manivasagan, P.; Jang, E.-S. Methylene Blue-Loaded Mesoporous Silica-Coated Gold Nanorods on Graphene Oxide for Synergistic Photothermal and Photodynamic Therapy. *Pharmaceutics* **2022**, *14*, 2242. https://doi.org/10.3390/pharmaceutics14102242

Academic Editor: Nataša Skalko-Basnet

Received: 25 August 2022
Accepted: 18 October 2022
Published: 20 October 2022

1. Introduction

Cancer is a leading cause of human death globally, and exploiting photo-nanotheranostics is an urgent requirement for cancer treatment [1,2]. Photo-nanotheranostics is the integration of near-infrared (NIR) light-triggered diagnostics and therapeutics, which are combined into a new all-in-one nanomaterial that has great potential for accurate cancer diagnosis and effective phototherapy [3]. As an emerging photo-triggered diagnostic technique, surface-enhanced Raman scattering (SERS) imaging has been increasingly considered a promising diagnostic method for the development of theranostic systems because of its high sensitivity, enhanced photostability, and multiplexing abilities [4]. SERS has emerged as an effective imaging method for accurate cancer detection [5]. As emerging phototherapeutic strategies, photothermal therapy (PTT) and photodynamic therapy (PDT) have attracted increasing clinical attention in recent years, which is beneficial for enhanced cancer therapy because of their synergistic effects [6]. PTT has attracted considerable interest in cancer treatment because of its minimal invasiveness, high efficiency, low toxicity, and deep tissue penetration [7,8]. In PTT, a photothermal nanoagent directly converts the absorbed NIR light into heat-to-kill cancer cells [9]. PDT has attracted much attention as a less invasive therapeutic modality for treating cancer. It employs photosensitizers (PSs) and light irradiation at certain wavelengths of photon energy to produce reactive oxygen species (ROS), including singlet oxygen ($^1O_2$), to destroy cancer cells [7]. Therefore, the integration of SERS imaging, PTT, and PDT treatment into one system could provide a new photo nanotheranostic platform for precise cancer diagnosis and therapy [6].

To date, various nanoparticles (NPs) with strong NIR absorption properties and an effective heat transfer mechanism have been developed as photo-nanotheranostic agents for SERS imaging-guided phototherapy, including polymer NPs [10], carbon NPs [11], copper sulfide NPs [12], and gold nanostructures [13]. Among these photothermal nanomaterials, gold nanorods (GNRs) have emerged as interesting nanomaterials for theranostic applications, including drug delivery, PTT, SERS imaging, and photoacoustic imaging, because of their excellent small size, biocompatibility, ease of synthesis, ease of surface modification, and strong tunable extinction in the NIR region (600–1100 nm) [14]. The seed-mediated method has become one of the most powerful and versatile methods for the synthesis of GNRs due to its advantages, such as a simple process and high yield of GNRs [15,16]. GNRs have proven to be appropriate candidates for SERS imaging-guided PTT because of their strong SERS effects and high photothermal conversion properties [17]. However, GNRs with nonporous structures have the poor dye-loading capacity, which seriously limits their efficacy in PDT [18]. The Stöber method is a widely used method for obtaining silica-coated NPs due to its advantages, such as ease of synthesis, low cost, mild reaction conditions, and a broad range of achievable particle sizes [19,20]. Mesoporous silica ($mSiO_2$) is highly suitable as a coating material for GNRs [21]. Mesoporous silica-coated gold nanorods (GNR@$mSiO_2$) show great potential for theranostic applications because of their high specific surface area, high accessibility, tunable pore size and volume, high dye loading capability, high stability, well-defined surface properties, and excellent biocompatibility [22]. Methylene blue (MB) is a water-soluble, Food and Drug Administration (FDA)-approved, and photosensitizer (PS) drug that is frequently used in PDT to treat cancerous and non-cancerous diseases [23]. MB has high levels of ROS, such as singlet oxygen ($^1O_2$) quantum yield, under laser light exposure [24]. MB was directly loaded with high loading capability inside the pores of $mSiO_2$ to obtain MB-GNR@$mSiO_2$ [25]. Most importantly, the integration of GNRs with other NPs, such as graphene oxide (GO) is being actively pursued because of its enhanced photothermal conversion property when compared with single NPs [3]. GO is a sheet-like two-dimensional layer of carbon nanomaterials and is the most attractive material for the PTT of cancer because of its excellent biocompatibility, conductivity, large surface area, stability, facile surface functionalization, and high optical absorption coefficient in the NIR region [26]. Therefore, the integration of MB-GNR@$mSiO_2$ with GO is a promising strategy for achieving high PTT efficacy because of its synergistic effect [9]. In this study, we developed multifunctional MB-GNR@$mSiO_2$-GO as a novel photo-nanotheranostic agent for SERS imaging-guided synergistic PTT/PDT.

2. Materials and Methods

2.1. Materials

All the materials and reagents were purchased from Sigma-Aldrich (St. Louis, MO, USA) and used without further purification.

2.2. Synthesis of MB-GNR@$mSiO_2$-GO

Gold nanorods (GNRs) were prepared in an aqueous cetyltrimethylammonium bromide (CTAB) solution using a seed-mediated method [15]. For the mesoporous silica ($mSiO_2$) coating on the surface of GNRs (GNR@$mSiO_2$), GNR@$mSiO_2$ was synthesized according to a previously published procedure, with minor modifications [27,28]. Briefly, 10 mL of the purified GNR solution was added to 5 mL of absolute ethanol and 5 mL of deionized water (DW), and the mixture was stirred for 3 h at ambient temperature. Subsequently, 200 μL of NaOH solution (0.1 M) was slowly added dropwise with stirring. Subsequently, 60 μL of tetraethyl orthosilicate (TEOS; 20%) in methanol solution was added three times at 30 min intervals with gentle stirring. The mixture was allowed to react for 2 days at 30 °C with vigorous stirring. Mesoporous silica-coated GNRs (GNR@$mSiO_2$) were collected by centrifugation at 15,000 rpm for 30 min and washed thrice with ethanol. The purified GNR@mSiO_2 samples obtained were redispersed in 10 mL of ethanol for further experiments. Methylene blue (MB) loaded onto GNR@mSiO_2 was prepared according to previous reports [25]. MB (0.1 mg) was added to 10 mL of purified GNR@mSiO_2, and the reaction solution was then stirred for 2 days at 25 °C, which was covered with aluminum foil to avoid light exposure and to protect the fluorescent dyes from photodegradation. The reaction solutions (MB-GNR@mSiO_2) were separated by centrifugation and washed repeatedly with DW until the supernatant changed from dark blue to colorless.

For the conjugation of amine functional groups on the surface of MB-GNR@mSiO_2, 1 mL of 3-aminopropyltriethoxysilane (APTES) was mixed with 10 mL of MB-GNR@mSiO_2 and stirred for 6 h at ambient temperature. The reaction solutions were centrifuged and washed thrice using DW. The aminated MB-GNR@mSiO_2-APTES was re-dispersed in DW for further experiments. GO was prepared using a modified Hummers' method [29]. The primary amino groups of MB-GNR@mSiO_2-APTES (5.0 mL) were reacted with the carboxyl groups of GO (0.1 mg) via a coupling reaction in the presence of 1-ethyl-3-(3-dimethylaminopropyl) carbodiimide hydrochloride (EDC; 1.0 mg) and *N*-hydroxysuccinimide (NHS; 1.0 mg) with overnight stirring at room temperature. The amino group of MB-GNR@mSiO_2-APTES can be easily cross-linked with the carboxyl group of GO, and MB-GNR@mSiO_2-GO can potentially prevent the leaching of MB from the mesopores of SiO_2. The reaction solutions (MB-GNR@mSiO_2-GO) were centrifuged, washed repeatedly using DW to eliminate unreacted reactants, and freeze-dried.

2.3. Characterization

The UV–vis spectra of GNR, GNR@mSiO_2, MB-GNR@mSiO_2, free GO, and MB-GNR@mSiO_2-GO were obtained using a Shimadzu (Kyoto, Japan) UV-2600 UV–vis spectrophotometer. The morphologies of the nanomaterials were characterized using a Cs-corrected field emission transmission electron microscope (Cs-corrected FE-TEM; JEM-ARM-200F, JEOL, Tokyo, Japan) equipped with energy-dispersive X-ray spectroscopy (EDX). The particle size range and zeta potential (ZP) of the nanomaterials were measured using a particle size and ZP analyzer (90 Plus, Brookhaven Instruments, Holtsville, NY, USA). Raman spectra at 785 nm (2 mW, spot area = 2–100 μm) were obtained using a Renishaw inVia Raman spectrometer (Gloucestershire, UK) equipped with a Leica DM 2500 microscope with 1200 g/mm grating. The final gold (Au) ion concentration of the nanomaterials in aqueous solutions was determined to be 36 μg Au/mL by inductively coupled plasma optical emission spectroscopy (ICP-OES, Santa Clara, CA, USA). The photothermal experiments were carried out using an NIR laser at 785 nm (Changchun New Industries Optoelectronics Technology, Changchun, China; power density: 0.8 W/cm^2, laser beam diameter: 4 mm, laser beam shape: round), and infrared (IR) thermal images were captured every 1 s using an FLIR CX-Series compact thermal imaging camera. A transparent cuvette was filled with a known concentration of NP solutions and the top surface was fixed with a laser pointer lens. An IR thermal camera was placed at a 30 cm distance to capture a lateral image of the cuvette and a laser.

2.4. Measurement of Photothermal Performance

One milliliter of each of the GNR, GNR@mSiO_2, MB-GNR@mSiO_2, GO, and MB-GNR@mSiO_2-GO aqueous solutions (36 μg/mL) was added to a quartz cuvette and exposed to a 785 nm NIR laser at 0.8 W/cm^2 for 25 min. The temperature profiles were recorded every 1 s using an FLIR CX-Series compact thermal imaging camera. Additionally, the gold concentration of the MB-GNR@mSiO_2-GO was calculated to be 36 μg Au/mL and different volumes (0–200 μL of 36 μg/mL) of the MB-GNR@mSiO_2-GO aqueous solutions (1 mL) were added to a quartz cuvette and exposed to a 785 nm laser at 0.8 W/cm^2 for 25 min. Temperature profiles and infrared (IR) thermal images were obtained using a FLIR thermal imaging camera. To evaluate the photothermal stability of MB-GNR@mSiO_2-GO, the MB-GNR@mSiO_2-GO (200 μL of 36 μg/mL) aqueous solution was filled in a quartz cuvette and then exposed to four cycles of laser on/off (785 nm, 0.8 W/cm^2). The UV–vis spectra of MB-GNR@mSiO_2-GO were recorded and characterized using TEM, DLS, and ZP after four cycles of laser irradiation. The photothermal conversion efficiency (η) of MB-GNR@mSiO_2-GO (200 µL of 36 µg/mL) was calculated according to the thermal balance using the following Equation (1), as described in previous reports [30,31].

$$\eta = \frac{hS(T_{Max} - T_{Sur}) - Q_{dis}}{I(1 - 10^{-A_{808}})} \quad (1)$$

2.5. Cells Culture

The mouse muscle fibroblast cell line (BLO-11) and human breast cancer cell line (MDA-MB-231 cells) were purchased from the Korean Cell Line Bank (Seoul, Korea). The BLO-11 and MDA-MB-231 cells were cultured in Dulbecco's modified Eagle's medium (DMEM) with 1% antibiotics (penicillin/streptomycin) and 10% fetal bovine serum (FBS) at 37 °C under 5% CO_2.

2.6. In Vitro PTT/PDT Synergistic Effect

The in vitro photothermal/photodynamic effects of GNR, MB-GNR@mSiO_2, GO, and MB-GNR@mSiO_2-GO against normal (BLO-11) and cancer (MDA-MB-231) cells were evaluated using the 3-(4,5-dimethylthiazol-2-yl)-2,5-diphenyltetrazolium bromide (MTT) assay. BLO-11 and MDA-MB-231 cells (7000 cells/well) were seeded in 96-well plates and cultured at 37 °C for 24 h. The cells were incubated with GNR, MB-GNR@mSiO_2, GO, and MB-GNR@mSiO_2-GO (200 µL of 36 µg/mL) for 12 h and 48 h and then irradiated with a 785 nm laser at 0.8 W/cm^2 for various time points (0, 10, 30, and 50 min). The cells were treated with all the nanomaterials and were incubated with a 785 nm laser irradiation as the control. All cells were further incubated for 6 h, and cell viability was evaluated using the MTT assay. The percentage of cell viability was determined by the following Formula (2):

$$\text{Percentage of cell viability } (\%) = \frac{\text{OD value of samples}}{\text{OD value of controls}} \times 100 \quad (2)$$

To further confirm the in vitro photothermal/photodynamic effect of GNR, MB-GNR@mSiO_2, and MB-GNR@mSiO_2-GO, the MDA-MB-231 cells were stained using calcein-AM and propidium iodide (PI) to visually identify live (green) and dead (red) cells. The MDA-MB-231 cells were cultured in 6-well plates and incubated overnight at 37 °C. After incubation, the cells were treated with GNR, MB-GNR@mSiO_2, and MB-GNR@mSiO_2-GO (200 µL of 36 µg/mL) for 12 h and irradiated with a 785 nm laser (0.8 W/cm^2) for various time points (0, 10, 30, and 50 min). Then, all cells were incubated for 6 h and stained using calcein-AM and PI for 30 min. Finally, the cells were imaged using a confocal fluorescence microscope (Nikon Confocal A1R; Nikon, Tokyo, Japan).

To quantify intracellular ROS, MDA-MB-231 cells were cultured in 6-well plates and incubated for 24 h at 37 °C. The MDA-MB-231 cells were then incubated with GNR, MB-GNR@mSiO_2, and MB-GNR@mSiO_2-GO (200 µL of 36 µg/mL) for 12 h and exposed to a 785 nm laser at 0.8 W/cm^2 for various time points (0, 10, 30, and 50 min). The cells were incubated for another 6 h and stained with 2,7-Dichlorofuorescin diacetate (DCFH-DA 10 µM) for 30 min. DCFH-DA is a non-fluorescent dye used to detect PDT-induced intracellular ROS production. The cells were washed with PBS and imaged using a confocal fluorescence microscope (Nikon Confocal A1R; Nikon, Tokyo, Japan).

2.7. SERS Imaging Measurement

The SERS imaging measurement was recorded with a Renishaw inVia Raman spectrometer (Gloucestershire, UK) equipped with a NIR laser (785 nm, 2 mW, spot area = 2–100 µm) and a Leica DM 2500 microscope with 1200 g/mm grating, which was focused for 40 s on the samples through a 20× microscope objective. MDA-MB-231 cells were grown on a chambered slide glass at 37 °C for 24 h and then cells were treated with GNR (control), MB-GNR@mSiO_2, and MB-GNR@mSiO_2-GO. After incubation, the slide glass was performed using a Renishaw inVia Raman microscope.

2.8. Statistical Analysis

Data were expressed as mean ± standard deviation (SD) in at least three trials. The significant difference between groups was performed using a Student's two-tailed *t*-test and one-way ANOVA with SPSS software 23.0. * $p < 0.05$ and ** $p < 0.01$ were statistically significant.

3. Results and Discussion

3.1. Synthesis and Characterization of MB-GNR@mSiO_2-GO

The preparation process of MB-GNR@mSiO_2-GO for PTT/PDT is illustrated in Scheme 1. GNRs were prepared in an aqueous CTAB solution using a seed-mediated growth method [15]. The synthesized GNRs were covered with a mesoporous layer of silica (mSiO_2) using a previously reported method, forming highly monodisperse GNR@mSiO_2 [27,28]. MB was directly loaded inside the pores of mSiO_2 with a high loading capacity to obtain MB-GNR@mSiO_2 [25]. Amine functionalization of the MB-GNR@mSiO_2 surfaces was performed using APTES as a silane coupling agent (MB-GNR@mSiO_2-APTES). GO was synthesized using a modified Hummers' method [29]. Finally, the carboxyl groups of GO were covalently conjugated to the free amino groups of MB-GNR@mSiO_2-APTES through a coupling reaction in the presence of EDC/NHS chemistry to form MB-GNR@mSiO_2-GO, which can potentially prevent leaching of the loaded MB molecules before MB-GNR@mSiO_2-GO is taken up by the cells [32].

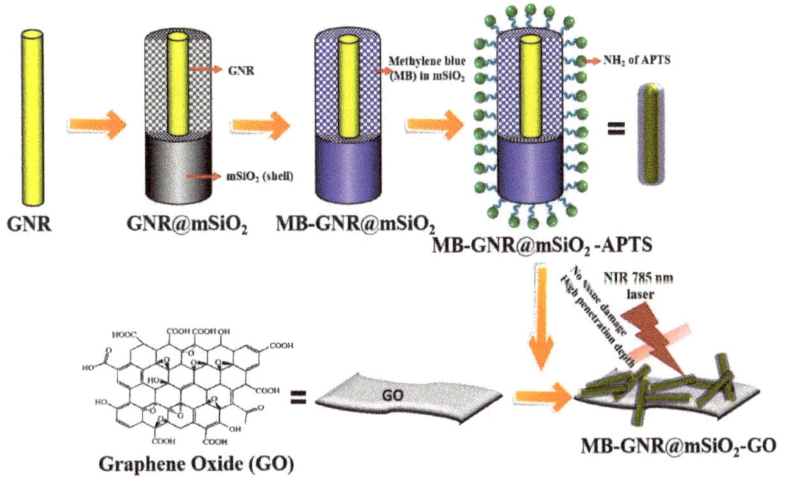

Scheme 1. Schematic illustration of the synthesis process of methylene blue-loaded mesoporous silica-coated gold nanorods on graphene oxide (MB-GNR@mSiO_2-GO) for photothermal and photodynamic therapy.

The morphologies of GNR, GNR@mSiO_2, MB-GNR@mSiO_2, GO, and MB-GNR@mSiO_2-GO were characterized using Cs-corrected FE-TEM (Figure 1). The GNRs have a uniform and well-defined rod-like structure with an average length of 47 ± 3.6 nm and a diameter of 12 nm ± 2.4 nm. The GNRs were coated with mSiO_2 to form highly monodispersed GNR@mSiO_2. The thickness of the mSiO_2 was 38 nm, and a nanoporous structure could be observed. Similar findings were reported by Xia et al., who demonstrated that the thickness of the GNR@mSiO_2 was 12 nm [33]. After loading with MB, MB-GNR@mSiO_2 exhibited a uniform size and morphology. The as-prepared GO was nanosheet-shaped with a smooth surface and wrinkled edges. MB-GNR@mSiO_2 was uniformly anchored on the

GO nanosheets (Figure S1). Additionally, Cs-corrected FE-TEM and the corresponding EDX mapping of MB-GNR@mSiO_2-GO revealed the presence of Au, Si, S, C, and O, confirming the presence of MB-GNR@mSiO_2-GO (Figure 2).

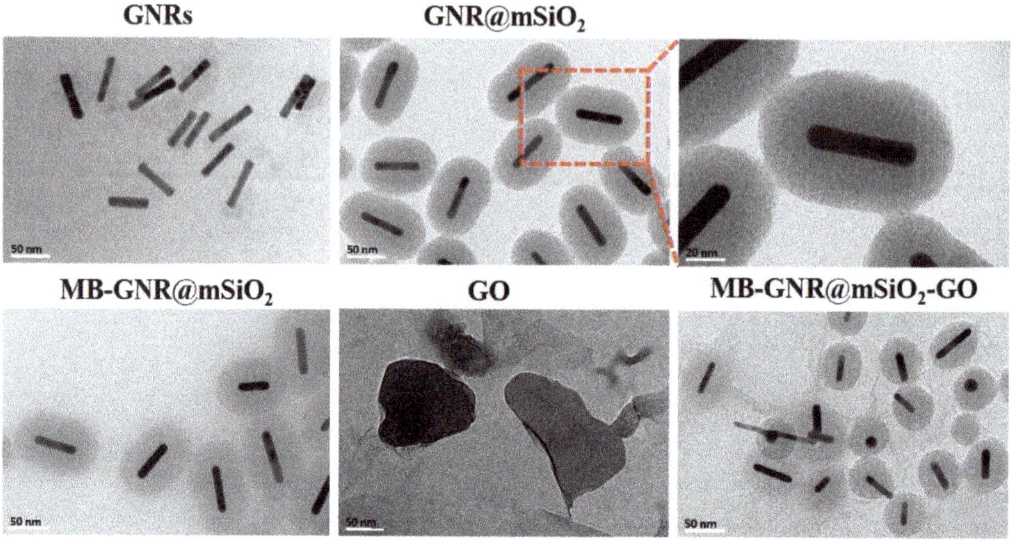

Figure 1. FETEM images of gold nanorods (GNRs), mesoporous silica-coated gold nanorods (GNR@mSiO_2), methylene blue-loaded mesoporous silica-coated gold nanorods (MB-GNR@mSiO_2), graphene oxide (GO), and methylene blue-loaded mesoporous silica-coated gold nanorods on graphene oxide (MB-GNR@mSiO_2-GO).

Figure 3a shows the UV–vis spectra of GNR, GNR@mSiO_2, MB-GNR@mSiO_2, GO, and MB-GNR@mSiO_2-GO. The as-prepared GNR has a strong longitudinal surface plasmon resonance (LSPR) peak at 800 nm, which is more suitable for PTT with an NIR laser [34]. After coating with SiO_2, the LSPR peak of GNR@mSiO_2 was slightly red-shifted to 810 nm, which could be attributed to an increase in the local refractive index of the surrounding medium after the creation of the silica shell layer by replacing the CTAB [35]. After loading with MB, the absorption spectra of MB-GNR@mSiO_2 showed a characteristic peak of MB at 655 nm and a strong LSPR peak at 800 nm, indicating that MB molecules were successfully loaded onto GNR@mSiO_2 and were highly effective in PDT, owing to their ability to generate ROS [36]. The absorption spectrum of GO spanned from the UV-vis range to the NIR range, which can effectively enhance the photothermal conversion efficiency [37]. The absorption spectra of MB-GNR@SiO_2-GO exhibited characteristic peaks of GO, MB, and GNR@SiO_2 at 280, 655, and 810 nm, respectively, confirming their strong photothermal and photodynamic properties.

The particle sizes of GNR, GNR@mSiO_2, MB-GNR@mSiO_2, GO, and MB-GNR@mSiO_2-GO were measured by dynamic light scattering (DLS). The GNR, GNR@mSiO_2, MB-GNR@mSiO_2, GO, and MB-GNR@mSiO_2-GO have average sizes of 43.67 $\pm$ 2.18 nm, 188.46 $\pm$ 3.42 nm, 196.57 $\pm$ 4.17 nm, 80.47 $\pm$ 4.02 nm, and 294.90 $\pm$ 2.74, respectively (Figure 3b). The average particle size of GNR@mSiO_2 increased to ~144.79 nm when compared with GNR, indicating that the mSiO_2 layer was effectively coated on the surface of the GNR. The average particle size of MB-GNR@mSiO_2 also increased to ~8.11 nm compared with GNR@mSiO_2, demonstrating that MB was successfully loaded onto GNR@mSiO_2. Additionally, the particle size of MB-GNR@mSiO_2-GO increased significantly to ~98.33 nm, implying that GO was covalently conjugated to MB-GNR@mSiO_2. As shown in Figure 3c, the zeta potentials (ZP) of GNR, GNR@mSiO_2, MB, MB-GNR@mSiO_2, GO, and MB-GNR@mSiO_2-GO were measured in PBS (pH 7.4). The ZP of GNR was +35.4 ± 2.0 mV due to the presence of positively charged CTAB on the surface of GNR. The ZP of GNR@mSiO_2 was −31.0 ± 1.5 mV because of the hydroxyl groups on the surface of GNR@mSiO_2. The ZP of MB was +7.5 ± 0.4 mV because of the positively charged blue dye. After MB loading, the ZP of MB-GNR@mSiO_2 was −31.4 ± 2.0 mV because the positively charged MB can be easily absorbed on the mSiO_2, which indicated the successful loading of MB and GNR@mSiO_2. The ZP of GO was −34.0 ± 2.0 mV because of the ionized carboxylic acid groups. The ZP of MB-GNR@mSiO_2-GO was −39.0 ± 2.0 mV which confirmed the successful conjugation of GO and MB-GNR@mSiO_2-GO [38].

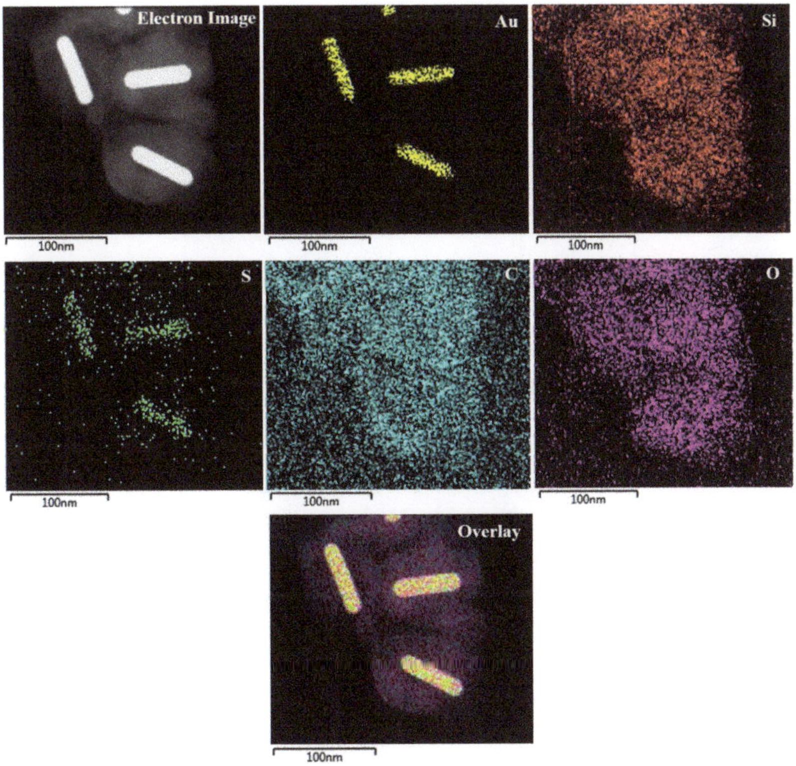

Figure 2. Cs-corrected field emission transmission electron microscope (Cs-corrected FE-TEM) image of methylene blue-loaded mesoporous silica-coated gold nanorods on graphene oxide (MB-GNR@mSiO_2-GO) and corresponding energy-dispersive X-ray (EDX) mapping of MB-GNR@mSiO_2-GO.

The micro-Raman spectra of the synthesized GNR, MB-GNR@mSiO_2, and MB-GNR@mSiO_2-GO were analyzed using an excitation laser at 785 nm (2 mW, spot area = 2–100 μm) and are shown in Figure 3d. Strong Raman signals were recorded for MB-GNR@mSiO_2 and MB-GNR@mSiO_2-GO, while GNR did not show any Raman signals. The Raman spectra of MB-GNR@mSiO_2 and MB-GNR@mSiO_2-GO showed intense bands at 246 cm^{-1}, 445 cm^{-1}, 771 cm^{-1}, 1317 cm^{-1}, 1397 cm^{-1}, and 1624 cm^{-1}, assigned to the Au–N stretching mode (ν(Au–N)), C–N–C skeletal deformation mode (δ(C–N–C), γ(C–H) stretching mode, ν(N–CH_3) mode, ν_{sym}(C–N) mode, and ν(C–C)$_{ring}$ mode, respectively. The Raman spectra of MB-GNR@mSiO_2-GO were much stronger than those of MB-GNR@mSiO_2, suggesting that GO has a high binding affinity for the MB-GNR@mSiO_2 surface [25,28,39].

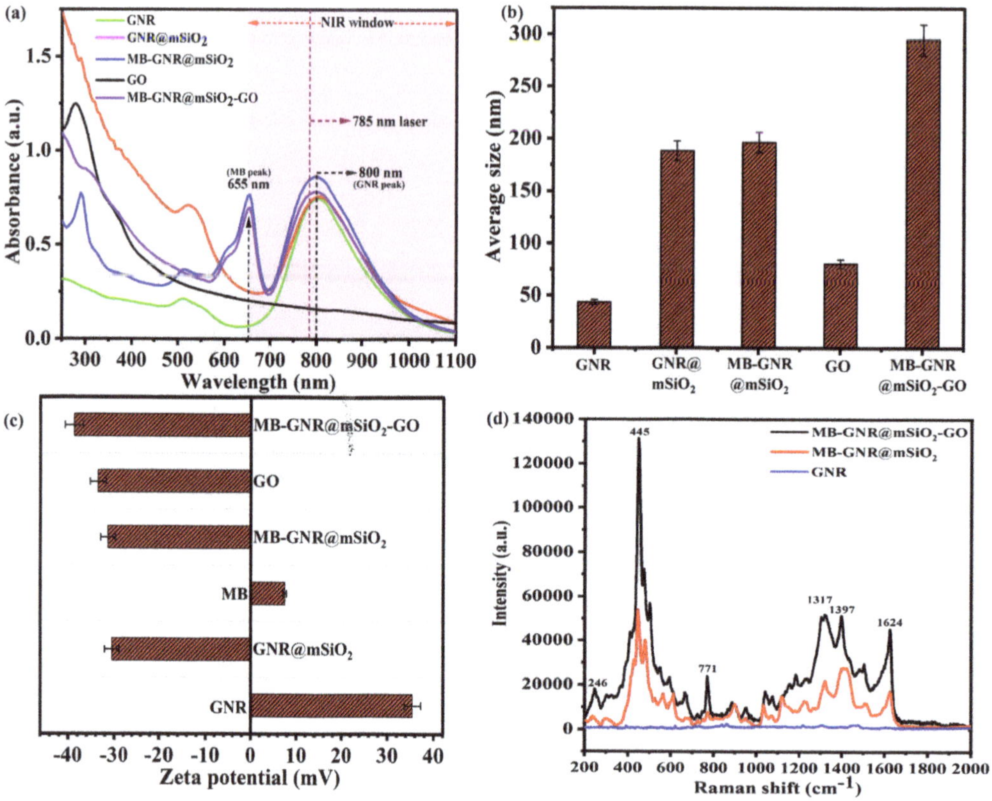

Figure 3. (a) UV–vis spectra of gold nanorods (GNRs), mesoporous silica-coated gold nanorods ($GNR@mSiO_2$), methylene blue-loaded mesoporous silica-coated gold nanorods ($MB-GNR@mSiO_2$), graphene oxide (GO), and methylene blue-loaded mesoporous silica-coated gold nanorods on graphene oxide ($MB-GNR@mSiO_2-GO$). (b) Particle size distribution histograms of GNR, $GNR@mSiO_2$, $MB-GNR@mSiO_2$, GO, and $MB-GNR@mSiO_2-GO$. (c) Zeta potential of GNR, $GNR@mSiO_2$, $MB-GNR@mSiO_2$, GO, and $MB-GNR@mSiO_2-GO$. (d) MicroRaman spectra of GNR, $MB-GNR@mSiO_2$, GO, and $MB-GNR@mSiO_2-GO$ solutions excited by a 785 nm laser (2 mW, spot area = 2–100 μm).

3.2. Measurement of Photothermal Performance

To further assess the photothermal efficacy of $MB-GNR@mSiO_2-GO$, we examined the temperature variations of GNR, $GNR@mSiO_2$, $MB-GNR@mSiO_2$, GO, and $MB-GNR@mSiO_2$-GO aqueous solutions (36 μg/mL) under a 785 nm laser irradiation at a power density of 0.8 W/cm^2 for 25 min (Figure 4a). The temperature variation of GNRs, $GNR@mSiO_2$, $MB-GNR@mSiO_2$, and GO aqueous solutions increased from 36.42 °C to 41.28 °C, which was insufficient to ablate the cancer cells. Meanwhile, the temperature variation of pure water only increased from 36.24 °C to 27.28 °C. In contrast, the temperature variation of the $MB-GNR@mSiO_2-GO$ aqueous solutions increased to 50.67 °C, which is sufficient for cancer therapy because cancer cells can be specifically eliminated at temperatures higher than hyperthermia (42–47 °C) [40]. Additionally, the Au concentration of $MB-GNR@mSiO_2-GO$ was calculated to be 36 μg Au/mL, and different volumes (0–200 μL of 36 μg/mL) of MB-$GNR@mSiO_2$-GO aqueous solutions (1 mL) were added to a quartz cuvette and exposed to a 785 nm laser at 0.8 W/cm^2 for 25 min. The temperature variation of $MB-GNR@mSiO_2-GO$ (0, 50, 100, 150, and 200 μL of 36 μg/mL) aqueous solutions were exposed to a 785 nm laser at 0.8 W/cm^2 for 25 min, resulting in temperature values of 37.4 °C, 45.79 °C, 48.72 °C, 49.71 °C, and 52.12 °C, respectively (Figure 4b). The photothermal effect of the MB-GNR@mSiO$_2$-GO (0, 50, 100, 150, and 200 µL of 36 µg/mL) aqueous solutions was further examined using a FLIR thermal imaging camera (Figure 4c).

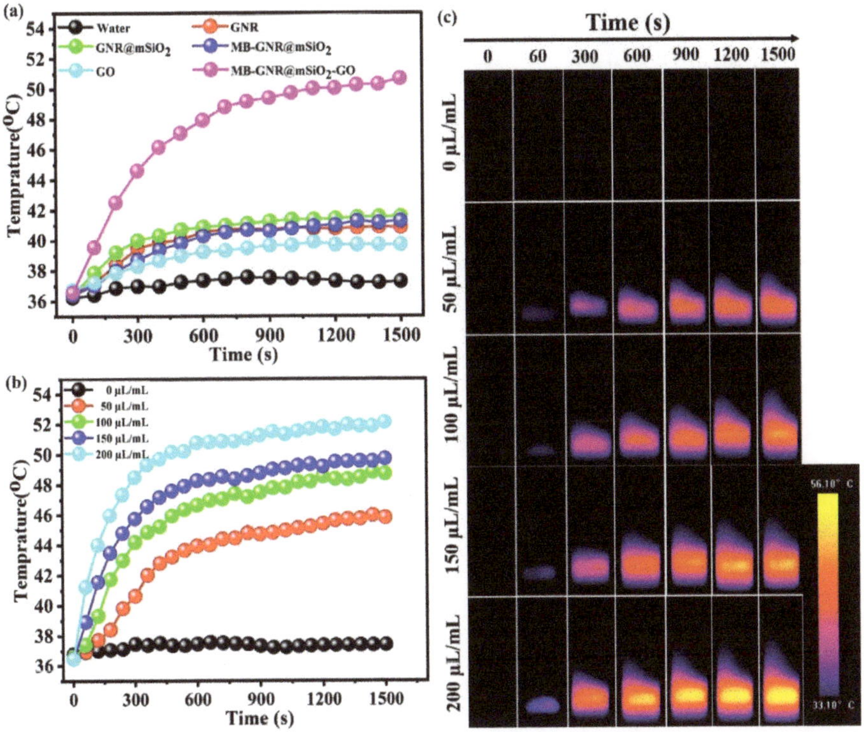

Figure 4. (**a**) The temperature variation curves of gold nanorods (GNRs), mesoporous silica-coated gold nanorods (GNR@mSiO$_2$), methylene blue-loaded mesoporous silica-coated gold nanorods (MB-GNR@mSiO$_2$), graphene oxide (GO), and methylene blue-loaded mesoporous silica-coated gold nanorods on graphene oxide (MB-GNR@mSiO$_2$-GO) aqueous solutions (36 µg/mL) under 785 nm laser irradiation at a power density of 0.8 W/cm^2 for 25 min. (**b**) The temperature variation curves of MB-GNR@mSiO$_2$-GO aqueous solutions with different volumes (0–200 µL of 36 µg/mL) under 785 nm laser irradiation at a power density of 0.8 W/cm^2 for 25 min. (**c**) IR thermal images of MB-GNR@mSiO$_2$-GO aqueous solutions with different volumes (0–200 µL of 36 µg/mL) under 785 nm laser irradiation at a power density of 0.8 W/cm^2 for 25 min.

The photothermal conversion efficiency (η) is a significant parameter used to quantify the ability of photothermal agents to convert light into heat [41]. To further demonstrate the photothermal conversion efficiency (η) of the MB-GNR@mSiO$_2$-GO aqueous solution, 1 mL of the MB-GNR@mSiO$_2$-GO (200 µL of 36 µg/mL) aqueous solution was exposed to a 785 nm laser at 0.8 W/cm^2 for 25 min, turned off for 25 min, and then left to cool down to room temperature (Figure 5a). The value of t_s was 398.81 s. The photothermal conversion efficiency (η) of MB-GNR@mSiO$_2$-GO (200 µL/mL) was calculated to be 48.93% (Figure 5b), confirming that MB-GNR@mSiO$_2$-GO has the best photothermal conversion capability, which was higher than that of commonly reported photothermal materials, such as GNRs (21%) [42], GO nanosheets (18%) [9], Fe_3O_4@CuS NPs (19.2%) [43], $Cu_{2-x}S$ nanocrystals (16.3%) [44], and Fe_3O_4 nanoclusters (20.8%) [45]. More interestingly, the photothermal conversion efficiency of MB-GNR@mSiO$_2$-GO was significantly higher (48.93%) than that of either GNRs (21%) alone or GO nanosheets (18%) alone, which strongly supports the idea that the dual plasmonic photothermal agents have a synergistic plasmonic effect [9,46].

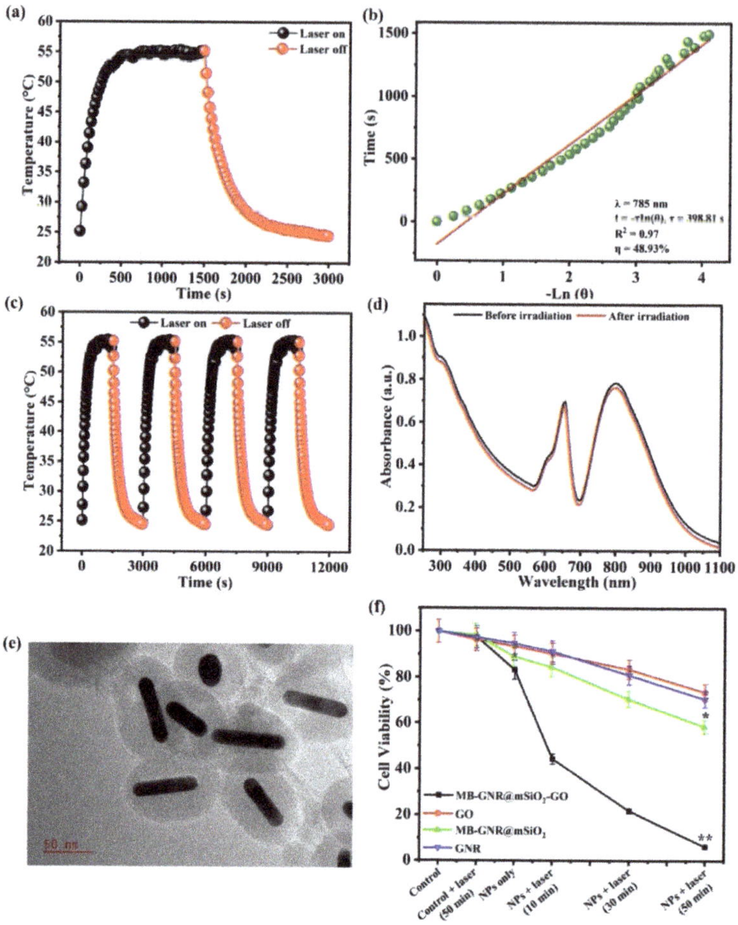

Figure 5. (**a**) The temperature variation curves of methylene blue-loaded mesoporous silica-coated gold nanorods on the graphene oxide (MB-GNR@mSiO$_2$-GO) aqueous solution (200 μL of 36 μg/mL) under 785 nm laser irradiation at 0.8 W/cm^2 for 25 min, followed by natural cooling for 25 min. (**b**) The linear relationship of time data versus –In θ obtained from the cooling period of the panel. (**c**) The temperature variation curves of the MB-GNR@mSiO$_2$-GO aqueous solution (200 μL of 36 μg/mL) under four 785 nm laser on/off cycles (0.8 W/cm^2). (**d**) UV–vis spectra of MB-GNR@mSiO$_2$-GO before and after four cycles of laser irradiation. (**e**) Field emission transmission electron microscope (FETEM) image of MB-GNR@mSiO$_2$-GO before and after four cycles of laser irradiation. (**f**) Viability of MDA-MB-231 cells treated with gold nanorods (GNRs), methylene blue-loaded mesoporous silica-coated gold nanorods (MB-GNR@mSiO$_2$), graphene oxide (GO), and methylene blue-loaded mesoporous silica-coated gold nanorods on graphene oxide (MB-GNR@mSiO$_2$-GO) (200 μL of 36 μg/mL) for 12 h of incubation with or without 785 nm laser irradiation at 0.8 W/cm^2 for various time points (0, 10, 30, and 50 min). Data were expressed as mean ± S.D. (n = 3, * significant $p < 0.05$, ** significant $p < 0.01$).

It is widely known that GNR and GO have excellent heat-conducting properties, which are crucial for enhancing the photothermal stability of MB-GNR@mSiO$_2$-GO (200 μL of

36 μg/mL) [47]. To further evaluate the photothermal stability of MB-GNR@mSiO_2-GO, the solution was placed in a quartz cuvette and then exposed to four cycles of laser on/off at 0.8 W/cm^2. As shown in Figure 5c, no significant changes in the temperature variation were observed for MB-GNR@mSiO_2-GO after four laser on/off cycles, indicating that MB-GNR@mSiO_2-GO has good photothermal stability. Additionally, the UV–vis spectra of MB-GNR@mSiO_2-GO were recorded after four cycles of laser irradiation (Figure 5d). The UV–vis spectra demonstrated that MB-GNR@mSiO_2-GO exhibited no obvious changes in absorbance after four cycles of laser irradiation. MB-GNR@mSiO_2-GO was also characterized using TEM, DLS, and ZP after four cycles of laser irradiation. The FETEM image of MB-GNR@mSiO_2-GO also showed a uniform size and morphology after four cycles of laser irradiation (Figure 5e). The particle size of MB-GNR@mSiO_2-GO after four cycles of laser irradiation was characterized by DLS as ~292.75 ± 2.92 nm. The ZP of MB-GNR@mSiO_2-GO was −36.81 ± 3.31 mV. These results indicate that MB-GNR@mSiO_2-GO has excellent photothermal stability after four cycles of laser irradiation [47].

3.3. *In Vitro PTT/PDT Synergistic Effect*

To evaluate the in vitro photothermal/photodynamic synergistic efficacy of GNR, MB-GNR@mSiO_2, GO, and MB-GNR@mSiO_2-GO, the BLO-11, and MDA-MB-231 cells were incubated with GNR, MB-GNR@mSiO_2, GO, and MB-GNR@mSiO_2-GO (200 μL of 36 μg/mL) for 12 h and 48 h, and then were exposed to a 785 nm laser at 0.8 W/cm^2 for various time points (0, 10, 30, and 50 min). All cells were further incubated for 6 h, and their viability was determined by the MTT assay. As shown in Figure S2a,b, BLO-11 cells showed no obvious cytotoxicity without laser irradiation and the cell viability of BLO-11 cells with laser irradiation for 10, 30, and 50 min was above 80% when the high concentration of GNR, MB-GNR@mSiO_2, and GO was 200 μL of 36 μg/mL for 12 h and 48 h, indicating that GNR, MB-GNR@mSiO_2, and GO have excellent biocompatibility. The cell viability with laser irradiation for 10 and 30 min was above 80% at a higher concentration of MB-GNR@mSiO_2-GO (200 μL of 36 μg/mL) and the cell viability with laser irradiation for 50 min was more than 75% at a high concentration of MB-GNR@mSiO_2-GO up to 200 μL of 36 μg/mL, suggesting very little cytotoxicity and good biocompatibility for the MB-GNR@mSiO_2-GO. As shown in Figures 5f and S3, no noticeable cytotoxicity was observed in any of the treated groups for 12 h and 48 h of incubation without laser irradiation, suggesting that the as-prepared GNR, MB-GNR@mSiO_2, GO, and MB-GNR@mSiO_2-GO were highly biocompatible. However, upon irradiation, GNR, MB-GNR@mSiO_2, and GO showed noticeable anticancer activity against MDA-MB-231 cells, which was 70.58%, 58.55%, and 73.57% for 12 h, and was 64.41%, 56.32%, and 71.42% for 48 h at the high concentration of GNR, MB-GNR@mSiO_2, and GO up to 200 μL of 36 μg/mL and exposed to a 785 nm laser at 0.8 W/cm^2 for 50 min. The cells were treated with MB-GNR@mSiO_2-GO (200 μL of 36 μg/mL) for 12 h and 48 h and then were irradiated with a 785 nm laser at 0.8 W/cm^2 for 50 min, which showed significantly reduced cell viability (6.32% and 4.46%), demonstrating the possibility that laser irradiation could boost the cytotoxic effects of MB-GNR@mSiO_2-GO [34]. As shown in Figures 5f and S3, no significant difference was observed after 12 h and 48 h incubation with GNR, MB-GNR@mSiO_2, GO, and MB-GNR@mSiO_2-GO(200 μL of 36 μg/mL) with or without laser irradiation, confirming that all the materials have very little toxicity to cells without laser irradiation; in addition, all the materials with laser irradiation for 50 min showed significant cytotoxicity to MDA-MB-231 cells because these materials have excellent photothermal conversion properties.

The photothermal effects of GNR, MB-GNR@mSiO_2, and MB-GNR@mSiO_2-GO on MDA-MB-231 cells were further confirmed using calcein-AM and PI co-staining to visually differentiate live (green) and dead (red) cells. As shown in Figure 6, there were no cellular deaths in the control only, control with laser irradiation, GNR only, GNR with laser irradiation, MB-GNR@mSiO_2 only, MB-GNR@mSiO_2 with laser irradiation, and MB-GNR@mSiO_2-GO only groups because all of the cells showed green fluorescence and negligible cell deaths in these groups. In contrast, MB-GNR@mSiO_2-GO with laser irradiation showed strong red fluorescence, and a large red area was observed, suggesting that MB-GNR@mSiO$_2$-GO with laser irradiation induced complete cellular destruction [48].

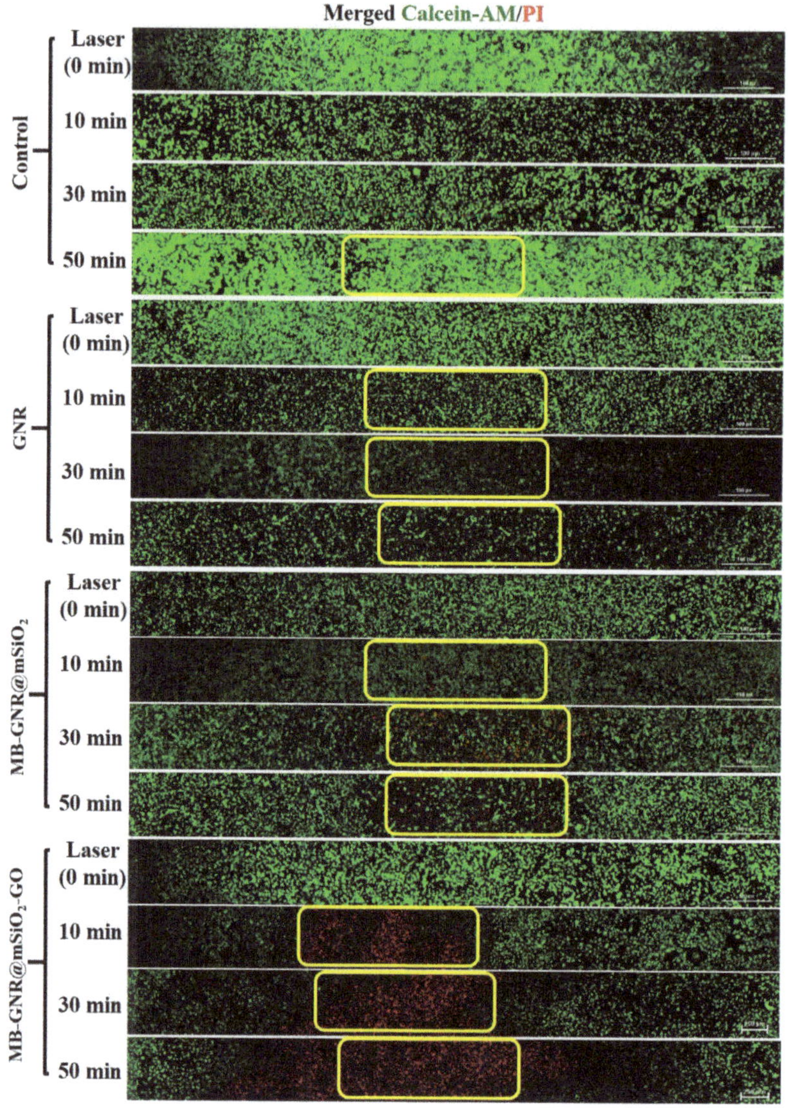

Figure 6. Merged confocal fluorescence images of MDA-MB-231 cells treated with gold nanorods (GNRs), methylene blue-loaded mesoporous silica-coated gold nanorods (MB-GNR@mSiO$_2$), and methylene blue-loaded mesoporous silica-coated gold nanorods on graphene oxide (MB-GNR@mSiO$_2$-GO) (200 µL of 36 µg/mL) for 12 h with or without 785 nm laser irradiation at 0.8 W/cm^2 for various time points (0, 10, 30, and 50 min). The cells were stained using calcein-AM (live: green) and PI (dead: red) (20× magnification). The yellow color indicates the laser spot area.

To evaluate the photodynamic effect of GNR, MB-GNR@mSiO$_2$, and MB-GNR@mSiO$_2$-GO on intracellular ROS generation, DCHF-DA was used as a fluorescent dye to monitor changes in the intracellular ROS levels. As shown in Figure 7, no strong green signal was detected in the control only, control with laser irradiation, GNR only, GNR with laser irradiation, MB-GNR@mSiO$_2$ only, MB-GNR@mSiO$_2$ with laser irradiation, and MB-GNR@mSiO$_2$-GO only groups, which did not damage the cancer cells. In contrast, MB-GNR@mSiO$_2$-GO produced strong green fluorescence after 785 nm laser irradiation for 50 min, which could damage biomolecules and cellular organs such as the mitochondria and nucleus, implying the generation of ROS [49].

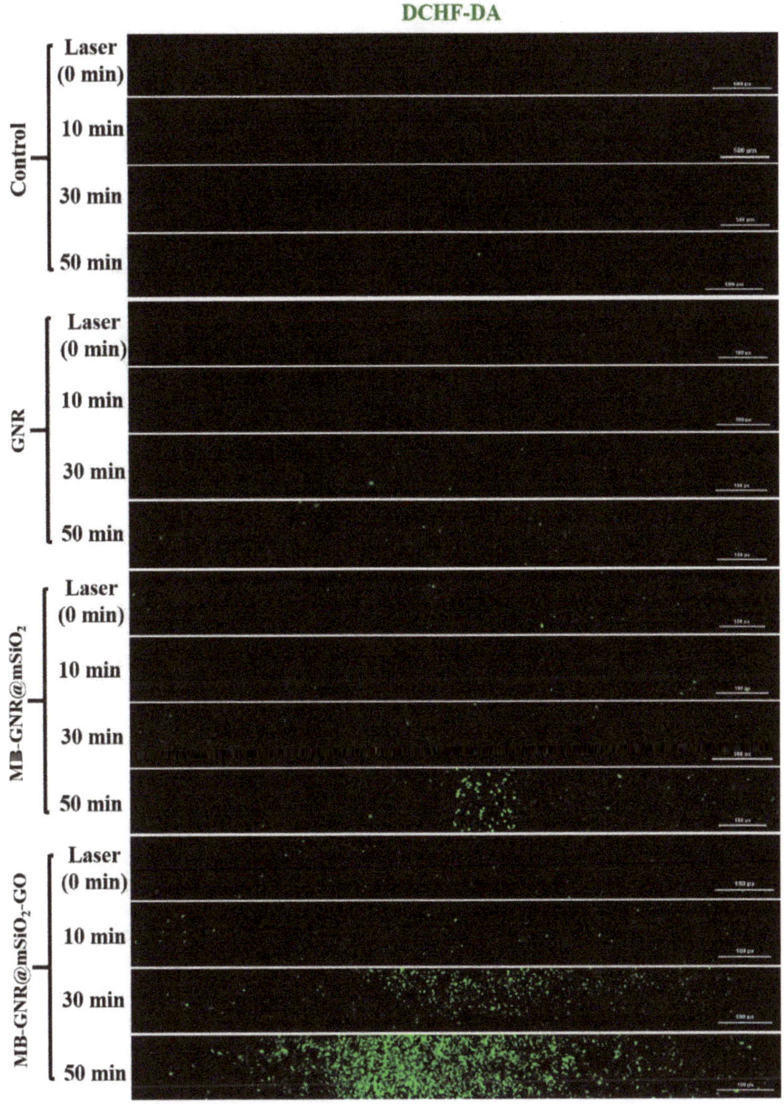

Figure 7. DCHF-DA staining confocal fluorescence images of MDA-MB-231 cells treated with gold nanorods (GNRs), methylene blue-loaded mesoporous silica-coated gold nanorods (MB-GNR@mSiO$_2$), and methylene blue-loaded mesoporous silica-coated gold nanorods on graphene oxide (MB-GNR@mSiO$_2$-GO) (200 μL of 36 μg/mL) with or without 785 nm laser irradiation at 0.8 W/cm^2 for various time points (0, 10, 30, and 50 min) (20× magnification).

3.4. Surface-Enhanced Raman Scattering (SERS) Imaging

Surface-enhanced Raman scattering (SERS) imaging is a technique for boosting the Raman signals of molecules using nanomaterials, such as metal colloids [50]. SERS has drawn much interest in the field of biological research and has emerged as an effective imaging method for early cancer detection [51]. The bright-field images and SERS spectra of the GNR (control), MB-GNR@mSiO_2, and MB-GNR@mSiO_2-GO-treated MDA-MB-231 cells under an excitation laser of 785 nm are shown in Figure 8a–c. Strong SERS peaks were observed for MB-GNR@mSiO_2 and MB-GNR@mSiO_2-GO, whereas no SERS peaks were observed for the control. The SERS peaks of MB-GNR@mSiO_2 and MB-GNR@mSiO_2-GO were clearly observed at 445 cm^{-1}, 1397 cm^{-1}, and 1624 cm^{-1}, which were assigned to the C–N–C skeletal deformation mode (δ(C–N–C), ν_{sym}(C–N), and $\nu(C–C)_{ring}$ modes, indicating that the SERS peaks of MB-GNR@mSiO_2-GO were sufficient to detect cancer at an early stage.

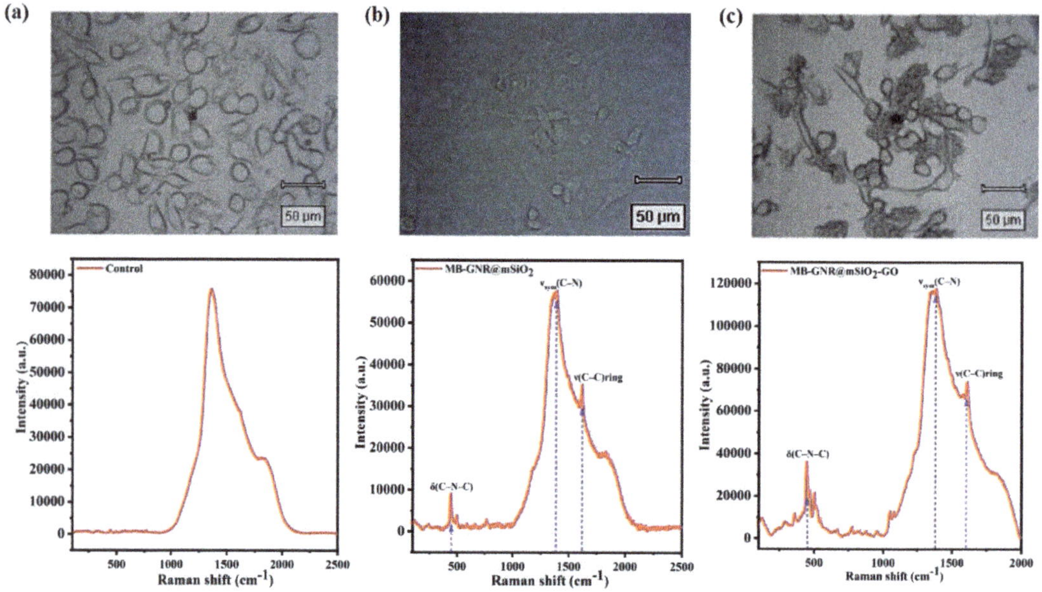

Figure 8. Bright-field images and micro-Raman spectra of MDA-MB-231 cells treated with control (**a**), methylene blue-loaded mesoporous silica-coated gold nanorods (MB-GNR@mSiO_2) (**b**), and methylene blue-loaded mesoporous silica-coated gold nanorods on graphene oxide (MB-GNR@mSiO_2-GO) (**c**). The bright-field images indicate the laser-activated area (λ = 785 nm, 2 mW, spot area = 2–100 μm).

4. Conclusions

In summary, we successfully designed and developed multifunctional MB-GNR@mSiO_2-GO as a new phototheranostic agent for intracellular SERS imaging-guided synergistic PTT/PDT of cancer. MB-GNR@mSiO_2-GO showed strong SPR absorption in the NIR region, high MB loading capability, high photothermal conversion efficiency (48.93%), superior photothermal stability, and a strong SERS effect. The in vitro cytotoxicity of MB-GNR@mSiO_2-GO had excellent biocompatibility in the absence of laser irradiation. Most intriguingly, the strong SERS effect of MB-GNR@mSiO_2-GO permits accurate cancer cell detection by SERS imaging, and the cells treated with MB-GNR@mSiO_2-GO and laser irradiation led to significant cell death, confirming that MB-GNR@mSiO_2-GO has a strong SERS effect and superior photothermal efficiency under laser irradiation. The in vitro photodynamic effect of MB-GNR@mSiO_2-GO showed enhanced cell apoptosis under laser irradiation, which could damage the mitochondria and nucleus, indicating that MB-GNR@mSiO$_2$-GO has excellent ROS generation capability with the help of 785 nm laser irradiation. We believe that this multifunctional MB-GNR@mSiO$_2$-GO has great potential for further clinical applications in the diagnosis and therapy of cancer.

Supplementary Materials: The following supporting information can be downloaded at: https://www.mdpi.com/article/10.3390/pharmaceutics14102242/s1, Figure S1: FETEM images of methylene blue-loaded mesoporous silica-coated gold nanorods on graphene oxide (MB-GNR@mSiO$_2$-GO); Figure S2: Viability of BLO-11 cells treated with gold nanorod (GNR), methylene blue-loaded mesoporous silica-coated gold nanorod (MB-GNR@mSiO$_2$), graphene oxide (GO), and methylene blue-loaded mesoporous silica-coated gold nanorods on graphene oxide (MB-GNR@mSiO$_2$-GO) (200 μL/mL) for 12 h (a) and 48 h (b) with or without 785 nm laser irradiation at 0.8 W/cm^2 for various time points (0, 10, 30, and 50 min). Data were expressed as mean ± S.D (n = 3); Figure S3: Viability of MDA-MB-231 cells treated with gold nanorod (GNR), methylene blue-loaded mesoporous silica-coated gold nanorod (MB-GNR@mSiO$_2$), graphene oxide (GO), and methylene blue-loaded mesoporous silica-coated gold nanorods on graphene oxide (MB-GNR@mSiO$_2$-GO) (200 μL/mL) for 48 h with or without 785 nm laser irradiation at 0.8 W/cm^2 for various time points (0, 10, 30, and 50 min). Data were expressed as mean ± S.D. (n = 3, * significant $p < 0.05$, ** significant $p < 0.01$).

Author Contributions: The manuscript was written through the contributions of all authors. Conceptualization, S.-H.S. and A.J.; investigation, A.J. and H.-W.H.; formal analysis, S.-H.S. and H.-W.H.; data curation, S.-H.S., writing—original draft preparation, P.M.; data curation, E.-S.J.; funding acquisition, E.-S.J. All authors have read and agreed to the published version of the manuscript.

Funding: E.-S.J. is especially grateful for the research grant from the Korean Ministry of Education, Science & Technology (2016R1D1A3B0201175615) and the Grand Information Technology Research Center Program (IITP-2022-2020-0-01612) through the Institute of Information & Communications Technology Planning & Evaluation (IITP) funded by the Ministry of Science and ICT (MSIT), Korea. This research was supported by the Basic Science Research Program through the National Research Foundation of Korea (NRF), funded by the Ministry of Education (NRF-2022R1I1A1A01068693).

Institutional Review Board Statement: Not applicable.

Informed Consent Statement: Not applicable.

Data Availability Statement: Not applicable.

Conflicts of Interest: The authors declare no conflict of interest.

References

1. Shao, W.; Yang, C.; Li, F.; Wu, J.; Wang, N.; Ding, Q.; Gao, J.; Ling, D. Molecular design of conjugated small molecule nanoparticles for synergistically enhanced PTT/PDT. *Nano-Micro Lett.* **2020**, *12*, 147. [CrossRef] [PubMed]
2. Ferrari, M. Cancer nanotechnology: Opportunities and challenges. *Nat. Rev. Cancer* **2005**, *5*, 161–171. [CrossRef] [PubMed]
3. Wu, C.; Li, D.; Wang, L.; Guan, X.; Tian, Y.; Yang, H.; Li, S.; Liu, Y. Single wavelength light-mediated, synergistic bimodal cancer photoablation and amplified photothermal performance by graphene/gold nanostar/photosensitizer theranostics. *Acta Biomater.* **2017**, *53*, 631–642. [CrossRef] [PubMed]
4. Li, M.; Xu, J.; Romero-Gonzalez, M.; Banwart, S.A.; Huang, W.E. Single cell Raman spectroscopy for cell sorting and imaging. *Curr. Opin. Biotechnol.* **2012**, *23*, 56–63. [CrossRef]
5. Park, H.; Lee, S.; Chen, L.; Lee, E.K.; Shin, S.Y.; Lee, Y.H.; Son, S.W.; Oh, C.H.; Song, J.M.; Kang, S.H. SERS imaging of HER2-overexpressed MCF7 cells using antibody-conjugated gold nanorods. *Phys. Chem. Chem. Phys.* **2009**, *11*, 7444–7449. [CrossRef]
6. Cheng, L.; Wang, C.; Feng, L.; Yang, K.; Liu, Z. Functional nanomaterials for phototherapies of cancer. *Chem. Rev.* **2014**, *114*, 10869–10939. [CrossRef]
7. Manivasagan, P.; Bharathiraja, S.; Moorthy, M.S.; Oh, Y.-O.; Seo, H.; Oh, J. Marine biopolymer-based nanomaterials as a novel platform for theranostic applications. *Polym. Rev.* **2017**, *57*, 631–667. [CrossRef]
8. Deng, K.; Liu, D.; Wang, Z.; Zhou, Z.; Chen, Q.; Luo, J.; Zhang, Y.; Hou, Z.; Lin, J. Surface-Functionalized NdVO4: Gd3+ Nanoplates as Active Agents for Near-Infrared-Light-Triggered and Multimodal-Imaging-Guided Photothermal Therapy. *Pharmaceutics* **2022**, *14*, 1217. [CrossRef] [PubMed]
9. Younis, M.R.; An, R.B.; Yin, Y.-C.; Wang, S.; Ye, D.; Xia, X.-H. Plasmonic nanohybrid with high photothermal conversion efficiency for simultaneously effective antibacterial/anticancer photothermal therapy. *ACS Appl. Bio Mater.* **2019**, *2*, 3942–3953. [CrossRef] [PubMed]

10. Li, J.; Rao, J.; Pu, K. Recent progress on semiconducting polymer nanoparticles for molecular imaging and cancer phototherapy. *Biomaterials* **2018**, *155*, 217–235. [CrossRef] [PubMed]
11. Gong, H.; Peng, R.; Liu, Z. Carbon nanotubes for biomedical imaging: The recent advances. *Adv. Drug Deliv. Rev.* **2013**, *65*, 1951–1963. [CrossRef] [PubMed]
12. Li, Y.; Lu, W.; Huang, Q.; Li, C.; Chen, W. Copper sulfide nanoparticles for photothermal ablation of tumor cells. *Nanomedicine* **2010**, *5*, 1161–1171. [CrossRef] [PubMed]
13. Narayanan, N.; Kim, J.H.; Santhakumar, H.; Joseph, M.M.; Karunakaran, V.; Shamjith, S.; Saranya, G.; Sujai, P.T.; Jayasree, R.S.; Barman, I. Nanotheranostic Probe Built on Methylene Blue Loaded Cucurbituril [8] and Gold Nanorod: Targeted Phototherapy in Combination with SERS Imaging on Breast Cancer Cells. *J. Phys. Chem. B* **2021**, *125*, 13415–13424. [CrossRef] [PubMed]
14. Huang, X.; Neretina, S.; El-Sayed, M.A. Gold nanorods: From synthesis and properties to biological and biomedical applications. *Adv. Mater.* **2009**, *21*, 4880–4910. [CrossRef]
15. Nikoobakht, B.; El-Sayed, M.A. Preparation and growth mechanism of gold nanorods (NRs) using seed-mediated growth method. *Chem. Mater.* **2003**, *15*, 1957–1962. [CrossRef]
16. Kim, H.J.; Wang, W.; Bu, W.; Hossen, M.M.; Londono-Calderon, A.; Hillier, A.C.; Prozorov, T.; Mallapragada, S.; Vaknin, D. Salt mediated self-assembly of poly (ethylene glycol)-functionalized gold nanorods. *Sci. Rep.* **2019**, *9*, 20349. [CrossRef]
17. Terentyuk, G.; Panfilova, E.; Khanadeev, V.; Chumakov, D.; Genina, E.; Bashkatov, A.; Tuchin, V.; Bucharskaya, A.; Maslyakova, G.; Khlebtsov, N. Gold nanorods with a hematoporphyrin-loaded silica shell for dual-modality photodynamic and photothermal treatment of tumors in vivo. *Nano Res.* **2014**, *7*, 325–337. [CrossRef]
18. Tsai, M.-F.; Chang, S.-H.G.; Cheng, F.-Y.; Shanmugam, V.; Cheng, Y.-S.; Su, C.-H.; Yeh, C.-S. Au nanorod design as light-absorber in the first and second biological near-infrared windows for in vivo photothermal therapy. *ACS Nano* **2013**, *7*, 5330–5342. [CrossRef]
19. Stöber, W.; Fink, A.; Bohn, E. Controlled growth of monodisperse silica spheres in the micron size range. *J. Colloid Interface Sci.* **1968**, *26*, 62–69. [CrossRef]
20. Khatiri, R.; Reyhani, A.; Mortazavi, S.; Hossainalipour, M. Immobilization of serum albumin on the synthesized three layers core–shell structures of super-paramagnetic iron oxide nanoparticles. *J. Ind. Eng. Chem.* **2013**, *19*, 1642–1647. [CrossRef]
21. Huang, P.; Zeng, B.; Mai, Z.; Deng, J.; Fang, Y.; Huang, W.; Zhang, H.; Yuan, J.; Wei, Y.; Zhou, W. Novel drug delivery nanosystems based on out-inside bifunctionalized mesoporous silica yolk–shell magnetic nanostars used as nanocarriers for curcumin. *J. Mater. Chem. B* **2016**, *4*, 46–56. [CrossRef]
22. Zhang, Z.; Wang, L.; Wang, J.; Jiang, X.; Li, X.; Hu, Z.; Ji, Y.; Wu, X.; Chen, C. Mesoporous silica-coated gold nanorods as a light-mediated multifunctional theranostic platform for cancer treatment. *Adv. Mater.* **2012**, *24*, 1418–1423. [CrossRef]
23. Tardivo, J.P.; Del Giglio, A.; De Oliveira, C.S.; Gabrielli, D.S.; Junqueira, H.C.; Tada, D.B.; Severino, D.; de Fátima Turchiello, R.; Baptista, M.S. Methylene blue in photodynamic therapy: From basic mechanisms to clinical applications. *Photodiagn. Photodyn. Ther.* **2005**, *2*, 175–191. [CrossRef]
24. Kohle, F.F.; Li, S.; Turker, M.Z.; Wiesner, U.B. Ultrasmall PEGylated and targeted core–shell silica nanoparticles carrying methylene blue photosensitizer. *ACS Biomater. Sci. Eng.* **2019**, *6*, 256–264. [CrossRef]
25. Seo, S.-H.; Kim, B.-M.; Joe, A.; Han, H.-W.; Chen, X.; Cheng, Z.; Jang, E.-S. NIR-light-induced surface-enhanced Raman scattering for detection and photothermal/photodynamic therapy of cancer cells using methylene blue-embedded gold nanorod@ SiO_2 nanocomposites. *Biomaterials* **2014**, *35*, 3309–3318. [CrossRef]
26. Zhang, Z.; Shi, J.; Song, Z.; Zhu, X.; Zhu, Y.; Cao, S. A synergistically enhanced photothermal transition effect from mesoporous silica nanoparticles with gold nanorods wrapped in reduced graphene oxide. *J. Mater. Sci.* **2018**, *53*, 1810–1823. [CrossRef]
27. Gorelikov, I.; Matsuura, N. Single-step coating of mesoporous silica on cetyltrimethyl ammonium bromide-capped nanoparticles. *Nano Lett.* **2008**, *8*, 369–373. [CrossRef]
28. Han, H.-W.; Joe, A.; Jang, E.-S. Reduced cytotoxicity of CTAB-templated silica layer on gold nanorod using fluorescence dyes and its application in cancer theranostics. *J. Ind. Eng. Chem.* **2021**, *96*, 202–212. [CrossRef]
29. Hummers, W.S., Jr.; Offeman, R.E. Preparation of graphitic oxide. *J. Am. Chem. Soc.* **1958**, *80*, 1339. [CrossRef]
30. Roper, D.K.; Ahn, W.; Hoepfner, M. Microscale heat transfer transduced by surface plasmon resonant gold nanoparticles. *J. Phys. Chem. C* **2007**, *111*, 3636–3641. [CrossRef]
31. Liu, X.; Li, B.; Fu, F.; Xu, K.; Zou, R.; Wang, Q.; Zhang, B.; Chen, Z.; Hu, J. Facile synthesis of biocompatible cysteine-coated CuS nanoparticles with high photothermal conversion efficiency for cancer therapy. *Dalton Trans.* **2014**, *43*, 11709–11715. [CrossRef] [PubMed]
32. Yang, Y.-H.; Liu, C.-H.; Liang, Y.-H.; Lin, F.-H.; Wu, K.C.-W. Hollow mesoporous hydroxyapatite nanoparticles (hmHANPs) with enhanced drug loading and pH-responsive release properties for intracellular drug delivery. *J. Mater. Chem. B* **2013**, *1*, 2447–2450. [CrossRef] [PubMed]
33. Xia, H.-X.; Yang, X.-Q.; Song, J.-T.; Chen, J.; Zhang, M.-Z.; Yan, D.-M.; Zhang, L.; Qin, M.-Y.; Bai, L.-Y.; Zhao, Y.-D. Folic acid-conjugated silica-coated gold nanorods and quantum dots for dual-modality CT and fluorescence imaging and photothermal therapy. *J. Mater. Chem. B* **2014**, *2*, 1945–1953. [CrossRef]
34. Manivasagan, P.; Jun, S.W.; Truong, N.T.P.; Hoang, G.; Mondal, S.; Moorthy, M.S.; Kim, H.; Phan, T.T.V.; Doan, V.H.M.; Kim, C.-S. A multifunctional near-infrared laser-triggered drug delivery system using folic acid conjugated chitosan oligosaccharide encapsulated gold nanorods for targeted chemo-photothermal therapy. *J. Mater. Chem. B* **2019**, *7*, 3811–3825. [CrossRef]

35. Liu, X.; Liu, L.; Hu, X.; Zhou, S.; Ankri, R.; Fixler, D.; Xie, Z. Multimodal bioimaging based on gold nanorod and carbon dot nanohybrids as a novel tool for atherosclerosis detection. *Nano Res.* **2018**, *11*, 1262–1273. [CrossRef]
36. Planas, O.; Bresolí-Obach, R.; Nos, J.; Gallavardin, T.; Ruiz-González, R.; Agut, M.; Nonell, S. Synthesis, photophysical characterization, and photoinduced antibacterial activity of methylene blue-loaded amino-and mannose-targeted mesoporous silica nanoparticles. *Molecules* **2015**, *20*, 6284–6298. [CrossRef] [PubMed]
37. Xing, C.; Jing, G.; Liang, X.; Qiu, M.; Li, Z.; Cao, R.; Li, X.; Fan, D.; Zhang, H. Graphene oxide/black phosphorus nanoflake aerogels with robust thermo-stability and significantly enhanced photothermal properties in air. *Nanoscale* **2017**, *9*, 8096–8101. [CrossRef]
38. Fang, S.; Lin, J.; Li, C.; Huang, P.; Hou, W.; Zhang, C.; Liu, J.; Huang, S.; Luo, Y.; Fan, W. Dual-stimuli responsive nanotheranostics for multimodal imaging guided trimodal synergistic therapy. *Small* **2017**, *13*, 1602580. [CrossRef] [PubMed]
39. Shim, K.D.; Jang, E.S. SERS Signal Enhancement of Methylene Blue-embedded Agglomerated Gold Nanorod@ SiO_2 Core@ Shell Composites. *Bull. Korean Chem. Soc.* **2018**, *39*, 936–940. [CrossRef]
40. Manivasagan, P.; Jun, S.W.; Hoang, G.; Mondal, S.; Kim, H.; Doan, V.H.M.; Kim, J.; Kim, C.-S.; Oh, J. Anti-EGFR antibody conjugated thiol chitosan-layered gold nanoshells for dual-modal imaging-guided cancer combination therapy. *J. Control. Release* **2019**, *311*, 26–42. [CrossRef]
41. Yu, X.; Yang, K.; Chen, X.; Li, W. Black hollow silicon oxide nanoparticles as highly efficient photothermal agents in the second near-infrared window for in vivo cancer therapy. *Biomaterials* **2017**, *143*, 120–129. [CrossRef] [PubMed]
42. Hessel, C.M.; Pattani, V.P.; Rasch, M.; Panthani, M.G.; Koo, B.; Tunnell, J.W.; Korgel, B.A. Copper selenide nanocrystals for photothermal therapy. *Nano Lett.* **2011**, *11*, 2560–2566. [CrossRef]
43. Wu, Z.C.; Li, W.P.; Luo, C.H.; Su, C.H.; Yeh, C.S. Rattle-Type Fe_3O_4@ CuS Developed to Conduct Magnetically Guided Photoinduced Hyperthermia at First and Second NIR Biological Windows. *Adv. Funct. Mater.* **2015**, *25*, 6527–6537. [CrossRef]
44. Wang, S.; Riedinger, A.; Li, H.; Fu, C.; Liu, H.; Li, L.; Liu, T.; Tan, L.; Barthel, M.J.; Pugliese, G. Plasmonic copper sulfide nanocrystals exhibiting near-infrared photothermal and photodynamic therapeutic effects. *ACS Nano* **2015**, *9*, 1788–1800. [CrossRef] [PubMed]
45. Huang, C.-C.; Chang, P.-Y.; Liu, C.-L.; Xu, J.-P.; Wu, S.-P.; Kuo, W.-C. New insight on optical and magnetic Fe_3O_4 nanoclusters promising for near infrared theranostic applications. *Nanoscale* **2015**, *7*, 12689–12697. [CrossRef] [PubMed]
46. Manivasagan, P.; Joe, A.; Han, H.-W.; Thambi, T.; Selvaraj, M.; Chidambaram, K.; Kim, J.; Jang, E.-S. Recent advances in multifunctional nanomaterials for photothermal-enhanced Fenton-based chemodynamic tumor therapy. *Mater. Today Bio* **2022**, *13*, 100197. [CrossRef] [PubMed]
47. Qi, Z.; Shi, J.; Zhang, Z.; Cao, Y.; Li, J.; Cao, S. PEGylated graphene oxide-capped gold nanorods/silica nanoparticles as multifunctional drug delivery platform with enhanced near-infrared responsiveness. *Mater. Sci. Eng. C* **2019**, *104*, 109889. [CrossRef] [PubMed]
48. Wang, H.; Zhang, M.; Zhang, L.; Li, S.; Li, L.; Li, X.; Yu, M.; Mou, Z.; Wang, T.; Wang, C. Near-infrared light and pH-responsive Au@ carbon/calcium phosphate nanoparticles for imaging and chemo-photothermal cancer therapy of cancer cells. *Dalton Trans.* **2017**, *46*, 14746–14751. [CrossRef] [PubMed]
49. Bharathiraja, S.; Moorthy, M.S.; Manivasagan, P.; Seo, H.; Lee, K.D.; Oh, J. Chlorin e6 conjugated silica nanoparticles for targeted and effective photodynamic therapy. *Photodiagn. Photodyn. Ther.* **2017**, *19*, 212–220. [CrossRef] [PubMed]
50. Moisoiu, V.; Socaciu, A.; Stefancu, A.; Iancu, S.D.; Boros, I.; Alecsa, C.D.; Rachieriu, C.; Chiorean, A.R.; Eniu, D.; Leopold, N. Breast cancer diagnosis by surface-enhanced Raman scattering (SERS) of urine. *Appl. Sci.* **2019**, *9*, 806. [CrossRef]
51. Sun, C.; Gao, M.; Zhang, X. Surface-enhanced Raman scattering (SERS) imaging-guided real-time photothermal ablation of target cancer cells using polydopamine-encapsulated gold nanorods as multifunctional agents. *Anal. Bioanal. Chem.* **2017**, *409*, 4915–4926. [CrossRef] [PubMed]

pharmaceutics

Article

Gold Nanoparticles Synthesized with Common Mullein (*Verbascum thapsus*) and Castor Bean (*Ricinus communis*) Ethanolic Extracts Displayed Antiproliferative Effects and Induced Caspase 3 Activity in Human HT29 and SW480 Cancer Cells

Karen M. Soto [1,*], Ivan Luzardo-Ocampo [2], José M. López-Romero [1], Sandra Mendoza [3], Guadalupe Loarca-Piña [3], Eric M. Rivera-Muñoz [4] and Alejandro Manzano-Ramírez [1,*]

1 Centro de Investigaciones y de Estudios Avanzados del I. P. N. Unidad Querétaro, Queretaro 76230, Mexico
2 Instituto de Neurobiología, Universidad Nacional Autónoma de México (UNAM-Campus Juriquilla), Queretaro 76230, Mexico
3 Research and Graduate Program in Food Science, Universidad Autónoma de Querétaro, Queretaro 76010, Mexico
4 Centro de Física Aplicada y Tecnología Avanzada (CFATA), Universidad Nacional Autónoma de México (UNAM-Campus Juriquilla), Queretaro 76230, Mexico
* Correspondence: karen.soto@cinvestav.mx (K.M.S.); amanzano@cinvestav.mx (A.M.-R.)

Citation: Soto, K.M.; Luzardo-Ocampo, I.; López-Romero, J.M.; Mendoza, S.; Loarca-Piña, G.; Rivera-Muñoz, E.M.; Manzano-Ramírez, A. Gold Nanoparticles Synthesized with Common Mullein (*Verbascum thapsus*) and Castor Bean (*Ricinus communis*) Ethanolic Extracts Displayed Antiproliferative Effects and Induced Caspase 3 Activity in Human HT29 and SW480 Cancer Cells. *Pharmaceutics* **2022**, *14*, 2069. https://doi.org/10.3390/pharmaceutics14102069

Academic Editors: Manivasagan Panchanathan and Eue-Soon Jang

Received: 5 September 2022
Accepted: 24 September 2022
Published: 28 September 2022

Abstract: Gold nanoparticles (AuNPs) are promising nanomaterials exhibiting anti-cancer effects. Green AuNPs synthesis using plant extracts can be used to achieve stable and beneficial nanoparticles due to their content of bioactive compounds. This research aimed to synthesize and evaluate the antiproliferative and caspase-3 activity induction of green AuNPs synthesized with common mullein (*V. thapsus*) flowers (AuNPsME) and castor bean (*R. communis*) leaves (AuNPsCE) ethanolic extracts in human HT29 and SW480 colorectal cancer cells. Their effect was compared with chemically synthesized AuNPs (AuNPsCS). The extracts mainly contained *p*-coumaric acid (71.88–79.93 μg/g), ferulic acid (19.07–310.71 μg/g), and rutin (8.14–13.31 μg/g). The obtained nanoparticles presented typical FT-IR bands confirming the inclusion of polyphenols from *V. thapsus* and *R. communis* and spherical/quasi-spherical morphologies with diameters in the 20.06–37.14 nm range. The nanoparticles (20–200 μg/mL) showed antiproliferative effects in both cell lines, with AuNPsCE being the most potent (IC_{50} HT29: 110.10 and IC_{50}SW480: 64.57 μg/mL). The AuNPsCS showed the lowest intracellular reactive oxygen species (ROS) generation in SW480 cells. All treatments induced caspase 3/7 activity to a similar or greater extent than 30 mM H_2O_2-treated cells. Results indicated the suitability of *V. thapsus* and *R. communis* extracts to synthesize AuNPs, displaying a stronger antiproliferative effect than AuNPsCS.

Keywords: antiproliferative effect; castor bean (*R. communis*); colorectal cancer; common mullein (*V. thapsus*); gold nanoparticles; reactive oxygen species

1. Introduction

Gold nanoparticles (AuNPs), also called "gold colloids", are usually particles with different morphologies, including nanospheres, nano bars, nanolayers, and nanocubes, and small sizes ranging from 1 to 100 nm. These nanoparticles represent one of the most promising nanomaterials because of various advantages, such as their ease of synthesis, bio-molecular coupling abilities or easy functionalization with almost every type of electron-donating molecule, and multiple properties, allowing their use in different applications, e.g., sensors, photodynamic therapies, bioimaging, drug carriers, and others [1–4].

One of the most important applications of AuNPs is their use as therapeutic molecules for the medical treatment of cancer. Gold nanoparticles exhibit cytotoxic properties against certain types of cancer cell lines like Breast cancer cells (MDA-MB-231) [5], liver cancer cells (HepG2) [6], pancreatic cancer cells (PANC-1) [7], and others. The mechanism of action of the cytotoxic properties has not yet been fully described so far. However, some of the possible mechanisms include interaction with the cell membrane and internalization in the cell by different charges on the surface of cells and AuNPs showing cytotoxic activity via ROS (Reactive Oxygen Species) production, DNA and mitochondrial damage, activation of caspase cascade of apoptosis, cell rupture and inhibition of cell growth, and loss of membrane stability [8–10]. Colorectal cancer (CRC) is a common malignancy and one of the leading causes of cancer death worldwide; the second and third most frequently occurring cancer in women and men, respectively, is the third most common cancer, with 1.8 million new cases diagnosed in 2018. Colon cancer cell lines, e.g., HT29 and SW480, are fundamental tools for analyzing the role of antitumor compounds and treatments [11,12].

Several methods can be used to synthesize AuNPs. The most used is the synthesis involving the reducing of a precursor agent ($HAuCl_4$) with reducing agents, such as sodium borohydride ($NaBH_4$). However, generating toxic compounds harmful to humans and the environment requires searching for new sources of reducing agents. In this sense, plant extracts represent a suitable alternative with multiple advantages over other green synthesis methods. Plant extracts have a high quantity of diverse phytochemicals, such as proteins, polyphenols, and related multifunctional-electron-rich species serving as dual reducing and stabilizing agents, encapsulating gold atoms through the phytochemical crown on the AuNPs surface, increasing their stability [10,13]. Moreover, AuNPs synthesis does not need microbial cultures, making them more straightforward to further scale-up.

Considering the wide variety of medicinal plants, our research group has previously examined underutilized plant species and derived wastes for manufacturing nanoparticles [10,14]. Underutilized plant species, castor bean (*R. communis*), and common mullein (*V. thapsus*) are traditionally used to treat several conditions. Castor bean is a plant found in various regions of the world, mainly in tropical areas, and is widely used to produce castor oil. Castor bean leaves have been used as a medicinal plant due to the presence of phytochemicals, mainly phenolic compounds such as flavonoids (rutin, quercetin, and (+)-catechin), hydroxycinnamic acids (caffeic, chlorogenic, *p*-Coumaric, and sinapic acids), and hydroxybenzoic acids (gallic and ellagic acids) [15]. As a result of these compounds' presence, the leaves and leaves extracts have been used for their anti-inflammatory [16] and anti-cancer effects [17]. On the other hand, common mullein (*V. thapsus*) is a plant used in traditional medicine to treat inflammatory diseases, bronchial congestion, coughs, diarrhea, and snakebite [18]. *R. communis* leaf extract was used for the green synthesis of spherical gold nanoparticles with diameters between 40 and 80 nm and probed against Hela and HepG2 cells, obtaining 58.64% and 42.74% inhibitory effects, respectively [19], while *V. thapsus* was used for the synthesis of metallic nanoparticles, e.g., AgNPs for antibacterial activity [20], CuO nanoparticles for photocatalytic degradation of methylene blue [21], and Iron nanoparticles for Cr (VI) reduction [22].

As one of the most critical applications of green gold, nanoparticles are used for their anti-cancer effects, and plant extracts contain bioactive compounds displaying anti-cancer benefits and are functionally optimal as reducing agents to synthesize stable AuNPs. Their green synthesis using either castor bean or common mullein provides a technological activity with medicinal value against several types and stages of cancer [23,24]. Therefore, this research aimed to synthesize, characterize, and evaluate the antiproliferative effect of green AuNPs made with castor bean leaves and common mullein flowers' ethanolic extracts against two cell lines of colorectal cancer, one of the most critical sicknesses. It is important to mention that this is the first report of the use of *V. thapsus* for green gold synthesis and the first report of the antiproliferative against CRC effect of *R. communis* gold nanoparticles. In this study, the AuNPs obtained with chemical synthesis and green synthesis were compared to observe the differences that may occur in the biological activity and in the mechanism of antiproliferative action, which could be affected by the extracts present in the plants. It is important to mention that this is the first report of the use of *V. thapsus* for green gold synthesis and the first report of the antiproliferative effect of *R. communis* gold nanoparticles against CRC.

2. Materials and Methods

2.1. Chemicals and Reagents

Solvents, chloroauric acid ($HAuCl_4$), sodium; 1,1-diphenyl-2-picrylhydrazyl (DPPH); 2,2′-azinobis-(3-ethylbenzothiazoline-6-sulfonic acid) (ABTS), $NaBH_4$, 2′,7′-dichlorodihydrofluorescein (H_2DCFDA), H_2O_2, DMSO, HPLC standards, and potassium persulfate were purchased from Sigma Chemical Co. (St. Louis, MO, USA). Dulbecco's Modified Eagle's Medium (DMEM), fetal bovine serum (FBS), and antibiotic antimycotic solution were from Gibco, Waltham, MA, USA. The EnzChek® Caspase-3 Assay was purchased from ThermoFisher Scientific (Waltham, MA, USA).

2.2. Extract Obtention and Characterization

Common mullein (*V. thapsus*) (World Checklist of Selected Plant Families, WCSP record in review: Sp. Pl. 177 1753) flowers and castor bean (*R. communis*) (WCSP record: 178867) leaves were collected in Queretaro (Mexico) and deposited in the "Dr. Jerzy Rzedowski" (QMEX) herbarium of Universidad Autónoma de Querétaro (Juriquilla, Querétaro, Mexico). The selected plant parts were washed with deionized water and dried in a convection oven at 60 °C for 6 h. The dried residues were then powdered to fine particles using a mixer grinder (KRUPS GX4100, Solingen, Germany) and passed through a 60-mesh screen (250 µm particle size). Ten grams of the powdered castor and mullein plants were added to 100 mL ethanol and left to stand for 4 h. The extracts were filtered (Whatman Paper no. 40) and stored at 4 °C for further analysis. These extracts were termed ME and CE for common mullein and castor bean, respectively.

To spectrophotometrically analyze phenolic compounds, total phenolic compounds (TPC) were determined using the Folin-Ciocalteu technique [25], and total phenolics were expressed as mg of gallic acid equivalents per gram of dry weight (mg GAE/g DW). For the in vitro antioxidant capacity (AOX), the DPPH [26] and $ABTS^{\bullet+}$ [27] radical scavengers were tested, and the results were indicated as mg Trolox equivalents per gram of dried weight (mg TE/g DW).

The identification and quantification of individual phenolic compounds were performed through high-performance liquid chromatography coupled to a diode array detector (HPLC-DAD) [28]. Briefly, an Agilent 1100 Series HPLC System (Agilent Technologies, Palo Alto, CA, USA) with a DAD detector (Agilent Technologies), and using a thermostatically controlled (35 ± 0.6 °C) Zorbax Eclipse XDB-C18 column (Agilent Technologies, 9.4 × 250 mm, 5.0 µm of granule size) was used. The samples were eluted using a linear gradient of concentration from water containing 0.1% of acetic acid (Solvent A) to 100% of acetonitrile (solvent B) as follows: 80–83% A for 7 min, 83–60% A for 5 min, 60–50% A for 1 min, and 50–85% A for 2 min (The solvent B complete the 100%). An acquisition speed of 1 min with a volume injection of 20 µL was used, and the samples were analyzed in triplicate. The quantification was carried out using external HPLC-grade chemical standards, and curves were plotted. Compounds were expressed in mg equivalents of each one per gram of dry weight (mg eq/g DW).

2.3. Green (AuNPsME and AuNPsCE) and Chemical (AuNPsCS) Synthesis of Gold Nanoparticles

For the chemical synthesis, AuNPsCS were prepared with the methodology presented by Deraedt et al. [29]. For this, 1.5 mg of $HAuCl_4$ (Mw = 339.7 g/mol, n = 4.4×10^{-3} mol) were dissolved in 32 mL of water to obtain [Au] = 1.38×10^{-1} mM. After 15 min of stirring, 1 mL of a water solution of $NaBH_4$ (10 equivalents of Au) was added quickly.

For the green synthesis of AuNPs, 10 mL of an aqueous solution of $HAuCl_4$ (0.5 mM) were mixed with 50 and 100 µL of the ethanolic extract of mullein flowers and castor leaves, referred to as AuNPsME and AuNPsCE, respectively. The reaction mixture was incubated at 60 °C for 5 min, the reduction of Au^{3+} ions was monitored using color changes, and the UV−vis spectrum was recorded from 300 to 800 nm with baseline correction using water as the blank on a Spectra Max Tunable Microplate Reader (Molecular Devices Co., Sunnyvale, CA, USA).

2.4. AuNPs Characterization

The morphology and diameters of AuNPs were determined using a scanning transmission electron microscope (STEM SU8230, Hitachi, Tokyo, Japan). The diameters of the AuNPs were obtained by measuring at least 50 particles with Image J software [30]. DLS and Z potential were performed using a Zetasizer Nano ZS (Malvern Panalytical, Worcestershire, UK). Infrared absorption spectra of dried AuNPs and lyophilized extracts were recorded using Fourier-transformed infrared (FT-IR) equipment (Spectrometer Spectrum Two, PerkinElmer, Waltham, MA, USA). The X-ray pattern of the AuNPs was obtained with an X-ray diffractometer (Dmax 2100 Rigaku Americas, The Woodlands, TX, USA) that has a CuKα radiation generator (k = 1.5418° A), from 5 to 50 °, on a 2θ scale, with a step size of 0.02° [31]. The thermal analysis of the AuNPs was carried out using a TGA-50, Shimzadu thermogravimetric analyzer (Shimzadu Corp., Kyoto, Japan) by increasing the temperature from 25 to 800 °C at the rate of 5 °C/min, and nitrogen gas flow rate of 10 mL/min.

2.5. Cell Culture

Human HT-29 (ATCC HTB-38) and SW-480 [SW-480] (ATCC CCL-228) colon adenocarcinoma cells were acquired from American Type Culture Collection. Cells were cultured using Dulbecco's Modified Eagle's Medium (DMEM, Gibco, Waltham, MA, USA) supplemented with 2.2 g/L $NaHCO_3$, 10% fetal bovine serum (FBS, Gibco), and 1% antibiotic antimycotic solution (100×, Gibco). The cells were maintained in a humidified 5% CO_2 atmosphere at 37 °C.

2.5.1. Cell Viability Assay by 3-(4,5-Dimethylthiazol-2-yl)-2,5-Diphenyl Tetrazolium Bromide (MTT) Assay

The assessment of cell viability was conducted using the MTT assay. The cells (3.125×10^4 cells/cm^2) were seeded in 96-well plates for 24 h. The media (DMEM + 10% FBS) was then replaced with several concentrations of the green gold nanoparticles AuNPsCE and AuNPsME (20, 50, 100, 150, and 200 µg/mL AuNPs) dissolved in DMEM + 0.5% bovine seric albumin (BSA). To avoid excessive cell growth and maintain a similar cell population that was present at the beginning of the experiment. After 24 h, a 5 mg/mL MTT solution dissolved in PBS 1× was added, and the cells were left to stand for 3 h. The media was then replaced with a sodium dodecyl sulfate (SDS)-HCl solution (14 g SDS dissolved in 140 mL distilled water + 112 µL HCl), and the plate was placed in a rocket shaker overnight (40 cycles/min). The absorbance was read at 595 nm in a Varioskan plate reader (ThermoFisher Scientific, Waltham, MA, USA), and the cell viability was expressed as a percentage of the negative control cells (cells without treatments: DMEM + 10% FBS). Cells treated with 30 µM H_2O_2 were used as the positive control. The half inhibitory concentration (IC_{50}) was calculated using a biological model adjusted equation provided by GraphPad Prism v. 8.2 (Dotmatics, Boston, MA, USA). Three independent experiments using six replicates were used.

2.5.2. Intracellular Reactive Oxygen Species (ROS) Quantification

The cells (2.50×10^4 cells/cm^2) were seeded in 96-well plates for 24 h. Afterward, the media was replaced with several concentrations of the treatments (20, 50, 100, 150, and 200 µg/mL AuNPs) prepared as indicated in Section 2.5.1., for 24 h [32,33]. Once the media was removed, the cells were then washed twice with PSB 1×, and 100 µM 2′,7′-dichlorodihydrofluorescein (H_2DCFDA) solution (D399, ThermoFisher Scientific, Waltham, MA, USA), prepared as indicated by the manufacturer in DMSO, was added (final DMSO concentration: 5%). Subsequently, the cells were incubated at 30 min at room temperature

(25 ± 1 °C). The fluorescence was read at 485/525 nm (excitation/emission) in a Varioskan plate reader (Thermo Fisher Scientific, Waltham, MA, USA) [34]. The results were expressed as relative fluorescence units (RFU). DMEM + 10% FBS and 30 μM H_2O_2-treated cells were used as positive and negative controls, respectively. Two independent experiments in triplicates were conducted.

2.5.3. Caspase 3/7 Activity Assay

The EnzChek® Caspase-3 Assay kit #2 (E13183, Invitrogen™, ThermoFisher Scientific, Waltham, MA, USA)) was used to assess apoptosis detection by assaying increases in the caspase-3 and other Asp-Glu-Val-Asp (DEVD)-specific protease activities such as Caspase 7. The cells (1.05×10^5 cells/cm²: 1×10^6 cells/well in 6-well plates) were seeded for 24 h. Then, the media was replaced with 1 mL of the IC_{50} concentration of each extract (IC_{50} AuNPsME: 116.7 and 153 μg/mL; IC_{50} AuNPsCE: 110.10 and 64.57 μg/mL, and IC_{50} AuNPsCS: 151.80 and 126.30 μg/mL, for HT29 and SW480 cells, respectively) for 24 h. Subsequently, the media was removed, cells were washed twice with PBS 1×, trypsinized (0.25% Trypsin-EDTA, ThermoFisher Scientific) for 15 min, and trypsin activity was stopped with 2% FBS-PBS 1×. Then, the cells were centrifuged to collect them (350× *g*, 10 min), and frozen at −80 °C. To quantify the Caspase-3 activity, the manufacturer's instructions were followed using the collected cells. A caspase-3 inhibitor provided by the kit was also used to treat the cells (1 mM Ac-DEVD-CHO) to confirm that the observed results were due to caspase-3 activity and no other proteases (results not shown). The fluorescence was measured at 496/520 nm of excitation/emission fluorescence, and results were expressed in RFU. Untreated cells and 30 mM H_2O_2-treated cells were used as negative and positive controls, respectively. Two independent experiments in duplicates were conducted for each treatment.

2.6. *Statistical Analysis*

Unless indicated, the results were expressed as the mean ± SD of at least two independent experiments in duplicates. An analysis of variance (ANOVA) followed by a post-hoc Tukey–Kramer's test was used, considering significance at $p < 0.05$. The analysis was carried out using JMP v. 16.0 software (SAS, Cary, NC, USA).

3. Results and Discussion

3.1. *Extraction and Characterization of Ethanolic Extracts*

The total phenolic compounds, the antioxidant capacity (ABTS and DPPH radical scavenging), and the individual phenolic composition attained using HPLC-DAD are shown in Figure 1. The best antioxidant activity was obtained with castor bean ethanolic leave extract (CE) with 726.65 ± 10.72 and 429.89 ± 2.01 mg TE/g DW for the $ABTS^{\bullet+}$ and DPPH radical scavenging, respectively, being almost ten times higher compared to mullein extract (Figure 1A). These results could be closely related to the total content of total phenolic compounds (TPC), which are known to provide antioxidant activity. Hence, the castor extract presented the highest range of total phenolic compounds with 101.32 ± 0.81 mg GAE/g DW. The obtained ME-TPC was consistent with previous reports of *V. thapsus* aerial and flower extracts (44.84–60.13 mg GAE/g DW or mL extract) [35,36], although Taleb et al. [36] did not specify the amount of powdered extract used, and the flowers were subjected to several ethanolic extractions (96% *v*/*v*) by maceration.

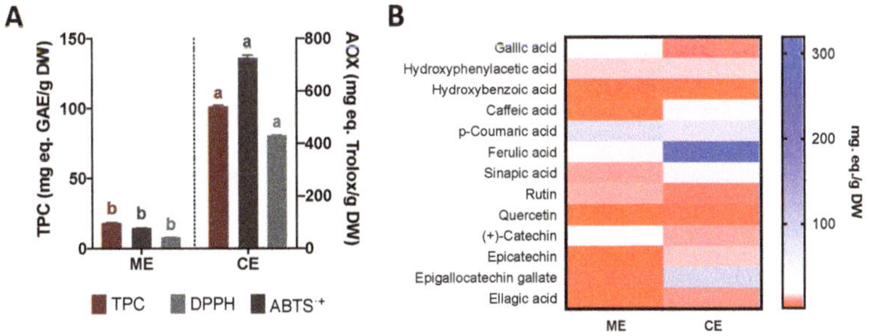

Figure 1. Polyphenolic composition of common mullein (*V. thapsus*) (ME) and castor bean (R. *communis*) (CE) extracts. (**A**) Content of total phenolic compounds (TPC) and antioxidant capacity (AOX), measured by ABTS and DPPH methods of *V. thapsus* (ME) and *R. communis* (CE) extracts. (**B**) Individual phenolic compounds identified by HPLC-DAD for ME and CE extracts. The results were expressed as the mean ± SD of three independent experiments in triplicates. Different letters in (**A**) (a, b) represent significant differences by Student's *t*-test ($p < 0.05$) between ME and CE.

Castor bean ethanolic extracts also yielded a higher TPC (Figure 1A) than reported values for methanolic and aqueous extracts of castor bean leaves (~30–50 mg GAE/g DW) [15]. The values are significantly lower than TPC contents from five *R. communis* Tunisian populations (174.25–623.7 mg GAE/g DW) [37]. However, the authors conducted a methanol-maceration extraction that was further concentrated in a rotary evaporator, which could explain almost 10-times the TPC than this study.

It has been reported that the DPPH and FRAP antioxidant capacities from *V. thapsus* aqueous and ethyl-acetate extracts are positively correlated with (+)-quercetin, rutin, ferulic acid, harpagoside, protocatechuic acid, rosmarinic acid, and salicylic acid, similarly to some of the identified and quantified compounds found in ME [35]. The richness in phenolic compounds of *V. thapsus* exhibiting a high ABTS$^{•+}$ and DPPH radical scavenging was demonstrated by Mihailović et al. [38], which showed that *V. thapsus* methanolic (8.11% *v*/*v* methanol) and aqueous (10.03% water) extracts from their aerial parts displayed some of the lowest IC_{50} values of radical scavenging in comparison with other *Verbascum* species. Considering CE, polar solvents have been validated as some of the most efficient extracting antioxidant compounds from *R. communis* leaves, displaying a high antiradical activity by ABTS$^{•+}$ and DPPH methods [39].

The identification and content of 13 phenolic compounds conducted using HPLC-DAD are presented in Figure 1B as a composition heatmap. Three hydroxybenzoic acid derivatives and benzaldehydes, four hydroxycinnamic acids and derivatives, two flavonoids, three flavones, and one hydrolyzable tannin (Table S1) were identified and quantified in the extracts. The extracts presented different concentrations and very noticeable differences in certain compounds. The main compounds in ME were *p*-coumaric acid, gallic acid, (+)-catechin, ferulic acid, and rutin. Mihailović et al. [38] reported similar caffeic acid concentrations from methanolic and water *V. thapsus* extracts (0.81–1.53 µg/g DW). Selseleh et al. [35] reported lower concentrations of *p*-coumaric acid (0.11 µg/g DW), ferulic acid (1.33 µg/g), rutin (2.90 µg/g), and similar levels of quercetin (1.35 µg/g) from methanolic fractions of *V. thapsus* flowers.

Regarding CE, the main compounds are ferulic acid, epigallocatechin gallate, coumaric acid, and caffeic acid (Figure 1B, Table S1). Vasco-Leal et al. [15] reported similar contents of gallic acid (10–25–35.81 µg/g DW), *p*-coumaric acid (0.63–517.12 mg/g DW), rutin (1.04–5.91 µg/g DW), and (+)-catechin (6–147.65 µg/g) than those found in this study. Caffeic acid was positively correlated ($p < 0.05$) with DPPH radical scavenging in *R. communis* extracts. In both samples, phenolic acids, whose structure is shorter than flavonoids, are presented as major compounds, and their possible participation in the synthesis and stabilization of metallic nanoparticles such as silver has been verified [40].

3.2. Synthesis and Characterization of AuNPs

The synthesis of AuNPs was monitored with color change and UV-Vis spectra. Figure 2A shows the UV-Vis spectra of AuNPs synthesized with the ethanolic extracts and $NaBH_4$ (chemical synthesis: AuNPsCS). AuNPsME presented a red wine color and a maximum absorption band of 550 nm. AuNPsCE showed a red wine coloration after the synthesis and two absorption bands at 540 nm and 675 nm. The synthesis of particles can explain these two bands with different diameters or nanorods. In contrast, the AuNPsCS showed a purple hue and maximum absorption at 548 nm. Generally, metal nanoparticles have free electrons, which give surface plasmon resonance (SPR) absorption band due to the combined vibration of electrons of metal nanoparticles in resonance with a light wave. The sharp bands of gold colloids for the plasmon resonance appeared in the visible range at 540–550 nm, dependent on the morphology and size distribution [41]. Rahman et al. [42] obtained dark brown AuNPsCE made with 5-fold more concentrated CE than those reported in this research. Moreover, an absorption peak at 550 nm was observed, close to the reported 540 nm from this research.

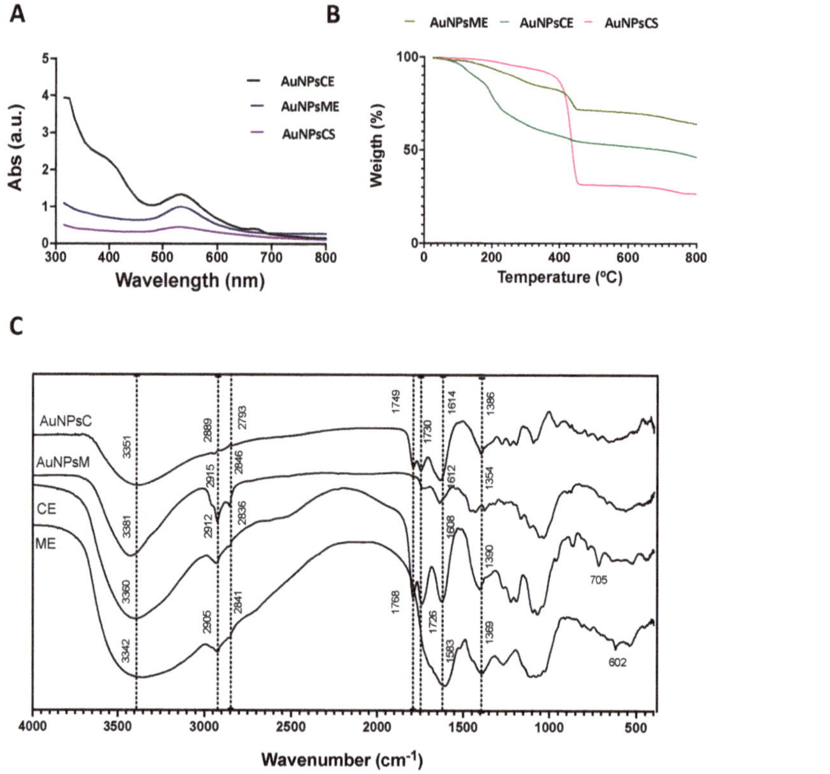

Figure 2. Synthesis and characterization of AuNPs. (**A**) Color change monitoring by UV-Vis spectra; (**B**) Thermogravimetric analysis (TGA) of AuNPs; (**C**) FT-IR spectra of the synthesized AuNPs. AuNPsCE: Gold nanoparticles synthesized with castor bean (*R. communis*) leaves extract; AuNPsCS: Gold nanoparticles produced by chemical synthesis; AuNPsME: Gold nanoparticles synthesized with common mullein (*V. thapsus*) flowers ethanolic extract; CE: Castor bean (*R.* communis) leaves ethanolic extract; ME: Common mullein (*V. thapsus*) flowers ethanolic extract.

3.2.1. Thermogravimetric Analysis

Thermogravimetric analysis (TGA) is presented in Figure 2B. TGA was conducted to determine the phytochemical compounds (phenolic and small proteins) that capped the surface of AuNPs. Green gold nanoparticles synthesized with mullein and castor extracts showed a two-stage process of losing weight. AuNPsME had a 23% weight loss in a 221–455 °C range and a total weight loss of 36% at 800 °C. AuNPsCE presented a weight loss of 27% in the 115–460 °C range and a final weight loss of 53.2%. The first weight loss can be attributed to the capped organic materials on the surface of particles. After 400 °C, the degradation of resistant aromatic compounds present in the extract compounds is presented, and the final weight loss refers to the presence of pure AuNPs [43,44]. In chemical synthesis, only one stage was observed at 305–800 °C, with a final weight loss of 83.3%. These results indicate that the bioactive compounds present in the extract, principally polyphenolic compounds, were capped on the synthesized AuNPs and were degraded due to the high temperature, which is not observed in chemically synthesized AuNPs.

3.2.2. Infrared Spectroscopy

Figure 2C presents the FT-IR spectra of pure ethanolic mullein, castor extracts, and synthesized gold nanoparticles. The FT-IR method was used to identify the responsible functional groups involved in reducing, synthesizing, and capping the gold nanoparticles. All the spectra exhibited similar bands: between 3800–3200 cm^{-1} can be assigned to the phenolic hydroxyl and aliphatic hydroxyl groups (-OH); bands in 2900–2600 cm^{-1} correspond to symmetric stretching of methyl (CH_3) and methylene (CH_2) group; absorption bands in 1760–1730 cm^{-1} indicate C=O stretching, suggesting the presence of high quantities of phenolic acids; the band at 1600–1580 cm^{-1} represents the presence of C=C stretching in aromatic compounds; bands in 1390–1350 cm^{-1} indicate alkane (C-H) bending and asymmetric stretching of the glycosidic bond (C-O-C), and the 700–600 band corresponded to the CH out of plane bending in carbonyl compounds [15,42,45]. Overall, the fingerprint region between 800 and 1200 cm^{-1} indicates the area for carbohydrate-based biopolymers. The band corresponding to the hydroxyl groups became narrower after synthesis, suggesting that hydroxyl groups play an important role in gold ion reduction and the formation of hydroxyquinones that capped the AuNPs surface [46,47].

3.2.3. Morphology and Size Distribution

The morphology and size distribution of the AuNPs was examined by STEM and DLS analysis (Figure 3). Most nanoparticles presented a quasi-spherical morphology, and only in green synthesis exhibited triangular morphologies. AuNPsME mainly presented spherical morphologies with 20.1 ± 0.2 nm. For AuNPsCE, quasi-spherical morphologies with a median diameter of 37.1 ± 0.4 nm predominated. On the contrary, AuNPsCS exhibited spherical and flower-like morphologies with the highest diameters (102.3 ± 2.1 nm). In green synthesis, the particles with triangular morphologies present a size larger than the quasi-spherical morphologies, AuNPsME presents an average diameter of 30.8 ± 12.7, and AuNPsCE of 41.3 ± 16.2. Although the mean diameters are not so different from the other morphologies, they present larger standard errors, indicating a greater dispersion.

Results from green synthesis are consistent because hierarchical structures, such as triangular ones, are formed by smaller particles, generally spherical, considered growth seeds. Due to their size and shape, these structures are less stable for aggregation. On the other hand, flower-like nanoparticles synthesized with $NaBH_4$, presented larger diameters corresponding to the aggregation. The growth of nanoparticles may be assisted by the capping agents, limiting the size and controlling the shape. The non-specific binding on all the exposed surfaces of gold in green synthesis may lead to anisotropic growth. In addition, the different compounds present in the plant extracts may prevent aggregation as capping agents [48]. The hydrodynamic diameter of DLS follows a similar trend to STEM diameter as the smallest particles are synthesized with ME with 49.41 nm and a PDI (polydispersity index) of 0.437, followed by the AuNPsCE with 56.2 nm and PDI of 0.391, and the largest are those of chemical synthesis with 92.12 nm and 0.143 of PDI. Generally, the hydrodynamic diameters are more prominent because they include the compounds on the surface and cannot differentiate two particles very close together. The PDI values greater than 0.1 may imply polydisperse particle size distributions, which may be due to the variable compound composition of the extracts that cap the surface of AuNPs [49,50].

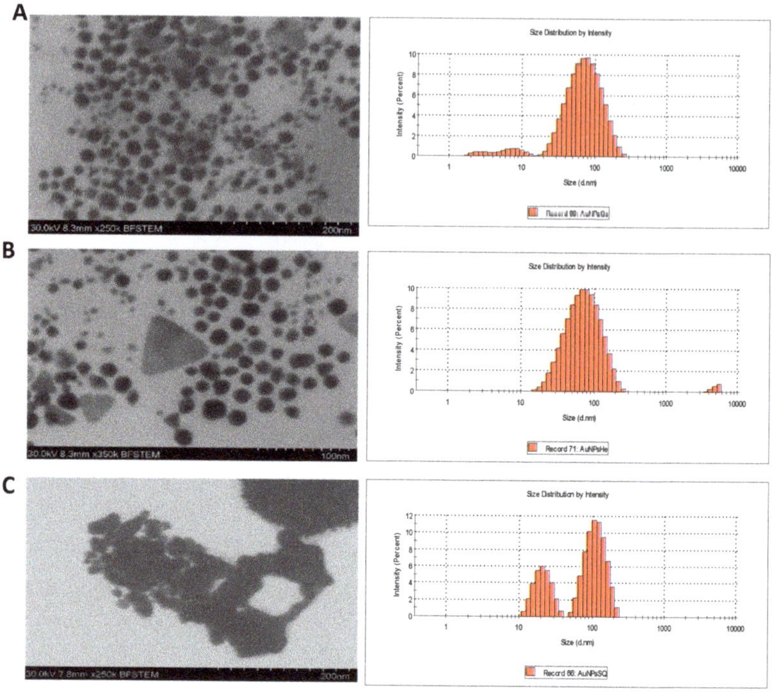

Figure 3. Scanning transmission electron microscope (STEM) and dynamic light scattering (DLS) analyses from (**A**) AuNPsME; (**B**) AuNPsCE; and (**C**) AuNPsCS. AuNPsCE: Gold nanoparticles synthesized with a castor (*R. communis*) leaves extract; AuNPsCS: Gold nanoparticles produced by chemical synthesis; AuNPsME: Gold nanoparticles synthesized with common mullein (*V. thapsus*) flowers ethanolic extract.

Surface Z potential was measured to record the surface charge on green and chemical AuNPs. Generally, nanoparticles are considered to be stable if the surface potential values are between −30 mV and +30 mV [51]. The zeta values of AuNPsME, AuNPsCE, and AuNPsSQ are 25.43 ± 0.66, 34.96 ± 0.79, and 63.86 ± 1.18 respectively. These negative zeta potentials confirmed the formation of negative charges on the surface of the AuNPs and predict that the particles made with green synthesis have greater stability compared to the particles made by chemical synthesis. The anticancer effect of AuNPs was found to be highly dependent on a range of factors related to their physical characteristics, such as surface coating, shape, size, and superficial charge. Regarding the size, particles with small size and negative charge, like that obtained in green synthesis, are capable of taking advantage of the enhanced permeability and retention effect.

3.3. Ethanolic Extracts-Based AuNPs Are More Cytotoxic than AuNPsCS

MTT analysis was used to assess the cytotoxic activity of different concentrations of green and chemical AuNPs in two colon cancer cell lines, SW480 and HT29, which vary in terms of epithelial-mesenchymal transition (EMT) markers, WNT activity, and stemness signatures. SW480 is derived from a primary Duke's stage B adenocarcinoma. It has a high expression of EMT markers, with a migratory capacity phenotype, while the HT29 derived from Duke's C stage human colon adenocarcinoma cell line has been ranked lower in terms of EMT and humanized intestinal stem cell signature, challenging the system to differentiate between different stages of the same disease [12,52].

3.3.1. Cellular Viability of AuNPsME

Figure 4 shows the impact of AuNPsME on HT29 and SW480 cell viability. AuNPsME extracts >100 μg/mL are needed to decrease HT29 cell viability <80% (Figure 4A), whereas overall lower cell viability was observed in SW480 cells (Figure 4B). However, since a major impact was shown between 100 and 150 μg/mL for HT29, their calculated IC_{50} values were lower than those for SW480 (−23.72%).

Representative pictures for the effect of AuNPsME in HT29 and SW480 cells (Figure 4E,F) showed high levels of cells unsticking at AuNPsME > 150 μg/mL, which coincided with the observed cell viability tests. As indicated by the IC_{50} values and the composition of the extracts shown in Table S1, the obtained concentrations mainly contained 4.60–8.79 μg/mL p-coumaric acid, 2.10–19.88 μg/mL ferulic acid, 0.57–5.62 μg/mL epigallocatechin gallate, 0.36–2.40 μg/mL caffeic acid, and 1.03–1.73 μg/mL hydroxyphenylacetic acid, among other compounds. Only a few reports have tested the safety of ME in selected models. For instance, Dragičevic et al. [53] suggested that *V. thapsus* flowers aqueous extracts, prepared at 1:10 flower:water maceration, further rotary evaporator-mediated concentration up to 36 μg/mL, and dilutions starting from 1000× (0.036 μg/mL), did not induce malformations in zebrafish embryo, but viabilities in the range of 57.9–67.3% in human lung epithelial cells (BEAS-2B) suggested certain toxic effects that merit further research. Other authors reported that 50–200 μg/mL *V. thapsus* aqueous leaves extracts are safe in mice RAW264.7 macrophages and human adult chondrocytes (HC) [54]. Aerial parts of *V. thapsus* at 100, 300, and 500 mg/kg body weight, which are much higher than the doses used in this research, did not produce any cytotoxic effects expressed in the increase of frequency of micronucleated polychromatic erythrocytes (MNPCE) in BALB/c mice after a 24 h administration [55]. There are no reports on the impact of *V. thapsus* or even *V. thapsus*-based nanoparticles on HT29 and SW480 cells. However, AuNPs made from *V. thapsus* leaf extracts (20 μg/mL) showed IC50 of 287–308 μg/mL against differentiated bronchogenic carcinoma HLC-1 and moderately differentiated lung adenocarcinoma cells [56].

3.3.2. Cellular Viability of AuNPsCE

The impact of AuNPsCE on HT29 and SW480 cell viability is shown in Figure 5. A dose–response trend was observed for both cell lines (Figure 5A,B), with the effects being stronger on SW480 cells since even the lowest concentration showed cell viabilities <80%. The results could be confirmed in the 1.7-fold lower IC50 values for SW480 compared to HT29 cells (Figure 5C,D). The representative pictures showed a high degree of viability loss at the highest concentration for HT29 cells (Figure 5E), whereas results for SW480 cells are similar between 50 and 200 μg/mL (Figure 5F). Previously, our research group tested CE extracts on RAW264.7 macrophages, and concentrations in the range of 10–50% v/v (equivalents to 2–5 mg/mL aqueous CE) were not cytotoxic (cell viability > 80%) (results not shown) [15]. Gold nanoparticles made from *R. communis* leaf extracts showed inhibition values ranging 30–70% from 25 to 200 μg/mL AuNPsCE on normal rat splenic cells, although no differences were shown from 50 to 200 μg/mL AuNPsCE [42], suggesting that the doses assayed in this research can be considered safe within the 20–100 μg/mL range. Particularly for human hepatic (HepG2) and cervical (HeLa) cell lines, AuNPsCE inhibited 58.64% and 42.74% cell growth, respectively, but no mechanistic effects were provided. Although there are no reports on the effect of AuNPsCE on colon cancer models in vitro or in vivo, CE extracts from several castor bean parts have been assayed in several cancer types. An ethanolic extract of the fruit demonstrated reduced cytotoxicity (0.005–1 μg/mL) in human MCF-7 and MDA-MB-231 breast cancer cell lines [57], and chloroform extracts from the oil demonstrated antiproliferative effects on human SK-MEL-28 skin melanoma cells [58]. Taken together, *R. communis* extracts have increasingly been assayed for their anticancer effects, and encapsulation methods aiming to protect its bioactive molecules ensure an enhanced effect against cancer cells.

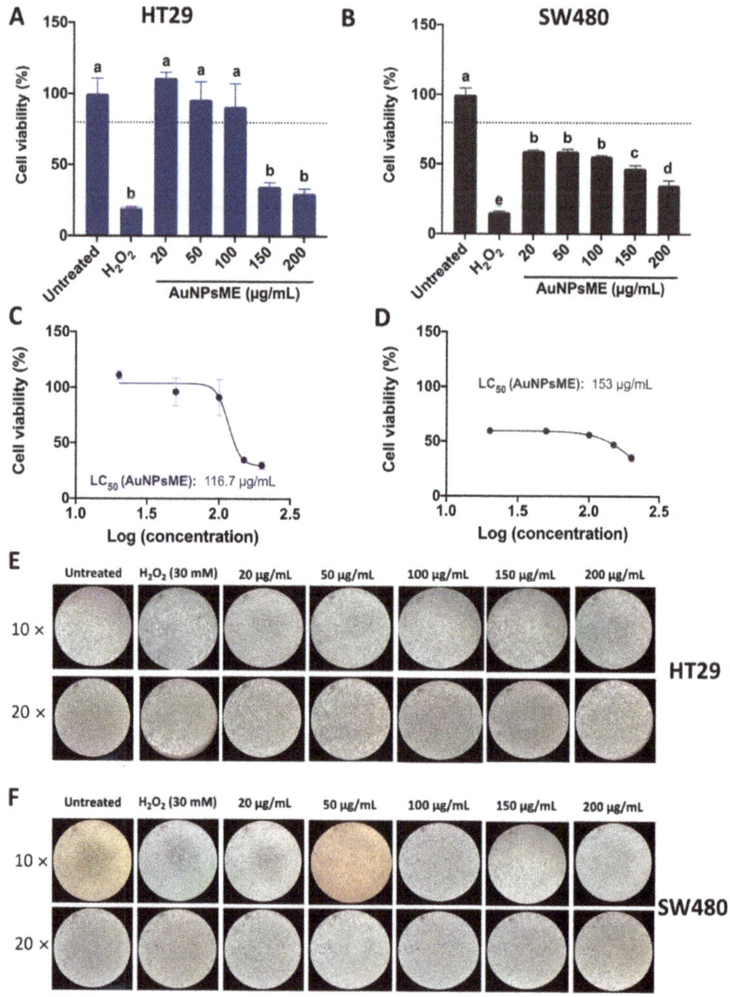

Figure 4. Assessment of HT29 and SW480 cell viability after treatment with common mullein (*V. thapsus*) flowers ethanolic extracts-gold nanoparticles (AuNPsME) treatments. (**A**) Evaluation of the impact of AuNPsME (20–200 µg/mL) and controls (untreated cells, 30 mM H_2O_2) in HT29 cells viability; (**B**) Effect of AuNPsME in SW480 cells viability; (**C**) Calculated IC_{50} of AuNPsME treatments on HT29 cells; (**D**) Calculated IC_{50} of AuNPsME treatments on SW480 cells; (**E**) Representative pictures (10× and 20×) of AuNPsME effects on HT29 cells; (**F**) Representative pictures (10× and 20×) of AuNPsME effects on SW480 cells. Hydrogen peroxide (H_2O_2)-treated cells (30 mM) were used as a positive control. Untreated cells (DMEM + 0.5% ASB) were used as a negative control. Pictures from untreated controls were the same as AuNPsCE and AuNPsCS. In Figure 4A,B, different letters (a, b, c, d, e) express significant differences ($p < 0.05$) by Tukey-Kramer's test. The dashed line in (**A**,**B**) represents 80% cell viability.

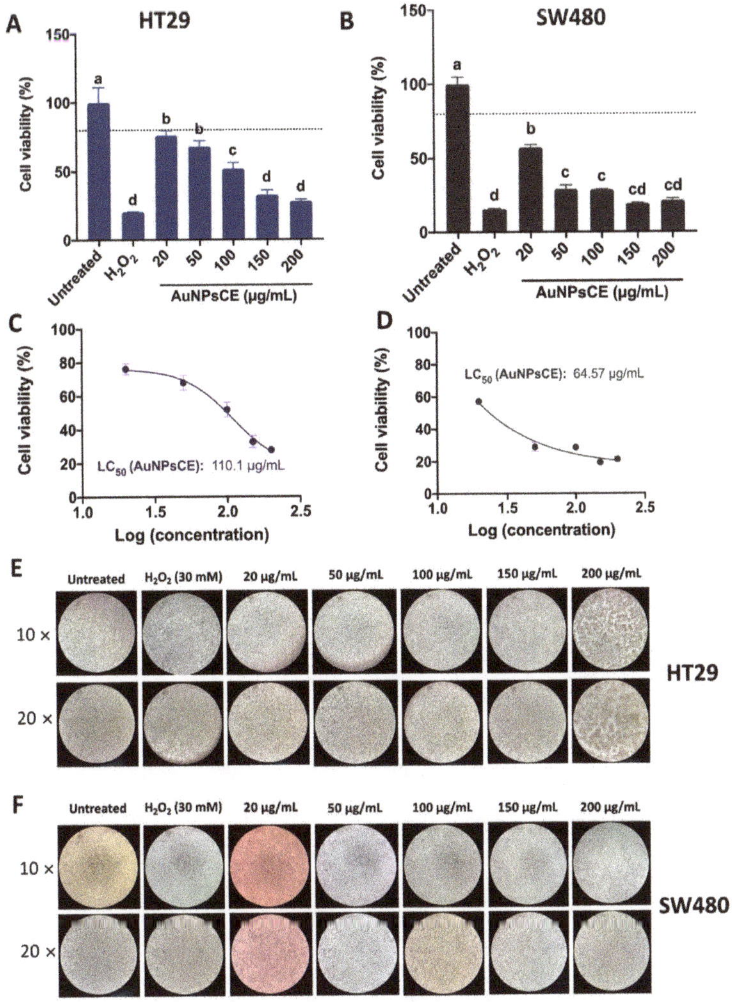

Figure 5. Impact of castor bean (*R. communis*) leaves ethanolic extracts-based gold nanoparticles (AuNPsCE) on HT29 and SW480 cells viability. (**A**) Effect of AuNPsCE in HT29 cells viability; (**B**) Influence of AuNPsCE in SW480 cells viability; (**C**) Quantified IC_{50} of AuNPsCE treatments on HT29 cells; (**D**) Quantified IC_{50} of AuNPsCE treatments on SW480 cells; (**E**) Representative pictures (10× and 20×) of AuNPsCE impact on HT29 cells; (**F**) Representative pictures (10× and 20×) of AuNPsCE impact on SW480 cells. Hydrogen peroxide (H_2O_2)-treated cells (30 mM) were used as a positive control. Untreated cells (DMEM + 0.5% ASB) were used as a negative control. Pictures from untreated controls were the same as AuNPsME and AuNPsCS. For Figure 5A,B different letters (a, b, c, d) express significant differences ($p < 0.05$) by Tukey-Kramer's test. The dashed line in (**A**,**B**) represent 80% cell viability.

3.3.3. Cellular Viability of AuNPsCS

Figure 6 depicts the effects on the HT29 (Figure 6A) and SW480 (Figure 6B) cell viability of AuNPsCS. Compared to AuNPsME (Figure 4) and AuNPsCE (Figure 5), AuNPsCS displayed a stronger inhibitory effect ($p < 0.05$) for 20–50 µg/mL, the same effect as AuNPsCE at 100 µg/mL, and the same effect as the other nanoparticles for 150 and 200 µg/mL in the HT29 cell line (Figure S1A). For the SW480 cells, a lower effect than AuNPsME or

AuNPsCE was observed for 20 μg/mL, a similar effect to AuNPsME at 50–100 μg/mL, and lower than AuNPsME at 150–200 μg/mL (Figure S1B). Based on the calculated IC_{50} for HT29 (Figure 6C) and SW480 (Figure 6D), AuNPsCS are less effective against HT29 cells but have a better effect than AuNPsME in inhibiting SW480 cell proliferation. The representative pictures showed the strongest effects in the 100–200 μg/mL concentrations, agreeing with the results for cell viability.

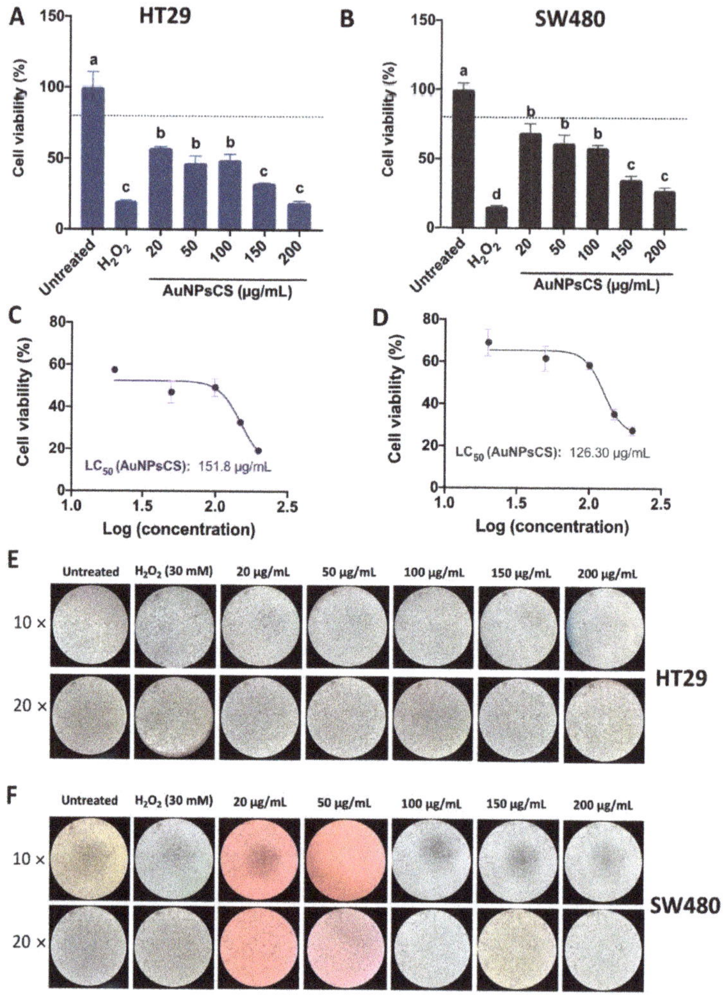

Figure 6. Impact of gold nanoparticles made by chemical synthesis (AuNPsCS) on HT29 and SW480 cells viability. (**A**) Effect of AuNPsCS in HT29 cells viability; (**B**) Influence of AuNPsCS in SW480 cells viability; (**C**) Quantified IC_{50} of AuNPsCS treatments on HT29 cells; (**D**) Quantified IC_{50} of AuNPsCS treatments on SW480 cells; (**E**) Representative pictures (10× and 20×) of AuNPsCS impact on HT29 cells; (**F**) Representative pictures (10× and 20×) of AuNPsCS impact on SW480 cells. Hydrogen peroxide (H_2O_2)-treated cells (30 mM) were used as a positive control. Untreated cells (DMEM + 0.5% ASB) were used as a negative control. Pictures from untreated controls were the same as AuNPsME and AuNPsCE. For Figure 6A,B, different letters (a, b, c, d) express significant differences ($p < 0.05$) by Tukey-Kramer's test. The dashed line in (**A**,**B**) represent 80% cell viability.

AuNPsCS have been tested in colorectal cancer cell lines, exhibiting viabilities <80% in HT29 and SW480 cells at 1–200 μM concentrations, although HT29 cells displayed lower cell viabilities than SW480 [59], similar to the results presented in this research. It has been suggested that both HT29 and SW480 present a relatively high AuNPsCS uptake [59], suggesting that AuNPs-based nanoparticles could be an effective strategy aiming to deliver bioactive compounds in these cells.

In general, we can observe that the gold particles with the best results were those made with the ethanolic extract of castor oil with a dose-dependent effect, and it is well known that the size and capping agent of AuNPs plays an essential role in cellular viability and proliferation. We can observe this in the results because, compared to the chemical synthesis particles, the green ones presented a smaller particle size, and in the case of AuNPsCE this extract presents a greater antioxidant capacity and a higher content of phenolic compounds capping the surface of particles, which would lead to more significant activity. It is important to emphasize that the particles had a greater effect on SW480 cells that represent an early stage of colon cancer.

3.4. A Stronger ROS Generation Was Shown in the Ethanolic-Based AuNPs Compared to AuNPsCS

Reactive oxygen species (ROS) are a specific type of oxygen-containing reactive molecules that play an important role in different cellular processes and are essential for cell proliferation at basal levels. However, in high concentrations, they present a cytotoxic effect, causing necrosis or apoptosis and cell damage by ionizing DNA and other cellular molecules. This effect is important for different therapeutic applications, like the AuNPs anticancer effect that increases the production of ROS [60,61].

The effect of the evaluated nanoparticles on ROS generation in HT29 and SW480 cells is presented in Figure 7. Both green-synthesized nanoparticles showed a similar trend, where the highest concentration (200 μg/mL) surpassed the positive control (30 mM H_2O_2). Notably, AuNPsME (Figure 7A) displayed a higher ROS generation in 100 μg/mL than 150 μg/mL, a trend also shown for AuNPsCE (Figure 7C) but without differences with the 20–150 μg/mL range. Despite showing a similar behavior in SW480 cells, 200 μg/mL AuNPsME displayed a higher effect (Figure 7B) than AuNPsCE (Figure 7D). For AuNPsCS, similar effects were presented in HT29 (Figure 7E) to SW480 (Figure 7F).

The ability with which metallic nanoparticles can destroy cancer cells partially depends on the stimulation of ROS in the damaged components of the cells [62,63]. The generation of a ROS response in both cell lines after AuNPsME treatments supports other findings indicating that the sole ME treatment alters the production of superoxide anion in selected cell lines [53]. Verbascum calvum flower extracts were assayed against human lung A549 and breast MCF-7 cancer cells, showing antiproliferative effects attributed to ROS generation and pro-inflammatory effects [64]. Despite no reports for green synthesized AuNPsCE inducing ROS in colorectal cancer cells, phytochemicals from R. communis leaves such as p-coumaric acid have been linked to ROS generation, inducing further apoptosis and a fall in the mitochondrial membrane potential [57]. This effect has been observed in SW480 cells treated with phenolic-rich berries extracts [65]. *R. communis* leaves extracts induce ROS in LPS-stimulated murine RAW264.7 macrophages, acting as critical mediators of several pro-inflammatory responses, such as the canonical pathway or even NLRP3 inflammasome activation [34,39].

Overall, AuNPsCS were less effective in inducing ROS in the cancer cells, particularly at 200 μg/mL, compared to AuNPsME and AuNPsCE (Figure S2), which agrees with the reported low ability of AuNPs to produce ROS in immune cells [66], but effects depend on the cell line and the exposition time. Hence, AuNPs are considered ideal radiosensitizers in cancer radiotherapy since the high atomic number of Au enhances the radiation dose absorbed by tumors emitting secondary ionizing radiations, resulting in ROS production and increased oxidative stress [67].

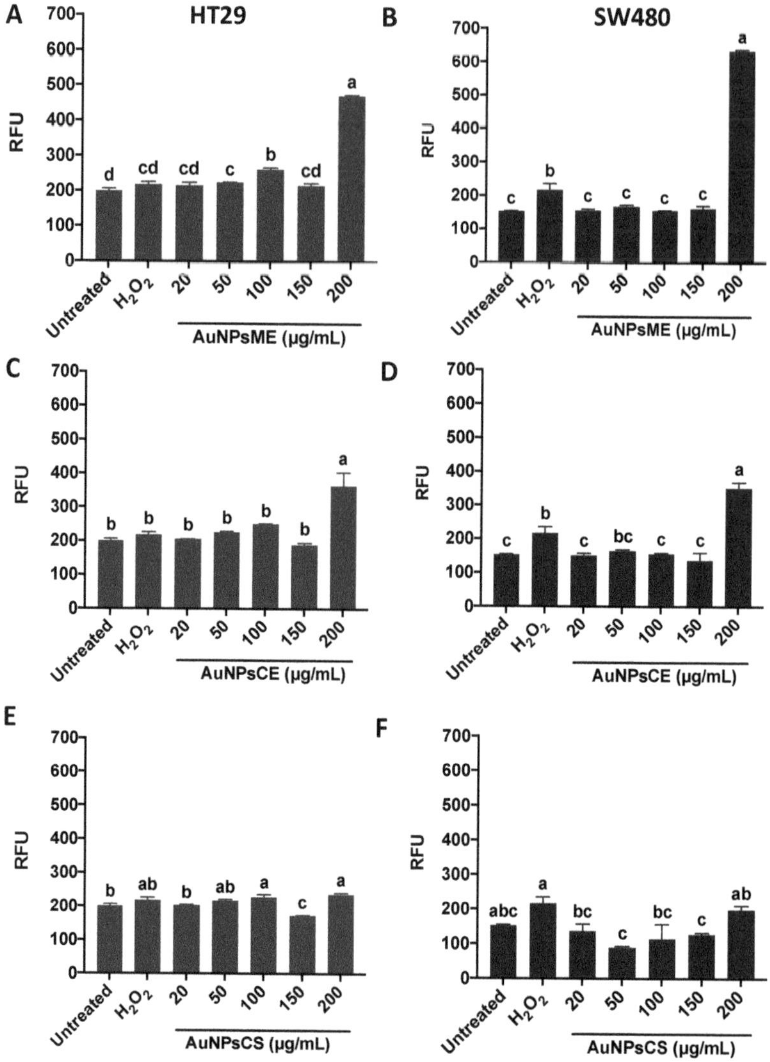

Figure 7. Intracellular reactive oxygen species (ROS) generation in HT29 and SW480 cells after treatments with common mullein (*V. thapsus*) ethanolic extracts-based gold nanoparticles (AuNPsME) (**A**,**B**); castor bean (*R. communis*) ethanolic extracts-based gold nanoparticles (AuNPsCE) (**C**,**D**); and gold nanoparticles made by chemical synthesis (AuNPsCS) (**E**,**F**). Different letters (a, b, c, d) in Figure 7A–F express significant differences ($p < 0.05$) by Tukey-Kramer's test. AuNPsCE: Gold nanoparticles synthesized with a castor bean (*R. communis*) leaves extract; AuNPsCS: Gold nanoparticles produced by chemical synthesis; AuNPsME: Gold nanoparticles synthesized with common mullein (*V. thapsus*) flowers ethanolic extract; RFU: Relative fluorescence units. Hydrogen peroxide (H_2O_2)-treated cells (30 mM) were used as a positive control. Untreated cells (DMEM + 0.5% ASB) were used as a negative control.

3.5. Caspase-3/7 Activity Is Increased Due to AuNPs Treatments

Caspase-3 is a relatively small protein consisting of two subunits, 12- and 17-kDa subunits that contain three and five thiol functions, respectively, members of a family of Cysteine-Aspartic proteases, best known for their ability to mediate the cleavage of specific target proteins. Caspases are all produced initially as inactive zymogens (called procaspases) that are then subject to activation by a wide range of specific internal and/or external signals, such as a high production of ROS [67]. As observed in Figure 8, the treatments (IC_{50}AuNPs) were less effective in inducing caspase 3/7 activity on HT29 (Figure 8A) than on SW480 cells (Figure 8B). For HT29, there were no differences between the positive control (30 mM H2O2) and AuNPs treatments, but the induction by AuNPsCE and AuNPsCS was higher ($p < 0.05$) than the untreated cells. Only AuNPsME and AuNPsCS displayed higher ($p < 0.05$) pro-caspase 3/7 effects than the positive control in SW480 cells, but all treatments were higher than the untreated cells. Obtained results indicated that, except for AuNPsME on HT29 cells, all treatments were able to induce Caspase 3/activity in both cell lines.

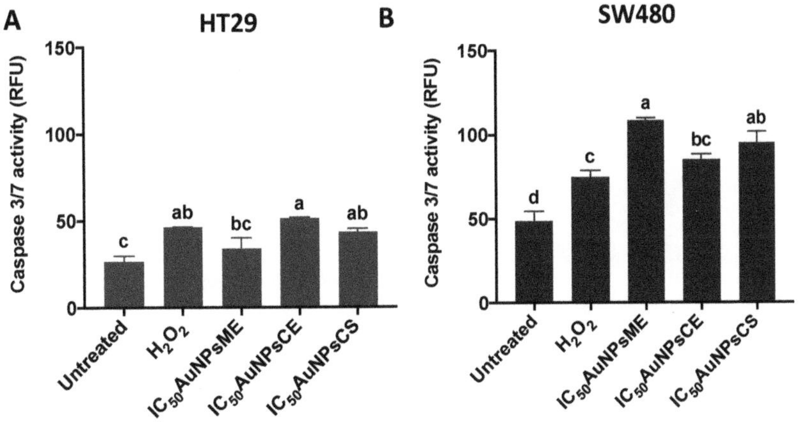

Figure 8. Evaluation of caspase 3/7 activity after treatment of HT29 (**A**,**B**) SW480 cells with the half inhibitory concentrations (IC_{50}) of AuNPsME, AuNPsCE, and AuNPsCS. Different letters (a, b, c, d) in Figure 8A,B indicate significant differences ($p < 0.05$) by Tukey-Kramer's test. AuNPsCE: Gold nanoparticles synthesized with a castor bean (R. communis) leaves extract; AuNPsCS: Gold nanoparticles produced by chemical synthesis; AuNPsME: Gold nanoparticles synthesized with common mullein (V. thapsus) flowers ethanolic extract; RFU: Relative fluorescence units. Hydrogen peroxide (H_2O_2) treated cells (30 mM) were used as a positive control. Untreated cells (DMEM + 0.5% ASB) were used as a negative control.

Within the radiosensitization provided by AuNPs, DNA damage induction and the development of pro-apoptotic mechanisms represent one of the most studied cytotoxic effects of the AuNPs-based nanoformulations in colon cancer [67]. Caspase-3 is a crucial driver of the intrinsic apoptotic pathway, potentially activated by ROS generation, leading to further cell death. Since the assayed AuNPs partially induced ROS and Caspase 3/7 activity in both cell lines, their relationship could be hypothesized. However, further in vitro and in vivo assays evaluating additional pro-apoptotic proteins are required to fully elucidate their mechanism of action, and not all the treatments were successful in inducing ROS, suggesting additional mechanisms inducing Caspase 3/7. It was hypothesized that gold nanoparticles synthesized with *V. thapsus* leaf extracts could induce apoptosis in several lung cancer cell lines [56], but no pro-apoptotic proteins were evaluated. *R. communis* leaves extracts-based nanoparticles evaluated in HepG2 and HeLa cell lines were presumably linked to DNA fragmentation due to ROS generation [39,42].

4. Conclusions

The antioxidant activity of the extracts can be related to the synthesis of gold nanoparticles since compounds capable of donating electrons can act as reducing agents of gold ions, capping the surface to stabilize them. Results suggested that gold nanoparticles obtained by green synthesis using *V. thapsus* flowers and *R. communis* leaves ethanolic extracts are suitable nanomaterials exhibiting antiproliferative effects through ROS generation and activation of caspase 3/7 activity in human HT29 and SW480 colorectal cancer cells. Compared to AuNPs obtained by chemical synthesis, AuNPsME and AuNPsCE displayed a more robust antiproliferative response in SW480 cells, and higher ROS generation in both cancer cell lines, with no differences in caspase 3/7 activation. The cytotoxic activity of AuNPs has a dose–response effect that depends on the particle size and capping compounds on the surface of AuNPs. The best cytotoxic activity was obtained with AuNPsCE, which presents the major phenolic compounds in the extract. We conclude that the capping compounds, morphology, and size are important factors in the cytotoxic activity of green gold nanoparticles, and this cytotoxicity is modulated by reactive oxygen species production and Caspases 3/7 activation. Despite the large size nanoparticles obtained with the chemical synthesis, they presented cytotoxic activity through both pathways studied, which can be related to the release of ions or particles that did not aggregate. To understand specific cancer-related mechanisms of action, studies on in vivo models are suggested.

Supplementary Materials: The following supporting information can be downloaded at: https://www.mdpi.com/article/10.3390/pharmaceutics14102069/s1, Table S1: Individual phenolic compounds' composition by HPLC-DAD or ME and CE extracts; Figure S1: Impact of AuNPs on HT29 (A) and SW480 (B) cells viability. Figure S2. Effect of AuNPs on HT29 (A) and SW480 (B) ROS generation. Figure S3. UV-Vis spectra of the effect of pH in the synthesis of AuNPs.

Author Contributions: Conceptualization, K.M.S., I.L.-O. and A.M.-R.; methodology, K.M.S. and I.L.-O.; validation, K.M.S., I.L.-O., S.M., G.L.-P. and A.M.-R.; formal analysis, K.M.S., I.L.-O. and E.M.R.-M.; investigation, A.M.-R., G.L.-P., S.M., E.M.R.-M. and I.L.-O.; data curation, K.M.S. and I.L.-O.; writing—original draft preparation, K.M.S. and I.L.-O.; writing—review and editing, K.M.S., I.L.-O. and A.M.-R.; visualization, K.M.S. and I.L.-O.; supervision, J.M.L.-R. and A.M.-R.; project administration, K.M.S. and A.M.-R.; funding acquisition, A.M.-R. and J.M.L.-R. All authors have read and agreed to the published version of the manuscript.

Funding: This research was funded by the Science and Technology Council of the State of Queretaro, project number 45/ESP701/2022.

Institutional Review Board Statement: Not applicable.

Informed Consent Statement: Not applicable.

Data Availability Statement: Not applicable.

Acknowledgments: Karen M. Soto thanks Consejo Nacional de Ciencia y Tecnología (CONACyT-Mexico) for her postdoctoral fellowship (grant ID: 483977). Author I. Luzardo acknowledges Programa de Becas Posdoctorales de la UNAM (DGAPA-CIC) for his postdoctoral fellowship (grant number: 5267). Authors thank LIDTRA at CINVESTAV, Unidad Querétaro, and The National Laboratory for Characterization in materials at CFATA, UNAM for the facilities provided for the characterization of nanoparticles, M.C. Manuel Aguilar Franco for his support in STEM and processing of the photomicrographs, and Beatriz Millan Malo for her support in X-ray diffraction procedure.

Conflicts of Interest: The authors declare no conflict of interest.

References

1. Kyzioł, A.; Łukasiewicz, S.; Sebastian, V.; Kuśtrowski, P.; Kozieł, M.; Majda, D.; Cierniak, A. Towards Plant-Mediated Chemistry–Au Nanoparticles Obtained Using Aqueous Extract of *Rosa damascena* and Their Biological Activity in Vitro. *J. Inorg. Biochem.* **2021**, *214*, 111300. [CrossRef] [PubMed]
2. Chen, Y.; Feng, X. Gold Nanoparticles for Skin Drug Delivery. *Int. J. Pharm.* **2022**, *625*, 122122. [CrossRef] [PubMed]
3. Mikhailova, E.O. Gold Nanoparticles: Biosynthesis and Potential of Biomedical Application. *J. Funct. Biomater.* **2021**, *12*, 70. [CrossRef] [PubMed]
4. Medici, S.; Peana, M.; Coradduzza, D.; Zoroddu, M.A. Gold Nanoparticles and Cancer: Detection, Diagnosis and Therapy. *Semin. Cancer Biol.* **2021**, *76*, 27–37. [CrossRef]

5. Benedec, D.; Oniga, I.; Cuibus, F.; Sevastre, B.; Stiufiuc, G.; Duma, M.; Hanganu, D.; Iacovita, C.; Stiufiuc, R.; Lucaciu, C.M. Origanum Vulgare Mediated Green Synthesis of Biocompatible Gold Nanoparticles Simultaneously Possessing Plasmonic, Antioxidant and Antimicrobial Properties. *Int. J. Nanomed.* **2018**, *13*, 1041–1058. [CrossRef]
6. Kalimuthu, K.; Lubin, B.C.; Bazylevich, A.; Gellerman, G.; Shpilberg, O.; Luboshits, G.; Firer, M.A. Gold Nanoparticles Stabilize Peptide-Drug-Conjugates for Sustained Targeted Drug Delivery to Cancer Cells. *J. Nanobiotechnology* **2018**, *16*, 34. [CrossRef]
7. Wang, L.; Xu, J.; Yan, Y.; Liu, H.; Karunakaran, T.; Li, F. Green Synthesis of Gold Nanoparticles from *Scutellaria Barbata* and Its Anticancer Activity in Pancreatic Cancer Cell (PANC-1). *Artif. Cells Nanomed. Biotechnol.* **2019**, *47*, 1617–1627. [CrossRef]
8. Ferreira, D.; Fontinha, D.; Martins, C.; Pires, D.; Fernandes, A.R.; Baptista, P.V. Gold Nanoparticles for Vectorization of Nucleic Acids for Cancer Therapeutics. *Molecules* **2020**, *25*, 3489. [CrossRef]
9. Donga, S.; Bhadu, G.R.; Chanda, S. Antimicrobial, Antioxidant and Anticancer Activities of Gold Nanoparticles Green Synthesized Using *Mangifera Indica* Seed Aqueous Extract. *Artif. Cells Nanomed. Biotechnol.* **2020**, *48*, 1315–1325. [CrossRef]
10. Soto, K.M.; Mendoza, S.; López-Romero, J.M.; Gasca-Tirado, J.R.; Manzano-Ramírez, A. Gold Nanoparticles: Synthesis, Application in Colon Cancer Therapy and New Approaches-Review. *Green Chem. Lett. Rev.* **2021**, *14*, 663–676. [CrossRef]
11. Nalli, M.; Puxeddu, M.; la Regina, G.; Gianni, S.; Silvestri, R. Emerging Therapeutic Agents for Colorectal Cancer. *Molecules* **2021**, *26*, 7463. [CrossRef] [PubMed]
12. Gala de Pablo, J.; Armistead, F.J.; Peyman, S.A.; Bonthron, D.; Lones, M.; Smith, S.; Evans, S.D. Biochemical Fingerprint of Colorectal Cancer Cell Lines Using Label-Free Live Single-Cell Raman Spectroscopy. *J. Raman Spectrosc.* **2018**, *49*, 1323–1332. [CrossRef] [PubMed]
13. Thipe, V.C.; Karikachery, A.R.; Çakılkaya, P.; Farooq, U.; Genedy, H.H.; Kaeokhamloed, N.; Phan, D.-H.; Rezwan, R.; Tezcan, G.; Roger, E.; et al. Green Nanotechnology—An Innovative Pathway towards Biocompatible and Medically Relevant Gold Nanoparticles. *J. Drug Deliv. Sci. Technol.* **2022**, *70*, 103256. [CrossRef]
14. Soto, K.M.; Quezada-Cervantes, C.T.; Hernández-Iturriaga, M.; Luna-Bárcenas, G.; Vazquez-Duhalt, R.; Mendoza, S. Fruit Peels Waste for the Green Synthesis of Silver Nanoparticles with Antimicrobial Activity against Foodborne Pathogens. *LWT* **2019**, *103*, 293–300. [CrossRef]
15. Vasco-Leal, J.F.; Cuellar-Nuñez, M.L.; Luzardo-Ocampo, I.; Ventura-Ramos, E.; Loarca-Piña, G.; Rodriguez-García, M.E. Valorization of Mexican *Ricinus communis* L. Leaves as a Source of Minerals and Antioxidant Compounds. *Waste Biomass Valorization* **2020**, *12*, 2071–2088. [CrossRef]
16. Luzardo-Ocampo, I.; Vasco-Leal, J.; Cuellar-Nuñez, M.L.; Quintero-Castaño, V.; Pérez-Serrano, R.; Luzardo-Ocampo, I. Castor Bean (*Ricinus communis* L.) Polyphenolic Extracts Exhibited Anti-Inflammatory Effects on LPS-Stimulated RAW 264.7 Macrophages. *Curr Dev. Nutr.* **2022**, *6*, 309. [CrossRef]
17. Abdul, W.M.; Hajrah, N.H.; Sabir, J.S.M.; Al-Garni, S.M.; Sabir, M.J.; Kabli, S.; Saini, K.S.; Bora, R.S. Therapeutic Role of *Ricinus communis* L. and Its Bioactive Compounds in Disease Prevention and Treatment. *Asian Pac. J. Trop. Med.* **2018**, *11*, 177. [CrossRef]
18. Turker, A.U.; Gurel, E. Common Mullein (*Verbascum thapsus* L.): Recent Advances in Research. *Phytother. Res.* **2005**, *19*, 733–739. [CrossRef]
19. Ghramh, H.A.; Khan, K.A.; Ibrahim, E.H.; Setzer, W.N. Synthesis of Gold Nanoparticles (AuNPs) Using *Ricinus Communis* Leaf Ethanol Extract, Their Characterization, and Biological Applications. *Nanomaterials* **2019**, *9*, 765. [CrossRef]
20. Gonca, S.; Arslan, H.; Isik, Z.; Özdemir, S.; Dizge, N. The Surface Modification of Ultrafiltration Membrane with Silver Nanoparticles Using *Verbascum thapsus* Leaf Extract Using Green Synthesis Phenomena. *Surf. Interfaces* **2021**, *26*, 101291. [CrossRef]
21. Weldegebrieal, G.K. Photocatalytic and Antibacterial Activityof CuO Nanoparticles Biosynthesized Using *Verbascum thapsus* Leaves Extract. *Optik* **2020**, *204*, 164230. [CrossRef]
22. Mohamed, S.; Zela, I.; Yusuf, A.; Hudaverdi, A.; Mutlu, Y.; Dizge, N. Green Synthesis of Zero Valent Iron Nanoparticles Using *Vervascum thapsus* and Its Cr (VI) Reduction Activity. *Bioresour. Technol. Rep.* **2022**, *13*, 100637. [CrossRef]
23. Darweesh, R.S.; Ayoub, N.M.; Nazzal, S. Gold Nanoparticles and Angiogenesis: Molecular Mechanisms and Biomedical Applications. *Int. J. Nanomed.* **2019**, *14*, 7643–7663. [CrossRef] [PubMed]
24. Cai, W.; Gao, T.; Hong, H.; Sun, J. Applications of gold nanoparticles in cancer nanotechnology. *Nanotechnol. Sci. Appl.* **2008**, *1*, 17–32. [CrossRef]
25. Singleton, V.L.; Orthofer, R.; Lamuela-Raventós, R.M. Analysis of Total Phenols and Other Oxidation Substrates and Antioxidants by Means of Folin-Ciocalteu Reagent. *Methods Enzymol.* **1999**, *299*, 152–178. [CrossRef]
26. Brand-Williams, W.; Cuvelier, M.E.; Berset, C. Use of a Free Radical Method to Evaluate Antioxidant Activity. *LWT-Food Sci. Technol.* **1995**, *28*, 25–30. [CrossRef]
27. Re, R.; Pellegrini, N.; Proteggente, A.; Pannala, A.; Yang, M.; Rice-Evans, C. Antioxidant Activity Applying an Improved ABTS Radical Cation Decolorization Assay. *Free Radic. Biol. Med.* **1999**, *26*, 1231–1237. [CrossRef]
28. Ramírez-Jiménez, A.K.; Reynoso-Camacho, R.; Mendoza-Díaz, S.; Loarca-Piña, G. Functional and Technological Potential of Dehydrated *Phaseolus Vulgaris* L. Flours. *Food Chem.* **2014**, *161*, 254–260. [CrossRef]
29. Deraedt, C.; Salmon, L.; Gatard, S.; Ciganda, R.; Hernandez, R.; Ruiz, J.; Astruc, D. Sodium Borohydride Stabilizes Very Active Gold Nanoparticle Catalysts. *Chem. Commun.* **2014**, *50*, 14194–14196. [CrossRef]
30. Schneider, C.A.; Rasband, W.S.; Eliceiri, K.W. NIH Image to ImageJ: 25 Years of Image Analysis. *Nat. Methods* **2012**, *9*, 671–675. [CrossRef]

31. Torres-Vargas, O.L.; Luzardo-Ocampo, I.; Hernandez-Becerra, E.; Rodríguez-García, M.E. Physicochemical Characterization of Unripe and Ripe Chontaduro (Bactris Gasipaes Kunth) Fruit Flours and Starches. *Starch Stärke* **2021**, *73*, 2000242. [CrossRef]
32. Lan, H.; Yuan, H.; Lin, C. Sulforaphane Induces P53-Deficient SW480 Cell Apoptosis via the ROS-MAPK Signaling Pathway. *Mol. Med. Rep.* **2017**, *16*, 7796–7804. [CrossRef] [PubMed]
33. Sheikh, B.Y.; Sarker, M.R.; Kamarudin, M.N.A.; Mohan, G. Antiproliferative and Apoptosis Inducing Effects of Citral via P53 and ROS-Induced Mitochondrial-Mediated Apoptosis in Human Colorectal HCT116 and HT29 Cell Lines. *Biomed. Pharmacother.* **2017**, *96*, 834–846. [CrossRef]
34. Luzardo-Ocampo, I.; Loarca-Piña, G.; de Mejia, E.G. Gallic and Butyric Acids Modulated NLRP3 Inflammasome Markers in a Co-Culture Model of Intestinal Inflammation. *Food Chem. Toxicol.* **2020**, *146*, 111835. [CrossRef] [PubMed]
35. Selseleh, M.; Ebrahimi, S.N.; Aliahmadi, A.; Sonboli, A.; Mirjalili, M.H. Metabolic Profiling, Antioxidant, and Antibacterial Activity of Some Iranian *Verbascum* L. Species. *Ind. Crops. Prod.* **2020**, *153*, 112609. [CrossRef]
36. Taleb, S.; Saeedi, M. The Effect of the *Verbascum thapsus* on Episiotomy Wound Healing in Nulliparous Women: A Randomized Controlled Trial. *BMC Complement. Med. Ther.* **2021**, *21*, 166. [CrossRef]
37. Wafa, G.; Amadou, D.; Larbi, K.M.; Héla, E.F.O. Larvicidal Activity, Phytochemical Composition, and Antioxidant Properties of Different Parts of Five Populations of *Ricinus communis* L. *Ind. Crops. Prod.* **2014**, *56*, 43–51. [CrossRef]
38. Mihailović, V.; Kreft, S.; Benković, E.T.; Ivanović, N.; Stanković, M.S. Chemical Profile, Antioxidant Activity and Stability in Stimulated Gastrointestinal Tract Model System of Three *Verbascum* Species. *Ind. Crops. Prod.* **2016**, *89*, 141–151. [CrossRef]
39. Nemudzivhadi, V.; Masoko, P. In Vitro Assessment of Cytotoxicity, Antioxidant, and Anti-Inflammatory Activities of *Ricinus communis* (Euphorbiaceae) Leaf Extracts. *Evid. Based Complementary Altern. Med.* **2014**, *2014*, 1–8. [CrossRef]
40. Alegria, E.; Ribeiro, A.; Mendes, M.; Ferraria, A.; do Rego, A.; Pombeiro, A. Effect of Phenolic Compounds on the Synthesis of Gold Nanoparticles and Its Catalytic Activity in the Reduction of Nitro Compounds. *Nanomaterials* **2018**, *8*, 320. [CrossRef]
41. Amendola, V.; Pilot, R.; Frasconi, M.; Maragò, O.M.; Iatì, M.A. Surface Plasmon Resonance in Gold Nanoparticles: A Review. *J. Phys. Condens. Matter* **2017**, *29*, 203002. [CrossRef] [PubMed]
42. Rahman, T.J.; Anwar, M.; Zeb, M.A.; Liaqat, W. Green synthesis, characterization, antibacterial activity of metal nanoparticles and composite oxides using leaves extract of *Ocimum basilicum* L. *Microsc. Res. Tech.* **2022**, *85*, 2857–2865. [CrossRef] [PubMed]
43. Botteon, C.E.A.; Silva, L.B.; Ccana-Ccapatinta, G.V.; Silva, T.S.; Ambrosio, S.R.; Veneziani, R.C.S.; Bastos, J.K.; Marcato, P.D. Biosynthesis and Characterization of Gold Nanoparticles Using Brazilian Red Propolis and Evaluation of Its Antimicrobial and Anticancer Activities. *Sci. Rep.* **2021**, *11*, 1974. [CrossRef] [PubMed]
44. Aljabali, A.; Akkam, Y.; Al Zoubi, M.; Al-Batayneh, K.; Al-Trad, B.; Abo Alrob, O.; Alkilany, A.; Benamara, M.; Evans, D. Synthesis of Gold Nanoparticles Using Leaf Extract of *Ziziphus Zizyphus* and Their Antimicrobial Activity. *Nanomaterials* **2018**, *8*, 174. [CrossRef]
45. Quilaqueo, M.; Millao, S.; Luzardo-Ocampo, I.; Campos-Vega, R.; Acevedo, F.; Shene, C.; Rubilar, M. Inclusion of Piperine in β-Cyclodextrin Complexes Improves Their Bioaccessibility and in Vitro Antioxidant Capacity. *Food Hydrocoll.* **2019**, *91*, 143–152. [CrossRef]
46. Alghuthaymi, M.A.; Rajkuberan, C.; Santhiya, T.; Krejcar, O.; Kuča, K.; Periakaruppan, R.; Prabukumar, S. Green Synthesis of Gold Nanoparticles Using *Polianthes Tuberosa* L. Floral Extract. *Plants* **2021**, *10*, 2370. [CrossRef]
47. Wang, B.; Yang, G.; Chen, J.; Fang, G. Green Synthesis and Characterization of Gold Nanoparticles Using Lignin Nanoparticles. *Nanomaterials* **2020**, *10*, 1869. [CrossRef]
48. Fang, A.; Sun, Y.; Feng, D.; Ma, M.; Xu, Z.; Zhang, T.; Shi, F. Flower-like Gold Nanoparticles Labeled and Silver Deposition Rapid Vertical Flow Technology for Highly Sensitive Detection of Brucella Antibodies. *Analyst* **2021**, *146*, 5362–5368. [CrossRef]
49. Clayton, K.N.; Salameh, J.W.; Wereley, S.T.; Kinzer-Ursem, T.L. Physical Characterization of Nanoparticle Size and Surface Modification Using Particle Scattering Diffusometry. *Biomicrofluidics* **2016**, *10*, 054107. [CrossRef]
50. Takechi-Haraya, Y.; Ohgita, T.; Demizu, Y.; Saito, H.; Izutsu, K.I.; Sakai-Kato, K. Current Status and Challenges of Analytical Methods for Evaluation of Size and Surface Modification of Nanoparticle-Based Drug Formulations. *AAPS PharmSciTech* **2022**, *23*, 150. [CrossRef]
51. Al Saqr, A.; Khafagy, E.-S.; Alalaiwe, A.; Aldawsari, M.; Alshahrani, S.; Anwer, K.; Khan, S.; Lila, A.; Arab, H.; Hegazy, W. Synthesis of Gold Nanoparticles by Using Green Machinery: Characterization and in Vitro Toxicity. *Nanomaterials* **2021**, *11*, 808. [CrossRef] [PubMed]
52. Nowakowska, M.; Pospiech, K.; Lewandowska, U.; Piastowska-Ciesielska, A.W.; Bednarek, A.K. Diverse Effect of WWOX Overexpression in HT29 and SW480 Colon Cancer Cell Lines. *Tumor Biology* **2014**, *35*, 9291–9301. [CrossRef] [PubMed]
53. Dragicevic, S.; Kovacevic, D.; Divac-Rankov, A.; Nikolic, A.; Radojkovic, D.; Radovic, S. Evaluation of Toxicity and Antioxidative Effects of *Tussilago Farfara* and *Verbascum thapsus* Water Extracts in Zebrafish and in Bronchial Epithelial Cells. *Arch. Biol. Sci.* **2019**, *71*, 409–416. [CrossRef]
54. Calabrese, G.; Zappalà, A.; Dolcimascolo, A.; Acquaviva, R.; Parenti, R.; Malfa, G.A. Phytochemical Analysis and Anti-Inflammatory and Anti-Osteoarthritic Bioactive Potential of *Verbascum thapsus* L. (Scrophulariaceae) Leaf Extract Evaluated in Two In Vitro Models of Inflammation and Osteoarthritis. *Molecules* **2021**, *26*, 5392. [CrossRef] [PubMed]
55. Escobar, F.M.; Sabini, M.C.; Zanon, S.M.; Cariddi, L.N.; Tonn, C.E.; Sabini, L.I. Genotoxic Evaluation of a Methanolic Extract of *Verbascum thapsus* Using Micronucleus Test in Mouse Bone Marrow. *Nat. Prod. Commun.* **2011**, *6*, 989–991. [CrossRef]

56. Lu, K.; Ou, Y.; Jiang, L.; Xiao, Y.; Chinnathambi, A.; Alahmadi, T.A.; Wainwright, M. Novel Formulation, Cytotoxicity, Antioxidant, and Anti-Human Lung Cancer Properties of Gold Nanoparticles Containing *Verbascum thapsus* L. Leaf Aqueous Extract. *Arch. Med. Sci.* **2021**, 1–24. [CrossRef]
57. Majumder, M.; Debnath, S.; Gajbhiye, R.L.; Saikia, R.; Gogoi, B.; Samanta, S.K.; Das, D.K.; Biswas, K.; Jaisankar, P.; Mukhopadhyay, R. *Ricinus communis* L. Fruit Extract Inhibits Migration/Invasion, Induces Apoptosis in Breast Cancer Cells and Arrests Tumor Progression in Vivo. *Sci. Rep.* **2019**, *9*, 14493. [CrossRef]
58. Darmanin, S.; Wismayer, P.S.; Camilleri Podesta, M.T.; Micallef, M.J.; Buhagiar, J.A. An Extract from *Ricinus communis* L. Leaves Possesses Cytotoxic Properties and Induces Apoptosis in SK-MEL-28 Human Melanoma Cells. *Nat. Prod. Res.* **2009**, *23*, 561–571. [CrossRef]
59. Hosseinzadeh, H.; Atyabi, F.; Varnamkhasti, B.S.; Hosseinzadeh, R.; Ostad, S.N.; Ghahremani, M.H.; Dinarvand, R. SN38 Conjugated Hyaluronic Acid Gold Nanoparticles as a Novel System against Metastatic Colon Cancer Cells. *Int. J. Pharm.* **2017**, *526*, 339–352. [CrossRef]
60. Tabatabaie, F.; Franich, R.; Feltis, B.; Geso, M. Oxidative Damage to Mitochondria Enhanced by Ionising Radiation and Gold Nanoparticles in Cancer Cells. *Int. J. Mol. Sci.* **2022**, *23*, 6887. [CrossRef]
61. Minai, L.; Yeheskely-Hayon, D.; Yelin, D. High Levels of Reactive Oxygen Species in Gold Nanoparticle-Targeted Cancer Cells Following Femtosecond Pulse Irradiation. *Sci. Rep.* **2013**, *3*, srep02146. [CrossRef] [PubMed]
62. Venugopal, K.; Rather, H.A.; Rajagopal, K.; Shanthi, M.P.; Sheriff, K.; Illiyas, M.; Rather, R.A.; Manikandan, E.; Uvarajan, S.; Bhaskar, M.; et al. Synthesis of Silver Nanoparticles (Ag NPs) for Anticancer Activities (MCF 7 Breast and A549 Lung Cell Lines) of the Crude Extract of Syzygium Aromaticum. *J. Photochem. Photobiol. B* **2017**, *167*, 282–289. [CrossRef] [PubMed]
63. Lin, J.; Gulbagca, F.; Aygun, A.; Tiri, R.N.E.; Xia, C.; Van Le, Q.; Gur, T.; Sen, F.; Vasseghian, Y. Phyto-Mediated Synthesis of Nanoparticles and Their Applications on Hydrogen Generation on $NaBH_4$, Biological Activities and Photodegradation on Azo Dyes: Development of Machine Learning Model. *Food Chem. Toxicol.* **2022**, *163*, 112972. [CrossRef]
64. M Amin, H.I.; Hussain, F.H.S.; Gilardoni, G.; Thu, Z.M.; Clericuzio, M.; Vidari, G. Phytochemistry of *Verbascum* Species Growing in Iraqi Kurdistan and Bioactive Iridoids from the Flowers of *Verbascum* Calvum. *Plants* **2020**, *9*, 1066. [CrossRef] [PubMed]
65. Arango-Varela, S.S.; Luzardo-Ocampo, I.; Reyes-Dieck, C.; Yahia, E.M.; Maldonado-Celis, M.E. Antiproliferative Potential of Andean Berry (Vacciniu m Meridionale Swartz) Juice in Combination with Aspirin in Human SW480 Colon Adenocarcinoma Cells. *J. Food Biochem.* **2021**, *45*, e13760. [CrossRef] [PubMed]
66. Alkilany, A.M.; Murphy, C.J. Toxicity and Cellular Uptake of Gold Nanoparticles: What We Have Learned so Far? *J. Nanoparticle Res.* **2010**, *12*, 2313–2333. [CrossRef] [PubMed]
67. Aldahhan, R.; Almohazey, D.; Khan, F.A. Emerging Trends in the Application of Gold Nanoformulations in Colon Cancer Diagnosis and Treatment. *Semin. Cancer Biol.* 2021, in press. [CrossRef]

pharmaceutics MDPI

Article

Gold Nanoclusters-Based NIR-II Photosensitizers with Catalase-like Activity for Boosted Photodynamic Therapy

Qing Dan [1,2], Zhen Yuan [3], Si Zheng [1,2], Huanrong Ma [1], Wanxian Luo [1], Li Zhang [1], Ning Su [1], Dehong Hu [2], Zonghai Sheng [2,*] and Yingjia Li [1,*]

[1] Department of Medicine Ultrasonics, Nanfang Hospital, Southern Medical University, Guangzhou 510515, China
[2] Shenzhen Institute of Advanced Technology, Chinese Academy of Sciences, Shenzhen 518055, China
[3] Faculty of Health Sciences, University of Macau, Taipa, Macau SAR, China
* Correspondence: zh.sheng@siat.ac.cn (Z.S.); lyjia@smu.edu.cn (Y.L.)

Abstract: Photodynamic therapy (PDT) under fluorescence imaging as a selective and non-invasive treatment approach has been widely applied for the therapy of cancer and bacterial infections. However, its treatment efficiency is hampered by high background fluorescence in the first near-infrared window (NIR-I, 700–900 nm) and oxygen-dependent photosensitizing activity of traditional photosensitizers. In this work, we employ gold nanoclusters (BSA@Au) with the second near-infrared (NIR-II, 1000–1700 nm) fluorescence and catalase-like activity as alternative photosensitizers to realize highly efficient PDT. The bright NIR-II fluorescence of BSA@Au enables the visualization of PDT for tumor with a high signal-to-background ratio (SBR = 7.3) in 4T1 tumor-bearing mouse models. Furthermore, the catalase-like activity of BSA@Au endows its oxygen self-supplied capability, contributing to a five-fold increase in the survival period of tumor-bearing mice receiving boosted PDT treatment compared to that of the control group. Moreover, we further demonstrate that BSA@Au-based PDT strategy can be applied to treat bacterial infections. Our studies show the great potential of NIR-II BSA@Au as a novel photosensitizer for boosted PDT against cancer and bacterial infections.

Keywords: photodynamic therapy; NIR-II fluorescence; gold nanoclusters; hypoxia; nanozyme; cancer theranostics

Citation: Dan, Q.; Yuan, Z.; Zheng, S.; Ma, H.; Luo, W.; Zhang, L.; Su, N.; Hu, D.; Sheng, Z.; Li, Y. Gold Nanoclusters-Based NIR-II Photosensitizers with Catalase-like Activity for Boosted Photodynamic Therapy. *Pharmaceutics* **2022**, *14*, 1645. https://doi.org/10.3390/pharmaceutics14081645

Academic Editors: Manivasagan Panchanathan and Eue-Soon Jang

Received: 22 June 2022
Accepted: 3 August 2022
Published: 7 August 2022

1. Introduction

Photodynamic therapy (PDT) is acknowledged as a clinical treatment modality for cancer and bacterial infections, which is selective, non-invasive, and feasible to combine with imaging. It generally employs a visible or near-infrared laser to excite photosensitizers to generate fluorescence (FL) for guiding therapy. Meanwhile, it generates singlet oxygen ($^1O_2$) in the presence of oxygen to kill cancer cells and bacteria [1–4]. To date, organic dyes (i.e., porphyrins, indocyanine dyes, BODIPYs, and aggregation-induced emission dyes) [5–8], inorganic nanoparticles (i.e., quantum dots, black phosphorus, and TiO_2) [9–11], polymer nanoparticles (i.e., polyfluorene and polythiophene) [12,13], and other nanomaterials (i.e., noble metal complexes and metallic nanomaterials) [14,15], have been developed for PDT. However, the FL emission of these photosensitizers is limited in visible (400–700 nm) and the first near-infrared (NIR-I, 700–900 nm) windows, resulting in intravital imaging with high background signal. Moreover, the $^1O_2$ generation of photosensitizers gradually decreases within tissue due to the pre-existing hypoxia and consumption of oxygen during the PDT process, leading to unsatisfactory PDT outcomes [16]. Therefore, it is of great significance to develop novel photosensitizers with long emission wavelength and self-supplied oxygen for boosted PDT against tumors and infectious diseases.

In recent years, several dyes with photosensitive activity in the second near-infrared (NIR-II, 1000–1700 nm) window have been explored, which exhibits unique merits over the

NIR-I window, such as a reduction in photon absorption and scattering, as well as improvement of signal-to-background ratio (SBR) [17–19]. However, these NIR-II photosensitizers generally show low PDT efficiency due to the oxygen insufficiency within laser irradiation area. Recently, NIR-II gold nanoclusters (NCs), consisting of several to about a hundred gold atoms with a size below 3 nm, are of particular interest to researchers owing to their intrinsic FL emission, renal clearance, and high biocompatibility [20]. Gold NCs with the NIR-II emission have been developed for the bright in vivo NIR-II FL imaging of malignant tumors, bones, and brain vessels [21–25]. In addition to FL imaging capability, gold NCs could also generate toxic $^1O_2$ under NIR laser irradiation [26]. Furthermore, our previous studies reported that gold NCs showed catalase-like activity, which could decompose endogenous hydrogen peroxide (H_2O_2) within tumor tissue to O_2 for enhancing PDT [27]. However, to our best knowledge, there has no reports on the utilization of the three features of gold NCs, including NIR-II FL, $^1O_2$ generation, and catalase-like activity, for boosted PDT against tumors and bacterial infections.

Herein, we developed the NIR-II gold NCs (BSA@Au) as novel photosensitizers for boosted PDT. As shown in Scheme 1, BSA@Au was synthesized by one-step biomineralization. The facile synthesis process is advantageous for biological applications. The bright NIR-II FL of BSA@Au could realize high-performance in vivo bioimaging for the optimization of PDT. Meanwhile, the superb catalase-like activity of BSA@Au could catalyze the decomposition of excess H_2O_2 to produce O_2 within the microenvironment of the tumor and bacterial-infected tissues for enhancing PDT. Our studies provide a paradigm of using gold NCs as efficient photosensitizers for boosted PDT.

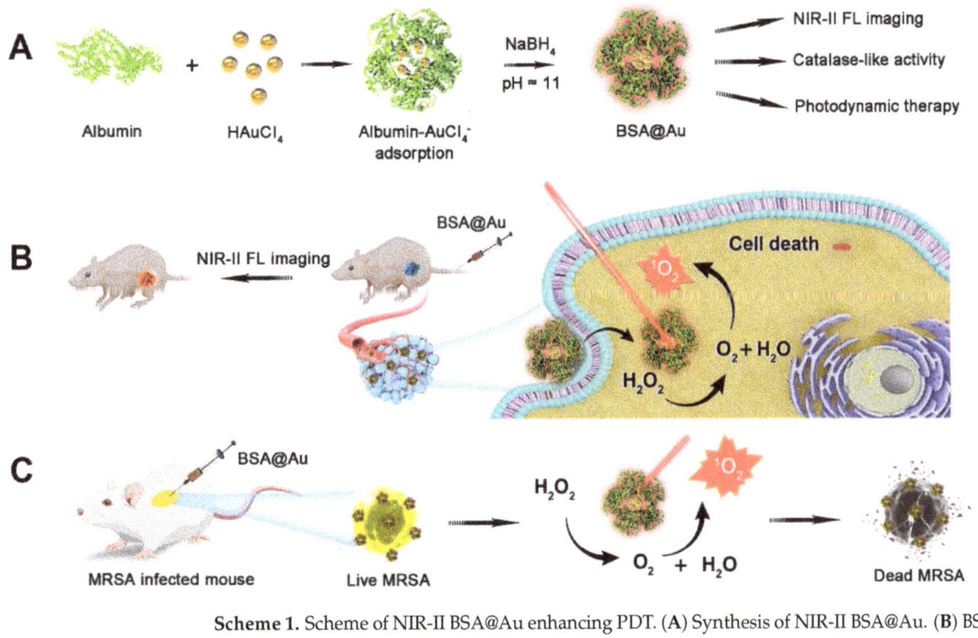

Scheme 1. Scheme of NIR-II BSA@Au enhancing PDT. (**A**) Synthesis of NIR-II BSA@Au. (**B**) BSA@Au combining with PDT against breast cancer. (**C**) BSA@Au combining with PDT against methicillin-resistant *Staphylococcus aureus* (MRSA) infection.

2. Materials and Methods

2.1. Materials

Gold (III) chloride trihydrate ($HAuCl_4 \cdot 3H_2O$), bovine serum albumin (BSA), sodium borohydride ($NaBH_4$), and 1,3-Diphenylisobenzofuran (DPBF) were provided by Sigma-

Aldrich. The Cell counting kit-8 (CCK-8) was supplied by Dojindo. Pimonidazole HCl and IgG_1 mouse antibody (conjugated with FITC) were purchased from Hypoxyprobe.

2.2. Synthesis of BSA@Au

BSA@Au was prepared adapted from previous reports [28]. Briefly, all glassware were soaked with aqua regia (HCl:HNO_3, volume ratio 1:3), then washed with ultrapure water. $HAuCl_4$ solution (10 mM, 1.25 mL) was mixed with BSA solution (0.76 mM, 2.5 mL). Then, NaOH solution (1 M, 0.25 mL) and $NaBH_4$ (0.1 M, 0.1 mL) were introduced. The whole process was under vigorously stirring. When the mixture turned into deep brown, it was incubated in a shaker (180 rpm, 37 °C) for 12 h to obtain BSA@Au nanoclusters. The as-prepared BSA@Au was purified by dialysis (molecular weight cutoff: 8–14 KD) with 0.1 M phosphate solution (pH $\approx$ 7.4) and stored in the dark at 4 °C for further investigation.

2.3. In Vivo FL Imaging

Mice bearing a 4T1 tumor received an intravenous (i.v.) injection of BSA@Au (0.46 mM, 200 µL) and were imaged at 0, 2, 4, 10, and 24 h. At 10 h, the main organs (including heart, liver, spleen, lung, and kidneys) and tumor were collected for ex vivo FL imaging. NIR-I FL imaging was conducted with excitation wavelength at 675 nm, and emission wavelength at 800 nm. The NIR-II FL imaging conditions were set as follows: excitation wavelength: 808 nm; long-pass filter: 1000 nm; and exposure time: 800 ms.

2.4. In Vivo PDT for Tumor

To evaluate the PDT efficacy, we randomly divided Balb/C mice bearing 4T1 tumor into 4 groups (n = 5): (1) Control; (2) Laser; (3) BSA@Au; (4) BSA@Au + Laser. At 10 h post i.v. injection of BSA@Au, the 808 nm laser (0.3 W/cm^2, 30 min) was applied and the temperature of tumors in group (4) was recorded to avoid the thermal effect. Since the single dose treatment, the tumor volumes and body weights of mice were recorded every 2 days for further analysis. When tumor grew to ~800 mm^3, the mouse was defined as dead to obtain the survival curve.

2.5. In Vitro Antibacterial Assay

For antibacterial experiments, the amounts of live MRSA after PDT treatment were quantified by plate counting method. Firstly, MRSA suspensions were mixed with PBS, BSA@Au, and H_2O_2 (200 µM). Then, the suspensions ($OD_{600nm} \approx 0.1$) were placed in 96-well plates. After incubation at 37 °C for 30 min, the samples were further illuminated by 808 nm laser (0.3 W/cm^2, 30 min). Then, the suspensions were diluted ($1:1 \times 10^5$) and transferred to LB agar plates for further cultivation at 37 °C. After 24 h, the colonies were counted by ImageJ software. In addition, bacterial morphology details of MRSA after different treatments were observed by scanning electron microscopy (SEM).

2.6. In Vivo PDT for MRSA Infection

To investigate in vivo antimicrobial performance, the infected mice were randomly divided into 4 groups (n = 5): (1) Control; (2) Laser; (3) BSA@Au; (4) BSA@Au + Laser. Mice of groups (3) and (4) were locally treated with BSA@Au solution (2.5 mM, 20 µL) at wound sites. The wound areas and body weights were recorded every 3 days. After 12 days of treatments, the wounds were excised for H&E staining analysis.

2.7. Statistical Analysis

Data are shown as the mean $\pm$ SD. Statistical significance was analyzed by student's t-test analysis. $p < 0.05$ was considered as statistical significance.

3. Results and Discussion

3.1. Preparation and Characterization of NIR-II BSA@Au

In this work, NIR-II emissive BSA@Au was synthesized by the biomineralization method. BSA was used as a template and protector for the growth of BSA@Au, and $NaBH_4$ was introduced as the critical reductant to modulate the NIR-II FL emission intensity. As shown in the high-resolution transmission electron microscope (HRTEM) image, the core size of BSA@Au was ~2 nm (Figure 1A), which was below the glomerular filtration cutoff (~6 nm) for renal clearance after i.v. injection [29,30]. The hydrodynamic diameter of BSA@Au was 10.1 $\pm$ 1.4 nm and stayed highly stable after four weeks of storage in water or fetal bovine serum (FBS) (Figure 1B,C). The size of BSA@Au measured by dynamic light scattering (DLS) was larger than that determined by HRTEM because DLS detected an average hydrodynamic size whereas the HRTEM images exhibited the dehydration morphology of BSA@Au [31].

Generally, gold-based nanoparticles exhibit FL emission in both NIR-I and NIR-II windows. We used the NIR FL spectrum instrument and NIR-II FL imaging system to verify the FL property of BSA@Au. When prepared under various conditions, it demonstrated different fluorescent performance. We noted that the NIR-II FL intensity of BSA@Au was closely related to the molar ratio of BSA: $HAuCl_4$ and NaOH: $NaBH_4$, which were optimized to be 0.15 and 25, respectively. Particularly, BSA@Au did not show NIR-II FL emission without $NaBH_4$ (Table S1, Figures 1D,E and S1A), which indicated the important reduction role of $NaBH_4$ to transfer Au^{3+} to Au to form the gold nucleus [32,33]. Furthermore, it was observed that NIR-II BSA@Au were formed rapidly within 0.5 h after adding $NaBH_4$, and the FL intensity gradually became stronger with the reaction time, reaching the peak value at 12 h, before decreasing (Figure S1B,C). Therefore, BSA (0.76 mM, 2.5 mL), $HAuCl_4$ (10 mM, 1.25 mL), NaOH (1 M, 0.25 mL), and $NaBH_4$ (0.1 M, 0.1 mL) were applied to synthesize NIR-II emissive BSA@Au under 37 °C for 12 h for further study.

To obtain deep insights into the optical properties of BSA@Au, its spectra properties were investigated. As shown in Figure 1F, the UV–Vis absorption spectrum of BSA@Au showed strong absorption at ~280 nm and decayed exponentially towards ~1100 nm. Moreover, the fluorescence emission spectrum exhibited two characteristic emission peaks at ~960 nm and ~1010 nm, reaching the NIR-II region. The mechanism of NIR-II emission of BSA@Au might involve complex electron transfer between BSA and gold cores [34]. In addition, we obtained the FL spectra of BSA@Au under various excitation wavelengths (450~808 nm). It showed an excitation wavelength-dependent FL emission peak, and the optimized FL emission spectrum was under 808 nm excitation (Figure S1D). Such a broadband-excitable bright FL emission was beneficial for diverse applications, and 808 nm excitation was chosen for further investigation.

Then, we used IR-26 as the reference and measured the fluorescent QY of the BSA@Au. Results showed that the QY of BSA@Au was measured to be ~3.5% (Figure S2), which was much higher than the QY of other NIR-II gold NCs (Table 1) and enabled it to realize bright FL imaging performance. Next, we tested the penetration depth of BSA@Au. Firstly, in the NIR-I window, the Eppendorf tube containing BSA@Au could not be detected when covered by chicken tissues (>4 mm thickness). In contrast, bright images of BSA@Au were obtained in NIR-II window at a penetration depth of more than 10 mm with high SBR (Figure S3A). This indicated the potential that BSA@Au could achieve high-performance in vivo imaging. Next, we assessed the photostability of BSA@Au compared with the clinical NIR dye indocyanine green (ICG). It was found that, under NIR-II FL imaging, BSA@Au was still bright under 808 nm laser irradiation (0.3 W/cm^2, 30 min) whereas the FL intensity of its ICG counterpart nearly decreased to ~0%, demonstrating that BSA@Au exhibited excellent photostability in contrast with ICG (Figure S4), which was owing to the role of BSA to protect BSA@Au from bleaching [35]. Taken together, the as-synthesized NIR-II emissive BSA@Au with a size of ~2 nm showed good water solubility, hydrodynamic size stability, photostability, and high QY of ~3.5%, demonstrating promising biomedical applications.

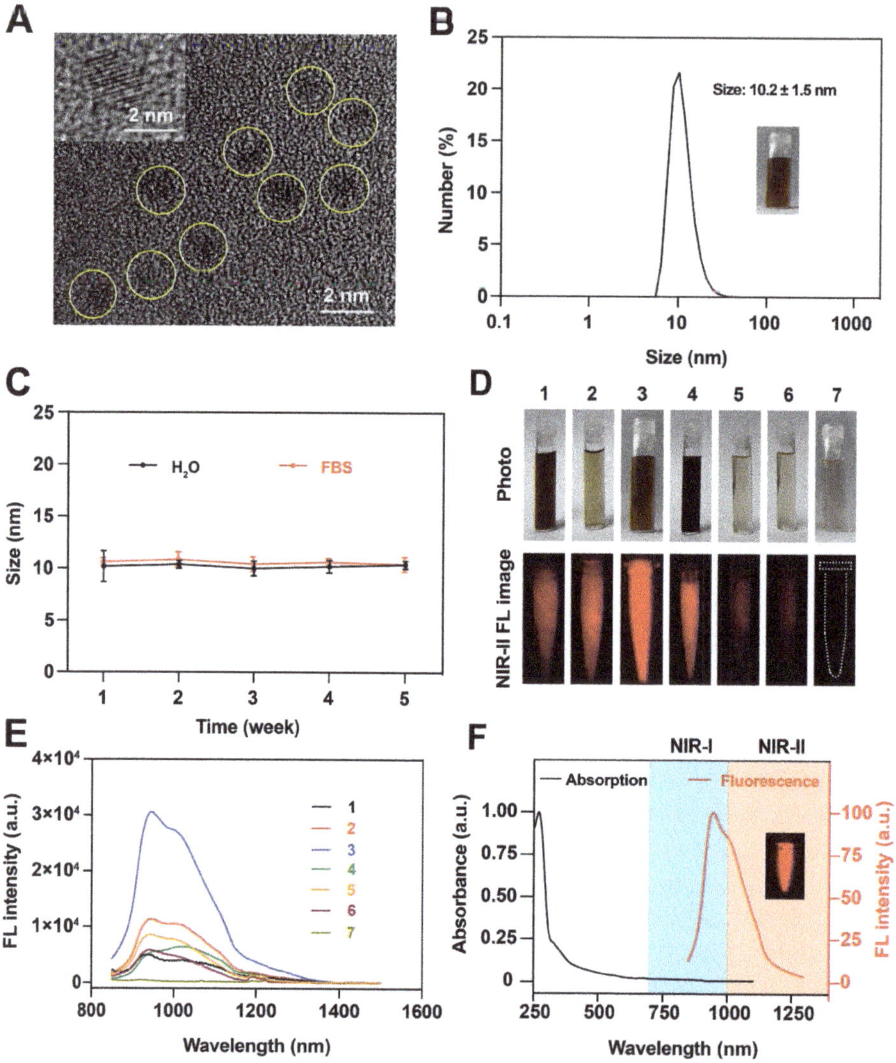

Figure 1. Characterization of BSA@Au. (**A**) The HRTEM image of BSA@Au. The inset was the lattice structure of BSA@Au. (**B**) Hydrodynamic size of BSA@Au. The inset was the digital photo of BSA@Au. (**C**) Colloidal stability of BSA@Au in H_2O or FBS after 4 weeks storage at 4 °C. (**D**) Digital photos and NIR-II FL images of BSA@Au samples 1–7 synthesized under different conditions. (**E**) The FL emission spectra of BSA@Au samples 1–7 under 808 nm excitation. In (**D**,**E**), 1: BSA@Au (50 mg/mL, 2.5 mL), $HAuCl_4$ (10 mM, 2.5 mL), NaOH (1 M, 0.25 mL), and $NaBH_4$ (0.1 M, 0.25 mL); 2: BSA@Au (50 mg/mL, 2.5 mL), $HAuCl_4$ (10 mM, 0.75 mL), NaOH (1 M, 0.25 mL), and $NaBH_4$ (0.1 M, 0.25 mL); 3: BSA@Au (50 mg/mL, 2.5 mL), $HAuCl_4$ (10 mM, 1.25 mL), NaOH (1 M, 0.25 mL), and $NaBH_4$ (0.1 M, 0.1 mL); 4: BSA@Au (50 mg/mL, 0.75 mL), $HAuCl_4$ (10 mM, 2.5 mL), NaOH (1 M, 0.25 mL), and $NaBH_4$ (0.1 M, 0.1 mL); 5: BSA@Au (50 mg/mL, 2.5 mL), $HAuCl_4$ (10 mM, 0.75 mL), NaOH (1 M, 0.5 mL), and $NaBH_4$ (0.1 M, 0.05 mL); 6: BSA@Au (50 mg/mL, 2.5 mL), $HAuCl_4$ (10 mM, 0.75 mL), NaOH (1 M, 1 mL), and $NaBH_4$ (0.1 M, 0.05 mL); 7: BSA@Au (50 mg/mL, 2.5 mL), $HAuCl_4$ (10 mM, 1.25 mL), and NaOH (1 M, 0.25 mL). (**F**) Absorption spectrum and FL spectrum of BSA@Au. The inset was the NIR-II FL image of BSA@Au.

Table 1. Overview of NIR-II emissive gold NCs.

Gold NCs	Core Size (nm)	QY (%)	Applications	Ref.
CD-Au NCs	1.85 ± 0.24	0.11	FL imaging of cancer	[21]
$Au_{25}(SG)_{18}$	3.2	0.67	FL imaging of brain injury and tumor metastasis	[22]
$Au_{25}(SG)_{18}$	3.30 ± 0.82	0.27	FL imaging of bones	[23]
$Au_{25}(SG)_{18}$-Pt	3.2	0.04	FL imaging guided chemotherapy	[24]
RNase-A@AuNCs	2.2 ± 0.1	1.19	FL imaging of gastrointestinal tract	[25]
$Au_{25}(SR_1)_x(SR_2)_{18-x}$	1.8	0.54	Antibacterial infections	[36]
Ln@AuNCs	35	/	In vivo FL imaging of H_2S	[37]
NR@Min-23@AuNCs	6.4	0.21	PDT theranostics against breast cancer	[38]
BSA@Au	2	3.53	Phototheranostics of cancer and bacterial infections	This work

3.2. Catalase-like Activity and Singlet Oxygen Generation

Tumor and infection microenvironments generally show high H_2O_2 concentration and low O_2 level [39,40]. We hypothesized that the as-prepared BSA@Au could serve as a catalase mimic to alleviate hypoxic microenvironment within tumor and infected tissues (Figure 2A). In a typical experiment, after BSA@Au was added into H_2O_2 solution, we observed obvious gas bubbles generated from the mixed solution (Figure 2B). Furthermore, when BSA@Au reacted with H_2O_2 monitored by an ultrasound imaging system, it showed significantly enhanced imaging performance both on B mode and contrast-enhanced ultrasound mode (Figure 2C). More interestingly, clear gas echo signals were observed in 4T1 tumor-bearing mice injected with BSA@Au under an ultrasound imaging system (Figure S5A), demonstrating that BSA@Au could catalyze H_2O_2 to produce O_2.

To gain deep understanding of the catalase-like activity of BSA@Au, we investigated the apparent kinetic parameters according to Michaelis-Menten method and Lineweaver-Burk equation [41]. We found that O_2 production was dependent on the concentration of BSA@Au and H_2O_2 (Figure 2D,E). Hence, it was believed that with more BSA@Au accumulation at the tumor sites or bacteria-infected lesions, more O_2 would generate in situ. In Figure S5B, the critical enzyme kinetic parameters of the Michaelis-Menten constant (K_m) and maximum initial velocity (V_{max}) were obtained and compared with other nanozymes. In this study, the apparent K_m reflected that BSA@Au showed high affinity towards H_2O_2, and V_{max} indicated the good catalytic ability of BSA@Au, which were superior to other nanozymes (Table S2). However, compared with our previous study, the V_{max} of BSA@Au was lower than that of Au NCs-ICG (with a size of ~1 nm) by an order of magnitude, which might be attributed to the relatively larger size and less turnover number of BSA@Au [27]. Moreover, in the future, the size, shape, and ligand effects should be considered to improve the catalytic performance of BSA@Au.

Next, we evaluated the $^1O_2$ generation ability of BSA@Au using 1,3-Diphenylisobenzofuran (DPBF) probe under 808 nm irradiation. Compared with the clinical photosensitizer ICG ($^1O_2$ generation efficiency, $\Phi_{\Delta ICG} = 0.2\%$), the $\Phi_{\Delta BSA@Au}$ was determined to be about 0.4% (Figure 2F), which was 2.2-fold higher than that of ICG. More inspiringly, our results showed that the generated O_2 from H_2O_2 degradation could enhance the generation of $^1O_2$, which could overcome the pre-existing hypoxia and O_2 depletion during the PDT process. In brief, upon 808 nm laser irradiation, the $^1O_2$ generation of BSA@Au was measured by DPBF. As shown in Figure 2G, the DPBF was quenched by BSA@Au combining with 808 nm laser irradiation, and significantly decreased after the addition of H_2O_2. This suggested that BSA@Au could serve as a self-sufficient O_2 supplying photosensitizer within H_2O_2-rich regions and thus enhance in vivo PDT for cancer and bacteria-infected diseases.

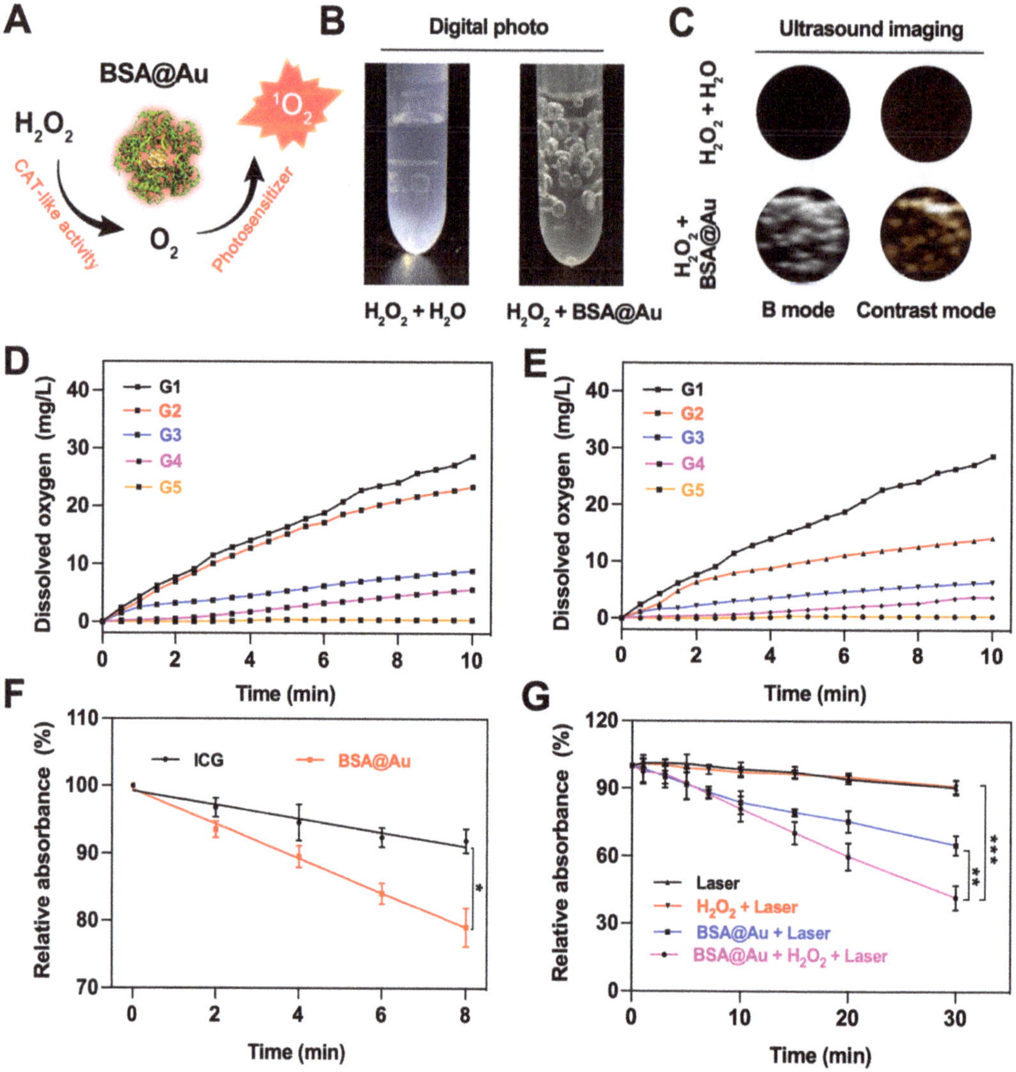

Figure 2. Catalase (CAT)-like activity and $^1O_2$ generation ability of BSA@Au. (**A**) Scheme of BSA@Au worked as a catalase mimic and photosensitizer in cascade reactions. (**B**) Photograph of bubbles generated from reaction between BSA@Au and H_2O_2. (**C**) Bubbles generation imaged by ultrasound device on B mode and contrast mode. (**D**) O_2 production from H_2O_2 of different concentrations decomposed by BSA@Au. G1: H_2O_2 (12.5 mM) + BSA@Au (0.23 mM); G2: H_2O_2 (9.37 mM) + BSA@Au (0.23 mM); G3: H_2O_2 (6.25 mM) + BSA@Au (0.23 mM); G4: H_2O_2 (1.25 mM) + BSA@Au (0.23 mM); G5: H_2O_2 (12.5 mM) + H_2O. (**E**) O_2 production from H_2O_2 catalyzed by BSA@Au of various concentrations. G1: H_2O_2 (12.5 mM) + BSA@Au (0.23 mM); G2: H_2O_2 (12.5 mM) + BSA@Au (0.12 mM); G3: H_2O_2 (12.5 mM) + BSA@Au (0.06 mM); G4: H_2O_2 (12.5 mM) + BSA@Au (0.03 mM); G5: H_2O_2 (12.5 mM) + H_2O. (**F**) $^1O_2$ production efficiency of BSA@Au measured by DPBF in absorbance at 414 nm under 808 nm laser irradiation with ICG as a reference. (**G**) $^1O_2$ generation measured by DPBF at 414 nm against time in the presence of different solutions. * $p < 0.5$, ** $p < 0.01$, *** $p < 0.005$.

3.3. Toxicity Evaluation and In Vitro PDT

The cytotoxicity of BSA@Au was investigated using the CCK-8 assay. The 4T1 cells were incubated with BSA@Au of various concentrations. Compared with the control groups, there was almost no obvious cell viability decrease in 4T1 cells treated with BSA@Au (Figure 3A), which indicated the low cytotoxicity of BSA@Au. Similarly, BSA@Au exhibited no harm to red blood cells in hemolysis assay (Figure 3B). Based on the above results, we confirmed the negligible in vivo toxicity of BSA@Au. Two weeks after i.v. injection of the BSA@Au with high concentration (5 mM, 100 μL), the blood samples and main organs, including heart, liver, spleen, lung, and kidneys, were collected from mice for blood routine examinations and H&E analysis, respectively. This showed that the tissue sections and blood indicators of BSA@Au group were not significantly altered compared to normal references (Figure S6, Table S3). The superior biocompatibilities of the BSA@Au might be attributed to the natural carrier of BSA and the ultrasmall size of BSA@Au, which together supported the in vivo application of BSA@Au for imaging and therapy.

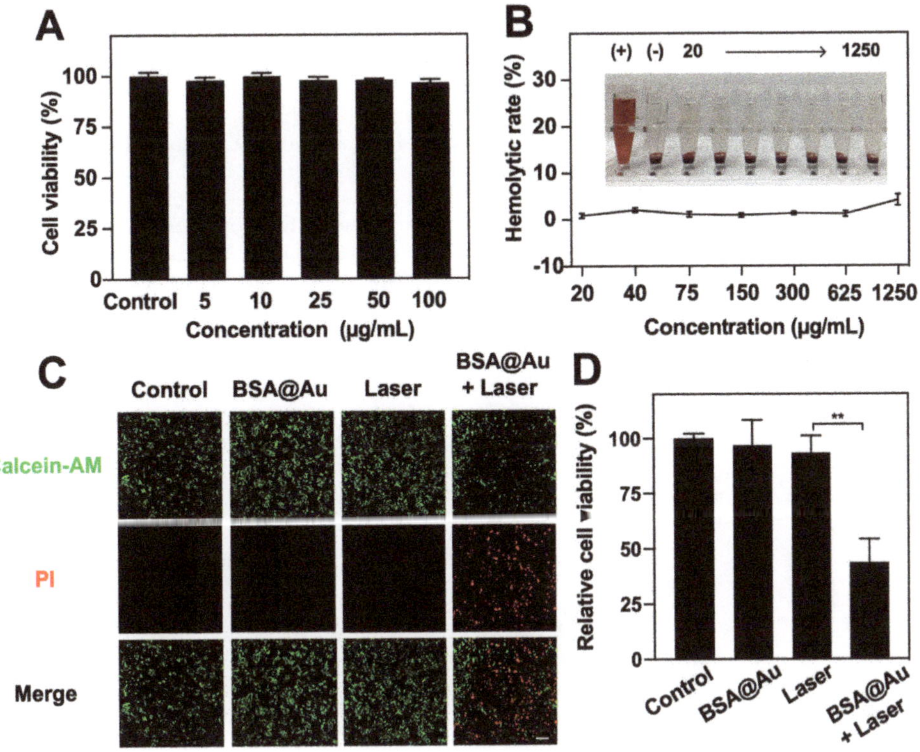

Figure 3. Cytotoxicity and in vitro anti-tumor PDT. (**A**) Cytotoxicity evaluation on 4T1 cells incubated with BSA@Au. (**B**) Hemolytic assay of BSA@Au on red blood cells (RBCs) from Balb/C mice. (+) represented RBCs solution mixed with ultrapure water; (−) indicated RBCs solution mixed with PBS. (**C**) Live/dead 4T1 cells visualization post PDT. The scale bar was 100 μm. (**D**) 4T1 cells viability post PDT. ** $p < 0.01$.

Given the high $^1O_2$ generation ability of BSA@Au, we tested the PDT efficacy against 4T1 cells. As shown in Figure 3C, Calcein-AM and PI dyes were used to visually evaluate the live/dead 4T1 cells after PDT. Nearly no red FL signals were detected in the control, BSA@Au, and Laser groups, while intense red FL signals in the BSA@Au + Laser group were observed. Cell viability after PDT was further quantified by CCK-8 method, revealing that only ~43.7% cells survived when treated by BSA@Au and laser irradiation. However, control groups did not show obvious damage to 4T1 cells (Figure 3D). These results demonstrated the efficient PDT potential of BSA@Au.

3.4. In Vivo Imaging

Inspired by in vitro FL imaging performance of BSA@Au, the in vivo NIR-I and NIR-II FL imaging of BSA@Au was evaluated in 4T1 tumor-bearing mice. As illustrated in Figure 4A, images in the NIR-I window roughly outlined the tissue, due to the absorption and scattering of NIR-I photons by biological tissues. However, the imaging quality was well improved in the NIR-II window, showing the superiority of NIR-II FL imaging. Clearly, the FL signals of the tumor region were gradually increased with time and reached the accumulation peak at 10 h post-injection, and then gradually declined. The SBR was quantitatively analyzed. The tumor FL signal shows about 7.3-fold higher than that of background at 10 h post injection (Figure 4B), which could precisely discriminate and localize the tumor tissues. In addition, the metabolism and biodistribution of BSA@Au in major organs and tumor were investigated by ex vivo NIR-II FL imaging. As shown in Figures 4C and S7A, at 10 h post i.v. injection, the heart, lung, spleen, and kidneys exhibited nearly no or weak FL signals, while the liver and tumor showed strong FL intensity, suggesting the BSA@Au preferred to accumulate in these organs. The effective accumulation of BSA@Au in tumor could reduce the nanoparticles injection dosage to minimize the system hazards. Furthermore, different metabolic behaviors of BSA@Au in liver and kidneys were observed. The liver gradually exhibited FL signals within 10 h post injection of BSA@Au, while the kidneys showed bright FL signals at 0.5 h and 4 h post-injection, and the signal became weak at 10 h (Figure S7B,C). In addition, the signal of tumor and main organs (heart, liver, spleen, lung, and kidneys) remarkably decreased at 48 h post injection (Figure S7D,E). This suggested that BSA@Au could be cleared out from liver and kidneys because of the small size. Taken together, these important features enabled BSA@Au to be a promising medical probe for in vivo imaging to guide therapy.

3.5. In Vivo Anti-Tumor Therapy

Due to the excellent $^1O_2$ generation ability and catalytic activity, we explored the anti-cancer potential of BSA@Au on 4T1 tumor-bearing mice. Firstly, we evaluated the hypoxia modulation performance of BSA@Au. After i.v. injection of BSA@Au, the obtained immunofluorescence images of tumor sections showed that the hypoxic areas of tumor tissue markedly decreased from ~46.2% to ~28.2% (Figure 5A,B). This revealed that BSA@Au had desirable catalase-like activity to decompose the overexpressed H_2O_2 within the tumor region to produce O_2 for alleviating the hypoxic tumor microenvironment.

Subsequently, we investigated the in vivo PDT outcomes of BSA@Au. Mice bearing 4T1 tumor were divided into four groups with random (n = 5): (1) Control; (2) Laser; (3) BSA@Au; (4) BSA@Au + Laser. For group (4), 10 h after i.v. injection of BSA@Au, the mice were irradiated by 808 nm laser (0.3 W/cm^2, 30 min). Notably, all mice from groups (2)–(4) only received a single-dose injection and/or single-dose laser irradiation during treatment. Moreover, we monitored the temperature changes of irradiated regions during the treatment. No obvious thermal increase was observed (Figure S8), which suggested that only the PDT effect (without photothermal effect) was involved in the killing of cancer cells. The tumor volume variation and body weights were recorded every 2 days. Results showed that the tumor growth was significantly inhibited in the group (4) whereas the tumor volumes increased quickly in groups (1)–(3), indicating the superior antitumor efficiency of BSA@Au in combination with laser irradiation (Figures 6C and S9). Based on the efficient anti-cancer ability, mice from BSA@Au + Laser group survived for a much longer period (60 days) compared with the other three groups (Figure 5D).

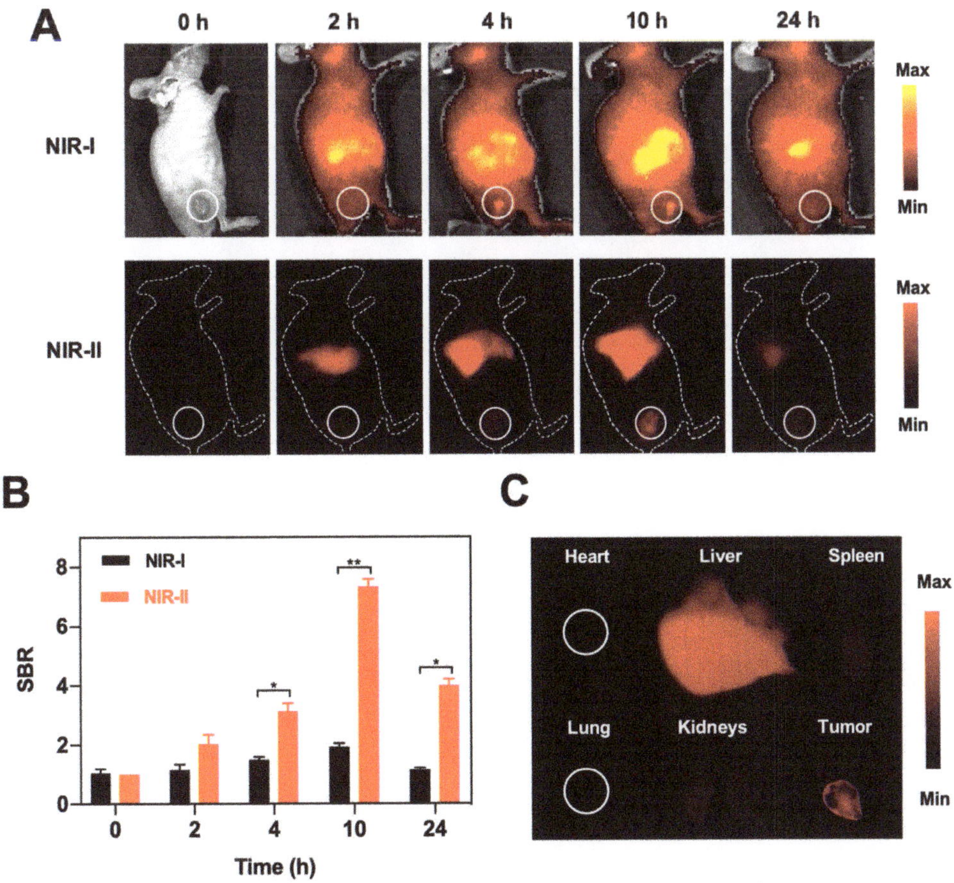

Figure 4. NIR-I and NIR-II FL imaging. (**A**) NIR-I and NIR-II FL images of 4T1 tumor-bearing mice at different time points (0, 2, 4, 10, and 24 h) post i.v. injection of BSA@Au. (**B**) SBR comparison between in vivo NIR-I FL and NIR-II FL images of 4T1 tumor-bearing mice. (**C**) The NIR-II FL images of main organs (heart, liver, spleen, lung, and kidneys) and tumor harvested from 4T1 tumor-bearing mice at 10 h post i.v. injection of BSA@Au. * $p < 0.05$, ** $p < 0.01$.

Meanwhile, tumor tissues from all groups were collected on the next day after various treatments for H&E staining and TUNEL assay. Obvious cell apoptosis and necrosis were found in BSA@Au + Laser group; in contrast, nearly no abnormality was detected in Control, Laser, and BSA@Au groups (Figure 5E). These results further showed that single-dose BSA@Au injection combined with single laser irradiation possessed great damage to cancer tissues. In addition, during the whole therapy process, no body fluctuation was recorded (Figure S10), indicating the high biosafety of BSA@Au combined with laser for PDT. The results together demonstrated that BSA@Au could remarkably modulate hypoxia and realize efficient in vivo anti-tumor outcomes under NIR-II FL imaging guidance.

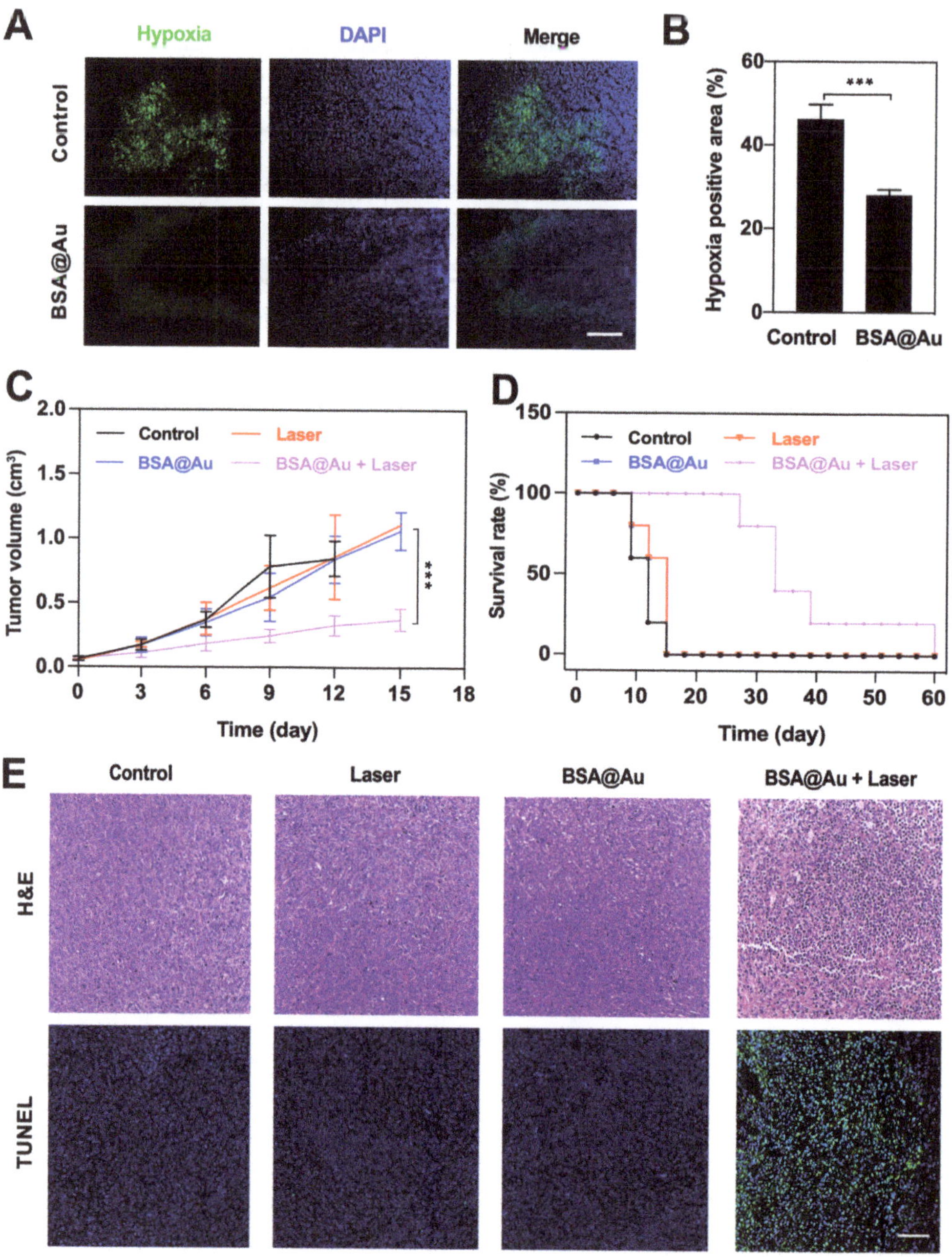

Figure 5. In vivo anti-tumor effect. (**A**) Representative immunofluorescence images of positive hypoxia area of tumor tissues after i.v. injection of PBS or BSA@Au. The scale bar was 100 μm. (**B**) The semi-quantification of positive hypoxia area shown in (**A**). (**C**) Tumor volumes variations over time after different treatments. (**D**) Survival rate of mice received different treatments. (**E**) H&E and TUNEL assay of tumor tissue sections from mice received different treatments. The scale bar was 100 μm. *** $p < 0.005$.

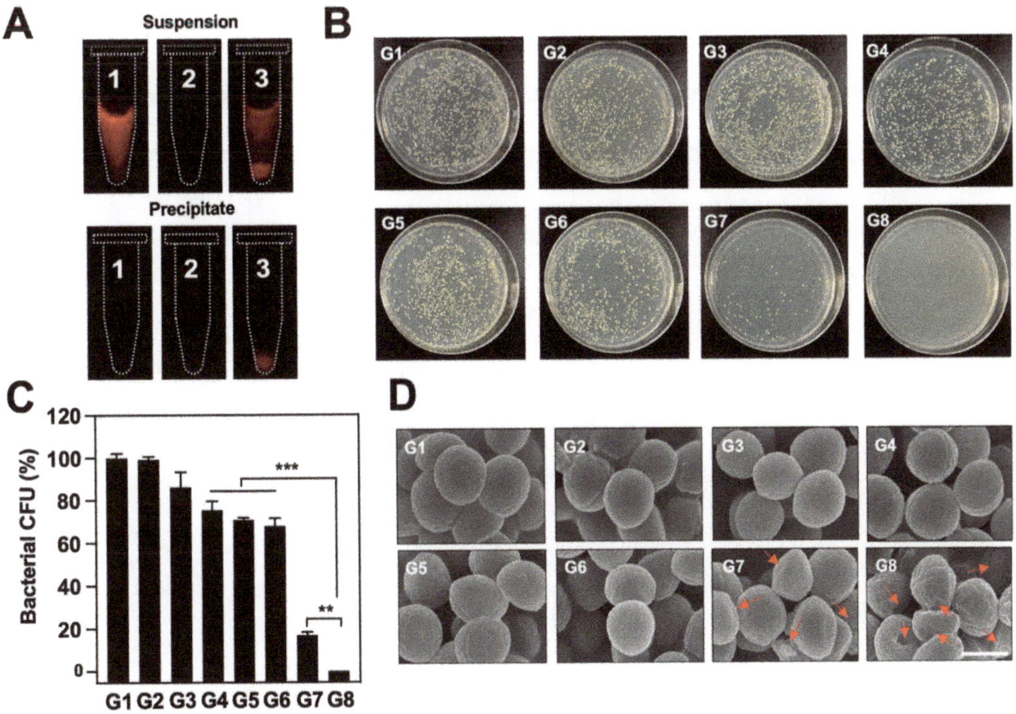

Figure 6. The anti-bacterial efficacy of BSA@Au. (**A**) Observation of BSA@Au binding with MRSA by NIR-II FL imaging. Tube 1, 2, and 3 contained the BSA@Au solution, MRSA solution, and the mixed suspension of BSA@Au and MRSA, respectively. (**B**) MRSA bacterial colonies test on LB agar plates after different treatments. (**C**) Quantification of survival rate of MRSA cells after different treatments. (**D**) Representative SEM images of MRSA exposed to different treatments. The red arrow indicates the crimple, hollow, and crack of MRSA. The scale bar was 1 µm. In (**C**,**D**), G1: PBS; G2: Laser; G3: BSA@Au; G4: H_2O_2; G5: BSA@Au + H_2O_2; G6: H_2O_2 + Laser; G7: BSA@Au + Laser; G8: BSA@Au + H_2O_2 + Laser. ** $p < 0.01$, *** $p < 0.005$.

3.6. *In Vitro Antibacterial Efficacy*

Bacterial infection have been considered as a key element of cancer management [42]. Given that the infection microenvironment is similar to tumor microenvionment, which features low O_2 and high expression of H_2O_2 [39,40,43], we investigated the anti-bacterial capability of BSA@Au by boosted PDT effect. MRSA, a known type of drug-resistant bacterium, was involved for further evaluation. Inspired by previous reports that human serum albumin has a unique affinity with MRSA [44], we investigated the binding capability of the BSA@Au towards MRSA. Tube 1 was set as the control group, only containing BSA@Au solution; in tubes 2 and 3, MRSA was incubated with PBS or BSA@Au for 4 h, respectively. The three samples were imaged under the NIR-II FL imaging system after centrifugation (5000 rpm, 5 min). The intense signal in tube 3 displayed that BSA@Au could tightly bind with MRSA. Notably, the FL signal of the suspensions was from free BSA@Au solutions, and the FL signal of the precipitates was from the BSA@Au binding with MRSA because BSA@Au could not be spun down at a low centrifugation speed unless they were bound with MRSA (Figure 6A). Then, we evaluated the in vitro anti-bacterial efficacy of BSA@Au in combination with 808 nm laser irradiation. MRSA suspensions were divided into eight groups: (1) Control; (2) Laser; (3) BSA@Au; (4) H_2O_2; (5) H_2O_2 + Laser; (6) BSA@Au + H_2O_2; (7) BSA@Au + Laser; and (8) BSA@Au + H_2O_2 + Laser. Compared with other treatments, the group (7) killing 70% of the bacteria because of the good toxic $^1O_2$ generation ability of BSA@Au (Figure 6B, C). Considering the hypoxic microenvironment within infectious tissues, which lacks O_2 and overexpresses H_2O_2, we added H_2O_2 (200 μM) to simulate the bacteria-infected microenvironment. Excitingly, no live bacteria colonies (~0%) was observed in group (8), revealing enhanced antibacterial effect after O_2 enrichment within infectious microenvironment. More visibly, as shown in the SEM images, the MRSA cells remained intact with grape-like morphology in groups (1)–(6). However, the MRSA bacteria were distorted and denatured in group (7)–(8). Moreover, the obvious shrinking and cracking of bacteria in group (8) were recorded (Figure 6D) due to $^1O_2$ generated from BSA@Au and H_2O_2 under 808 nm laser irradiation.

3.7. *In Vivo Antibacterial Efficacy*

To further verify the in vivo anti-bacterial effect, we established a MRSA-infected wound model on Balb/C mice, and then recorded the wound closure and bacterial residue after different treatments. At first, we observed the in vivo NIR-II FL imaging of MRSA-infected wound by local injection of BSA@Au. The bright FL signal was collected, and the SBR was determined to be about 66.8 (Figure S11), demonstrating the potential of BSA@Au as an NIR-II FL probe for in vivo imaging. The wounds treated by BSA@Au + Laser showed a significantly faster healing speed than those of the group (1)–(3), and almost completely healed within 12 days, whereas the wound areas of control group were still about 85.5% of the original wounds (Figure 7A,B). In addition, at the end of the treatment period, we collected the wound tissues from healthy mouse, control group, and BSA@Au + Laser group for continuing cultivation in LB culture medium. As expected, a large number of live MRSA cells were observed in the control group, whereas no bacteria remained in the healthy mouse and BSA@Au + Laser groups because of the excellent anti-bacterial effect of the treatment (Figure 7C). The skin tissues were also used for histology analysis. H&E staining assay displayed that obvious effusion, loose dermis, and inflammatory cells were found in the infected tissue, whereas less necrosis, fewer granulated cells, and more fibrous tissues were observed in the cured wound (Figure 7D). During the whole treatment, no obvious changes in body weight were observed, suggesting PDT for MRSA therapy was safe and biocompatible (Figure S12). Moreover, during the PDT process, the skin temperature did not increase (Figure S13), which excluded the photothermal effect. MRSA-infection as a drug-resistant infectious disease showed little response to clinical antibiotics. However, PDT, as a novel treatment strategy for bacterial infections, could achieve the superb anti-MRSA effect and avoid treatment resistance without using antibiotics. Hence, in the future, BSA@Au might be utilized to detect and treat the deep-seated infections in NIR-II phototheranostic platform.

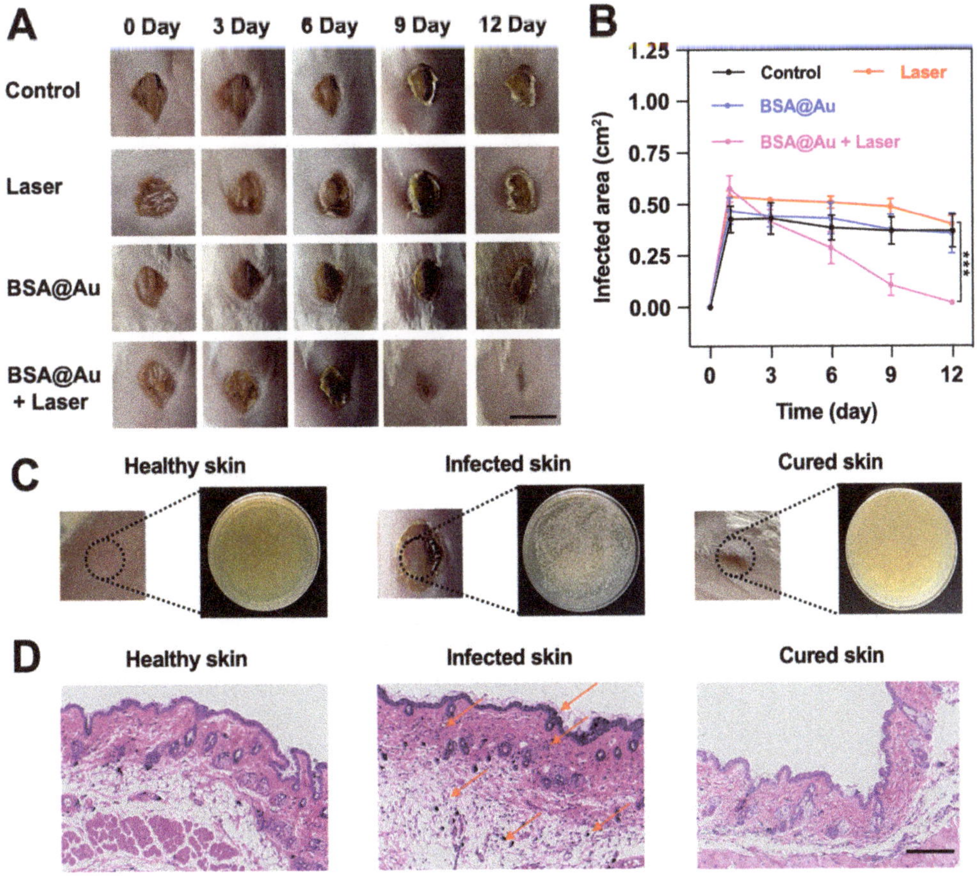

Figure 7. In vivo anti-bacterial effect. (**A**) Photographs of the infectious wounds after in vivo treatments. The scale bar was 1 cm. (**B**) Quantitative analysis of wound areas over time. (**C**) The MRSA colonies on LB-agar plates from the healthy skin tissue, the infected skin tissue, and the cured skin tissue, respectively. (**D**) H&E staining analysis of wounds from the healthy skin tissue, the infected skin tissue, and the cured skin tissue, respectively. The scale bar was 200 μm, and the red arrows represented tissue inflammation. *** $p < 0.005$.

4. Conclusions

In summary, we provided proof-of-principle evidence for the application of an O_2 self-sufficient photosensitizer for boosted cancer/infection PDT under the guidance of NIR-II FL imaging. In this system, BSA@Au was endowed with strong NIR-II FL emission, superb fluorescent QY (~3.5%), strong photobleaching resistance, excellent catalase-like activity, as well as good biocompatibility to realize imaging-guided PDT for cancer and bacteria-infected lesions. With bright NIR-II imaging performance, BSA@Au visualized the tumor location with high SBR (~7.3) and biodistribution, as well as the metabolism pathway itself, enabling high-performance imaging guided PDT. Moreover, BSA@Au served as a catalase mimic decomposing H_2O_2 to O_2 for alleviating a hypoxic microenvironment to augment O_2-dependent PDT. With single 808 nm laser irradiation, it achieved efficient tumor growth inhibition and MRSA-infection curation. Therefore, the developed BSA@Au might fulfill the unmet need for phototheranostics of cancer and bacterial infections.

Supplementary Materials: The following supporting information can be downloaded at: https://www.mdpi.com/article/10.3390/pharmaceutics14081645/s1, Table S1: The preparation optimization of BSA@Au; Table S2: Comparison of kinetic parameters among nanozymes; Table S3: Blood sample examination of mice; Figure S1: Optimization of BSA@Au; Figure S2: Fluorescent QY of BSA@Au; Figure S3: Penetration depth of NIR-I and NIR-II FL imaging; Figure S4: Photostability of BSA@Au; Figure S5: Catalase-like activity of BSA@Au; Figure S6: Biosafety of BSA@Au; Figure S7: Ex vivo FL images; Figure S8: Temperature of mice during cancer PDT; Figure S9: Tumor inhibition of mice received different treatments; Figure S10: Body weights of mice during cancer PDT; Figure S11: In vivo NIR-II FL imaging for MRSA-infected mice; Figure S12: Body weights of mice during MRSA PDT; Figure S13: Temperature of mice during MRSA PDT. References [45–52] are cited in the Supplementary Materials.

Author Contributions: Conceptualization, Z.S. and Y.L.; data curation, D.H.; formal analysis, Q.D.; funding acquisition, H.M., W.L., L.Z., Z.S. and Y.L.; investigation, Z.Y.; methodology, Q.D., S.Z. and N.S.; project administration, Z.S. and Y.L.; resources, Z.Y., Z.S. and Y.L.; software, Q.D.; supervision, Z.S. and Y.L.; validation, S.Z., N.S. and Y.L.; visualization, Q.D.; writing—original draft, Q.D.; writing—review & editing, Z.S. and Y.L. All authors have read and agreed to the published version of the manuscript.

Funding: This work is supported by the Natural Science Foundation of China (92159304, 82102081, 82171958, 81901812, 81971638, 82027803, 81927807, 82071949, 81871371, 81801720, 81701711, 81671709, 81827807), Science and Technology Planning Project of Guangzhou City (201804010106), and President Foundation of Nanfang Hospital, Southern Medical University (2020A001), CAS Key Laboratory of Health Informatics (2011DP173015), the Science and Technology Key Project of Shenzhen (JCYJ20190812163614809, JCYJ20200109114612308, JCYJ20210324120011030), Guangdong Basic and Applied Basic Research Fund (2020A1515110011, 2020A1515010395, 2022A1515010384), Key Laboratory for Magnetic Resonance and Multimodality Imaging of Guangdong Province (2020B1212060051), and the Key Technology and Equipment R&D Program of Major Science and Technology Infrastructure of Shenzhen (202100102, 202100104).

Institutional Review Board Statement: The animal study protocol was approved by the Institutional Animal Care and Use of Shenzhen Institutes of Advanced Technology (protocol code: SIAT-IACUC-20190304-YGS-YXZX-SZH-01, and date of approval: 20190304).

Informed Consent Statement: Not applicable.

Data Availability Statement: Data supporting reported results beyond what is reported in this manuscript are available upon reasonable request from the corresponding authors.

Acknowledgments: The authors appreciate the contribution of Zihan Mei (Shanghai Ruijin Hospital) to provide the method to establish MRSA-infected mouse model.

Conflicts of Interest: The authors declare no conflict of interest.

References

1. Hong, G.; Antaris, A.L.; Dai, H. Near-Infrared Fluorophores for Biomedical Imaging. *Nat. Biomed. Eng.* **2017**, *1*, 10. [CrossRef]
2. Zhu, S.; Tian, R.; Antaris, A.L.; Chen, X.; Dai, H. Near-Infrared-II Molecular Dyes for Cancer Imaging and Surgery. *Adv. Mater.* **2019**, *31*, 1900321. [CrossRef] [PubMed]
3. Lin, H.; Lin, Z.; Zheng, K.; Wang, C.; Lin, L.; Chen, J.; Song, J. Near-Infrared-II Nanomaterials for Fluorescence Imaging and Photodynamic Therapy. *Adv. Opti. Mater.* **2021**, *9*, 2002177. [CrossRef]
4. Dai, H.; Shen, Q.; Shao, J.; Wang, W.; Gao, F.; Dong, X. Small Molecular NIR-II Fluorophores for Cancer Phototheranostics. *Innovation* **2021**, *2*, 100082. [CrossRef] [PubMed]
5. Turan, I.S.; Yildiz, D.; Turksoy, A.; Gunaydin, G.; Akkaya, E.U. A Bifunctional Photosensitizer for Enhanced Fractional Photodynamic Therapy: Singlet Oxygen Generation in the Presence and Absence of Light. *Angew. Chem.* **2016**, *128*, 2925–2928. [CrossRef]
6. Zheng, Z.; Zhang, T.; Liu, H.; Chen, Y.; Kwok, R.T.K.; Ma, C.; Zhang, P.; Sung, H.H.Y.; Williams, I.D.; Lam, J.W.Y.; et al. Bright Near-Infrared Aggregation-Induced Emission Luminogens with Strong Two-Photon Absorption, Excellent Organelle Specificity, and Efficient Photodynamic Therapy Potential. *ACS Nano* **2018**, *12*, 8145–8159. [CrossRef]
7. Thomas, A.P.; Palanikumar, L.; Jeena, M.T.; Kim, K.; Ryu, J.-H. Cancer-Mitochondria-Targeted Photodynamic Therapy with Supramolecular Assembly of HA and a Water Soluble NIR Cyanine Dye. *Chem. Sci.* **2017**, *8*, 8351–8356. [CrossRef]

8. Schmitt, J.; Heitz, V.; Sour, A.; Bolze, F.; Ftouni, H.; Nicoud, J.; Flamigni, L.; Ventura, B. Diketopyrrolopyrrole-Porphyrin Conjugates with High Two-Photon Absorption and Singlet Oxygen Generation for Two-Photon Photodynamic Therapy. *Angew. Chem. Int. Ed.* **2015**, *54*, 169–173. [CrossRef]
9. Ge, J.; Lan, M.; Zhou, B.; Liu, W.; Guo, L.; Wang, H.; Jia, Q.; Niu, G.; Huang, X.; Zhou, H.; et al. A Graphene Quantum Dot Photodynamic Therapy Agent with High Singlet Oxygen Generation. *Nat. Commun.* **2014**, *5*, 4596. [CrossRef]
10. Mou, J.; Lin, T.; Huang, F.; Chen, H.; Shi, J. Black Titania-Based Theranostic Nanoplatform for Single NIR Laser Induced Dual-Modal Imaging-Guided PTT/PDT. *Biomaterials* **2016**, *84*, 13–24. [CrossRef]
11. Wang, H.; Yang, X.; Shao, W.; Chen, S.; Xie, J.; Zhang, X.; Wang, J.; Xie, Y. Ultrathin Black Phosphorus Nanosheets for Efficient Singlet Oxygen Generation. *J. Am. Chem. Soc.* **2015**, *137*, 11376–11382. [CrossRef] [PubMed]
12. Feng, G.; Fang, Y.; Liu, J.; Geng, J.; Ding, D.; Liu, B. Multifunctional Conjugated Polymer Nanoparticles for Image-Guided Photodynamic and Photothermal Therapy. *Small* **2017**, *13*, 1602807. [CrossRef] [PubMed]
13. Han, H.-H.; Wang, C.-Z.; Zang, Y.; Li, J.; James, T.D.; He, X.-P. Supramolecular Core-Glycoshell Polythiophene Nanodots for Targeted Imaging and Photodynamic Therapy. *Chem. Commun.* **2017**, *53*, 9793–9796. [CrossRef] [PubMed]
14. Vankayala, R.; Huang, Y.-K.; Kalluru, P.; Chiang, C.-S.; Hwang, K.C. First Demonstration of Gold Nanorods-Mediated Photodynamic Therapeutic Destruction of Tumors via Near Infra-Red Light Activation. *Small* **2014**, *10*, 1612–1622. [CrossRef]
15. Huang, H.; Yu, B.; Zhang, P.; Huang, J.; Chen, Y.; Gasser, G.; Ji, L.; Chao, H. Highly Charged Ruthenium (II) Polypyridyl Complexes as Lysosome-Localized Photosensitizers for Two-Photon Photodynamic Therapy. *Angew. Chem. Int. Ed.* **2015**, *54*, 14049–14052. [CrossRef]
16. Zhao, J.; Zhong, D.; Zhou, S. NIR-I-to-NIR-II Fluorescent Nanomaterials for Biomedical Imaging and Cancer Therapy. *J. Mater. Chem. B* **2018**, *6*, 349–365. [CrossRef]
17. Zhang, Z.; Xu, W.; Kang, M.; Wen, H.; Guo, H.; Zhang, P.; Xi, L.; Li, K.; Wang, L.; Wang, D.; et al. An All-Round Athlete on the Track of Phototheranostics: Subtly Regulating the Balance between Radiative and Nonradiative Decays for Multimodal Imaging-Guided Synergistic Therapy. *Adv. Mater.* **2020**, *32*, 2003210. [CrossRef]
18. Tang, C.; Song, C.; Wei, Z.; Liang, C.; Ran, J.; Cai, Y.; Dong, X.; Han, W. Polycyclic Naphthalenediimide-Based Nanoparticles for NIR-II Fluorescence Imaging Guided Phototherapy. *Sci. China Chem.* **2020**, *63*, 946–956. [CrossRef]
19. Bian, H.; Ma, D.; Zhang, X.; Xin, K.; Yang, Y.; Peng, X.; Xiao, Y. Tailored Engineering of Novel Xanthonium Polymethine Dyes for Synergetic PDT and PTT Triggered by 1064 nm Laser toward Deep-Seated Tumors. *Small* **2021**, *17*, 2100398. [CrossRef]
20. Jin, R. Quantum Sized, Thiolate-Protected Gold Nanoclusters. *Nanoscale* **2010**, *2*, 343–362. [CrossRef]
21. Song, X.; Zhu, W.; Ge, X.; Li, R.; Li, S.; Chen, X.; Song, J.; Xie, J.; Chen, X.; Yang, H. A New Class of NIR-II Gold Nanocluster-Based Protein Biolabels for in vivo Tumor-Targeted Imaging. *Angew. Chem. Int. Ed.* **2021**, *60*, 1306–1312. [CrossRef] [PubMed]
22. Liu, H.; Hong, G.; Luo, Z.; Chen, J.; Chang, J.; Gong, M.; He, H.; Yang, J.; Yuan, X.; Li, L.; et al. Atomic-Precision Gold Clusters for NIR-II Imaging. *Adv. Mater.* **2019**, *31*, 1901015. [CrossRef] [PubMed]
23. Li, D.; Liu, Q.; Qi, Q.; Shi, H.; Hsu, E.; Chen, W.; Yuan, W.; Wu, Y.; Lin, S.; Zeng, Y.; et al. Gold Nanoclusters for NIR-II Fluorescence Imaging of Bones. *Small* **2020**, *16*, 2003851. [CrossRef]
24. Yang, Y.; Yu, Y.; Chen, H.; Meng, X.; Ma, W.; Yu, M.; Li, Z.; Li, C.; Liu, H.; Zhang, X.; et al. Illuminating Platinum Transportation While Maximizing Therapeutic Efficacy by Gold Nanoclusters via Simultaneous Near-Infrared-I/II Imaging and Glutathione Scavenging. *ACS Nano* **2020**, *14*, 13536–13547. [CrossRef] [PubMed]
25. Wang, W.; Kong, Y.; Jiang, J.; Xie, Q.; Huang, Y.; Li, G.; Wu, D.; Zheng, H.; Gao, M.; Xu, S.; et al. Engineering the Protein Corona Structure on Gold Nanoclusters Enables Red-Shifted Emissions in the Second Near-infrared Window for Gastrointestinal Imaging. *Angew. Chem. Int. Ed.* **2020**, *59*, 22431–22435. [CrossRef] [PubMed]
26. Lillo, C.R.; Calienni, M.N.; Rivas Aiello, B.; Prieto, M.J.; Rodriguez Sartori, D.; Tuninetti, J.; Toledo, P.; del Valle Alonso, S.; Moya, S.; Gonzalez, M.C.; et al. BSA-Capped Gold Nanoclusters as Potential Theragnostic for Skin Diseases: Photoactivation, Skin Penetration, in Vitro, and in vivo Toxicity. *Mat. Sci. Eng. C* **2020**, *112*, 110891. [CrossRef]
27. Dan, Q.; Hu, D.; Ge, Y.; Zhang, S.; Li, S.; Gao, D.; Luo, W.; Ma, T.; Liu, X.; Zheng, H.; et al. Ultrasmall Theranostic Nanozymes to Modulate Tumor Hypoxia for Augmenting Photodynamic Therapy and Radiotherapy. *Biomater. Sci.* **2020**, *8*, 973–987. [CrossRef]
28. Xie, J.; Zheng, Y.; Ying, J.Y. Protein-Directed Synthesis of Highly Fluorescent Gold Nanoclusters. *J. Am. Chem. Soc.* **2009**, *131*, 888–889. [CrossRef]
29. Du, B.; Yu, M.; Zheng, J. Transport and Interactions of Nanoparticles in the Kidneys. *Nat. Rev. Mater.* **2018**, *3*, 358–374. [CrossRef]
30. Loynachan, C.N.; Soleimany, A.P.; Dudani, J.S.; Lin, Y.; Najer, A.; Bekdemir, A.; Chen, Q.; Bhatia, S.N.; Stevens, M.M. Renal Clearable Catalytic Gold Nanoclusters for in vivo Disease Monitoring. *Nat. Nanotechnol.* **2019**, *14*, 883–890. [CrossRef]
31. Chen, D.; Li, J. Ultrasmall Au Nanoclusters for Bioanalytical and Biomedical Applications: The Undisclosed and Neglected Roles of Ligands in Determining the Nanoclusters' Catalytic Activities. *Nanoscale Horiz.* **2020**, *5*, 1355–1367. [CrossRef] [PubMed]
32. Shang, Y.; Min, C.; Hu, J.; Wang, T.; Liu, H.; Hu, Y. Synthesis of Gold Nanoparticles by Reduction of $HAuCl_4$ under UV Irradiation. *Solid State Sci.* **2013**, *15*, 17–23. [CrossRef]
33. Lin, C.; Tao, K.; Hua, D.; Ma, Z.; Zhou, S. Size Effect of Gold Nanoparticles in Catalytic Reduction of P-Nitrophenol with $NaBH_4$. *Molecules* **2013**, *18*, 12609–12620. [CrossRef] [PubMed]
34. Zhu, M.; Aikens, C.M.; Hollander, F.J.; Schatz, G.C.; Jin, R. Correlating the Crystal Structure of a Thiol-Protected Au_{25} Cluster and Optical Properties. *J. Am. Chem. Soc.* **2008**, *130*, 5883–5885. [CrossRef]

35. Poderys, V.; Jarockyte, G.; Bagdonas, S.; Karabanovas, V.; Rotomskis, R. Protein-Stabilized Gold Nanoclusters for PDT. ROS and Singlet Oxygen Generation. *J. Photochem. Photobiol. B Biol. B* **2020**, *204*, 111802. [CrossRef]
36. Pang, Z.; Li, Q.; Jia, Y.; Yan, W.; Qi, J.; Guo, Y.; Hu, F.; Zhou, D.; Jiang, X. Controlling the Pyridinium-Zwitterionic Ligand Ratio on Atomically Precise Gold Nanoclusters Allowing for Eradicating Gram-Positive Drug-Resistant Bacteria and Retaining Biocompatibility. *Chem. Sci.* **2021**, *12*, 14871–14882. [CrossRef]
37. Li, S.; Ma, Q.; Wang, C.; Yang, K.; Hong, Z.; Chen, Q.; Song, J.; Song, X.; Yang, H. Near-Infrared II Gold Nanocluster Assemblies with Improved Luminescence and Biofate for In Vivo Ratiometric Imaging of H_2S. *Anal. Chem.* **2022**, *94*, 2641–2647. [CrossRef]
38. Kong, Y.; Santos-Carballal, D.; Martin, D.; Sergeeva, N.N.; Wang, W.; Liu, G.; Johnson, B.; Bhayana, B.; Lin, Z.; Wang, Y.; et al. A NIR-II-Emitting Gold Nanocluster-Based Drug Delivery System for Smartphone-Triggered Photodynamic Theranostics with Rapid Body Clearance. *Mater. Today* **2021**, *51*, 96–107. [CrossRef]
39. Hockel, M.; Vaupel, P. Tumor Hypoxia: Definitions and Current Clinical, Biologic, and Molecular Aspects. *JNCI J. Natl. Cancer Inst.* **2001**, *93*, 266–276. [CrossRef]
40. Devraj, G.; Beerlage, C.; Brüne, B.; Kempf, V.A.J. Hypoxia and HIF-1 Activation in Bacterial Infections. *Microbes Infect.* **2017**, *19*, 144–156. [CrossRef]
41. Johnson, K.A.; Goody, R.S. The Original Michaelis Constant: Translation of the 1913 Michaelis–Menten Paper. *Biochemistry* **2011**, *50*, 8264–8269. [CrossRef] [PubMed]
42. Elsland, D.; Neefjes, J. Bacterial Infections and Cancer. *EMBO Rep.* **2018**, *19*. [CrossRef] [PubMed]
43. Schaffer, K.; Taylor, C.T. The Impact of Hypoxia on Bacterial Infection. *FEBS J.* **2015**, *282*, 2260–2266. [CrossRef] [PubMed]
44. Chan, P.-H.; Chen, Y.-C. Human Serum Albumin Stabilized Gold Nanoclusters as Selective Luminescent Probes for Staphylococcus Aureus and Methicillin-Resistant Staphylococcus Aureus. *Anal. Chem.* **2012**, *84*, 8952–8956. [CrossRef]
45. Murphy, J.E.; Beard, M.C.; Norman, A.G.; Ahrenkiel, S.P.; Johnson, J.C.; Yu, P.; Mićić, O.I.; Ellingson, R.J.; Nozik, A.J. PbTe Colloidal Nanocrystals: Synthesis, Characterization, and Multiple Exciton Generation. *J. Am. Chem. Soc.* **2006**, *128*, 3241–3247. [CrossRef]
46. Wang, Q.; Xu, J.; Geng, R.; Cai, J.; Li, J.; Xie, C.; Tang, W.; Shen, Q.; Huang, W.; Fan, Q. High Performance One-for-All Phototheranostics: NIR-II Fluorescence Imaging Guided Mitochondria-Targeting Phototherapy with a Single-Dose Injection and 808 nm Laser Irradiation. *Biomaterials* **2020**, *231*, 119671. [CrossRef]
47. Mu, J.; Zhang, L.; Zhao, M.; Wang, Y. Catalase Mimic Property of Co_3O_4 Nanomaterials with Different Morphology and Its Application as a Calcium Sensor. *ACS Appl. Mater. Interfaces* **2014**, *6*, 7090–7098. [CrossRef]
48. Fan, J.; Yin, J.-J.; Ning, B.; Wu, X.; Hu, Y.; Ferrari, M.; Anderson, G.J.; Wei, J.; Zhao, Y.; Nie, G. Direct Evidence for Catalase and Peroxidase Activities of Ferritin–Platinum Nanoparticles. *Biomaterials* **2011**, *32*, 1611–1618. [CrossRef]
49. Wang, X.; Zhang, Y.; Li, T.; Tian, W.; Zhang, Q.; Cheng, Y. Generation 9 Polyamidoamine Dendrimer Encapsulated Platinum Nanoparticle Mimics Catalase Size, Shape, and Catalytic Activity. *Langmuir* **2013**, *29*, 5262–5270. [CrossRef]
50. Deng, H.; Shen, W.; Peng, Y.; Chen, X.; Yi, G.; Gao, Z. Nanoparticulate Peroxidase/Catalase Mimetic and Its Application. *Chem. Eur. J.* **2012**, *18*, 8906–8911. [CrossRef]
51. Su, H.; Liu, D.-D.; Zhao, M.; Hu, W.-L.; Xue, S.-S.; Cao, Q.; Le, X.-Y.; Ji, L.-N.; Mao, Z.-W. Dual-Enzyme Characteristics of Polyvinylpyrrolidone-Capped Iridium Nanoparticles and Their Cellular Protective Effect against H_2O_2-Induced Oxidative Damage. *ACS Appl. Mater. Interfaces* **2015**, *7*, 8233–8242. [CrossRef] [PubMed]
52. Liu, C.-P.; Wu, T.-H.; Liu, C.-Y.; Chen, K.-C.; Chen, Y.-X.; Chen, G.-S.; Lin, S.-Y. Self-Supplying O_2 through the Catalase-Like Activity of Gold Nanoclusters for Photodynamic Therapy against Hypoxic Cancer Cells. *Small* **2017**, *13*, 1700278. [CrossRef] [PubMed]

pharmaceutics

MDPI

Article

Saporin Toxin Delivered by Engineered Colloidal Nanoparticles Is Strongly Effective against Cancer Cells

Lucia Salvioni †, Filippo Testa †, Linda Barbieri, Marco Giustra, Jessica Armida Bertolini, Giulia Tomaino, Paolo Tortora, Davide Prosperi and Miriam Colombo *

Department of Biotechnology and Bioscience, University of Milano-Bicocca, Piazza della Scienza 2, 20126 Milano, Italy; lucia.salvioni@unimib.it (L.S.); f.testa@campus.unimib.it (F.T.); l.barbieri17@campus.unimib.it (L.B.); m.giustra2@campus.unimib.it (M.G.); jessica.bertolini@unimib.it (J.A.B.); g.tomaino3@campus.unimib.it (G.T.); paolo.tortora@unimib.it (P.T.); davide.prosperi@unimib.it (D.P.)

* Correspondence: miriam.colombo@unimib.it; Tel.: +39-02-6448-3388

† These authors contributed equally to this work.

Abstract: Ribosome-inactivating proteins, including Saporin toxin, have found application in the search for innovative alternative cancer therapies to conventional chemo- and radiotherapy. Saporin's main mechanism of action involves the inhibition of cytoplasmic protein synthesis. Its strong theoretical efficacy is counterbalanced by negligible cell uptake and diffusion into the cytosol. In this work, we demonstrate that by immobilizing Saporin on iron oxide nanoparticles coated with an amphiphilic polymer, which promotes nanoconjugate endosomal escape, a strong cytotoxic effect mediated by ribosomal functional inactivation can be achieved. Cancer cell death was mediated by apoptosis dependent on nanoparticle concentration but independent of surface ligand density. The cytotoxic activity of Saporin-conjugated colloidal nanoparticles proved to be selective against three different cancer cell lines in comparison with healthy fibroblasts.

Keywords: ribosome-inactivating proteins (RIPs); toxins; Saporin; colloidal nanoparticles; cancer therapy; breast cancer

Citation: Salvioni, L.; Testa, F.; Barbieri, L.; Giustra, M.; Bertolini, J.A.; Tomaino, G.; Tortora, P.; Prosperi, D.; Colombo, M. Saporin Toxin Delivered by Engineered Colloidal Nanoparticles Is Strongly Effective against Cancer Cells. *Pharmaceutics* **2022**, *14*, 1517. https://doi.org/10.3390/pharmaceutics14071517

Academic Editors: Manivasagan Panchanathan and Eue-Soon Jang

Received: 4 July 2022
Accepted: 18 July 2022
Published: 21 July 2022

1. Introduction

In the research into new drugs for tumor treatment, ribosome-inactivating proteins (RIPs) have recently emerged as promising agents of interest because of their potent cell-cycle-independent activity against cancer cells. RIPs are especially advantageous due to their enhanced thermostability, structural solidity against denaturation or alteration caused by proteolytic enzymes, and ease of handling [1,2].

RIP activity is traditionally attributed to rRNA N-glycosylase; this specific depurination leads to a permanent inactivation of the ribosome, irreversibly blocking protein synthesis and inducing DNA damage. Consequently, all RIPs cause cell death by apoptosis at very low dosages (less than a few milligrams/kg) [3,4].

Among the RIPs toxin family, Saporin (Sap, from the plant *Saponaria officinalis*) has attracted increasing attention in the last two decades thanks to its low systemic toxicity, due to its low propensity to enter cells in the absence of a specific carrier [5].

In vivo results have demonstrated the efficacy of immunotoxins (ITs) obtained by the conjugation of Saporin to monoclonal antibodies, which represents the ideal toxic moiety in cancer patients who are refractory to traditional therapies, including surgery, radiation therapy and chemotherapy [6,7].

Nevertheless, despite all of these significant benefits, the application of Saporin technology in current clinical practice is restricted to hematological malignancies and limited by several hurdles that need to be overcome related to the problems of immunogenicity, vascular leak syndrome due to endothelial toxicity and inefficient intracellular release [7–9]. The latter represents the main limitation on Saporin antitumor activity due to the very low protein uptake efficiency of cancer cells and negligible endosomal escape once internalized [9,10].

One possible approach to overcome this issue is to exploit Saporin conjugation to a suitable nanocarrier that improves its intracellular release [11]. In our ongoing effort to study the impact of surface coating of colloidal nanoparticles to achieve high uptake efficiency and optimal endosomal escape, we have developed biocompatible iron oxide nanoparticles coated with a pH-sensitive amphiphilic polymer [12,13]. This multidentate polymer, termed PMDA, obtained by modification of poly(isobutylene-alt-maleic anhydride) with 75% dodecylamines, demonstrated excellent endosomal escape capability upon endosome maturation [14]. PMDA-coated iron oxide nanoparticles, termed MYTS, proved to be non-toxic both in vitro and in vivo [12,15] and very easy to functionalize with different bioactive molecules, including proteins [16–19].

In the present work, we functionalized MYTS with Sap for the purpose of improving the intracellular release of toxins in cancer cells. Sap was reversibly conjugated to PMDA through a disulphide bond, sensitive to the cytosolic reducing environment. We used SK-BR-3 as a breast cancer cell model and NIH-3T3 fibroblasts as healthy control cells.

2. Materials and Methods

2.1. *Synthesis of MYTS-Sap-1, MYTS-Sap-2 and MYTS-Sap-3*

2.1.1. Chemicals and Materials

Twelve-nanometer iron oxide nanoparticles (NPs) were purchased from HiQ-Nano (Arnesano, Italy); 2,2′-(Ethylenedioxy)bis(ethylamine) (EDBE), Boric acid, $CHCl_3$, Dodecylamine (99%), EtOH, Poly(isobutylene-alt-maleic anhydride) (PMA) Mw ~6000 and N-(3-Dimethylaminopropyl)-N′-ethylcarbodiimide hydrochloride (EDC) were purchased from Sigma-Aldrich (Saint Louis, MO, USA); and sulfosuccinimidyl-6-(3′-(2-pyridyldithio)propionamido)hexanoate (sulfo-LC-SPDP) was purchased from CovaChem (Loves Park, IL, USA). Water used in all procedures was purified by passing through a MilliQ Millipore (Burlington, MA, USA) system.

2.1.2. Phase Transfer of Iron Oxide Cores

Toluene was removed from the iron oxide NPs using a Laborota 4000 efficient Rotary Evaporator (Heidolph Instrument, Schwabach, Germany) and NPs were resuspended in 2 mL $CHCl_3$ and quantified by UV–Vis spectroscopy, as previously described [20]. Polymer coating and phase transfer of iron oxide cores was performed using PMDA at a polymer/NP ratio of 150/nm^2. The synthesis of PMDA polymers and the ratio used were optimized in previous publications [21,22]. Specifically, for 2 mL of NPs 0.037 M, 347.2 µL of 0.5 M PMDA were added. The solvent was evaporated under reduced pressure and the obtained NPs were dispersed by sonication in aqueous 0.51 M Saline Borate Buffer (SBB), pH 9.0, for 20 min. The solution was then diluted in sterile MilliQ and washed by centrifugal filtration using an Amicon® Ultra centrifugal filter (100 kDa) (Merck Millipore, Bulington, MA, USA). The PMDA-coated NP (MYTS-PMDA) content was quantified as described above. All the synthetic procedures reported here and in the following paragraphs were conducted under sterile conditions.

2.1.3. EDBE Functionalization of MYTS-PMDA

A volume of 8 µL 0.1 M EDBE and a volume of 6 µL 0.05 M EDC were added to 10 mg of MYTS-PMDA (10 mg/mL). The mix was kept under stirring for 2 h at RT. Byproducts and the unconjugated EDBE were removed with an Amicon® Ultra centrifugal filter (50 kDa) using sterile MilliQ water. The final concentration of EDBE-conjugated MYTS (MYTS-EDBE) was quantified as described above.

2.1.4. Sulfo-LC-SPDP Functionalization of MYTS-EDBE

A volume of 506.6 µL 0.019 M sulfo-LC-SPDP was added to 10 mg of MYTS-EDBE (10 mg/mL). The mix was kept under stirring overnight at 4 °C. Byproducts and the unconjugated sulfo-LC-SPDP were removed with an Amicon® Ultra centrifugal filter (50 kDa) using sterile MilliQ water. The final concentration of sulfo-LC-SPDP-conjugated MYTS (MYTS-SPDP) was quantified as described above.

2.1.5. Saporin Conjugation

To achieve a specific conjugation to MYTS-SPDP through the formation of a disulfide bond, a recombinant version of Sap was produced, with the addition of a unique Cys residue located at the C-terminus of the protein (C-Sap; details about protein production are reported in the Supplementary Materials). However, the Cys residue could lead to the dimerization of the protein, making it unavailable for conjugation to the -SH group of MYTS-SPDP. To obtain monomeric Saporin with a free -SH group available for conjugation to MYTS-SPDP, the protein was activated with 1 mM EDTA and 50 mM DTT for 30 min at RT. The excess of DTT and EDTA were removed with a Zeba™ Spin desalting column (7 kDa) (Thermo Fisher Scientific, Waltham, MA, USA), using 50 mM potassium phosphate, pH 7.2, and 150 mM NaCl as final buffer.

MYTS-Sap-1, MYTS-Sap-2 and MYTS-Sap-3 were synthetized, using, respectively, 50, 100 and 200 μg of reacted Saporin per mg of MYTS-SPDP. The mixtures were kept under stirring for 18 h at 4 °C. Unreacted protein and byproducts were removed with an Amicon® Ultra centrifugal filter (100 kDa) using 50 mM potassium phosphate, pH 7.2, and 150 mM NaCl as washing buffer.

2.1.6. Saporin-Alexa Fluor™-647 Labelling

A Zeba™ Spin desalting column (7 kDa) was used to exchange Saporin buffer to 0.09 M $NaHCO_3$. Then, a volume of 11.65 μL Alexa Fluor™-647-NHS-ester (5 mg/mL in DMSO) (Thermo Fisher Scientific, Waltham, MA, USA) was added to 600 μg of Saporin (2 mg/mL), to obtain a molar dye/protein ratio of 2.5, and kept under stirring for 2 h at RT. At the end of the incubation, the unreacted dye was removed with 2 further washes with a Zeba™ Spin desalting column (7 kDa), using 50 mM potassium phosphate, pH 7.2, and 150 mM NaCl as final buffer.

The degree of labelling was calculated as dye concentration (M)/protein concentration (M), based on UV–Vis spectroscopy; 280 nm and 647 nm absorbances were used, respectively, for the quantification of the protein and the dye, using the following formulas that subtract the contribution of the dye at 280 nm:

$$\text{Degree of labelling} = \frac{\text{Abs}_{647\text{ nm}}}{\varepsilon_{\text{Alexa647}} \times \text{l} \times \text{protein concentration (M)}} \tag{1}$$

where $\varepsilon_{\text{Alexa647}}$ (molar extinction coefficient of Alexa Fluor™-647-NHS-ester) is 239,000 cm^{-1} M^{-1}, l (optical path length) is 1 cm and the protein concentration (M) was calculated as:

$$\text{Protein concentration (M)} = \frac{\text{Abs}_{280\text{ nm}} - 0.03 \times \text{Abs}_{647\text{ nm}}}{\varepsilon_{\text{Sap}} \times \text{l}} \tag{2}$$

where 0.03 is the correction factor recommended by the supplier to remove the contribution of Alexa Fluor™-647-NHS-ester from the 280 nm absorbance, l is 1 cm and ε_{Sap} is 23,380 (calculated by ProtParam [23]).

2.2. *NP Characterization*

2.2.1. Transmission Electron Microscopy (TEM)

The NP suspension was deposited onto a formvar-coated 200-mesh copper grid (Ted Pella, Redding, CA, USA) and allowed to dry before examination. NPs were then visualized with a Jeol JEM 2100Plus (Jeol, Tokyo, Japan) electron microscope equipped with a 9 MP complementary metal oxide superconductor (CMOS) Gatan Rio9 digital camera (Gatan

Inc., Pleasanton, CA, USA). The core diameter (n = 100) was calculated by processing the recorded images with Fiji/ImageJ software [24].

2.2.2. Dynamic Light Scattering (DLS) and ζ-Potential Analysis

NP hydrodynamic diameter and ζ-potential were analyzed on a Zetasizer Nano ZS ZEN3600 (Malvern Panalytical, Malvern, Worcestershire, UK), operating at a light source wavelength of 633 nm and a fixed scattering angle of 173°. For both DLS and ζ-potential analysis, the purified samples were diluted in distilled water to a concentration of 30 μg/mL and briefly sonicated prior to the analysis. The results were expressed as means ± standard deviations (SDs) of three measurements.

2.2.3. Conjugation Efficacy

The conjugation efficiency of Saporin was calculated after MYTS-Sap-Alexa-1, MYTS-Sap-Alexa-2 and MYTS-Sap-Alexa-3 synthesis as μg of conjugated Saporin/μg of reacted Saporin. The quantities (μg) of conjugated Saporin in the MYTS-treated samples were calculated via UV–Vis spectroscopy based on Alexa Fluor™-647-NHS-ester 647 nm absorbance and degree of protein labelling. Specifically, MYTS-SPDP nanoparticles were analyzed to calculate the Abs 550 nm/Abs 647 nm ratio. This ratio was used to determine the iron oxide core aspecific 647 nm contribution of MYTS-Sap-Alexa-1, MYTS-Sap-Alexa-2 and MYTS-Sap-Alexa-3. This value was subtracted from the 647 nm absorbance of each sample ($Abs_{647\ sub}$), and Saporin concentration (M) and conjugation efficiency (%) were calculated as follows:

$$\text{Saporin concentration (M)} = \frac{Abs_{647\ \text{nm sub}}}{\varepsilon_{647\ \text{nm}} \times 1 \times \text{degree of labelling}} \tag{3}$$

where $\varepsilon_{Alexa647}$ (molar extinction coefficient of Alexa Fluor™-647-NHS-ester) is 239,000 $cm^{-1}\ M^{-1}$ and l is 1 cm;

$$\text{Conjugation efficiency (\%)} = \frac{\text{Saporin concentration (mM)} \times \text{Protein MW} \times \text{Volume (mL)}}{\text{Reacted Saporin } (\mu g)} \tag{4}$$

where Saporin MW is 34,753 Da (calculated by ProtParam).

2.3. Cell Culture

SK-BR-3 cells were cultured in Dulbecco's Modified Eagle's Medium (DMEM) High Glucose/Ham's F12; U87, HeLa and NIH-3T3 cells were cultured in DMEM High Glucose. All media were supplemented with 2 mM L-glutamine, penicillin (50 U/mL), streptomycin (50 mg/mL) and 10% of either Fetal Bovine Serum (FBS) for SK-BR-3, HeLa and U87 or Calf Bovine Serum (CBS) for NIH-3T3. All cells were cultured at 37 °C in a humidified atmosphere containing 5% CO_2 and sub-cultured prior to confluence using trypsin/EDTA. All the cell media and supplements were purchased from Euroclone (Pero, Italy).

2.4. Cell Viability Assay (MTT Assay)

SK-BR-3, NIH-3T3, U87 and HeLa cells were seeded on a 96-well dish at a density of 1×10^4 (SK-BR-3 and U87) or 5×10^3 (NIH-3T3 and HeLa) cells/well and grown for 24 h in the appropriate medium. Cells were then incubated with MYTS-SPDP, MYTS-Sap-1, MYTS-Sap-2 and MYTS-Sap-3 at different concentrations (range: 1–50 μg/mL) and free Saporin at either the maximum concentration found in the MYTS-treated samples (corresponding to that of MYTS-Sap-3 at the highest concentration) for HeLa, U87 and NIH-3T3 cells or 2×, 3× and 4× the concentration for SK-BR-3. The equivalent Saporin nM concentrations for each sample can be found in Table S1. After 24 h, the medium was refreshed and cells were incubated for 4 h at 37 °C with 15 μL of 3-(4,5-dimethyl-2-thiazolyl)-2,5-diphenyl-2H-tetrazolium bromide (MTT) stock solution (Promega, Milano, Italy). At the end of the incubation, 0.1 mL of MTT solubilizing solution (Promega, Milano, Italy) was added to each well for 1 h at RT under shaking to solubilize the MTT formazan crystals. Absorbances were read with an EnSight™ Multimode Plate Reader (PerkinElmer, Waltham, MA, USA), using a test wavelength of 570 nm and a reference wavelength of 620 nm. Cellular viability was calculated by normalizing the treated sample absorbance against the absorbance of the untreated control sample. The results were expressed as the means ± standard deviations of four independent experiments.

2.5. Apoptosis Assay

SK-BR-3 and NIH-3T3 cells were seeded on a 12-well dish at a density of 2.5×10^5 cells/well and grown for 24 h. The cells were incubated with MYTS-SPDP, MYTS-Sap-1, MYTS-Sap-2 and MYTS-Sap-3 at concentrations of 75 μg/mL, with the highest concentration of free Saporin being found in the MYTS-treated samples (MYTS-Sap-3, equal to approximately 166.6 nM). After 24 h, floating cells were collected in flow cytometry tubes and adherent cells were detached with trypsin and pooled together with corresponding floating cells. Cells were centrifuged and washed once with ice-cold PBS, then resuspended in 100 μL Annexin-binding buffer (10 mM HEPES, 140 mM NaCl and 2.5 mM $CaCl_2$, pH 7.4) and incubated with 5 μL of Annexin V-Alexa Fluor™-488 solution (Thermo Fisher Scientific, Waltham, MA, USA) and 15 μg/mL Propidium Iodide (PI) for 15 min at RT in the dark. At the end of the incubation time, 200 μL of Annexin-binding buffer was added to the suspension and the cells were analyzed with a Gallios™ Flow Cytometer (Beckman Coulter Inc., Brea, CA, USA). For each analysis, 1×10^4 events were acquired after gating on single cells, and a sample of untreated cells was used to set the appropriate gate for the regions of positivity. The percentages of positive cells were reported as the mean ± SD of three individual experiments for SK-BR-3 and of a single experiment for NIH-3T3. Untreated cells were used as controls. The data were analyzed using FlowLogic™ Software (Inivai Technologies, Melbourne, Australia).

2.6. Protein Synthesis Inhibition Assay

SK-BR-3 and NIH-3T3 cells were seeded on a black 96-well dish at a density of 1×10^4 and 5×10^3 cells/well, respectively, and grown for 24 h. The cells were incubated with MYTS-SPDP, MYTS-Sap-1, MYTS-Sap-2 and MYTS-Sap-3 at a concentration of 50 μg/mL and with the highest concentration of free Saporin being found in the MYTS-treated samples (MYTS-Sap-3, equal to approximately 111.1 nM). After 6 h, the inhibition of protein synthesis was evaluated using a Click-iT® Plus OPP Alexa Fluor®-488 Protein Synthesis Assay Kit (Thermo Fisher Scientific, Waltham, MA, USA), according to the manufacturer's instructions. At the end of the procedure, cell green fluorescence signals were imaged with a PAULA microscope (Leica Microsystem, Buffalo Grove, IL, USA) and the images were analyzed with ImageJ software. Specifically, for each cell line, the original images were converted to greyscale 8-bit images, an automatic threshold was generated using the untreated sample with the "Auto Threshold" function and the same intensity value was used to generate the threshold for all the treated samples. Then, the "Watershed" function was used to separate any cell clusters into single cells and the "Measurement" function was used to measure the mean grey value of the above-threshold cells. The means and standard deviations of all cells mean grey values for each image were calculated and normalized against those of the untreated sample.

2.7. Cellular Uptake Assay

SK-BR-3 and NIH-3T3 cells were seeded on a 12-well dish at a density of 2.5×10^5 cells/well and grown for 24 h. The cells were incubated with MYTS-SPDP, MYTS-Sap-Alexa-1, MYTS-Sap-Alexa-2 and MYTS-Sap-Alexa-3 at an equal MYTS concentration of 10 μg/mL or at an equal Saporin concentration of 5.4 nM (found in MYTS-Sap-Alexa-1 10 μg/mL). The highest concentration of free Alexa-labelled Saporin used was found in the MYTS-treated samples, these being, respectively, 22.2 nM and 5.4 nM. After 4 h, the NP-containing medium was removed and cells were detached with trypsin, collected in flow cytometry tubes and washed twice with PBS. Cells were finally resuspended in

300 μL PBS 2 mM EDTA and analyzed using a Gallios Flow Cytometer (Beckman Coulter). For each analysis, 1×10^4 events were acquired after gating on single cells, and the median fluorescence intensities of samples were normalized against a sample of untreated cells. The data on the normalized medians of fluorescence intensity were reported as means ± standard deviations of three independent experiments. Untreated cells were used as a control. The data were analyzed using FlowLogic™ Software (Inivai Technologies, Melbourne, Australia).

2.8. Statistical Analysis

Statistical analysis was performed using a two-way ANOVA for multiple comparisons. The analyses were conducted using GraphPad Prism 6 (GraphPad Holdings, San Diego, CA, USA). Significance is indicated in the figures as: * $p < 0.05$, ** $p < 0.01$, *** $p < 0.001$, **** $p < 0.0001$.

3. Results

3.1. Production and Characterization of C-Cys-Engineered Saporin (C-Sap)

In order to be produced, purified and subsequently conjugated to colloidal nanoparticles, a highly potent mature form of Sap (SAP-3; [25]) was genetically modified with the insertion of a Cys residue at the C-terminus and a His Tag at the N-terminus (Figure S1A). Regarding the Cys insertion, it is worth noting that the conjugation strategy adopted for MYTS-Sap conjugation was selected for two main reasons: (1) disulfide bridges are selectively reduced in the cell cytoplasm, promoting the release of C-Sap in its site of action; (2) native Sap does not contain Cys residues, so the conjugation is highly specific and is not expected to affect the protein's structural integrity.

The recombinant protein was expressed in *E. coli* (strain BL21(DE3)) and isolated from the soluble fraction by affinity chromatography. The presence of Sap and the purity of the fractions containing His-Tag proteins were checked by SDS–PAGE after Coomassie blue staining and WB analysis (Figure S1B,C). The fractions containing most of the protein (lanes 3, 4 and 5 in Figure S1B,C) were collected together and the buffer exchanged via centrifugal filtration. This last step is crucial because an inefficient elimination of imidazole employed for the protein elution significantly compromises the freeze–thaw stability of C-Sap. The purification protocol used here was designed to allow the obtainment of around 2.8 mg of C-Sap/L culture, similar to what was previously reported [26].

3.2. Synthesis and Characterization of Saporin-Loaded Nanoparticles (MYTS-Sap)

Twelve-nanometer iron oxide nanocrystals (Figure 1A) were coated with a biocompatible amphiphilic polymer (PMDA), allowing the obtainment of a highly stable NP dispersion with surface carboxylic functional groups useful for bioconjugation (Figure 1B,C). MYTS-Sap NPs were synthesized according to an optimized procedure derived from Fiandra et al., with some adjustments [12] Briefly, MYTS-PMDA nanoparticles were modified with 2,2-(ethylenedioxy)bis(ethylamine) (EDBE) to introduce amino functionalities that allowed us to incorporate thiol-reactive groups onto the PMDA surface by reaction with Sulfosuccinimidyl-6-(3′-(2-pyridyldithio)propionamido)hexanoate (sulfo-LC-SPDP), leading to the formation of MYTS-SPDP. The functionalization steps did not alter the colloidal stability (measured by hydrodynamic size; Figure 1B, Table 1). The surface negative charge, determined by zeta potential (Table 1), was only slightly decreased upon conjugation with Sap, thus maintaining the original colloidal stability due to electrostatic repulsion. MYTS-SPDP particles were incubated with C-Sap at different concentrations to achieve toxin bioconjugates (Figure 1C). We optimized the reaction conditions to obtain three different batches of MYTS-Sap having ~1, 2 or 4 C-Sap molecules per NP, respectively, with a conjugation efficiency of around 40% (see Table 2). As expected, upon increasing the number of C-Sap molecules attached to the NPs, the hydrodynamic size of the nanoconjugates was augmented (Table 1). Western blot analysis performed in reducing and non-reducing conditions confirmed the presence of C-Sap in all three nanoconjugates and specific conjugation through disulfide bridging (Figure S2, Supplementary Materials). These three batches were designed to evaluate the effect of C-Sap concentration (i.e., the number of C-Sap molecules loaded on the NPs) on the cytotoxic activity of the nanoconjugates.

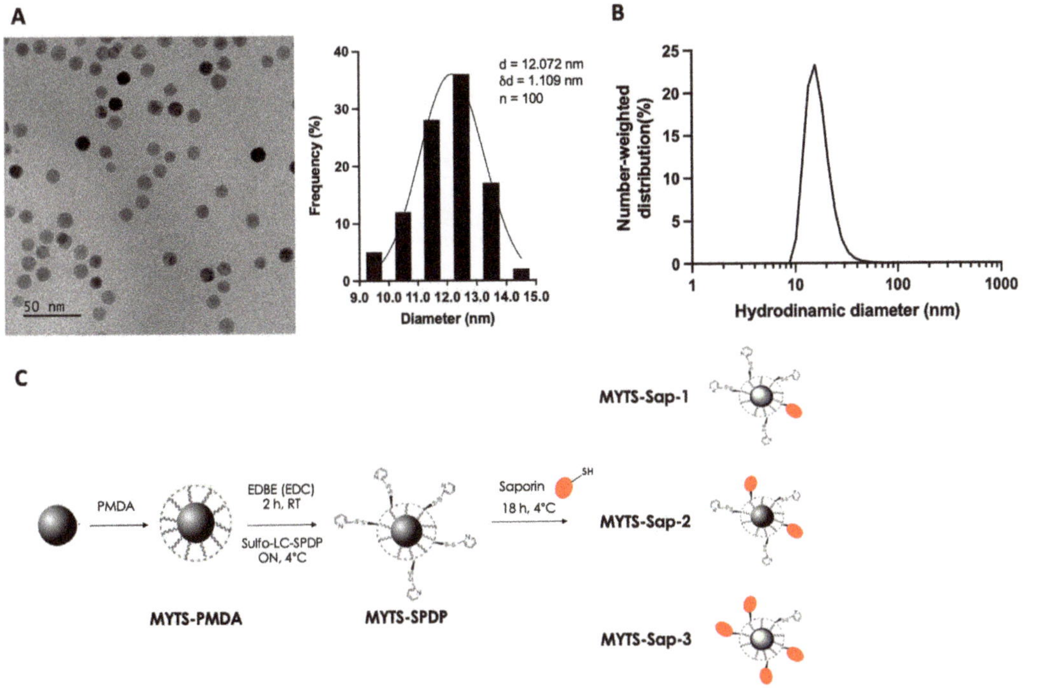

Figure 1. (**A**) TEM image of iron oxide cores. Scale bar: 50 nm; size distribution (n = 100). (**B**) Size distribution of MYTS-PMDA evaluated by DLS. (**C**) Schematic representation of MYTS-Sap synthesis reactions.

Table 1. Summary of DLS and zeta potential characterization of nanoparticles throughout the synthesis process. Data are expressed as the means of three measurements ± SDs.

Sample	Number-Weighted Hydrodynamic Diameter (nm)	Zeta Potential (mV)
MYTS-PMDA	17.2 ± 2.5	−46.0 ± 4.4
MYTS-SPDP	17.0 ± 4.6	−45.1 ± 3.2
MYTS-Sap-1	20.0 ± 1.5	−43.4 ± 2.9
MYTS-Sap-2	22.5 ± 6.8	−42.2 ± 0.7
MYTS-Sap-3	29.3 ± 4.7	−40.0 ± 0.9

Table 2. Quantification of Saporin conjugation in MYTS-Sap-1, MYTS-Sap-2 and MYTS-Sap-3. Conjugation efficiencies are expressed as the means of three independently synthesized batches ± SDs.

Sample	Reacted Saporin/MYTS (µg/mg)	Conjugation Efficiency (%)	Calculated Saporin/MYTS (Molar Ratio)
MYTS-Sap-1	50	37.4 ± 10.2	0.89
MYTS-Sap-2	100	37.0 ± 5.2	1.76
MYTS-Sap-3	200	38.6 ± 4.2	3.67

3.3. Antitumor Activity of MYTS-Sap

To assess the general cytotoxicity of MYTS-Sap, SK-BR-3 breast cancer cells were cultured and incubated with MYTS-SPDP and MYTS-Sap at increasing NP concentrations, i.e., 25, 37.5 and 50 µg/mL (Figure 2A). Although MYTS-SPDP nanoparticles were not toxic, MYTS-Sap particles exhibited dose-dependent activity. Considering the low protein uptake by cancer cells usually observed with the soluble toxin, we decided to increase the tested C-Sap concentration by up to four-fold compared to the highest amount loaded to the NPs (MYTS-Sap-3) (Table S3). Nevertheless, even at the highest concentration tested, unconjugated C-Sap was much less cytotoxic compared to MYTS-Sap. Apparently, since no difference was found between MYTS-Sap-1, -2 and -3, the number of C-Sap molecules attached to the NPs does not affect the antitumoral activity of MYTS-Sap. These results suggest that to achieve an effective cytotoxic effect, the principal objective is to improve Saporin internalization and endosomal escape through optimized carriers. Figure 2B shows that while healthy fibroblasts (NIH-3T3 murine fibroblasts) are generally more sensitive to non-functionalized NPs (i.e., MYTS-SPDP), these cells were not affected by MYTS conjugation with C-Sap—either MYTS-Sap-1, -2 or -3—all of which have the same toxicity level as MYTS-SPDP. Notably, unconjugated C-Sap was nontoxic at the highest concentration. Overall, the data reported in Figure 2A,B are encouraging, as MYTS-Sap proved to be selective toward SK-BR-3 breast cancer in a dose-dependent manner.

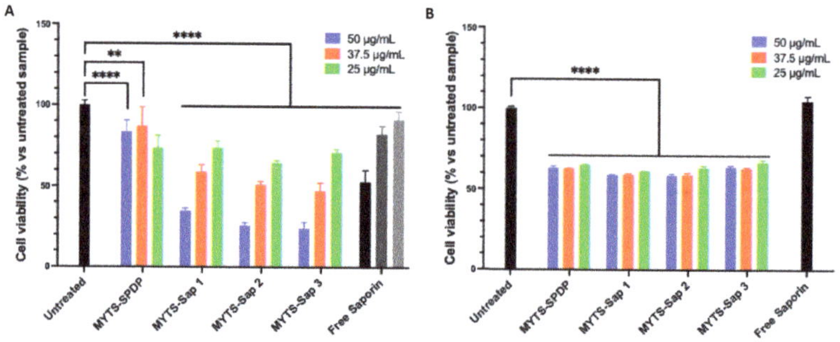

Figure 2. MTT assay performed incubating MYTS-SPDP, MYTS-Sap-1, MYTS-Sap-2 and MYTS-Sap-3 at concentrations of 25, 37.5 and 50 µg/mL with (**A**) SK-BR-3 cells and (**B**) for NIH-3T3 cells for 24 h. Corresponding Saporin concentrations (nM) are reported in Table S1. For free Saporin samples, approximately 2× (light gray), 3× (dark grey) and 4× (black) the maximum Saporin concentrations tested with MYTS-containing samples (50 µg/mL MYTS-Sap-3, 111.1 nM) were used for SK-BR-3, and the maximum concentration was used for NIH-3T3. **** $p < 0.0001$, ** $p < 0.01$ vs. untreated (black).

To attempt to understand the reason for the selective MYTS-Sap activity against breast cancer cells, the uptake of MYTS-Sap in SK-BR-3 was compared with that in NIH-3T3 under the same experimental conditions. The extent of MYTS uptake was assessed by flow cytometry analysis using C-Sap labeled with Alexa Fluor™-647-NHS-ester. Figure 3A shows that SK-BR-3 exhibited a remarkably higher uptake of MYTS-Sap compared to NIH-3T3. The NP concentration was kept constant and the increasing intensity of fluorescence from MYTS-Sap-1 to MYTS-Sap-3 was only due to the different number of C-Sap molecules per NP. Indeed, Figure 3B shows the result obtained with MYTS-Sap normalized to the number of toxin molecules. Although the level of MYTS-Sap uptake in NIH-3T3 was significantly lower than in SK-BR-3, it was not negligible. Combining these data with the absence of toxicity in NIH-3T3 (Figure 2B), the results of the uptake experiment suggested that a very low number of internalized NPs were able to escape the endosomes and release Sap molecules into the cell cytoplasm. Furthermore, despite the fact that the protein detected in SK-BR-3 increased in MYTS-Sap-1, -2 and -3 according to the relative proportions of C-Sap/NP (Figure 3A), the lack of substantial differences previously observed in the

MTT assay (Figure 2A) suggests that ligand density may somehow affect the efficiency of endosomal escape.

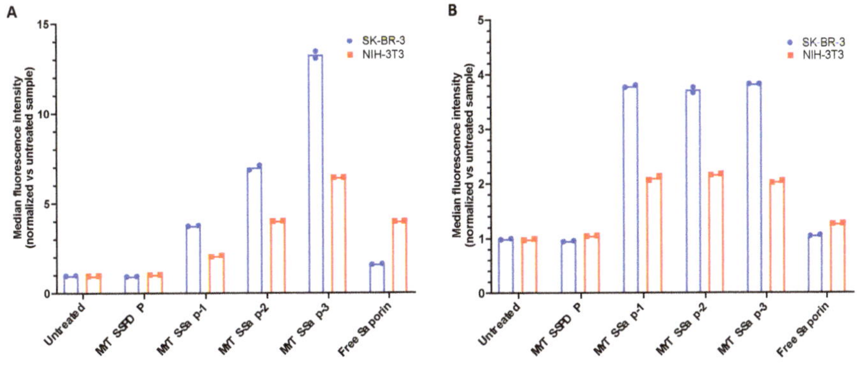

Figure 3. Cellular uptake of fluorescently labelled MYTS-Sap-1, MYTS-Sap-2 and MYTS-Sap-3 as evaluated by flow cytometry, with the incubation of an (**A**) equal MYTS concentration of 10 μg/mL or (**B**) an equivalent concentration of Saporin (5.4 nM) for 3 h. In both cases, free Saporin was used at the highest Saporin concentration tested (the one found in MYTS-Sap-3 equal to 22.2 nM and 5.4 nM, respectively).

3.4. Mechanism of MYTS-Sap Cytotoxic Activity

To confirm that the cytotoxic activity of MYTS-Sap in cancer cells could be indeed attributed to the effect of C-Sap release in the cell cytoplasm and not to an unspecific toxicity of the nanoconjugate, we investigated two specific molecular pathways of C-Sap activity. We first assessed the apoptosis/necrosis profile by Annexin-V-FITC/PI double staining assay, and, secondly, we determined the toxin-mediated protein synthesis inhibition [1].

Figure 4 shows the results of the flow cytometry analysis performed on SK-BR-3 (Figure 4A) and NIH-3T3 (Figure 4B) and the relevant histograms reflecting the relative ratios of viable, necrotic and apoptotic cells (Figure 4C,D, respectively).

Both in SK-BR-3 and NIH-3T3, unconjugated C-Sap did not affect the apoptosis/necrosis rate compared to the untreated control, in accordance with the results of the MTT assay. In contrast, following incubation with MYTS-Sap in SK-BR-3, we observed a remarkable increase in the number of apoptotic cells and, to a much lower extent, also of necrotic cells. Confirming the data from the MTT assay, the entity of apoptotic effect was independent of the number of C-Sap molecules bound to the MYTS NPs (Figure 4C). Once again, MYTS-Sap nanoparticles were ineffective toward NIH-3T3 cells (Figure 4D).

A protein synthesis inhibition assay was performed to test the generally accepted main mechanisms of intoxication and cell death induced by Saporin. Figure 5A,C shows that MYTS-Sap nanoparticles were able to significantly decrease the amount of newly translated proteins in SK-BR-3 cancer cells. In contrast, no evidence of proteins synthesis inhibition was observed for NIH-3T3 control cells (Figure 5B,C), confirming the selective action of MYTS-Sap in cancer cells.

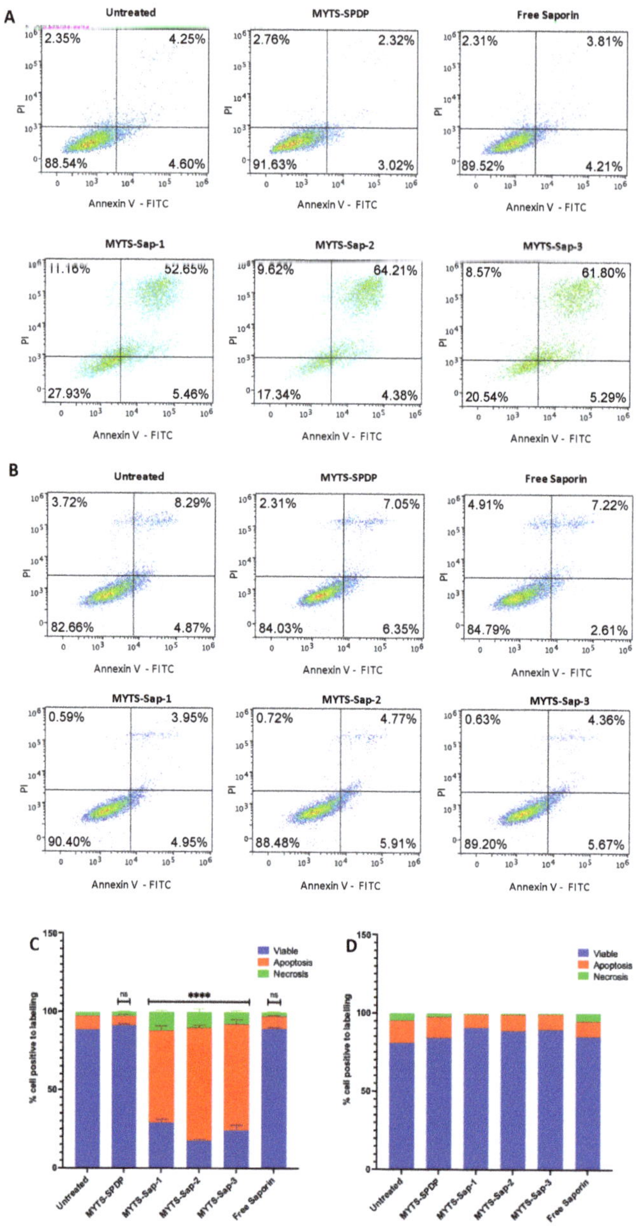

Figure 4. Annexin-V-FITC/PI double staining assay for the detection of apoptosis and/or necrosis performed on (**A**,**C**) SK-BR-3 cells and (**B**,**D**) NIH-3T3 cells, following 24 h of incubation with MYTS-SPDP, MYTS-Sap-1, MYTS-Sap-2 and MYTS-Sap-3 at concentrations of 75 µg/mL and free Saporin at the equivalent highest concentration found in the MYTS-treated samples (MYTS-Sap-3, approximately 166.6 nM). Viable: lower left quadrant; Apoptosis: lower + upper right quadrants; Necrosis: upper left quadrant. **** $p < 0.0001$; ns = non-significant vs. untreated.

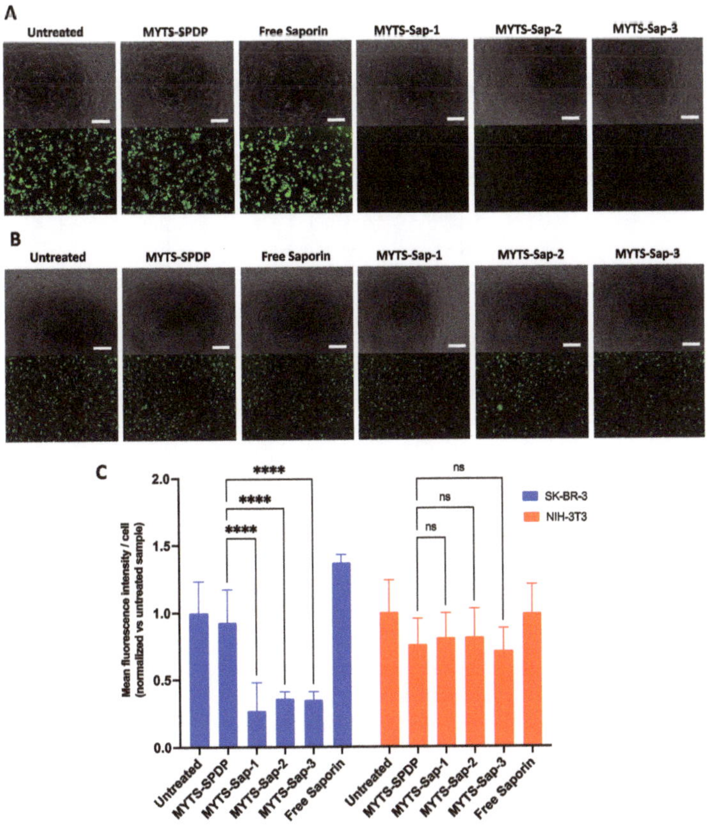

Figure 5. Protein synthesis inhibition assay performed on (**A**) SK-BR-3 cells and (**B**) NIH-3T3 cells, following 6 h incubation with MYTS-SPDP, MYTS-Sap-1, MYTS-Sap-2 and MYTS-Sap-3 at concentrations of 50 µg/mL and free Saporin and the equivalent highest concentration found in MYTS-treated samples (approximately 111.1 nM). Scale bars: 200 µm. (**C**) Quantification of fluorescence intensity/cell normalized against the untreated sample. **** $p < 0.0001$; ns = non-significant vs. MYTS-SPDP.

3.5. Dependence of MYTS-Sap Activity on Cancer Cell Lines

To explore the generalization of the cytotoxic activity of MYTS-Sap against cancer cells, MYTS-Sap were also tested in HeLa cervical cancer cells (Figure 6A) and in U87 glioblastoma cells (Figure 6B) through MTT assay. In all cancer cell cultures, dose-dependent cytotoxic activity of MYTS-Sap was observed. However, HeLa cells were much more sensitive to C-Sap nanoconjugates; thus, lower concentrations (in the range of 1–10 µg/mL) were required to investigate MYTS-Sap activity. Despite a reduction in cell viability being recorded in MYTS-SPDP-treated U87 cells, the effect displayed by Sap nanoconjugates was significantly higher. This is in contrast with what we observed with NIH-3T3, where no difference was found between bare and conjugated MYTS (Figure 2B).

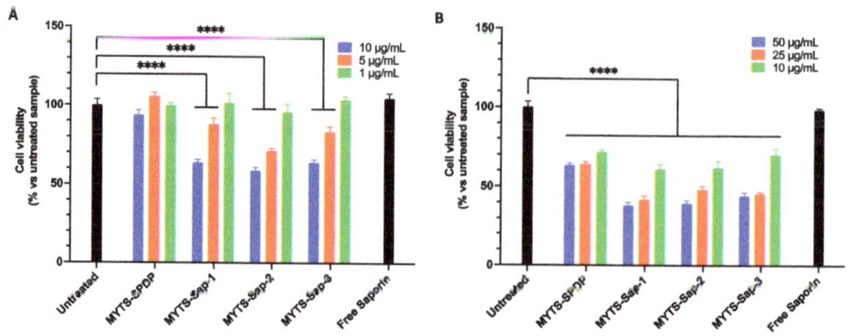

Figure 6. MTT assay performed incubating MYTS-SPDP, MYTS-Sap-1, MYTS-Sap-2 and MYTS-Sap-3 at concentrations of (**A**) 1, 5 and 10 µg/mL, with HeLa cells, and (**B**) 10, 25 and 50 µg/mL, with U87 cells, for 24 h. For free Saporin samples (black), the maximum Saporin concentration tested in MYTS-containing samples was used (10 µg/mL MYTS-Sap-3 for HeLa and 50 µg/mL MYTS-Sap-3 for U87). **** $p < 0.0001$ vs. untreated (black).

4. Conclusions

In summary, we have investigated the impact of nanoconjugation of a highly cytotoxic RIP, namely, Saporin, on therapeutic toxin activity against cancer cells. Despite the efforts dedicated to the development of new therapeutic options based on RIPs for the treatment of aggressive neoplasms, their clinical application remains elusive. Among the various limitations that have been identified in their application in cancer therapy, fundamental concerns have been raised regarding Sap stability against degradation from proteases and its capability to reach tumor tissue, enter the cell membrane and escape the endosome to interfere with ribosome activity and inhibit protein synthesis. In this work, we focused on exploring the potential of conjugating Sap to colloidal NPs suitably coated with an amphiphilic polymer which have the capacity to promote endosomal escape [14]. We prepared three batches of Sap-conjugated NPs, namely, MYTS-Sap-1, MYTS-Sap-2 and MYTS-Sap-3, bearing, respectively, 1, 2 and 4 Sap molecules on average. The treatment of SK-BR-3 (breast), HeLa (cervical) and U87 (brain) cancer cells with the three batches of MYTS-Sap nanoparticles resulted in dose-dependent cytotoxic activity. Surprisingly, despite protein internalization increasing with the number of conjugated Sap/NPs, no difference between the three batches was apparently recorded in any of the activity experiments. In contrast, control NIH-3T3 healthy fibroblasts did not exhibit sensitivity to Sap nanoconjugates. Evidence of the actual involvement of Sap in the mechanism of action of MYTS-Sap toxicity was provided by an uptake experiment combined with an assessment of protein synthesis inhibition in SK-BR-3 cells. The same experiments performed in NIH-3T3 cells confirmed the selective proapoptotic activity of the protein delivered by MYTS NPs.

Notably, the structural features of MYTS promoted cellular internalization in cancer cells selectively and mediated an efficient endosomal escape and Sap release in the cytoplasm. This unpredicted behavior suggests the potential of MYTS-Sap as a cancer-selective drug delivery system even in the absence of conjugation with targeting ligands, which is favorable for clinical translation. Such a strong advantage will deserve further mechanistic investigations.

Overall, the results of our study implied that the newly designed MYTS-Sap nanoparticles were highly cytotoxic against cancer cells, with broad applicability, inducing apoptotic effects mediated by inhibition of translocation of tRNA in the ribosome complex, thus blocking protein synthesis. Remarkably, cytotoxic activity of MYTS-Sap was highly selective against cancer cells in vitro, preserving healthy fibroblasts from apoptosis induction. This was achieved without the insertion of any targeting agent, highlighting the potential of this nanoformulation. In addition, MYTS-Sap cytotoxicity proved to be independent of the amount of Sap conjugated to NPs but dependent (only) on the concentration of Sap nanoconjugates. This may have been due to the influence of ligand density on the extent to which Sap reached the cytoplasm as well as the achievement of a toxin activity plateau in tested conditions. Additional investigations are required to corroborate either of the two hypotheses. This preliminary study encourages the investigation of the use of colloidally stable Sap nanomedicines for the development of innovative treatment approaches against solid tumors.

Supplementary Materials: The following supporting information can be downloaded at: https://www.mdpi.com/article/10.3390/pharmaceutics14071517/s1, Figure S1: Production of recombinant Saporin; Figure S2: Assessment of MYTS-Sap binding specificity. Table S1: Correlation between Sap and MYTS concentrations in MTT assay.

Author Contributions: Conceptualization, M.C. and D.P.; methodology, L.S., F.T., L.B., M.G., G.T. and J.A.B.; resources, M.C.; data curation, M.C. and L.S.; writing—original draft preparation, P.T. M.C. and D.P. writing—review and editing, L.S.; supervision, D.P. All authors have read and agreed to the published version of the manuscript.

Funding: This research was funded by AIRC (Associazione Italiana per la Ricerca sul Cancro) 'My First AIRC Grant' (MFAG 2014, ref. 16144 to M.C.).

Data Availability Statement: Not applicable.

Acknowledgments: We thank Tiziano Catelani for TEM analysis.

Conflicts of Interest: The authors declare no conflict of interest. The funders had no role in the design of the study; in the collection, analyses, or interpretation of data; in the writing of the manuscript, or in the decision to publish the results.

References

1. Polito, L.; Bortolotti, M.; Mercatelli, D.; Battelli, M.; Bolognesi, A. Saporin-S6: A Useful Tool in Cancer Therapy. *Toxins* **2013**, *5*, 1698–1722. [CrossRef] [PubMed]
2. Bolognesi, A.; Bortolotti, M.; Maiello, S.; Battelli, M.; Polito, L. Ribosome-Inactivating Proteins from Plants: A Historical Overview. *Molecules* **2016**, *21*, 1627. [CrossRef] [PubMed]
3. Sikriwal, D.; Ghosh, P.; Batra, J.K. Ribosome Inactivating Protein Saporin Induces Apoptosis through Mitochondrial Cascade, Independent of Translation Inhibition. *Int. J. Biochem. Cell Biol.* **2008**, *40*, 2880–2888. [CrossRef] [PubMed]
4. Lu, J.-Q.; Zhu, Z.-N.; Zheng, Y.-T.; Shaw, P.-C. Engineering of Ribosome-Inactivating Proteins for Improving Pharmacological Properties. *Toxins* **2020**, *12*, 167. [CrossRef]
5. Salvioni, L.; Zuppone, S.; Andreata, F.; Monieri, M.; Mazzucchelli, S.; di Carlo, C.; Morelli, L.; Cordiglieri, C.; Donnici, L.; de Francesco, R.; et al. Nanoparticle-Mediated Suicide Gene Therapy for Triple Negative Breast Cancer Treatment. *Adv. Ther.* **2020**, *3*, 2000007. [CrossRef]
6. Dosio, F.; Stella, B.; Cerioni, S.; Gastaldi, D.; Arpicco, S. Advances in Anticancer Antibody-Drug Conjugates and Immunotoxins. *Recent Pat. Anti-Cancer Drug Discov.* **2013**, *9*, 35–65. [CrossRef]
7. Giansanti, F.; Flavell, D.; Angelucci, F.; Fabbrini, M.; Ippoliti, R. Strategies to Improve the Clinical Utility of Saporin-Based Targeted Toxins. *Toxins* **2018**, *10*, 82. [CrossRef]
8. Li, M.; Liu, Z.-S.; Liu, X.-L.; Hui, Q.; Lu, S.-Y.; Qu, L.-L.; Li, Y.-S.; Zhou, Y.; Ren, H.-L.; Hu, P. Clinical Targeting Recombinant Immunotoxins for Cancer Therapy. *OncoTargets Ther.* **2017**, *10*, 3645–3665. [CrossRef]
9. Smith, W.; Johnston, D.; Holmes, S.; Wensley, H.; Flavell, S.; Flavell, D. Augmentation of Saporin-Based Immunotoxins for Human Leukaemia and Lymphoma Cells by Triterpenoid Saponins: The Modifying Effects of Small Molecule Pharmacological Agents. *Toxins* **2019**, *11*, 127. [CrossRef]
10. Alewine, C.; Hassan, R.; Pastan, I. Advances in Anticancer Immunotoxin Therapy. *Oncologist* **2015**, *20*, 176–185. [CrossRef]
11. Qiu, M.; Zhang, Z.; Wei, Y.; Sun, H.; Meng, F.; Deng, C.; Zhong, Z. Small-Sized and Robust Chimaeric Lipopepsomes: A Simple and Functional Platform with High Protein Loading for Targeted Intracellular Delivery of Protein Toxin in Vivo. *Chem. Mater.* **2018**, *30*, 6831–6838. [CrossRef]
12. Fiandra, L.; Colombo, M.; Mazzucchelli, S.; Truffi, M.; Santini, B.; Allevi, R.; Nebuloni, M.; Capetti, A.; Rizzardini, G.; Prosperi, D.; et al. Nanoformulation of Antiretroviral Drugs Enhances Their Penetration across the Blood Brain Barrier in Mice. *Nanomed. Nanotechnol. Biol. Med.* **2015**, *11*, 1387–1397. [CrossRef]
13. Marongiu, L.; Mingozzi, F.; Cigni, C.; Marzi, R.; di Gioia, M.; Garrè, M.; Parazzoli, D.; Sironi, L.; Collini, M.; Sakaguchi, R.; et al. Inositol 1,4,5-Trisphosphate 3-Kinase B Promotes Ca^{2+} Mobilization and the Inflammatory Activity of Dendritic Cells. *Sci. Signal.* **2021**, *14*, eaaz2120. [CrossRef]

14. Rotem, R.; Micale, A.; Rizzuto, M.A.; Migliavacca, M.; Giustra, M.; Salvioni, L.; Tasin, F.; Prosperi, D.; Colombo, M. Modeling the Interaction of Amphiphilic Polymer Nanoparticles with Biomembranes to Guide Rational Design of Drug Delivery Systems. *Colloids Surf. B Biointerfaces* **2020**, *196*, 111366. [CrossRef]
15. Colombo, M.; Mazzucchelli, S.; Montenegro, J.M.; Galbiati, E.; Corsi, F.; Parak, W.J.; Prosperi, D. Protein Oriented Ligation on Nanoparticles Exploiting O6-Alkylguanine-DNA Transferase (SNAP) Genetically Encoded Fusion. *Small* **2012**, *8*, 1492–1497. [CrossRef]
16. Truffi, M.; Colombo, M.; Sorrentino, L.; Pandolfi, L.; Mazzucchelli, S.; Pappalardo, F.; Pacini, C.; Allevi, R.; Bonizzi, A.; Corsi, F.; et al. Multivalent Exposure of Trastuzumab on Iron Oxide Nanoparticles Improves Antitumor Potential and Reduces Resistance in HER2-Positive Breast Cancer Cells. *Sci. Rep.* **2018**, *8*, 6563. [CrossRef]
17. Colombo, M.; Rizzuto, M.A.; Pacini, C.; Pandolfi, L.; Bonizzi, A.; Truffi, M.; Monieri, M.; Catrambone, F.; Giustra, M.; Garbujo, S.; et al. Half-Chain Cetuximab Nanoconjugates Allow Multitarget Therapy of Triple Negative Breast Cancer. *Bioconjugate Chem.* **2018**, *29*, 3817–3832. [CrossRef]
18. Mazzucchelli, S.; Colombo, M.; Verderio, P.; Rozek, E.; Andreata, F.; Galbiati, E.; Tortora, P.; Corsi, F.; Prosperi, D. Orientation-Controlled Conjugation of Haloalkane Dehalogenase Fused Homing Peptides to Multifunctional Nanoparticles for the Specific Recognition of Cancer Cells. *Angew. Chem. Int. Ed.* **2013**, *52*, 3121–3125. [CrossRef]
19. Colombo, M.; Mazzucchelli, S.; Collico, V.; Avvakumova, S.; Pandolfi, L.; Corsi, F.; Porta, F.; Prosperi, D. Protein-Assisted One-Pot Synthesis and Biofunctionalization of Spherical Gold Nanoparticles for Selective Targeting of Cancer Cells. *Angew. Chem. Int. Ed.* **2012**, *51*, 9272–9275. [CrossRef]
20. Hühn, J.; Carrillo-Carrion, C.; Soliman, M.G.; Pfeiffer, C.; Valdeperez, D.; Masood, A.; Chakraborty, I.; Zhu, L.; Gallego, M.; Yue, Z.; et al. Selected Standard Protocols for the Synthesis, Phase Transfer, and Characterization of Inorganic Colloidal Nanoparticles. *Chem. Mater.* **2017**, *29*, 399–461. [CrossRef]
21. Amigoni, L.; Salvioni, L.; Sciandrone, B.; Giustra, M.; Pacini, C.; Tortora, P.; Prosperi, D.; Colombo, M.; Regonesi, M.E. Impact of Tuning the Surface Charge Distribution on Colloidal Iron Oxide Nanoparticle Toxicity Investigated in Caenorhabditis Elegans. *Nanomaterials* **2021**, *11*, 1551. [CrossRef] [PubMed]
22. Mazzucchelli, S.; Sommaruga, S.; O'Donnell, M.; Galeffi, P.; Tortora, P.; Prosperi, D.; Colombo, M. Dependence of Nanoparticle-Cell Recognition Efficiency on the Surface Orientation of ScFv Targeting Ligands. *Biomater. Sci.* **2013**, *1*, 728. [CrossRef] [PubMed]
23. Gasteiger, E.; Hoogland, C.; Gattiker, A.; Duvaud, S.; Wilkins, M.R.; Appel, R.D.; Bairoch, A. Protein Identification and Analysis Tools on the ExPASy Server. In *The Proteomics Protocols Handbook*; John, M.W., Ed.; Humana Press: Totowa, NJ, USA, 2005; pp. 571–607.
24. Schindelin, J.; Arganda-Carreras, I.; Frise, E.; Kaynig, V.; Longair, M.; Pietzsch, T.; Preibisch, S.; Rueden, C.; Saalfeld, S.; Schmid, B.; et al. Fiji: An open-source platform for biological-image analysis. *Nat. Methods* **2012**, *9*, 676–682. [CrossRef]
25. Fabbrini, M.S.; Rappocciolo, E.; Carpani, D.; Solinas, M.; Valsasina, B.; Breme, U.; Cavallaro, U.; Nykjaer, A.; Rovida, E.; Legname, G.; et al. Characterization of a Saporin Isoform with Lower Ribosome-Inhibiting Activity. *Biochem. J.* **1997**, *322*, 719–727. [CrossRef] [PubMed]
26. Günhan, E.; Swe, M.; Palazoglu, M.; Voss, J.C.; Chalupa, L.M. Expression and Purification of Cysteine Introduced Recombinant Saporin. *Protein Expr. Purif.* **2008**, *58*, 203–209. [CrossRef] [PubMed]

pharmaceutics

MDPI

Article

Surface-Functionalized $NdVO_4:Gd^{3+}$ Nanoplates as Active Agents for Near-Infrared-Light-Triggered and Multimodal-Imaging-Guided Photothermal Therapy

Kerong Deng [1,2,3,†], Donglian Liu [1,†], Ziyan Wang [4,†], Zhaoru Zhou [2], Qianyi Chen [2], Jiamin Luo [2], Yaru Zhang [1,2,*], Zhiyao Hou [1,2,*] and Jun Lin [1,3,*]

1 The Sixth Affiliated Hospital of Guangzhou Medical University, Qingyuan People's Hospital, Qingyuan 511518, China; dengkr26@gmail.com (K.D.); ldl00987@163.com (D.L.)

2 Guangzhou Municipal and Guangdong Provincial Key Labrotory of Protein Modification and Degradation, School of Basic Medical Sciences, Guangzhou Medical University, Guangzhou 511436, China; 2019218158@stu.gzhmu.edu.cn (Z.Z.); 2019218157@stu.gzhmu.edu.cn (Q.C.); jiaminluo7@stu.gzhmu.edu.cn (J.L.)

3 State Key Laboratory of Rare Earth Resource Utilization, Changchun Institute of Applied Chemistry, Chinese Academy of Sciences, Changchun 130022, China

4 The First Clinical School of Guangzhou Medical University, Guangzhou Medical University, Guangzhou 510120, China; wangzy2018@stu.gzhmu.edu.cn

* Correspondence: zyr@gzhmu.edu.cn (Y.Z.); zyhou@gzhmu.edu.cn (Z.H.); jlin@ciac.ac.cn (J.L.)

† These authors contributed equally to this work.

Abstract: Development of nanotheranostic agents with near-infrared (NIR) absorption offers an effective tool for fighting malignant diseases. Lanthanide ion neodymium (Nd^{3+})-based nanomaterials, due to the maximum absorption at around 800 nm and unique optical properties, have caught great attention as potential agents for simultaneous cancer diagnosis and therapy. Herein, we employed an active nanoplatform based on gadolinium-ion-doped $NdVO_4$ nanoplates ($NdVO_4:Gd^{3+}$ NPs) for multiple-imaging-assisted photothermal therapy. These NPs exhibited enhanced NIR absorption and excellent biocompatibility after being grafted with polydopamine (pDA) and bovine serum albumin (BSA) layers on their surface. Upon expose to an 808 nm laser, these resulting NPs were able to trigger hyperthermia rapidly and cause photo-destruction of cancer cells. In a xenograft tumor model, tumor growth was also significantly inhibited by these photothermal agents under NIR laser irradiation. Owing to the multicomponent nanostructures, we demonstrated these nanoagents as being novel contrast agents for in vivo magnetic resonance (MR) imaging, X-ray computed tomography (CT), photoacoustic (PA) imaging, and second biological window fluorescent imaging of tumor models. Thus, we believe that this new kind of nanotherapeutic will benefit the development of emerging nanosystems for biological imaging and cancer therapy.

Keywords: near infrared light; $NdVO_4$; polydopamine; photothermal therapy; multimodal imaging

Citation: Deng, K.; Liu, D.; Wang, Z.; Zhou, Z.; Chen, Q.; Luo, J.; Zhang, Y.; Hou, Z.; Lin, J. Surface-Functionalized $NdVO_4:Gd^{3+}$ Nanoplates as Active Agents for Near-Infrared-Light-Triggered and Multimodal-Imaging-Guided Photothermal Therapy. *Pharmaceutics* **2022**, *14*, 1217. https://doi.org/10.3390/pharmaceutics14061217

Academic Editors: Manivasagan Panchanathan and Eue-Soon Jang

Received: 14 April 2022
Accepted: 4 June 2022
Published: 8 June 2022

1. Introduction

Although numerous treatment modalities are available for fighting cancer, photothermal therapy (PTT), a minimally noninvasive and highly efficient treatment modality, has still been an active field of research [1–5]. It currently relies on the precise energy delivery to target sites and the sensitivity of tumor area to temperature increase, resulting in irreversible cell damage for cancer ablation [6–8]. Considering the nonspecific thermal effects, nanoparticle-assisted PTT seems to be an area of intense interest recently. Taking advantage of laser-induced nanoparticles, it can be easier to localize heat generation inside the desire area to improve the efficacy of PTT and avoid potential damage to surrounding healthy cells [9–12]. So far, shallow light penetration in living tissue remains a major limitation for in vivo applications of conventional PTT, which has hindered its therapeutic efficiency.

Great efforts have been made to develop photothermal conducting nanoparticles active in the near infrared region (700–1400 nm), lying in the "biological window" [13–15]. Due to relatively lower scattering and absorption from the intrinsic tissue chromophores, NIR light provides maximal penetration of biological tissue [16–20]. Various inorganic nanoparticles with effective NIR absorption have been studied as photothermal agents for NIR light-mediated PTT [21–26], such as gold nanostructures and carbon-based nanostructures, and have achieved effective therapeutic outcomes in vivo. Nevertheless, single functionality for cancer treatment cannot meet the increasing needs for personalized treatment of early diseases. To fabricate photothermal nanoagents with advanced performance through the integration of imaging capability may bring about a great improvement in photothermal ablation therapy due to the real-time monitoring of therapeutic action with more accuracy and efficacy [27].

As a promising candidate for PTT, Nd^{3+}-based nanomaterials have attracted much attention recently. With high-absorption cross-section at around 800 nm and multiphonon relaxation of Nd^{3+}, these nanoparticles are increasingly attractive for photothermal agents to convert the absorbed radiation into heat as an 808 nm light-triggered nanoheater [28–32]. For example, ultrasmall $NdVO_4$ nanoparticles as heating agents to improve photothermal conversion efficiency have been explored [33]. More intriguingly, the main emission bands of Nd^{3+} at around 1050 nm and 1330 nm both lie in the "second biological window" (II-BW), which extends from 1000 nm up to 1400 nm [34]. The optical scattering in the II-BW has been proven to be reduced when compared to that in the I-BW, affording an improvement in resolution of sub-tissue images. Thus, these unique properties of Nd^{3+}-based nanomaterials may pave a new way to achieve multifunctional therapeutic agents with deep-tissue imaging capability. Even though multimodal imaging satisfies current requirements for clinical diagnosis, the development of appealing Nd^{3+}-based nanoagents for multimodal imaging guided phototherapy is minimally reported to our best knowledge.

In this work, we present a versatile nanoplatform based on unique $NdVO_4$:Gd^{3+} NPs for in vivo multimodal imaging and activatable NIR-induced photothermal therapy of cancer. In this design, Nd^{3+}-based nanomaterials have a strong absorption at 808 nm and could serve as a therapeutic agent for selective phototherapy using a low-energy NIR laser. The synthetic procedure and potential bioapplication of $NdVO_4$:Gd^{3+}-pDA@BSA (NV-p@BSA) NPs are illustrated in Scheme 1. Firstly, Gd^{3+}-doped $NdVO_4$ NPs were synthesized by the one-step method and then modified with polydopamine on their surface. Polydopamine modification not only facilitates the following hydrophilicity but also enhances the photo-to-heat response via its photothermal ability. Next, in order to enhance water-solubility and biological properties of these NPs, negatively charged BSA molecules were chosen as a model biological agent to conjugate on their surface to achieve NV-p@BSA NPs. Both in vitro and in vivo investigations indicated that NV-p@BSA NPs possess excellent biocompatibility and show effective photothermal therapeutic upon NIR irradiation. More importantly, these nanocrystals are capable of emitting luminescent signal around 1050 nm in the second window of the near-infrared (NIR-II) region, which is particularly suitable for deep tissue fluorescence imaging. Meanwhile, after administrated to tumor-bearing mice, these NPs also exhibit the ability of magnetic resonance (MR)/X-ray computed tomography (CT)/photoacoustic (PA)/NIR-II fluorescent multimodal imaging for visualizing the tumor areas. On the basis of the combination of these properties, we demonstrate the novel nanoplatform based on $NdVO_4$:Gd^{3+} NPs for multimodal-imaging-guided photothermal therapy.

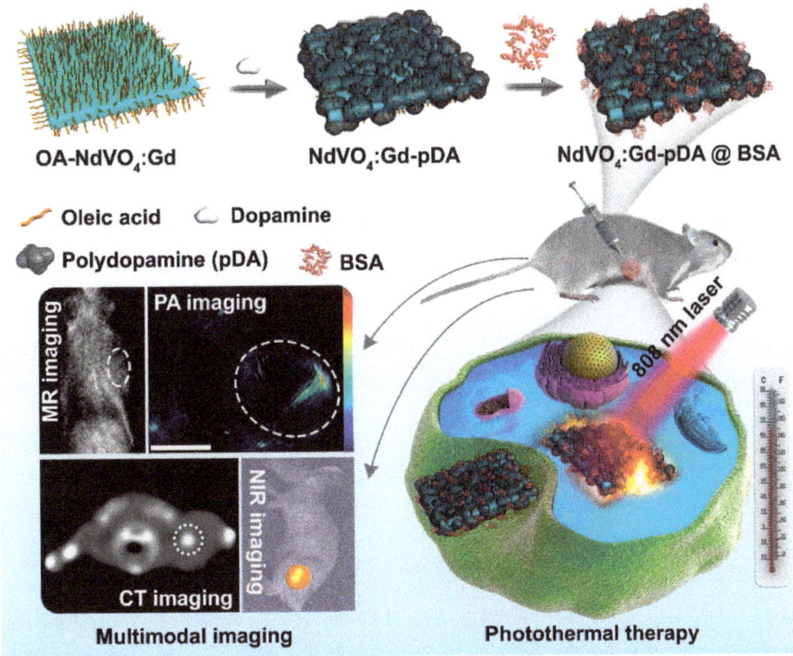

Scheme 1. Schematic illustration of the synthesis process of NV-p@BSA nanoplates and their potential bioapplication for in vivo multimodal-imaging-guided photothermal therapy under 808 nm laser irradiation.

2. Materials and Methods

2.1. Chemicals and Instrumentation

$Nd(NO_3)_3{\cdot}6H_2O$ and $Gd(NO_3)_3{\cdot}6H_2O$ were purchased from Beijing HWRK Co., LTD (Beijing, China). NH_4VO_3 was purchased from Sinopharm Chemical Reagent Co., Ltd. (Shanghai, China). Dopamine hydrochloride and polyethylene glycol (PEG) were acquired from Sigma-Aldrich Co. (Saint Louis, MO, USA). Dimethyl sulfoxide (DMSO) was obtained from Xilong Chemical Co., Ltd. (Swatow, China). Bovine serum albumin (BSA) was obtained from Shanghai Aladdin Biochemical Technology Co., Ltd. (Shanghai, China). All chemicals were used as received.

TEM images of NPs were acquired from a FEI Tecnai G2 S-Twin operating at an acceleration voltage of 200 kV. The X-ray powder diffraction was obtained using a D8 Advanced diffractometer (Bruker) with Cu Kα radiation (λ = 0.15405 nm). The UV–VIS absorption was recorded on U-3100 spectrophotometer (Hitachi). The Zeta potential was measured using a Malvern instrument Zetasizer Nano. The Confocal laser scanning microscopy images was performed on a LSM FV 1000 instrument (Olympus). Thermogravimetry data were recorded on a Netzsch Thermal Analyzer (STA 409) in an ambient environment with a heating rate of 10 °C min^{-1} from room temperature to 700 °C.

2.2. Synthesis of Oleic-Acid-Capped NdVO4:Gd^{3+} Nanoplates

Oleic-acid-capped $NdVO_4$ NPs were prepared via a facile hydrothermal method [35]. Briefly, 0.5 mmol NH_4NO_3 was added into 5 mL water containing 15 mmol NaOH to form a transparent solution. Then, 20 mL of a mixed solution of oleic acid and ethanol (v/v = 1:1) was added under strong stirring. Subsequently, 1 mL $Nd(NO_3)_3$ aqueous solution (1 mol L^{-1}) was added dropwise. After stirring for 10 min, the resulting solution was transferred into a Telfon-lined vessel and sealed, followed by treatment at 140 °C for

8 h. The as prepared NPs were collected after the autoclave cooled to room temperature naturally. For purification, these NPs were washed with cyclohexane and precipitated by ethanol twice. The 20% Gd^{3+}-doped $NdVO_4$ NPs were synthesized by the same procedure as described above, except we introduced 0.8 mL $Nd(NO_3)_3$ aqueous solution (1 mol L^{-1}) and 0.2 mL $Gd(NO_3)_3$ aqueous solution (1 mol L^{-1}) to the precursor mixture.

2.3. Synthesis of $NdVO_4$:Gd^{3+}-pDA@BSA NPs

The modification of $NdVO_4$:Gd^{3+} NPs was performed according to a previously published protocol [36]. Briefly, as-obtained $NdVO_4$:Gd^{3+} NPs were dispersed in 5 mL chloroform solution. Then, 20 mg dopamine hydrochloride was dissolved into 5 mL DMSO and added into the above solution under stirring to form a homogeneous solution. Subsequently, the mixture was heated to 70 °C for 1 h. After cooling down to room temperature, the nanoparticles were collected via centrifugation and washed with the mixture of DMSO and $CHCl_3$ (v/v = 1:1). Finally, 5 mL DMSO solution of $NdVO_4$:Gd^{3+}-pDA (20 mg) was slowly added to 10 mL BSA aqueous solution (1 mg mL^{-1}) to form a homogeneous solution with sonication. After stirring for 2 h, the mixture was separated by centrifugation and washed by water repeatedly to obtain $NdVO_4$:Gd^{3+}-pDA@BSA nanoplates. To obtain PEG-attached $NdVO_4$:Gd^{3+} nanoplates, a total of 5 mL of a chloroform dispersion of $NdVO_4$:Gd^{3+} nanoplates (0.1 mmol) was mixed with a chloroform solution (1 mL) containing 25 mg polyethylene glycol in a round bottom flask. Chloroform was then removed by evaporating slowly under nitrogen atmosphere for 24 h at room temperature. Then, the obtained mixed film was hydrated with deionized water (5 mL) after vigorous sonication. The products were collected by centrifugation at 12,000 rpm for 30 min and rinsed by deionized water at least three times.

2.4. In Vitro Photothermal Effects

A total of 200 μL of $NdVO_4$:Gd^{3+}-pDA aqueous suspension at various concentrations (31.25, 62.5, 125, 250, and 500 μg mL^{-1}) was placed in test wells. The solution was irradiated with 808 nm laser (1.0 W cm^{-2}, Changchun New Industries Optoelectronics Technology Co., Ltd., Changchun, China) for 5 min, and an infrared camera (FLIRT62101) was utilized to monitor the temperature change during this period. For comparison, the aqueous solution and PEG-modified $NdVO_4$:Gd^{3+} suspension was used as control. In this work, the particle concentrations were calculated on the basis of total mass of the resulting nanoplates. To keep the consistent particle concentration, the same calculation method was utilized through all the in vitro and in vivo experiments.

2.5. In Vitro Cytotoxicity and Phototoxicity Analyses

The in vitro cytotoxicity was tested with HeLa cells and L929 cells by 3-(4,5-dimethylthiazol-2-yl)-2,5-diphenyltetrazolium bromide (MTT) assay. The cultured HeLa cells and L929 cells were transferred, at a density of 6000 cells per well, to a 96-well plate and kept at 37 °C for 24 h. Then, $NdVO_4$:Gd^{3+}-pDA@BSA NP solutions were added to yield the final concentrations of 6.25, 12.5, 25, 50, 100, 200, 250, and 300 μg mL^{-1}, and the cells were then incubated with NPs for 24 h. Following this, the supernatant was removed, and 20 μL of MTT was added. After incubation, the medium was replaced with 150 μL of DMSO in each well. Subsequently, the plate was examined using a microplate reader, and cell viability was determined from absorbance at 490 nm relative to the controls.

To evaluate the photothermal cytotoxicity of NPs, HeLa cells were seeded onto 96-well plates for 24 h to adhere, and a series of concentrations of the NPs solution (15, 31, 62.5, 125, 250, and 300 μg mL^{-1}) were added for additional 4 h incubation. After that, the cells were exposed to a 808 nm laser for 5 min at a fixed power density of 1.0 W cm^{-2}. The cells were incubated for another 24 h. In parallel, the cells without any treatment were exposed to the laser light serving as a negative control. The phototoxicity of NPs was assessed by the MTT viability assay as described above.

2.6. Cellular Uptake of $NdVO_4$:Gd^{3+}-pDA@BSA NPs

HeLa cells were seeded in 6-well plate at a density of 8×10^4 cells per well and cultured overnight. Then, the medium was replaced with fresh culture medium containing NPs (150 μg mL^{-1}). After incubation for 4 h, the cells were washed with PBS and fixed with 10% paraformaldehyde. TEM images of fixed cells were taken on a Hitachi H-7650 TEM (Tokyo, Japan). For confocal laser scanning microscopy, FITC were used to label these NPs. Briefly, HeLa cells were incubated with FITC-labeled NPs at 37 °C for 4 h. After medium removal, the cells were washed with PBS and fixed in 4% paraformaldehyde for 10 min. For nucleus labeling, the cells were incubated with DAPI solution for 10 min. Then, the medium was removal and rinsed with PBS several times. The cells were examined with a confocal microscope.

2.7. Cell Lines and Animal Tumor Model

The human cervical cancer cells (HeLa), mouse hepatoma cells (H22), and mouse fibroblast cells (L929) were originally obtained from American Type Culture Collection (ATCC). HeLa cells and H22 cells were cultured in RMPI 1640 medium (Gibco) supplemented with 10% fetal bovine serum (Gibco) and 1% penicillin/streptomycin (Invitrogen) at 37 °C in 5% CO_2. L929 cells were cultured in DMEM medium supplemented with 10% fetal bovine serum (Gibco) and 1% penicillin/streptomycin (Invitrogen) at 37 °C in 5% CO_2. Healthy female Balb/c mice (six weeks old) were purchased from Guangdong Yaokang Biotechnology Co., Ltd. (Foshan, China). The mouse hepatoma cells H22 tumor model was established by subcutaneous injection with H22 cells (5×10^5 per mouse) into the left flank of healthy Balb/c mice. When the tumor volume reached an approximate volume of 100 mm^3, the in vivo studies were carried out. All the animal studies were fully compliant with the guidelines of National Institutes of Health Animal Care and Use Committee (NIHACUC) approved protocol.

2.8. In Vivo Multimodal Imaging

In vivo multimodal imaging experiments were performed immediately after intratumoral injection. For each imaging model, the time interval between injection and imaging could have been somewhat different due to the time required for image processing. Each imaging modality was separately performed on different mice, since the MR/CT/PA imaging equipment was located differently with restricted user access. For in vitro MR imaging, a series of NV-p@BSA aqueous solutions were prepared. The MR images were taken from a 1.2 T MR scanner for a small animal imaging system (Shanghai Huantong Science and Education Equipment Co., LTD. Shanghai, China). To perform in vivo MR imaging, 200 μL of NPs (20 mg kg^{-1}) were intratumorally injected into the anesthetized mice. The MR images were acquired before injection and after injection.

To perform X-ray CT imaging in vivo, tumor-bearing mice were first anesthetized with 10% chloral hydrate by intraperitoneal injection. A total of 100 μL of NV-p@BSA NPs was intratumorally injected into tumor-bearing mice in situ. Then, the mouse was scanned before injection and after injection. CT signal was recorded on a Philips 256-slice CT scanner (Philips Medical System).

For in vivo PA imaging, the mice bearing H22 xenograft tumors were anesthetized with 2% isofluorane and then located in prone position. Photoacoustic signals were detected at a wavelength of 808 nm. The PA images of the tumor area were monitored via a preclinical photoacoustic computed tomography scanner (MSOT InVision128) before and after injection of NPs (20 mg kg^{-1}).

For in vivo thermal imaging, 100 μL of saline, NV-PEG, and NV-p@BSA NPs were intratumorally injected into the tumors, individually. After injection, the tumor site was exposed to NIR laser at an output power of 1.0 W cm^{-2}, and IR images were recorded using infrared camera.

For in vivo NIR-II fluorescent imaging, 100 μL of NV-p@BSA NPs were intratumorally injected into the tumors. After injection, the tumor was exposed to NIR laser at an output power of 1.0 W cm^{-2}, and second biological windows fluorescent imaging was recorded using NIR II Fluorescence Imaging System (SWIR vision).

2.9. Phototherapeutic Efficacy

For antitumor activity in vivo, the H22-tumor-bearing mice were randomly distributed between 5 groups (n = 5) for different treatments: (1) control group (saline); (2) saline with NIR laser only; (3) NV-p@BSA NPs (20 mg kg^{-1}); (4) NV-PEG NPs (20 mg kg^{-1}) with NIR laser irradiation; (5) NV-p@BSA NPs (20 mg kg^{-1}) with NIR laser irradiation. For the laser treatment group, 808 nm light irradiation was conducted 6 h after injection at the power density of 1.0 W cm^{-2} for 10 min. During the experiment, tumor sizes were measured by a digital caliper, and body weights were recorded every 2 days. The in vivo treatment studies ended when the maximum tumor sizes of mice exceeded 20 mm. Tumor volume (V) was calculated by the following equation: V = (tumor length) × (tumor width)^2/2. Relative tumor volume was calculated as V/V_0 (V_0 is the initial tumor volume of each individual mouse). At the end of treatments, the mice were euthanized, and the tumors were separated and weighed.

2.10. Toxicology Evaluation

Healthy Balb/c mice were intravenously injected with saline (control group) and NV-p@BSA NPs at a dose of 20 mg kg^{-1} (test group). After injection at different times (days 1, 7, and 28), mice were euthanized and then the blood was collected for biochemistry analysis. In addition, the major organs (heart, liver, spleen, lung, and kidney) were harvested and dissected to make paraffin sections for further H&E staining. The tissue slices were examined on an inverted fluorescence microscope system (Nikon Ti–S).

3. Results and Discussion

3.1. Synthesis and Characterization

Oleic acid (OA)-stabilized $NdVO_4$ NPs were synthesized via a facile hydrothermal method, and these nanomaterials were structurally uniform with a square-like plate shape (Figure S1). With the modified synthesis route, Gd^{3+}-doped $NdVO_4$ ($NdVO_4$:Gd^{3+}) NPs can also be prepared. The resulting NPs were analyzed using a variety of techniques including transmission electron microscope (TEM), X-ray diffraction (XRD), UV–VIS–NIR absorption, and photoluminescence spectroscopy. Typical TEM images revealed that the as-obtained $NdVO_4$:Gd^{3+} NPs were of similar nanostructures as pristine $NdVO_4$ NPs, with an average length of 24 ± 6 nm (Figure 1A). To make these NPs suitable for bioapplication, an effective means was utilized to transfer these hydrophobic nanoagents to water phase. Polydopamine (pDA) were grafted on the surface of OA-$NdVO_4$:Gd^{3+} NPs through the polymerization of dopamine. As a capping agent, pDA layer had an evident influence on the color of $NdVO_4$:Gd^{3+} NPs, which changed from blue to brown (Figure 1B and inset). Power XRD patterns revealed that both $NdVO_4$:Gd^{3+} and $NdVO_4$:Gd^{3+}-pDA samples exhibited similar diffraction peaks in accordance with tetragonal $NdVO_4$ nanocrystals (Figure 1C). It can be inferred that the 20% Gd^{3+} doping in $NdVO_4$ nanocrystals did not affect their nanostructures. The UV–VIS–NIR absorption spectra showed that the as-obtained $NdVO_4$:Gd^{3+} and $NdVO_4$:Gd^{3+}-pDA had the similar characteristic peak of Nd^{3+} ions at around 808 nm (Figure 1D), providing evidence that the resulting $NdVO_4$:Gd^{3+}-pDA NP solution exhibits a strong absorption at 808 nm. Furthermore, the emission spectrum of these NPs under the excitation of 808 nm laser was examined. As depicted in Figure S2, the strong emission band at 1060 nm for $NdVO_4$:Gd^{3+}-pDA NPs could be observed clearly, ascribed to the energy transfer from $^4F_{3/2}$ level to $^4I_{11/2}$ level of Nd^{3+} ions. This result implies that $NdVO_4$:Gd^{3+}-pDA NPs have a great potential as effective NR-II imaging agents [37].

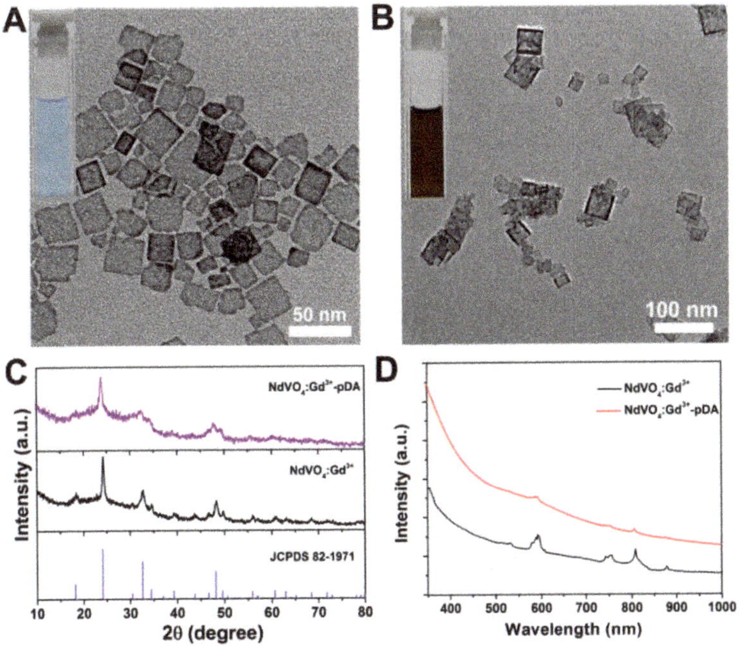

Figure 1. TEM images of $NdVO_4$:Gd^{3+} (**A**) and $NdVO_4$:Gd^{3+}-pDA nanoplates (**B**); inset: the digital photos of $NdVO_4$:Gd^{3+} in cyclohexane and $NdVO_4$:Gd^{3+}-pDA in DMSO, respectively. (**C**) XRD patterns of $NdVO_4$:Gd^{3+} samples, $NdVO_4$:Gd^{3+}-pDA nanoplates, and the standard JCPDS card no. 82-1971. (**D**) Emission spectrum of $NdVO_4$:Gd^{3+}-pDA nanoplates under excitation of 808 nm laser light.

To construct these NPs with good dispersity in aqueous solution, negatively charged BSA molecules were chosen as a model biological agent to conjugate on their surface. On the basis of the electrostatic interaction between BSA and amino groups of $NdVO_4$:Gd^{3+}-pDA NPs, BSA-modified $NdVO_4$:Gd^{3+}-pDA (NV-p@BSA) NPs were developed. As expected, the zeta-potential data showed the surface charge of $NdVO_4$:Gd^{3+}-pDA NPs (+30 mV) rapidly decreased to +6.7 mV after BSA coating (Figure S3). According to the thermogravimetric analysis (TGA) in Figure S4, the mass percent of the pDA layers in $NdVO_4$:Gd^{3+}-pDA nanoplates was evaluated as being around ≈14.5%. This modified strategy exhibited several advantages, including both coating layers (pDA and BSA) being biologically safe materials and the surface modification process being realized under a mild condition. Moreover, NV-p@BSA NPs were able to maintain their size after being dispersed in PBS solution for three days (Figure S5).

3.2. *Assessment of In Vitro Photothermal Effect*

The possibility of NV-p@BSA NPs as photothermal agents was evaluated. The temperature changes of aqueous medium containing NV-p@BSA NPs during irradiation of 808 nm laser were monitored by a NIR camera. Figure 2A shows the photothermal responses of these NPs at concentrations varying from 31.25 to 500 μg mL^{-1} during the NIR irradiation process. It can be seen that the as-prepared NV-p@BSA samples displayed an apparent concentration-dependent photo-to-heat effect. After 5 min of laser irradiation, the temperature of NP solutions at a low concentration (125 μg mL^{-1}) was able to increase from 25 °C to 55 °C, which is sufficient to destroy cancer cells. On the contrary, no significant temperature change was observed for deionized water under the same conditions, suggesting that the photothermal property originated from the as-prepared NPs. Moreover, to further study whether the pDA layer contributed to the photothermal efficiency of these NPs, PEGlyated-$NdVO_4:Gd^{3+}$ (NV-PEG) NPs were used as the control since the surface functionalization of PEG is a common strategy for phase transfer of OA-capped nanoparticles and PEG itself shows no signs of any photothermal properties. Compared with NV-PEG NPs, the resulting NV-p@BSA NPs exhibited superior light-to-heat conversion efficiency upon 808 nm laser excitation (Figure 2B). Therefore, it can be validated that the pDA layer can enhance the photothermal ability of $NdVO_4:Gd^{3+}$ NPs due to their own photothermal property, which is in line with previous reports [38–40].

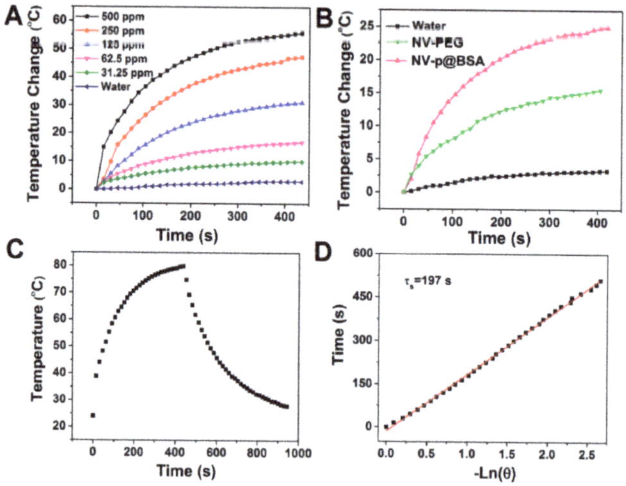

Figure 2. (**A**) Temperature elevation curves of NV-p@BSA solution with different concentrations when exposed to 808 nm NIR laser (1.0 W cm^{-2}) for 5 min. (**B**) Temperature changes of NV-p@BSA and NV-PEG solution at the same concentration under 808 nm NIR laser excitation. (**C**) Temperature changes of 500 μg mL^{-1} NV-p@BSA aqueous solution over one laser on/off cycle of NIR laser irradiation. (**D**) Linear plot of time data versus $-\ln(\theta)$ obtained from the cooling period of (**C**).

Accordingly, to analyze the photothermal conversion efficiency of these NV-p@BSA NPs, we measured the temperature changes of NPs solution (500 μg mL^{-1}) when irradiated with 808 nm laser (1.0 W cm^{-2}) and subsequently recorded the cooling process after switching off the NIR laser. As depicted in Figure 2C, the temperature elevation during this process was monitored to achieve a heat generation–dissipation curve. On the basis of the reported method, the photothermal transduction efficiency of samples can be expressed as the following Equation (1):

$$\eta = \frac{hS(T_{max} - T_{surr}) - Q_{dis}}{I(1 - 10^{-A_{808}})} \quad (1)$$

where T_{max} (K) is the equilibrium temperature, T_{surr} (K) is ambient temperature of the surroundings, Q_{dis} (W) is heat loss from light absorbed by the container, A_{808} is the absorbance of samples at 808 nm, h (W·cm^{-2}·K^{-1}) is heat transfer coefficient, and S (cm^2) represents the surface area of the container.

According to this calculation, the heat conversion efficiency of NV-p@BSA NPs was estimated to be 31.7% (Figure 2D). The photothermal conversion properties of NV-p@BSA did not show significant changes after the three cycles, indicating their good thermal stability (Figure S6). All the results suggested that the as-prepared NV-p@BSA NPs could be a promising 808 nm laser-driven photothermal agent.

3.3. *In Vitro Cytotoxicity Studies*

To examine the potential of these NPs as theranostic agents, in vitro dark cytotoxicity and NIR-induced photothermal cytotoxicity of NV-p@BSA NPs were evaluated using standard MTT assay. Two types of cells, normal cells (L929 cells) and tumor cells (HeLa cells), were used as model cells. After incubation with NV-p@BSA NPs in the concentration range of 0–300 μg mL^{-1} for 24 h, both cell viabilities were detected. As displayed in Figure 3A, no significant cytotoxicity in normal cells was observed in a high concentration of NPs (300 μg mL^{-1}), as well as in HeLa cells, suggesting that these NPs can serve as a biocompatible nanoplatform for cancer therapy in living cells. In addition, the photothermal effect of NV-p@BSA NPs was further assessed at the cellular level. In absence of NPs, NIR laser irradiation alone caused negligible living cell death (Figure 3B). In contrast, a noticeable concentration-dependent cytotoxicity against the cancerous cells was observed in the exposure of HeLa cells to NV-p@BSA NPs for 4 h followed by laser irradiation. As we can see, the cells viability significantly decreased to only 17% when HeLa cells were treated with a dose of 300 μg mL^{-1} NPs plus laser irradiation. These data infer that these NPs could effectively absorb NIR light and generate increased temperatures sufficient to induce irreversible damage to cancer cells.

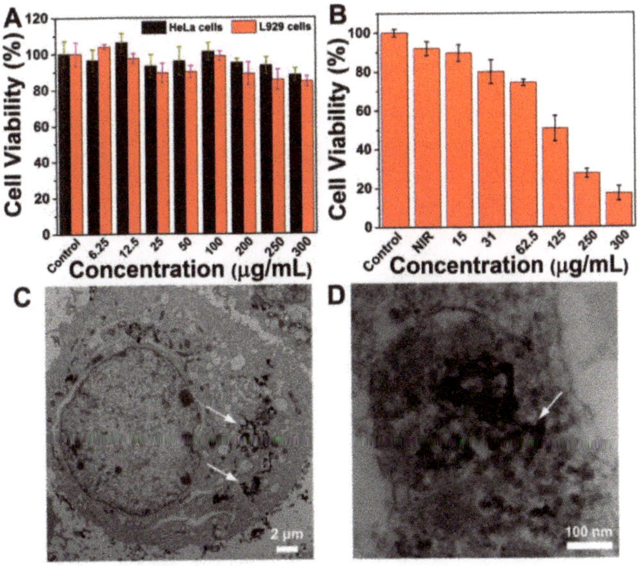

Figure 3. (**A**) In vitro cytotoxicity assays of NV-p@BSA NPs at various concentrations against HeLa and L929 cells. (**B**) Photo-induced cytotoxicity of NV-p@BSA samples plus the 808 nm laser irradiation (0.8 W cm^{-2}, 5 min) towards HeLa cells. (**C**,**D**) TEM images of HeLa cells incubated with NV-p@BSA NPs for 4 h. The white arrows indicate the locations of NV-p@BSA NPs within the cells.

Subsequently, the cellular uptake behavior of NV-p@BSA NP was investigated on HeLa cells. As shown in Figure S7, the green fluorescence from FITC-labeled NV-p@BSA NP and the blue fluorescence from DAPI-labeled cell nucleus exhibited effective overlay, verifying that NV-p@BSA NP could be taken up by HeLa cells. Meanwhile, biological TEM images of the treated cells were used to monitor their cellular location. As shown in Figure 3C,D, NV-p@BSA NPs were found to be effectively phagocytized by HeLa cells and were mainly located inside the endosomes. These data provide insight into the effective cellular uptake of NV-p@BSA NPs by tumor cells, which makes it possible for these NPs to eradicate cancer cells under the irradiation of an 808 nm laser.

3.4. In Vitro and In Vivo Multimodal Imaging

Versatile nanoplatforms with high contrast ability are urgently needed for guiding the therapeutic treatment in avoidance of damaging the surrounding healthy tissues. In this study, the in vivo multimodal imaging performance of as-prepared NV-p@BSA NPs was evaluated on a H22 xenograft model, including T_1-weighted MR, CT, PA, IR, and NIR-II imaging. Firstly, the in vitro MR nature of these Gd^{3+}-doped NPs was confirmed by measuring T_1 value as a function of their concentrations. Figure 4A demonstrates the T_1 MR signals were enhanced gradually with the increasing concentration of these solutions, which exhibited a noticeable brightening effect. The trend was well fitted by a linear line within the tested range of Gd^{3+} concentrations. The T_1 relaxivity coefficient value was estimated to be 1.726 Mm^{-1} s^{-1}. To investigate the in vivo MR imaging of NV-p@BSA NPs, MR images of tumor-bearing mice before and after the injection of NPs (100 μg mL^{-1}) were acquired. In the post-injection image, an MRI signal enhancement on the tumor site was observed.

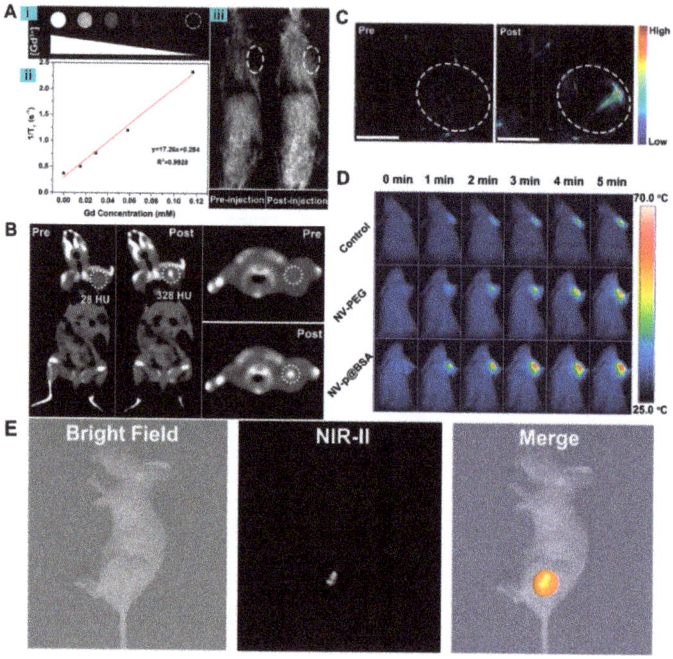

Figure 4. (**A**) Representative T_1-weighed MR contrast effects of NV-p@BSA NPs at different concentrations (i); graphs of relaxation rates R_1 versus different concentrations of NPs (ii); in vivo MR images of tumor-bearing mice before and after injection of NV-p@BSA NPs (iii). (**B**) In vivo CT images of tumor-bearing mice before and after injection of NV-p@BSA NPs. The white circles indicate the tumor regions. (**C**) Representative PA images at the tumor region before and after injection of NV-p@BSA NPs. Representative thermal images (**D**) and NIR-II fluorescence images (**E**) of mice subjected to 808 nm laser irradiation (1.0 W cm^{-2}).

X-ray computed tomographic (CT) imaging is proven with high resolution and deep penetration for clinical diagnoses. Figure 4B presents the in vivo CT images acquired before and after intratumoral injection of NV-p@BSA NPs (100 mL, 5 mg mL^{-1}) into tumor-bearing mice. Compared with the intrinsic tumor image pre-injection, the injected tumor area exhibited a brighter contrast effect, suggesting a significant increase in the CT signal. To demonstrate the feasibility of NV-p@BSA NPs for PA imaging, in vivo three-dimensional PA images of the tumor region were also conducted for NP-injected mice. For those treated with NV-p@BSA NPs, obvious PA amplitude for mice was revealed in comparison to the untreated tumor (Figure 4C). As shown in Figure 4D, the temperature of the tumor site in normal saline injection mice was not significantly increased in the control group, which confirmed that such a near infrared light irradiation condition had no additional damage to the tumor site. The surface temperature of the tumor increased rapidly after in situ injection of the same concentration of NV-PEG and NV-p@BSA, while the temperature of the latter material was significantly higher than that of the former under the same light irradiation conditions, indicating that the NV-p@BSA performed a higher heat production ability for tumor thermal ablation. The NIR emission peak of Nd^{3+} ions can be derived from radiative transitions of ${}^4F_{3/2}{\rightarrow}{}^4I_{11/2}$ (1060 nm), suggesting NV-p@BSA could be served as NIR-II fluorescence with favorable photostability. To evaluate the NIR-II fluorescence imaging feasibility of NV-p@BSA NPs in vivo, the whole-body imaging of tumor-bearing mice within the NIR-II region was conducted for NP intratumoral injection. For those treated with NV-p@BSA NPs, a strong NIR-II fluorescence signal for mice was detected in the tumor region (Figure 4E). Taken together, all these above data support that the NV-p@BSA NPs could be promising agents for synchronous MR/CT/PA/IR/NIR-II imaging, which may be useful for cancer diagnosis, owing to the excellent depth of visualization in tumor tissues.

3.5. *In Vivo Photothermal Effect*

The in vivo photothermal effect of as-obtained NV-p@BSA NPs were investigated via infrared thermal imaging and photothermal ablation of the tumor. Before in vivo NIR-light-triggered treatment, 200 μL of saline, NV-PEG, and NV-p@BSA solution was injected into the tumor-bearing mice. Under 808 nm laser irradiation, the cutaneous temperature of the tumor region was recorded by an infrared thermo-camera as a function of time. As described in Figure 4D, it produced a well localized hyperthermal and accelerated temperature response inside the tissue in the NV-PEG-NP-treated group, showing that the temperature increased to 50 °C. In comparison, the tumor temperature reached up 65 °C in the NV-p@BSA-treated mice. This implies that NV-p@BSA NPs exhibit more potent photothermal performance over NV-PEG NPs. Nevertheless, only inapparent thermal absorption by tissue was observed in NIR-laser-treated mice, suggesting the laser irradiation is safe for surrounding healthy tissues but enough to activate these NPs to induce photodamage to the tumor site. On the basis of these infrared thermal images, we found the heating spot was only limited to the tumor region, suggesting that the thermal damage could be constructed in the area of interest and that the surrounding normal cells were weakly affected.

Motivated by the outstanding in vitro photothermal performance and low cytotoxicity, the in vivo antitumor activity of this NV-p@BSA nanosystem was further investigated in tumor-bearing Balb/c mice that were inoculated subcutaneously with H22 hepatocellular carcinoma cells. These mice were randomly assigned into five groups and treated with an intratumoral injection of different samples, including saline, NV-PEG NPs, or NV-p@BSA NPs. In NIR-laser-treated groups, an 808 nm laser was focused on the tumor site of mice for 5 min (1.0 W cm^{-2}) after injection. The tumor growth was monitored within 28 days following the treatment. As displayed in Figure 5A, there was no significant difference in final tumor size for saline-, saline + NIR-, or NV-p@BSA-treated groups, indicating that these treatments had little effect on tumor inhibition. Once laser irradiation was applied, the tumor growth of mice treated with NV-PEG NPs was efficiently delayed. As expected, the tumor growth was greatly suppressed in the NV-p@BSA + NIR treated group, implying their pronounced photothermal performance (Figure 5B,C). These results also suggest that tHE pDA modification of these NPs further enhances the therapeutic efficacy when compared with NV-PEG NPs without the pDA layer, enabling them to be more effective in tumor photoablation.

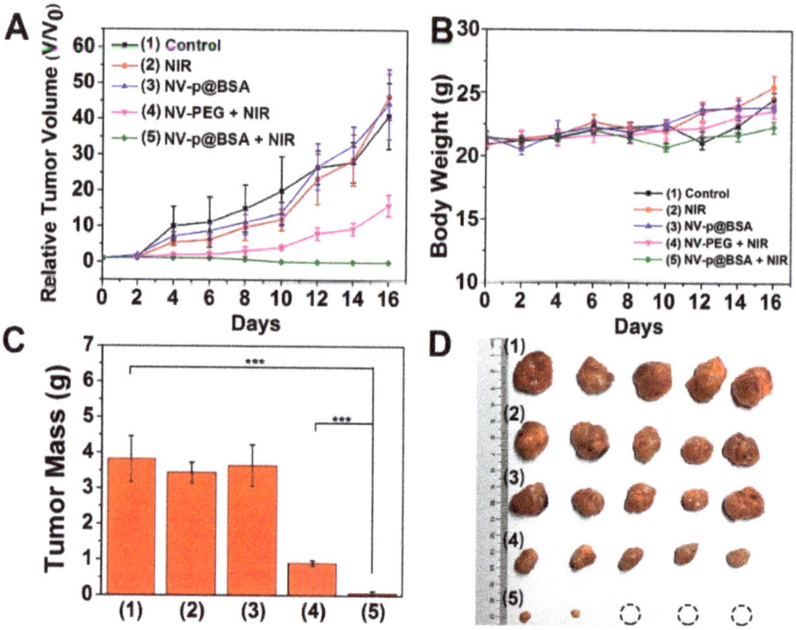

Figure 5. (**A**) Relative tumor volume changes and (**B**) corresponding body weight changes of tumor-bearing Balb/c mice after different treatment processes, as indicated. (**C**) Average weights of tumors collected from mice in each treatment group. (**D**) Photos of representative excised tumor tissues from different groups of mice at the end of treatments. p-values: *** $p < 0.001$.

3.6. In Vivo Biosafety Investigation

To assess the potential long-term toxicity of intratumorally injected NV-p@BSA NPs after photothermal ablation, a simulated in vivo toxicity assessment was performed by injecting NPs (20 mg kg^{-1}) into healthy Balb/c mice via the tail vein. This investigation could be achieved by monitoring blood biochemical index, body weight gain, and histological examinations. At the examined time points post-injection, the blood from NP-treated mice was collected and some important indicators were analyzed, including alanine aminotransferase (ALT), alkaline phosphatase (ALP), and aspartate aminotransferase (AST) for hepatic function, as well as blood urea (BUN) and serum creatinine (CREA) for renal function. As shown in Figure 6A–C, all these parameters were at a normal level and showed no visible differences compared with control group injected with saline. These biochemical data imply that no obvious abnormalities were associated with the treatment. In addition, all mice in the test group displayed similar changing profiles of body weight gain as the control group (Figure 6D), demonstrating that there were no acute side effects caused by our nanotherapeutic agent. To further assess the in vivo biocompatibility of these NPs, major organs of these injected mice were harvested on the 28th day post-injection and sliced for histology analysis. Hematoxylin and eosin (H&E)-stained sections suggest that no apparent lesion such as inflammatory response and necrocytosis was observed in the tissue sections (Figure 6E), further manifesting the low toxicity of the NV-p@BSA NPs during the experiment.

Overall, these preliminary results show that NV-p@BSA NPs had low side effects at the tested dose in vivo, suggesting their potential as a therapeutic nanoplatform for intravenous photothermal treatments. As a multifunctional nanoagent, NV-p@BSA NPs have potential application for multiple imaging-guided photothermal therapies in theranostic nanomedicine, although further in vivo toxicity studies and functional modification are still needed to be carried out.

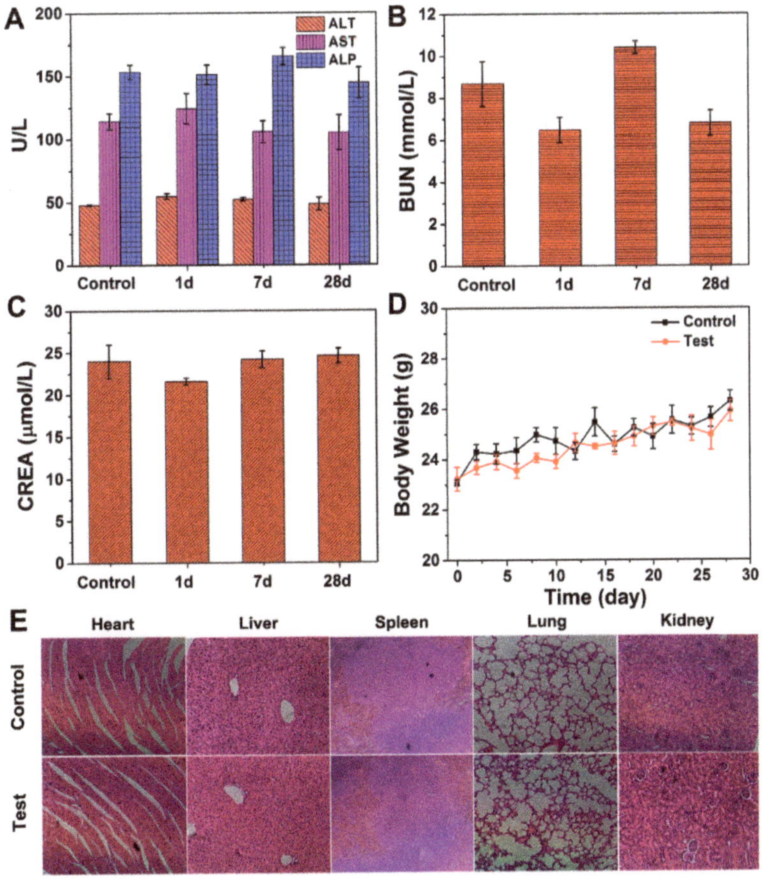

Figure 6. (**A–C**) Blood serum biochemistry data of healthy Balb/c mice with intravenous injection of NV-p@BSA NPs. The data were collected at different time points (1st, 7th, and 28th days post-injection). (**D**) Body weight change of Balb/c mice treated with NV-p@BSA NPs within 28 days compared to the control group without any injection. (**E**) H&E-stained histological slices from mice receiving no injection (control) or mice injected with NV-p@BSA NPs (test) at the 28th day post-injection.

4. Conclusions

In summary, this work reports a multifunctional NV-p@BSA nanoplatform for multimodal-imaging-guided photothermal therapy in vivo, owing to their excellent photothermal conversion capability and functional compositions. The NV-p@BSA nanosystem shows little appreciable dark toxicity and can serve as an efficient photothermal agent with irradiation time and dosage dependence when exposed to external NIR laser. With the effective cellular uptake by cancer cells, these NPs are proven to induce severe photo-cytotoxicity in vitro and exert significant tumor ablation post-injection in vivo. Remarkably, these NPs are capable of in vivo MR/CT/PA multimodal imaging. Therefore, we believe that these NV-p@BSA NPs hold great potential as theranostic agents for accurate diagnosis and efficient photothermal ablation of tumors.

Supplementary Materials: The following supporting information can be downloaded at: https://www.mdpi.com/article/10.3390/pharmaceutics14061217/s1, Figure S1: TEM image of as-obtained $NdVO_4$ nanoplates. Figure S2: NIR-II emission spectrum of as-obtained $NdVO_4$:Gd^{3+}-pDA solution. Figure S3: Surface charges of $NdVO_4$:Gd^{3+}-pDA (1) and NV-p@BSA (2) aqueous solution. Figure S4: Typical thermal gravimetric analysis curve of $NdVO_4$:Gd^{3+} and $NdVO_4$:Gd^{3+}-pDA nanoplates. Figure S5: The particle size of NV-p@BSA NPs in PBS solution over 72 h. Figure S6: Temperature increases by NV-p@BSA over three on/off cycles of 808 nm irradiation. Figure S7: The confocal microscope images of HeLa cells after being cultured with NV-p@BSA NPs for 24 h.

Author Contributions: The manuscript was written through contributions of all authors. Conceptualization, Z.H. and J.L. (Jun Lin); investigation, K.D., D.L., Z.W., Z.Z., Q.C. and J.L. (Jiamin Luo); formal analysis, K.D., Y.Z. and Z.W.; writing—original draft preparation, K.D. and D.L.; writing—review and editing, Y.Z.; Data curation, Z.Z.; funding acquisition, Z.H. and J.L. (Jun Lin). All authors have read and agreed to the published version of the manuscript.

Funding: This project was financially supported by the Science and Technology Cooperation Project between the Chinese and Australian Governments (2017YFE0132300); the National Natural Science Foundation of China (NSFC 52072082, 32101143, 51929201, 51672268, 51720105015, 51972138, 51872263, 51828202); the Science and Technology Development Planning Project of Jilin Province (20190201232JC); the CAS-Croucher Funding Scheme for Joint Laboratories (CAS18204); the Guangdong Natural Science Foundation (2019A1515012214); the Key Scientific Research Project of the Guangdong Provincial Department of Education (2020ZDZX2020); and the open research funds from the Sixth Affiliated Hospital of Guangzhou Medical University, Qingyuan People's Hospital (202011-105).

Institutional Review Board Statement: All animal studies were carried out following protocols approved by the Institutional Animal Care and Use Committee at Guangzhou Medical University and complied with all relevant ethical regulations (approval number 2021-069, date: 27 August 2021). Healthy female Balb/c mice (six weeks old) were purchased from the Center for Experimental Animals, Jilin University. The mouse hepatoma cell H22 tumor model was established by subcutaneous injection with H22 cells (5×10^5 per mouse) into the left flank of healthy Balb/c mice. When the tumor volume reached an approximate volume of 100 mm^3, the in vivo studies were carried out. All the animal studies were fully compliant with the guidelines of National Institutes of Health Animal Care and Use Committee (NIHACUC) approved protocol.

Informed Consent Statement: Not applicable.

Data Availability Statement: The results of this paper are entirely documented in the figures of this paper.

Conflicts of Interest: The authors declare no conflict of interest.

References

1. Lepock, J. Role of nuclear protein denaturation and aggregation in thermal radiosensitization. *Int. J. Hyperth.* **2004**, *20*, 115–130. [CrossRef] [PubMed]
2. Meredith, S.C. Protein denaturation and aggregation: Cellular responses to denatured and aggregated proteins. *Ann. N. Y. Acad. Sci.* **2006**, *1066*, 181–221. [CrossRef] [PubMed]
3. Kampinga, H.H.; Brunsting, J.F.; Stege, G.J.; Burgman, P.W.; Konings, A.W. Thermal protein denaturation and protein aggregation in cells made thermotolerant by various chemicals: Role of heat shock proteins. *Exp. Cell Res.* **1995**, *219*, 536–546. [CrossRef]
4. Liu, Y.; Bhattarai, P.; Dai, Z.; Chen, X. Photothermal therapy and photoacoustic imaging via nanotheranostics in fighting cancer. *Chem. Soc. Rev.* **2019**, *48*, 2053–2108. [CrossRef] [PubMed]
5. Liu, S.; Pan, X.; Liu, H. Two-dimensional nanomaterials for photothermal therapy. *Angew. Chem. Int. Ed.* **2020**, *132*, 5943–5953. [CrossRef]
6. Hu, J.J.; Cheng, Y.J.; Zhang, X.Z. Recent advances in nanomaterials for enhanced photothermal therapy of tumors. *Nanoscale* **2018**, *10*, 22657–22672. [CrossRef]
7. Jung, H.S.; Verwilst, P.; Sharma, A.; Shin, J.; Sessler, J.L.; Kim, J.S. Organic molecule-based photothermal agents: An expanding photothermal therapy universe. *Chem. Soc. Rev.* **2018**, *47*, 2280–2297. [CrossRef]
8. Wang, H.; Chang, J.; Shi, M.; Pan, W.; Li, N.; Tang, B. A dual-targeted organic photothermal agent for enhanced photothermal therapy. *Angew. Chem. Int. Ed.* **2019**, *131*, 1069–1073. [CrossRef]
9. Jaque, D.; Maestro, L.M.; del Rosal, B.; Haro-Gonzalez, P.; Benayas, A.; Plaza, J.; Rodríguez, E.M.; Solé, J.G. Nanoparticles for photothermal therapies. *Nanoscale* **2014**, *6*, 9494–9530. [CrossRef]

10. Beik, J.; Abed, Z.; Ghoreishi, F.S.; Hosseini Nami, S.; Mehrzadi, S.; Shakeri-Zadeh, A.; Kamrava, S.K. Nanotechnology in hyperthermia cancer therapy: From fundamental principles to advanced applications. *J. Control. Release* **2016**, *235*, 205–221. [CrossRef]
11. Lovell, J.F.; Jin, C.S.; Huynh, E.; Jin, H.; Kim, C.; Rubinstein, J.L.; Chan, W.C.; Cao, W.; Wang, L.V.; Zheng, G. Porphysome nanovesicles generated by porphyrin bilayers for use as multimodal biophotonic contrast agents. *Nat. Mater.* **2011**, *10*, 324–332. [CrossRef] [PubMed]
12. Lin, H.; Wang, Y.; Gao, S.; Chen, Y.; Shi, J. Theranostic 2D tantalum carbide (MXene). *Adv. Mater.* **2018**, *30*, 1703284. [CrossRef] [PubMed]
13. Smith, A.M.; Mancini, M.C.; Nie, S. Second window for in vivo imaging. *Nat. Nanotechnol.* **2009**, *4*, 710–711. [CrossRef] [PubMed]
14. Weissleder, R. A clearer vision for in vivo imaging. *Nat. Biotechnol.* **2001**, *19*, 316–317. [CrossRef] [PubMed]
15. Xiang, H.; Zhao, L.; Yu, L.; Chen, H.; Wei, C.; Chen, Y.; Zhao, Y. Self-assembled organic nanomedicine enables ultrastable photo-to-heat converting theranostics in the second near-infrared biowindow. *Nat. Commun.* **2021**, *12*, 218. [CrossRef]
16. Cao, F.; Chen, J.; Yu, D.; Wang, S.; Xu, X.; Liu, J.; Han, Z.; Huang, B.; Gu, Y.; Choy, K.L. Bionic detectors based on low-bandgap inorganic perovskite for selective NIR-I photon detection and imaging. *Adv. Mater.* **2020**, *32*, 1905362. [CrossRef]
17. Yang, J.; Zhao, Y.; Meng, Y.; Zhu, H.; Yan, D.; Liu, C.; Xu, C.; Zhang, H.; Xu, L.; Li, Y. Irradiation-free photodynamic therapy in vivo induced by enhanced deep red afterglow within NIR-I bio-window. *Chem. Eng. J.* **2020**, *387*, 124067. [CrossRef]
18. Yu, Z.; Chan, W.K.; Zhang, Y.; Tan, T.T.Y. Near-infrared-II activated inorganic photothermal nanomedicines. *Biomaterials* **2021**, *269*, 120459. [CrossRef]
19. Tang, Y.; Li, Y.; Hu, X.; Zhao, H.; Ji, Y.; Chen, L.; Hu, W.; Zhang, W.; Li, X.; Lu, X. "Dual lock-and-key"-controlled nanoprobes for ultrahigh specific fluorescence imaging in the second near-infrared window. *Adv. Mater.* **2018**, *30*, 1801140. [CrossRef]
20. Liu, Y.; Zhen, W.; Wang, Y.; Liu, J.; Jin, L.; Zhang, T.; Zhang, S.; Zhao, Y.; Song, S.; Li, C. One-dimensional Fe_2P acts as a Fenton agent in response to NIR II light and ultrasound for deep tumor synergetic theranostics. *Angew. Chem. Int. Ed.* **2019**, *131*, 2429–2434. [CrossRef]
21. Yang, T.; Wang, Q.; Lv, X.; Song, X.; Ke, H.; Guo, Z.; Huang, X.; Hu, J.; Li, Z.; Yang, P. Albumin-coordinated assembly of clearable platinum nanodots for photo-induced cancer theranostics. *Biomaterials* **2018**, *154*, 248–260.
22. Dreaden, E.C.; Alkilany, A.M.; Huang, X.; Murphy, C.J.; El-Sayed, M.A. The golden age: Gold nanoparticles for biomedicine. *Chem. Soc. Rev.* **2012**, *41*, 2740–2779. [CrossRef] [PubMed]
23. Li, X.; Shan, J.; Zhang, W.; Su, S.; Yuwen, L.; Wang, L. Recent advances in synthesis and biomedical applications of two-dimensional transition metal dichalcogenide nanosheets. *Small* **2017**, *13*, 1602660. [CrossRef] [PubMed]
24. Tan, C.; Cao, X.; Wu, X.J.; He, Q.; Yang, J.; Zhang, X.; Chen, J.; Zhao, W.; Han, S.; Nam, G.H. Recent advances in ultrathin two-dimensional nanomaterials. *Chem. Rev.* **2017**, *117*, 6225–6331. [CrossRef] [PubMed]
25. Huang, K.; Li, Z.; Lin, J.; Han, G.; Huang, P. Two-dimensional transition metal carbides and nitrides (MXenes) for biomedical applications. *Chem. Soc. Rev.* **2018**, *47*, 5109–5124. [CrossRef]
26. Chen, Y.W.; Su, Y.L.; Hu, S.H.; Chen, S.Y. Functionalized graphene nanocomposites for enhancing photothermal therapy in tumor treatment. *Adv. Drug Deliv. Rev.* **2016**, *105*, 190–204. [CrossRef]
27. Lin, J.; Wang, M.; Hu, H.; Yang, X.; Wen, B.; Wang, Z.; Jacobson, O.; Song, J.; Zhang, G.; Niu, G. Multimodal-imaging-guided cancer phototherapy by versatile biomimetic theranostics with UV and γ-irradiation protection. *Adv. Mater.* **2016**, *28*, 3273–3279. [CrossRef]
28. Yao, B.; Wu, K.; Zhang, C.; Zhang, H.; Wang, Z.; Wang, J.; Yu, H.; Yu, Y.; Jiang, M. Crystal growth and laser performance of neodymium-doped scandium orthovanadate. *J. Cryst. Growth* **2010**, *312*, 720–723. [CrossRef]
29. Xu, J.; Hu, C.; Liu, G.; Liu, H.; Du, G.; Zhang, Y. Synthesis and visible-light photocatalytic activity of $NdVO_4$ nanowires. *J. Alloys Compd.* **2011**, *509*, 7968–7972. [CrossRef]
30. Verma, S.; Gupta, R.; Bamzai, K. Effect of mixing Ce^{3+} and Nd^{3+} ions in equimolar ratio on structural, optical and dielectric properties on pure cerium orthovanadate and neodymium orthovanadate. *Mater. Res. Bull.* **2016**, *81*, 71–84. [CrossRef]
31. Deng, H.; Yang, S.; Xiao, S.; Gong, H.M.; Wang, Q.Q. Controlled synthesis and upconverted avalanche luminescence of cerium (III) and neodymium (III) orthovanadate nanocrystals with high uniformity of size and shape. *J. Am. Chem. Soc.* **2008**, *130*, 2032–2040. [CrossRef] [PubMed]
32. Monsef, R.; Ghiyasiyan-Arani, M.; Salavati-Niasari, M. Application of ultrasound-aided method for the synthesis of $NdVO_4$ nano-photocatalyst and investigation of eliminate dye in contaminant water. *Ultrason. Sonochem.* **2018**, *42*, 201–211. [CrossRef] [PubMed]
33. Del Rosal, B.; Pérez-Delgado, A.; Carrasco, E.; Jovanović, D.J.; Dramićanin, M.D.; Dražić, G.; de la Fuente, Á.J.; Sanz-Rodriguez, F.; Jaque, D. Neodymium-based stoichiometric ultrasmall nanoparticles for multifunctional deep-tissue photothermal therapy. *Adv. Opt. Mater.* **2016**, *4*, 782–789. [CrossRef]
34. Rocha, U.; Kumar, K.U.; Jacinto, C.; Villa, I.; Sanz-Rodríguez, F.; del Carmen Iglesias de la Cruz, M.; Juarranz, A.; Carrasco, E.; van Veggel, F.C.; Bovero, E. Neodymium-doped LaF_3 nanoparticles for fluorescence bioimaging in the second biological window. *Small* **2014**, *10*, 1141–1154. [CrossRef] [PubMed]
35. Liu, J.; Li, Y. General synthesis of colloidal rare earth orthovanadate nanocrystals. *J. Mater. Chem.* **2007**, *17*, 1797–1803. [CrossRef]
36. Quan, Q.; Xie, J.; Gao, H.; Yang, M.; Zhang, F.; Liu, G.; Lin, X.; Wang, A.; Eden, H.S.; Lee, S. HSA coated iron oxide nanoparticles as drug delivery vehicles for cancer therapy. *Mol. Pharm.* **2011**, *8*, 1669–1676. [CrossRef]

37. Welsher, K.; Sherlock, S.P.; Dai, H. Deep-tissue anatomical imaging of mice using carbon nanotube fluorophores in the second near-infrared window. *Proc. Natl. Acad. Sci. USA* **2011**, *108*, 8943–8948. [CrossRef]
38. Lin, L.S.; Cong, Z.X.; Cao, J.B.; Ke, K.M.; Peng, Q.L.; Gao, J.; Yang, H.H.; Liu, G.; Chen, X. Multifunctional Fe_3O_4@ polydopamine core-shell nanocomposites for intracellular mRNA detection and imaging-guided photothermal therapy. *ACS Nano* **2014**, *8*, 3876–3883. [CrossRef]
39. Chen, Y.; Ai, K.; Liu, J.; Ren, X.; Jiang, C.; Lu, L. Polydopamine-based coordination nanocomplex for T1/T2 dual mode magnetic resonance imaging-guided chemo-photothermal synergistic therapy. *Biomaterials* **2016**, *77*, 198–206. [CrossRef]
40. Liu, B.; Li, C.; Xing, B.; Yang, P.; Lin, J. Multifunctional UCNPs@PDA-ICG nanocomposites for upconversion imaging and combined photothermal/photodynamic therapy with enhanced antitumor efficacy. *J. Mater. Chem. B* **2016**, *4*, 4884–4894. [CrossRef]

pharmaceutics

Article

Monoclonal Antibody Functionalized, and L-lysine α-Oxidase Loaded PEGylated-Chitosan Nanoparticle for HER2/Neu Targeted Breast Cancer Therapy

Kandasamy Saravanakumar [1], Anbazhagan Sathiyaseelan [1], Soyoung Park [1], Song-Rae Kim [2], Veeraraghavan Vishnu Priya [3] and Myeong-Hyeon Wang [1,*]

1 Department of Bio Health Convergence, Kangwon National University, Chuncheon 200-701, Gangwon-do, Korea; saravana732@kangwon.ac.kr (K.S.); sathiyaseelan@kangwon.ac.kr (A.S.); 202116272@kangwon.ac.kr (S.P.)
2 Chuncheon Center, Korea Basic Science Institute (KBSI), Chuncheon 24341, Gangwon-do, Korea; ksr87@kbsi.re.kr
3 Department of Biochemistry, Saveetha Institute of Medical and Technical Sciences, Saveetha Dental College, Saveetha University, Chennai 600077, India; vishnupriya@saveetha.com
* Correspondence: mhwang@kangwon.ac.kr

Simple Summary: Breast cancer is one of the dominant cancers that threaten human beings worldwide. Moreover, the treatment of HER2+ breast cancer is challenging due to heterogeneity. The L-lysine α-oxidase (LO) enzyme is a well-known antitumor enzyme, but its clinical utility has been limited due to side effects, decreased stability, and inability to target tumor cells. To overcome the clinical challenges in delivery of LO enzymes and improve HER2+ breast cancer therapeutics, the present study developed the dual stimuli responsive nanocarrier system (CS-LO-PEG-HER NPs) for pH sensitive and HER2/neu targeted breast cancer therapy.

Abstract: Herein, we designed a nanocarrier to deliver the LO specifically to HER2+ breast cancer (BC) cells, where functionalization of mAb (anti-HER2+) with PEGylated chitosan enabled it to target the HER2+ BC cells. Taking advantage of overexpression of HER2+ in cancer cells, our nanocarrier (CS-LO-PEG-HER NPs) exhibited promising potency and selectivity against HER2+ BC cells (BT474). The CS-LO-PEG-HER NPs demonstrated the cytotoxicity in BT474 cells by promoting reactive oxygen species, mitochondrial membrane potential loss, and nucleus damage. The biocompatibility of CS-LO-PEG-HER NPs was evidenced by the hemolysis assay and H & E staining of major organs. The CS-LO-PEG-HER NPs showed anticancer potency against the BT474-xenograft tumor-bearing mice, as evident by the reduction of tumor size and cell density. These results indicate that CS-LO-PEG-HER NPs are biocompatible with mice while inhibiting tumor growth through alter the oxidative stress. Overall, this work provides a promising approach for the delivery of LO for good therapeutic effect in combination with mAb.

Keywords: L-lysine α-oxidase; HER2+; trastuzumab; breast cancer; PEGylated chitosan

Citation: Saravanakumar, K.; Sathiyaseelan, A.; Park, S.; Kim, S.-R.; Priya, V.V.; Wang, M.-H. Monoclonal Antibody Functionalized, and L-lysine α-Oxidase Loaded PEGylated-Chitosan Nanoparticle for HER2/Neu Targeted Breast Cancer Therapy. *Pharmaceutics* **2022**, *14*, 927. https://doi.org/10.3390/pharmaceutics14050927

Academic Editors: Manivasagan Panchanathan and Eue-Soon Jang

Received: 21 March 2022
Accepted: 21 April 2022
Published: 24 April 2022

1. Introduction

L-lysine α-oxidase (LO) was first discovered in *Trichoderma viride*, then in *T. harzianum* Rifai, in the 1980s [1,2]. This enzyme has been shown to inhibit the proliferation of cancer cells, such as human lung squamous cell carcinoma (RERF-LC-AI), prostate cancer (PC3), cervical cancer cell line (HeLa), ovarian cancer (SKOV3), breast cancer (MCF7), colon cancers (LS174T and HT29) and erythromyeloblastoid leukemia (K562), without causing any significant damage to normal human umbilical vein endothelial cells (HUVECs) [3,4]. Breast and ovarian cancer cells are more sensitive to LO than others [5]. The enzyme LO triggers H_2O_2 generation, which causes cytotoxicity through damaging the DNA and mitochondrial membrane [6]. Moreover, the cytotoxicity of LO is dependent on the concentration of L-lysine in the tissue [5]. The enzyme LO exhibits high antitumor activity due to the higher elevation of L-lysine in tumor tissue compared to normal tissue [4,7]. L-Lysine is an essential amino acid present in body tissue and blood plasma, which plays an important functional role in metabolism [8]. Therefore, deprivation of L-lysine is an important issue caused by LO administration to animals. Although LO is efficient in antitumor activity, an administration of 150 U/kg at a 48 h interval might cause toxicity, weight loss (>30%) and loss of L-lysine in blood plasma, which results in the death of mice [3]. The LO administration (75 U/Kg of four doses for 10 days at 2 days interval) suppresses the growth of cancer cells in Balb/c nude mice that were transplanted with cancer cells [5].

The LO is well known as a potential antitumor enzyme, but its overdose can cause toxicity to L-lysine in normal cells. Further, the therapeutic efficacy of LO is largely limited due to degradation and insufficient discharge at the cancer site. There is demand for the development of new strategies to release LO at cancerous sites for enhanced cancer therapy without causing any adverse effects to normal cells. Similarly, the overexpression of the human estrogen receptor (HER2+) efficiently stimulates breast cancer (BC) growth. Therefore, blocking the signal transfer between HER2 receptors would inhibit the proliferation of BC. Thus, trastuzumab (Herceptin) monoclonal antibody (mAb) is used to treat breast cancer by targeting the HER2/neu receptor [9]. In nanotechnology, herceptin (ligand) has gained attention for HER2+ BC targeted drug delivery due to its biocompatibility, safety, and high efficiency in binding to HER2+ receptors in BC. Although several works have reported the combination of drug and mAb delivery for enhanced cancer therapy, there is none of the work reported on antitumor enzyme delivery associated with mAb.

In this context, a nano-drug delivery system using chitosan and polyethylene glycol has the promise to release various drugs in the cancer microenvironment based on pH [10–12]. The efficacy of therapeutic enzymes is known to be increased by conjugation with antibody, folic acid, daidzein, liposomal nanoparticles, chitosan nanoparticles, gold nanoparticles, dendrimer, and nanotubes [13–18]. In addition, the chitosan nanoparticles are known to deliver monoclonal antibody, genes, and anti-cancer enzymes in the cancer sites [19–21]. To overcome the clinical challenges of higher toxicity and instability in the LO enzyme and increase its therapeutic efficacy, the present study was hypothesized to develop a PEGylated chitosan-based enzyme (LO) delivery system for HER2/neu targeted breast cancer therapy. A pH-dependent and herceptin (trastuzumab) functionalized L-lysine α-oxidase loaded PEGylated chitosan nanoparticle was engineered and characterized for HER2/neu targeted breast cancer therapy.

2. Materials and Methods

2.1. Chemicals, Cell Line, and Animals

Chitosan (CS; CAS No. 9012-76-4, low molecular weight −50,000–190,000 Da; 75–85% deacetylated chitin, poly (D-glucosamine), polyethylene glycol (PEG 6000), L-lysine α-oxidase (LO) from *Trichoderma viride*, ethidium bromide (EB), acridine orange (AO), hydrochloric acid, 1-ethyl-3-(3-dimethylaminopropyl) carbodiimide hydrochloride (EDC: HCl), rhodamine 123 (Rh123), N-hydroxysuccinimide (NHS), propodium iodine (PI), dichlorofluorescein diacetate (DCFH-DA) were purchased from Sigma-Aldrich (St. Louis, MO, USA). Antibody anti-HER2 (Herceptin, trastuzumab; A1046-100 Lot. 4E31A10460) was obtained from BioVision Inc., Milpitas, CA, USA. The cell viability assay kit CELLO MAXTM was procured from MediFab, Republic of Korea. Cell culture media (Roswell Park Memorial Institute Medium; RPMI), fetal bovine serum (FBS), SnakeSkin™ Dialysis Tubing (10 K MWCO, 35 mm), penicillin and streptomycin (P&S), and apoptosis kit (Annexin V FITC and PI) for flow cytometry were acquired from ThermoFisher Scientific (Waltham, MA, USA). Human breast cancer cell line (BT474) was purchased from Korean Cell Line Bank, (KCLB), Seoul, Korea. Seven-week-old BALB/c/nu/nu female mice weighing 20–22 g were purchased from Nara Biotech, Seoul, Korea.

2.2. Preparation of CS-LO NPs/CS-LO-PEG NPs

The characteristics of the chitosan used in the present study are reported in our previous work [22]. The CS-LO NPs were prepared by the TPP-ionotropic gelation method [23]. In brief, 5 mg of CS was dissolved in 10 mL of 0.5% (*v*/*v*) acetic acid and stirred at 50 °C for 4 h. The TPP solution (1 mg/mL) was prepared in deionized (DI) water and stirred for 2 h. The CS and TPP solutions were filtered through a syringe filter (0.45 μm). The pH of the CS solution was adjusted to 5.0 using 0.2 M NaOH. The 25 mL of TPP (1 mg/mL) was added to the 75 mL of CS (0.5 mg/mL) solution drop by drop and stirred at room temperature for 4 h. PEGylated CS nanoparticles were prepared according to the method reported earlier with slight modifications [24]. A total of 10 mg of the PEG 6000 was dissolved in CS (0.5 mg/mL) solution under stirrer condition for 1 h, and then the pH of the PEG-CS solution was adjusted to 5.0 using 0.2 M NaOH. Afterwards, the TPP (1 mg/mL) solution was added to the CS-PEG solution to form the nanoparticles. The different concentrations of LO (0.1–1 mg/mL) were incorporated into the CS solution or CS-PEG solution before the addition of TPP. The CS-LO NPs and CS-LO-PEG NPs were centrifuged at 14,000 rpm. The pellets were lyophilized and preserved at 4 °C in a refrigerator.

2.3. Functionalization of Antibody Anti-HER2 (Trastuzumab) in CS-LO-PEG NPs

The functionalization antibody in the NPs was carried out through covalent bonding performed according to the method reported elsewhere [25]. The COOH group of CS-LO-PEG NPs was activated by the addition of 10 μL of NHS (10 mM) and EDC (10 mM) and then added to CS-LO-PEG NPs (0.5 mg/2 mL) stirring at 150 rpm at ambient temperature for 8 h. Afterwards, the anti-HER2 (trastuzumab) (100 μg/mL of COOH group activated CS-LO-PEG NPs) was added and kept stirred using the magnetic stirrer at ambient temperature for 12 h. Then the anti-HER functionalized CS-LO-PEG NPs were named CS-LO-PEG-HER NPs and dialyzed using the SnakeSkin™ Dialysis Tubing (10 K MWCO, 35 mm) in phosphate buffer for 4 h, and then centrifuged at 20,000 rpm for 2 min, and the pelleted nanoparticles were stored at 4 °C for characterization, drug release, and anticancer experiments.

2.4. Enzyme Assay

Total protein was determined using the Bradford reagent (Bio–Rad). The enzyme (LO) was determined by measuring the H_2O_2 production in the reaction mixture at 555 nm using a spectrophotometer (SpectraMAx@ Plus 384 Microplate reader, Molecular devices) [5,26]. In brief, the enzyme solution was added to trisphosphate buffer (20 mM; pH 8.0) containing 2 mM of o-dianisidine, peroxidase (5 μg/mL), and substrate L-lysine (2 mM) and incubated at 30 °C for 5 min. Then absorbance was measured at 555 nm and the enzyme activity was expressed as IU [26].

2.5. Enzyme Loading and Entrapment Assay

Different concentrations of LO (0.1–1 mg/mL) were loaded onto CS-PEG-NPs. After completion of the LO loading producers, the unloaded LO was collected by centrifugation of the LO-CS-PEG NPs suspension at 12,000 rpm for 20 min. The supernatant was used as the LO source for the enzyme assay described in Section 2.4. The enzyme present in the supernatant was measured and subjected to the following equations to calculate the enzyme (LO) entrapment and loading efficiency: EEE (%) = (Ei − Ef)/Ei, where Ei: total input of LO, Ef is the amount of unloaded LO in the supernatant; and ELE = (Ei − Ef)/M, where Ei is the total amount of LO input, Ef is the amount of unloaded LO in the supernatant, and M: total weight of nanoparticle (LO-CS-PEG NPs).

2.6. Characterization of Nanoparticles

A series of analytical methods were used to confirm the successful formation of CS-LO-PEG-HER NPs. The loading of LO and, functionalization of mAb (Anti-HER) were analyzed by FTIR (Fourier-transform infrared spectroscopy; PerkinElmer Paragon 500,

Waltham, MA, USA) and NMR (Nuclear Magnetic Resonance, FT-NMR, Bruker, 600 MHz, MA, USA). For the FTIR analysis, the nanoparticle samples, including CS, PEG6000, LO, HER, CS-LO NPs, CS-LO-PEG NPs, and CS-LO-PEG-HER NPs, were prepared as KBr pellets using the standard sample preparation method, and the IR scanning was performed from the 400–4000 cm^{-1}. To determine the size and zeta potential of nanoparticles, DLS (Dynamic light scattering; zeta potential particle size analyzer Malvern, Eindhoven, The Netherland) and ELS (Electrophoretic light scattering) analysis were used. For the DLS/ELS analysis, 10 µg of each NPs was dissolved in 3 mL of DI water, sonicated for 2 min, and filtered using a 2 µm syringe filter. The filtered nanoparticle suspension was used for DLS analysis. Transmission electron microscopic (JEOL-JSM 1200EX, Tokyo, Japan) analysis was used to observe the morphology of nanoparticles. For TEM analysis, 10 µg of each NPs was dissolved in 1 mL of ethanol, sonicated for 2 min, and negative stained with 2% uranyl acetate then loaded into a copper grid for TEM observation [27].

2.7. *In Vitro Enzyme Release Assay*

In vitro release of the enzyme LO from the CS-LO-PEG-HER NPs was determined at three different pH values (5.0, 6.8, and 7.4). In brief, the 5 mg of the CS-LO-PEG-HER NPs were dissolved in 5 mL of different pH buffer solutions. The nanoparticle suspension was loaded in SnakeSkin™ Dialysis Tubing (35 mm) membrane bag. The nanoparticle loaded membrane was kept in a 250 mL beaker containing the 50 mL dissolution buffer solution (pH 5.0, 6.8, and 7.4) and allowed for magnetic stirring (50 rpm). Aliquots of 5 mL were collected at different time intervals (0–70 h) by replacing the same volume of dissolution buffer, and the enzyme release was measured using the enzyme assay methods described in Section 2.5. The experiments were performed three times, and pH-dependent release of the LO was calculated and expressed as the cumulative release (%).

2.8. *Cell Culture and Blood Compatibility Assay*

2.8.1. Cellular Internalization

The effect of the anti-HER functionalization in nanoparticles (CS-LO-PEG-HER NPs and CS-LO-PEG NPs) on cellular internalization was analyzed by in vitro cell culture experiments. In brief, the BT474 cells (1×10^4) were cultured in a 5 mL of RPMI medium incorporated with 10% of FBS and 1% antibiotic solution in six-well plates (costar) in 5% CO_2 atmosphere at 37 °C for 24 h. Afterward, 50 ng/mL of each nanoparticle (CS-LO-PEG-HER NPs and CS-LO-PEG NPs) was treated for 12 h, then the cells were collected using the aseptic cell scraper, dehydrated, and observed under TEM.

2.8.2. Hemolysis Assay

The blood compatibility of the nanoparticles (CS-LO-PEG-HER NPs and CS-LO-PEG NPs) was tested in sheep blood (Carlina, Seoul, Korea). In brief, 1 mL of sheep blood was dissolved in 10 mL of PBS (pH 7.4) and centrifuged at 2000 rpm for 10 min at 4 °C. After the centrifugation, the pellet containing the RBC was collected by washing three times with PBS (pH 7.4). The RBC suspension (4% *v*/*v*) was prepared in PBS for the hemolysis assay. For the assay, 200 µL of RBC suspension and 400 µL of different concentrations of each nanoparticle (12.5, 25, 50, 100, 150, 200 ng/mL) were added and incubated at 37 °C for 1 h and then centrifuged at 2000 rpm at 4 °C for 10 min. The absorbance of the supernatant was observed at 545 nm using the UV spectrophotometer. Here, 1% (*v*/*v*) of triton X-100 was used as a negative control while the PBS was used as a blank. The percentage of hemolysis was calculated by applying the formula: hemolysis (%) = (test/control) × 100.

2.8.3. Cytotoxicity

The cytotoxic effect of nanoparticles, enzymes, and antibody (anti-HER2) treatments in BT474 cells was analyzed by the WST assay. In detail, the BT474 cells (1×10^4) seeded in the 96 well plates containing the 100 µL of RBMI media (10% FBS and 1% antibiotic solution) and cultured at 37 °C at a 5% CO_2 atmosphere for 24 h. Then different concentrations

(25, 50, 100, 150, 200 ng/mL) of nanoparticles, enzyme, and antibody (anti-HER2) were treated at the same incubation condition for 12 h. After the treatment, 10 μL of WST solution was added and incubated for 45 min. The absorbance was read at 450 nm using the UV spectrophotometer. The cell viability (%) was calculated using the formula (OD of test/OD control) × 100.

2.8.4. Fluorescent Microscopic Assay

Effect of the nanoparticles, enzyme, and antibody (anti-HER2) treatments were determined in BT474 cells on mitochondrial membrane potential (MMP), nucleus, and reactive oxygen species (ROS) generation. In brief, the BT474 cells were cultured as mentioned in Section 2.8.1. After the confluence, the IC50 concentration of each sample was treated at 37 °C, in a 5% CO_2 atmosphere for 12 h. Then the MMP changes were observed by staining with Rh123, and the nucleus damage was visualized using DAPI staining. The ROS generation was noted through DCFH-DA staining, while the apoptosis stages were observed by AO/EB staining. All stain preparations and staining producers were carried out in accordance with the methods described in our previous works [28,29].

2.8.5. FITC, Annexin V Apoptosis Assay

The treatments induced apoptosis stages (live, early, late apoptosis, and necrosis) in BT474 cells were measured by flow cytometer using Annexin V, FITC staining according to the protocols described in the manufacture's instruction. In brief, the BT474 cells were cultured as mentioned in Section 2.8.1. The IC50 samples were treated for 12 h, and cells were collected using the aseptic cell scraper and washed with PBS (pH 7.4) by centrifugation at 4 °C for 4 min. Then the cells were fixed with 100 μL of binding buffer (1×) for 10 min at ambient temperature. Afterward, 1 μL of PI and 5μL of FITC-labeled annexin V were added and incubated for 15 min at room temperature. Finally, 400 μL of binding buffer (1×) was added, filtered, and analyzed using the flow cytometer (BD FACS aria II).

2.8.6. Cell Cycle Assay

The effect of nanoparticles, enzymes, and antibody (anti-HER2) treatments on the cell cycle of BT474 cells was analyzed by PI staining using a flow cytometer. The PBS was used to prepare the 70% ethanol and PI stocks. For the PI stock, 50 μg of PI and 100 μg of RNase A were dissolved in 1 mL of PBS (pH 7.2). The above-mentioned samples of treated cells were collected as described in Section 2.8.1 and fixed in 70% ethanol for 10 min. Then the cells were washed with PBS by centrifugation at 2000 rpm for 4 min at 4 °C. An equal volume of each treatment was adjusted using a hemocytometer and then stained with 500 μL of PI stock solution for 30 min. The cell cycle arrest was analyzed using the flow cytometer (BD FACS aria II).

2.9. In Vivo Animal Model

Seven-week-old athymic BALB/c female mice (20–22 g) were purchased from Nara Biotech, Seoul, Korea. Animal studies were performed according to the Kangwon National University Ethical Policy (KW-201019-1). The mice were housed (6 mice/cage) at a 12/12 h light/dark cycle and fed with a standard diet and water ad libitum. The BT-474 cells (1×10^6/mice) were subcutaneously injected into the right flank. The 0.1 mL of PBS was used as a cell vehicle and the implanted cells were allowed to grow for 10 days. Once tumor size reached 100–160 mm^3, the mice were randomly grouped for treatments, and each treatment group contained 4 mice, which is followed according to the Kangwon National University Ethical Policy: (i) saline water (PBS), (ii) LO (200 μg/kg), (iii) HER (200 μg/kg), (iv) CS-LO-PEG NPs (200 μg/kg), and (v) CS-LO-PEG-HER NPs (200 μg/kg). These samples were injected intravenously once a day for 12 days. During the experiment, the body weight and tumor volume were measured using the laboratory weighing balance and Vernier caliper. After the treatments, the mice were sacrificed by cervical dislocation, and organs and tumors were collected for histopathological analysis using Hematoxylin and Eosin staining according to the method reported earlier [27,30].

2.10. Statistical Analysis

Each experiment was repeated at least three times, and the results are expressed as a mean ± standard error. The significance of the factor was determined using analysis of variance (ANOVA) and a post-hoc test (Duncan's test). All statistical analysis is carried out via the use of a virtual tool (IBM SPSS statistical software version 20; IBM, New York NY, USA).

3. Results and Discussion

3.1. Synthesis and Characterization

3.1.1. DLS, Enzyme Entrapment, Loading Efficiency

The present study designed a nano-drug delivery system to deliver the LO specifically to HER2+ breast cancer (BC) cells, where functionalization of mAb (anti-HER2+) with PEGylated-chitosan enabled the targeting of the HER2+ BC cells. The physicochemical properties of NPs in response to the input of LO are shown in Table 1. The results revealed that the LO input significantly influenced the average size, zeta potential, polydispersity index (PDI), enzyme entrapment efficiency (EEE), and enzyme loading efficiency (ELF) (Table 1). Although these indicators varied in response to the LO input, all the nanoparticles exhibited the amicable properties that are favorable for biomedical drug delivery. Further, the size (138.53–182.60 nm) in line with PDI was 0.12–0.44, and the zeta potential was 26.73–36.60 mV (Table 1). The size of the nanoparticles has a vital role in cellular internalization and uptake of nanomedicine, as indicated by earlier reports [31], and the nanoparticles size of <200 nm is known to enhance cellular uptake and internalization through the EPR (permeability and retention) effect [32]. The results of the present work recorded that the input of 0.5 mg of LO had achieved CS-LO-PEG NPs with a size of 144.1 ± 2.96 nm with zeta potential of 35.86 ± 1.33 mV, and PDI of 0.28 ± 0.03, and it also exhibited the optimal EEE of 59.15 ± 0.44%, ELE of 14.26 ± 0.85%. These characteristics of nanoparticles are favorable for clinical application. Hence, they were selected for further characterization and anti-HER functionalization.

Table 1. Effect of different concentrations of enzyme input on size, zeta potential, polydispersity index (PDI), drug entrapment and drug loading in chitosan-PEGylated nanoparticles (CS-LO-PEG NPs). The results are presented as mean ± SE. Superscript letters on the values indicates significant differences between the data in the column ($p < 0.05$).

Initial Enzyme Input (mg)	Size (nm)	Zeta Potential (mV)	PDI	Enzyme Entrapment (%)	Enzyme Loading (%)
0.1	138.53 ± 4.68 [a]	26.73 ± 1.02 [a]	0.35 ± 0.04 [c]	75.77 ± 0.76 [d]	5.26 ± 0.31 [a]
0.2	143.76 ± 2.43 [b]	31.10 ± 0.92 [b]	0.44 ± 0.01 [d]	62.82 ± 0.82 [c]	10.59 ± 0.64 [b]
0.5	144.1 ± 2.96 [b]	35.86 ± 1.33 [c]	0.28 ± 0.03 [b]	59.15 ± 0.44 [b]	14.26 ± 0.85 [c]
1	182.60 ± 2.34 [c]	36.60 ± 0.30 [d]	0.12 ± 0.05 [a]	53.38 ± 0.57 [a]	17.49 ± 0.81 [d]

The DLS and ELS were again performed to show the zeta size, zeta potential, and PDI of CS-LO NPs, CS-LO-PEG NP, CS-LO-PEG-HER NPs (Figure S1). The results indicated that the size of the NPs increased with PEGylating and anti-HER2 functionalization. The CS-LO NPs exhibited a z-average size of 137.0 ± 1.02 nm, and a zeta potential of 36.4 ± 2.18 mV with PDI of 0.360 ± 0.04 (Figure S1a,b). The CS-LO-PEG NP displayed a z-average size of 143.0 ± 0.98 nm, and a zeta potential of 35.3 ± 0.52 mV with PDI of 0.357 ± 0.02 (Figure S1c,d), whereas CS-LO-PEG-HER NPs unveiled a z-average size of 165.9 ± 1.02 nm, and a zeta potential of 34.2 ± 1.60 mV with PDI of 0.300 ± 0.12 (Figure S1e,f). The results revealed that the 0.5 mg of LO-loaded nanoparticles (CS-LO NPs, CS-LO-PEG NP, and CS-LO-PEG-HER NPs) synthesized in the present work unveiled the optimal size, zeta potential, and PDI for drug delivery. The PDI value of a nanoparticle of <0.5 and a zeta potential of >30 mV indicates the mono-dispersion and stability of the NPs [33]. Furthermore, the positive charge of the NPs improved electrostatic interaction, cellular penetration, and uptake with the negatively charged cell membrane [34].

3.1.2. Morphological Characterization by TEM

The TEM results of CS-LO NPs, CS-LO-PEG NPs, and CS-LO-PEG-HER NPs are shown in Figure 1a–c. The morphology of the nanoparticles was found to be spherical and homogeneous spheres with size of <100 nm, which is smaller than the DLS measurement of size. The size difference between TEM and DLS was observed due to the differences in the sample preparation and hydrodynamic properties of NPs [22,35]. In addition, the DLS was performed to analyze the size of the particle in a liquid state according to the Stokes–Einstein principle, while TEM principally works in the dry state of samples under high vacuum pressure [36].

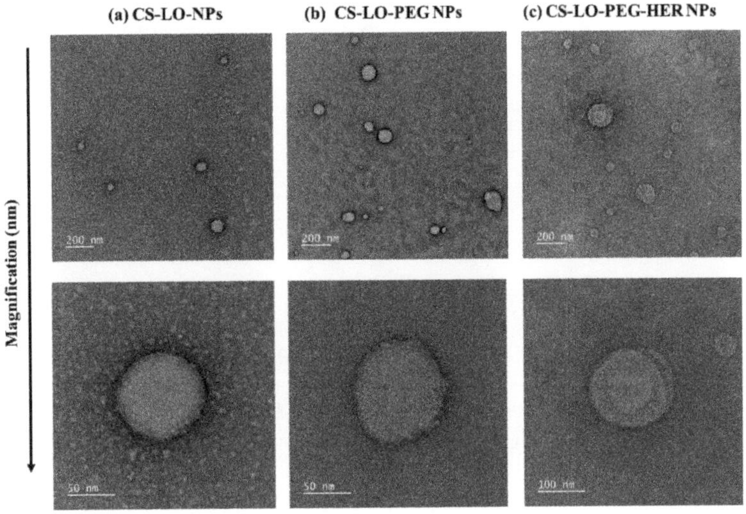

Figure 1. Transmission electron microscopic observation of HER functionalized L-lysine α-oxidase loaded chitosan-polyethylene glycol (PEG) nanoparticles. CS-LO NPs (**a**), CS-LO-PEG NPs (**b**) and CS-LO-PEG-HER NPs (**c**). Where CS is chitosan, LO is L-lysine α-oxidase, PEG is Polyethylene glycol 600, HER is Herceptin (Trastuzumab).

3.1.3. Surface Chemistry Analysis

The FTIR facilitated the analysis of the formation, loading, and functionalization of the LO and antibody (anti-HER2+) in the CS-LO-PEG-HER NPs in comparison with CS, PEG, HER, LO, and its conjugated NPs. To verify the loading of the LO, and functionalization of anti-HER, the IR spectrum of free LO, HER, CS-LO NPs, CS-LO PEG NPs were compared, and the results are shown in Figure 2a,b. The CS exhibited the characteristic vibration at 3351 cm^{-1}, 1648 cm^{-1}, and 1577 cm^{-1} assigned to the O-H stretching, C=O, amine I, and amine II respectively, while the 1419 cm^{-1}, 1375 cm^{-1}, 1314 cm^{-1} corresponding to O-H, and other characteristics peaks at 1060, 1026 assigned for C-N amine while peaks including 894 cm^{-1}, 555 cm^{-1}, 445 cm^{-1} accounted for C-H bending of CS [37,38]. The IR spectrum of the LO displayed primary protein absorption characteristic peaks at 3377 cm^{-1} accounting for N-H (primary amine), while 988 corresponded to C=C bending [39], Similarly, the protein absorption peaks for the lipase B and horseradish peroxidase are reported in the earlier work by using the IR spectrum [38,40,41]. The CS-LO NPs showed the overlapping and reappearance of characteristic peaks of CS and LO at 2882 cm^{-1}, 1577 cm^{-1}, and 961 cm^{-1}, respectively. Besides, some of the peaks related to the LO changed due to the interaction of amino groups with polymer surfaces [40]. In addition, the peaks of 1240 cm^{-1} and 899 cm^{-1} in CS-LO NPs indicate the formation of P=O and P-O-P bonds related to the interaction of an amino group of CS and the phosphate group of TPP mediated CS NPs formation [38,42].

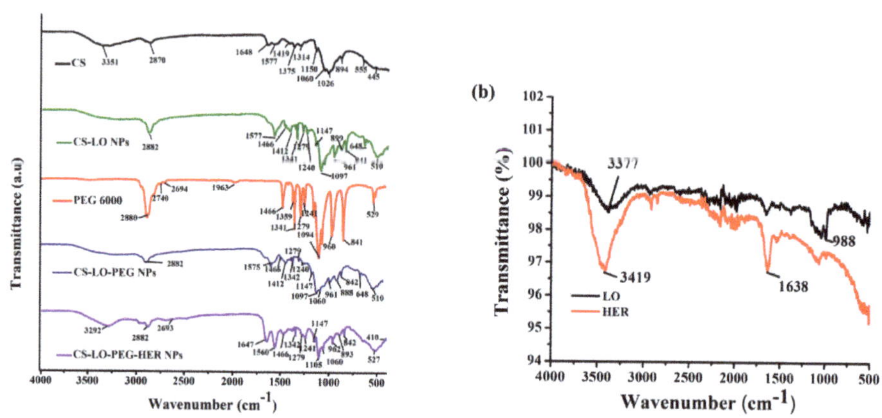

Figure 2. FTIR analysis of herceptin (trastuzumab) functionalized L-lysine α-oxidase loaded chitosan-polyethylene glycol (PEG) nanoparticles. Functional group changes in CS, CS-LO NPs, PEG600, CS-LO-PEG NPs and CS-LO-PEG-HER NPs are shown in (**a**), while LO and HER are shown in (**b**).

The PEG 6000 illustrated the characteristic peaks at 2880 cm^{-1}, 2740 cm^{-1}, 2694 cm^{-1}, corresponding to C-H, and 1466 cm^{-1} and 960 cm^{-1} accounted for by the ether bond [38]. The CS-LO PEG NPs exhibited the overlapping of the characteristic peaks, CS, LO and PEG, specifically the peaks such as 2882 cm^{-1}, 1575 cm^{-1}, 1466 cm^{-1}, 961 cm^{-1} that reappeared in the IR spectrum of CS-LO PEG NPs. Furthermore, the peaks that appeared at 2882 cm^{-1}, 1147 cm^{-1} and 1097 cm^{-1} evidenced the successful PEGylation with the amine group of CS and the formation of CS-LO PEG NPs [43]. The free anti-HER exhibited the peaks at 3419 cm^{-1} and 1638 cm^{-1}, which corresponded to the primary (N-H stretching) and secondary amine (N-H bending) groups [39]. The CS-LO-PEG-HER NPs showed the reappearance or overlapping of the IR peaks corresponding to CS (3292 cm^{-1}), LO (962 cm^{-1}), PEG (2693 cm^{-1}), HER (1647 cm^{-1}) with minor shifting from the original appearance, and this indicated the successful formation of CS-LO-PEG-HER NPs through TPP, EDC, and NHC mediated synthesis.

3.2. *In-Vitro Experiments*

3.2.1. Drug Release and Intracellular Distribution

The stability and pH-responsive release of LO from CS-LO-PEG-HER NPs were tested in three different pH buffer solutions (pH 5.0, 6.8, and 7.4). The results indicated that the LO release significantly varied with the time and pH value ($p < 0.05$), and the higher release of >80% was found at pH 5.0, while the lower ~15 was found at pH 7.4 after 64 h (Figure 3a). The pH-responsive diffusion and degradation properties of CS and PEG triggered the LO release at pH 5.0. Similarly, several works have been reported the pH responsive release of drug from PEGylated CS [44,45]. The pH of the cancer microenvironment is reported to be pH ~5.2–5.4 [46]. Furthermore, the enzyme LO is stable in a wide range of pH and temperature [4].

Cellular penetration, and uptake are considered important factors for successful nanomedicine. Therefore, we experimented to compare the effects of dual stimuli (pH, and electrostatic interaction) and multi stimuli (electrostatic, ligand, and antibody interaction and pH) on cellular uptake and penetration of NPs by TEM. The results are shown in Figure 3b–d. The data confirmed the higher cellular uptake and distribution of CS-LO-PEG

HER NPs than CS-LO PEG NPs. These results indicated that the Anti-HER2 (antibody–ligand) and HER2 (receptor) interaction triggered the higher cellular binding of formal CS-LO-PEG HER NPs than the bare CS-LO PEG NPs.

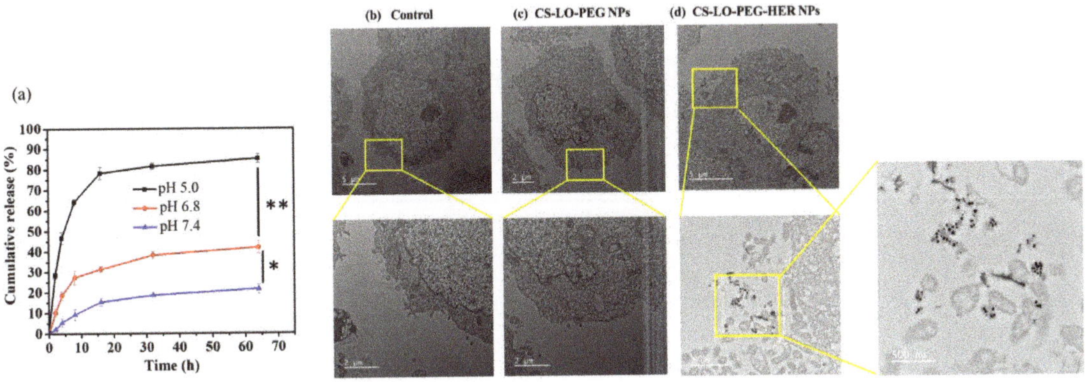

Figure 3. In vitro pH responsive enzyme release from the herceptin (trastuzumab) functionalized L-lysine α-oxidase loaded chitosan-polyethylene glycol (PEG) nanoparticles (CS-LO-PEG-HER NPs; (**a**)), cellular internalization without herceptin functionalized (CS-LO-PEG NPs) and herceptin functionalized (CS-LO-PEG-HER NPs) L-lysine α-oxidase loaded chitosan-polyethylene glycol (PEG) nanoparticles. (**b–d**) shown the nanoparticles internalization in high magnification. ** $p < 0.01$; * $p < 0.05$.

3.2.2. Blood Compatibility

Nanomedicines are favorable to intravenous injection for the efficient delivery of drugs without any toxicity to normal cells [47]. In addition, the enzyme LO is known to cause the toxicity of L-lysine in blood plasma [3]. To verify the blood toxicity of LO, a hemolysis assay was performed and compared the activity of CS-LO-PEG HER NPs and CS-LO PEG NPs. As shown in Figure 4a and Figure S2, both NPs exhibited a hemolysis rate of <5% even at the high concentration of 200 ng/mL, and hence, CS-LO-PEG HER NPs and CS-LO PEG NPs did not cause obvious RBC lysis, as shown in Figure S2. It has been reported that <5% of hemolysis can be considered blood compatible and used for drug delivery applications [48,49].

3.2.3. Cytotoxicity

The WST assay was applied to investigate the comparative cytotoxicity effects of LO, HER, CS-LO-PEG NPs and CS-LO-PEG HER-NPs in BT474 cells (Figure 4b). The cytotoxicity significantly varied between the samples or concentrations. The CS-LO-PEG HER-NPs induced higher cytotoxicity than the other samples. This higher cytotoxicity of CS-LO-PEG HER-NPs was attributed to the increased cellular uptake through the interaction of anti-HER2 (ligand) and surface receptor (HER2+). According to earlier work, the functionalization of anti-HER2 (trastuzumab) would support the greater uptake of the nanomedicine through receptor-mediated endocytosis [50–52]. The overexpression of HER2+ in the breast cancer cell line (BT474) enabled the higher interaction and bonding between the mAb (anti-HER2) and receptor (HER2+), which led to more efficient cytotoxicity than the other samples, including LO, HER, and CS-LO-PEG NPs. This finding is also supported by the cellular uptake and internalization studies.

3.2.4. Fluorescent Staining Assay

The cytotoxic effect of LO, HER, CS-LO-PEG NPs, and CS-LO-PEG HER-NPs in BT474 was further validated through a fluorescent staining assay (Figure 4c–e) Rh123 staining results revealed that the treatment of LO-CS-PEG-HER-NPs, HER, and LO caused a complete MMP loss, while CS-LO-PEG NPs exhibited moderate MMP loss as compared to control cells (Figure 4c). The loss of MMP was indicated through the loss of fluorescent intensity [53]. The nuclear stain DAPI is used to study the nucleus damage (nuclear fragmentation, chromatin condensation, margination). Moreover, DAPI would bind strongly with damaged cells as compared to healthy cells [54]. The results from DAPI staining revealed that the treatment of CS-LO-PEG HER-NPs caused higher cellular damage in the BT474 cells than in other samples (Figure 4d). The DCFH-DA is a dye used to measure the ROS generation under the excitation of fluorescents. The higher ROS generation was found in a decreasing order of sample treatments: LO-CS-PEG-HER NPs > CS-LO-PEG NPs > LO > HER (Figure 4e). These results agree with the general principle of LO-mediated cell toxicity through the production of H_2O_2 mediated oxidative stress [6]. Following this, the dual stain AO/EB assay was performed to observe the LO, HER, CS-LO-PEG NPs and CS-LO-PEG HER-NPs induced apoptosis stage in BT474 cells. The results showed more apoptosis (orange, chromatin damage) and necrosis (red, fragmented) cells than in other samples of treated cells (Figure 5a).

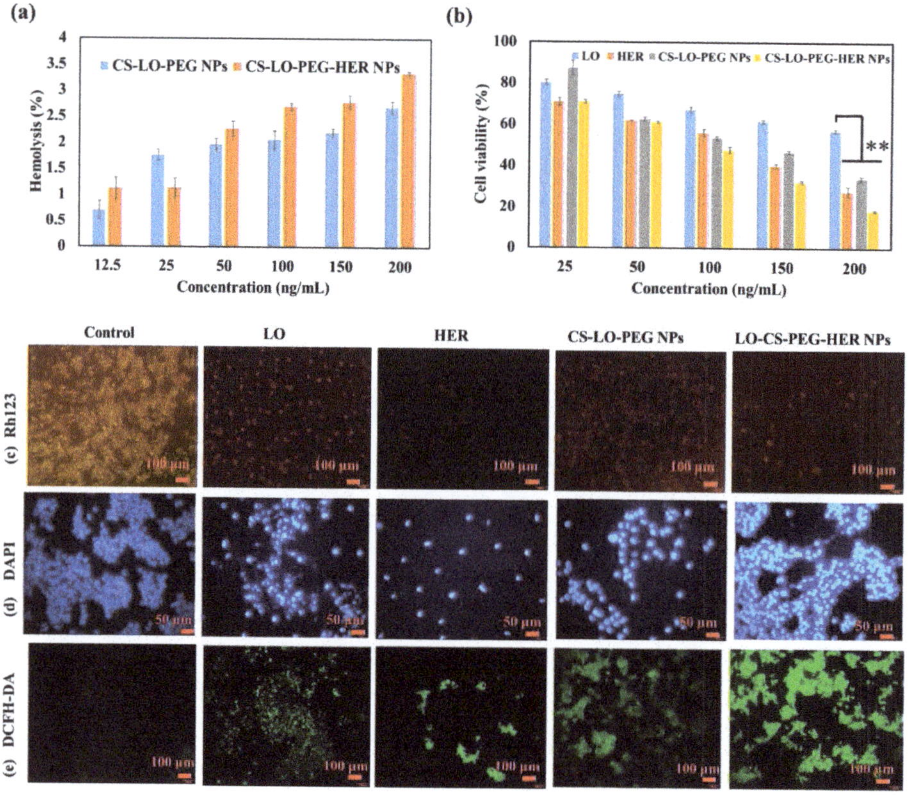

Figure 4. Blood compatibility of CS-LO-PEG-HER NPs and CS-LO-PEG- NPs (**a**), cytotoxicity of enzyme loaded nanoparticles in BT474 cell line (**b**), microscopic fluorescent analysis of enzyme or enzyme loaded nanoparticles induced mitochondrial membrane loss by Rh123 (**c**), nucleus damage by DAPI (**d**), reactive oxygen species generation by DCFH-DA (**e**). Where CS is chitosan, LO is L-lysine α-oxidase, PEG is polyethylene glycol 600, HER is herceptin (Trastuzumab). ** $p < 0.01$.

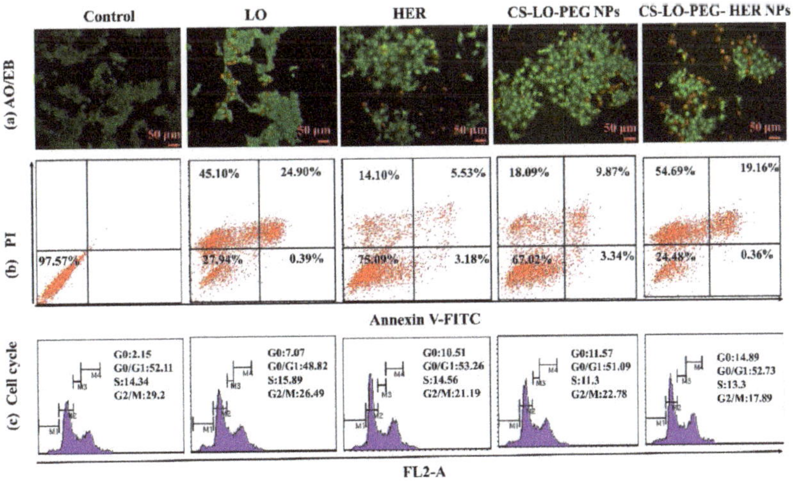

Figure 5. Effect of L-lysine α-oxidase, herceptin, and different nanoparticles treatment on apoptosis stages of BT474 cell line observed by AO/EB staining (**a**), FITC and annexin V staining based flow cytometry assay (**b**) and cell cycle stages by PI staining (**c**). Where CS is chitosan, LO is L-lysine α-oxidase, PEG is polyethylene glycol 600, HER is herceptin (Trastuzumab).

3.2.5. Measurement of Apoptosis and Cell Cycle Arrest by Flow Cytometer

A FITC annexin V, PI stain-based flow cytometry assay was executed to investigate the effect of LO, HER, CS-LO-PEG NPs, and CS-LO-PEG HER-NPs treatments in BT474 on apoptosis and cell cycle arrest (Figure 5b). Figure 5b displays the live cells in the left lower quadrants, early apoptosis in the right lower quadrants, and late apoptosis in the right upper quadrants, while the left upper quadrants showed the necrosis cells. The apoptosis or necrosis cells (%) were found higher in CS-LO-PEG HER-NPs (74.21%) and LO (70.39%) treated cells than in other treatments. Next, the cell cycle distributions due to the treatments in BT474 cells were investigated. The treatment of LO, HER, CS-LO-PEG NPs, and CS-LO-PEG HER-NPs showed a typical cell distribution pattern of G0, G0/G1, S, and G2/M phases in the cell cycle. The number of cells in each cell cycle stage significantly varied between the treatments ($p < 0.05$; Figure 5c).

The treated cells that exhibited higher cell populations in the G0/G1 phase were LO (48.82%), HER (53.26%), CS-LO-PEG NPs (51.09%), and CS-LO-PEG HER-NPs (52.73%). Moreover, the cell population was found high in the G0 phase (apoptosis) of the CS-LO-PEG-HER NPs (14.89%) compared to the control untreated cells. The present results revealed that the treatment of CS-LO-PEG-HER NPs exhibited cell arrest in the G0/G1 phase. Similarly, earlier work reported that the cell population in the G0 phase was related to the apoptosis cells [55]. Cell cycle arrest and apoptosis are regular processes to inhibit the abnormal growth of cells through activating signals related to the apoptosis-related pathway [56].

3.3. *In Vivo Experiments*

3.3.1. Survival, Tumor Size, and Body Weight

The anticancer activity of LO, HER, CS-LO-PEG NPs, and CS-LO-PEG HER-NPs was tested in vivo BT474 (breast cancer) tumor-bearing BALB/c female mice model (Figure 6a,b). The tumor volume and body weight were measured at different time intervals up to the end of the experiment. The tumor growth rapidly increased in the saline water treated (control group) ~558.25 ± 18.25 mm^3 at the end of the experiment. In contrast, LO, HER, or nanoparticles treated groups exhibited promising antitumor proliferation activity (Figure 6c,d). Moreover, the level of the tumor growth did not show any significance between the LO and HER treatments; however, the CS-LO-PEG HER-NPs exhibited significant tumor growth inhibition as compared to other treatments, including the control group (Figure 6c,d). Next, the bodyweight changes between the treatments were measured, and the results displayed significant difference between the CS-LO-PEG HER-NPs treatments and the control group (saline water), while other treatments did not show any significant differences from the control ($p < 0.05$). Also, the experiment observed a 100% survival rate in all the groups at the end of the experiment.

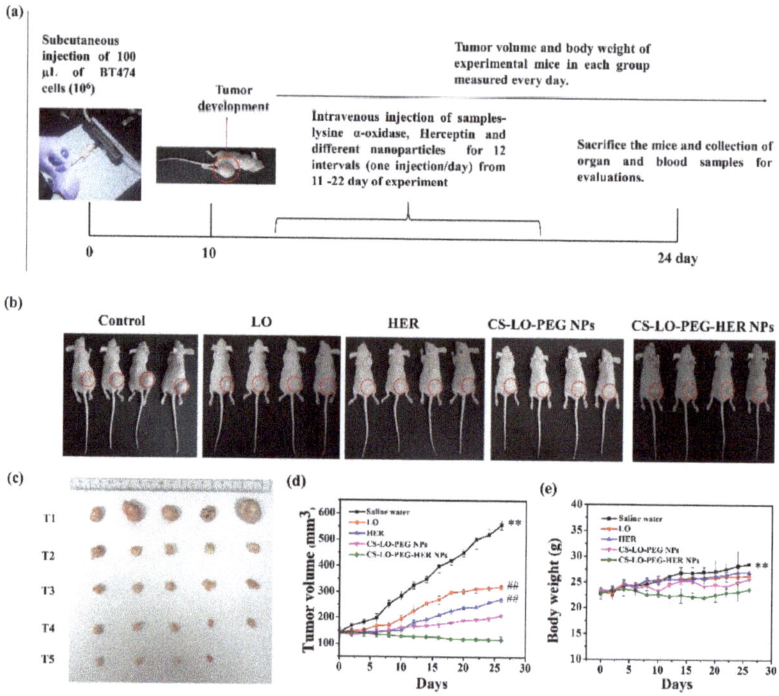

Figure 6. Effect of L-lysine α-oxidase, herceptin and different nanoparticles treatment on tumor bearing mice. Schematic illustration of in vivo antitumor study plan (**a**). Photographic visualization of tumor development (**b**), measurement of tumor volume in 24th day (**c**), tumor volume variation in relation to days (**d**) and body weight response to days (**e**). Where CS is chitosan, LO is L-lysine α-oxidase, PEG is polyethylene glycol 600, HER is herceptin (Trastuzumab). (T1) saline water (PBS), (T2) LO (200 µg/kg), (T3) HER (200 µg/kg), (T4) CS-LO-PEG NPs (200 µg/kg), and (T5) CS-LO-PEG-HER NPs (200 µg/kg). ** indicated the $p < 0.01$ significant variations between the untreated (saline water) mice with all other treated groups. ## indicated the non-significance with LO and HER while it exhibited the significance with other treatments and saline water ($p < 0.05$).

3.3.2. Histopathology

The effect of LO, HER, CS-LO-PEG NPs, and CS-LO-PEG HER-NPs treatments was observed by H and E staining in histopathological changes in the major organs such as kidney, liver, and spleen (Figure 7a–c). There was no apparent damage observed in the representative sections of the kidney, liver, and spleen when compared between the control and various treatments. These results also indicated mild histological changes in the HER (anti-HER2+) treated group, while all other drug formulations did not cause significant damage to the kidney, liver, and spleen. Furthermore, the present results evidenced the biocompatibility of drug formulations in vivo at the tested concentration. The nanomedicine-mediated tumor cell necrosis or apoptosis is often studied by the H and E staining assay [57]. Therefore, the changes in tumor cells were observed by H and E staining. As shown in Figure 7d, no apparent necrosis was observed in the saline water treated group while the necrosis was observed in the sample treated mice group in the decreasing order of CS-LO-PEG HER-NPs > CS-LO-PEG NPs > LO > HER. Similar observations of the non-necrosis in tumors treated with saline water has been reported [58]. Moreover, the density of the tumor tissue was found to be higher in the control group than in other treatments, while the CS-LO-PEG HER-NPs displayed a significant reduction of tumor cells. Overall, the in vivo anticancer experiment demonstrated that pH-responsive, HER2+ targeting property, CS-LO-PEG HER-NPs enhanced the utilization of LO efficiently to achieve a promising anticancer effect.

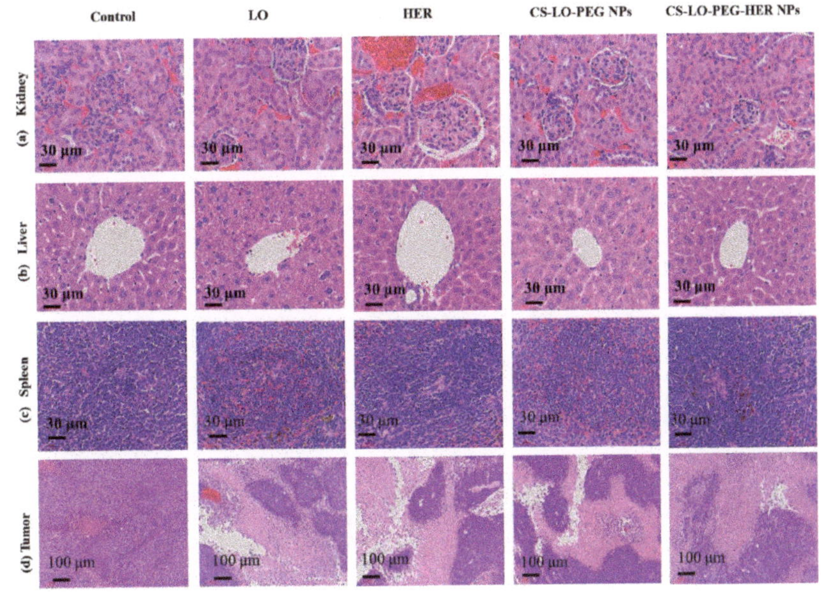

Figure 7. Effect of L-lysine α-oxidase, herceptin, and different nanoparticles treatment on histopathological changes in organs and tumor pathology. Kidney (**a**), liver (**b**), spleen (**c**), tumor (**d**). Where CS is chitosan, LO is L-lysine α-oxidase, PEG is polyethylene glycol 600, HER is herceptin (Trastuzumab).

4. Conclusions

In summary, the CS-LO-PEG NPs were prepared by the ionic gelation self-assembly method, and further functionalized with anti-HER2+ by covalent bonding. The formulation of these nanoparticles was confirmed by physicochemical characterization using FTIR, NMR, TEM, and zeta potential particle size analysis. The dual functions of CS-LO-PEG HER-NPs (i) pH-responsive release; and (ii) targeting the HER2+ breast cancer cells (BT474) were confirmed by comparing them to other nanoparticles used in the study. The cellular internalization of CS-LO-PEG HER-NPs was confirmed by TEM. It also increased the controlled release of LO at the cancer site by sensing the pH of the tumor site. Finally, this work proved that CS-LO-PEG HER-NPs exhibited significant anticancer potency in BT474-xenograft tumor mice, with promising biocompatibility as indicated by the experiments of cytotoxicity, hemolysis, and organ toxicity.

Supplementary Materials: The following supporting information can be downloaded at https://www.mdpi.com/article/10.3390/pharmaceutics14050927/s1: Figure S1. Determination of size (a,c,e) and zeta potential (b,d,f) of the nanoparticles. Z-average size and zeta potential of CS-LO NPs (a,b), CS-LO-PEG NPs (b–d) and CS-LO-PEG-HER NPs (e,f). Where CS is chitosan, LO is L-lysine α-oxidase, PEG is Polyethylene glycol 600, HER is Herceptin (Trastuzumab); Figure S2. Pictorial visualization of hemolysis activity of the nanoparticles CS-LO-PEG NPs and CS-LO PEG-HER NPs.

Author Contributions: Conceptualization; data curation; formal analysis; investigation; methodology; visualization; roles/writing-original draft; writing review and editing, K.S.; data curation; formal analysis; validation, A.S.; investigation; methodology, S.P.; formal analysis; validation, S.-R.K.; software; investigation, V.V.P.; funding acquisition; project administration; resources; software; supervision; validation, M.-H.W. All authors have read and agreed to the published version of the manuscript.

Funding: Brain Pool Program through the National Research Foundation of Korea (NRF) funded by the Ministry of Science, ICT and Future Planning (2021H1D3A2A01098535). This work was supported by the National Research Foundation of Korea (2021R1I1A1A01057742).

Institutional Review Board Statement: Not applicable.

Informed Consent Statement: Not applicable.

Data Availability Statement: Not applicable.

Acknowledgments: This work was supported by Brain Pool Program through the nnaNational Research Foundation of Korea (NRF) funded by the Ministry of Science, ICT and Future Planning (2021H1D3A2A01098535).

Conflicts of Interest: The authors declare no conflict of interest.

References

1. Kusakabe, H.; Kodama, K.; Kuninaka, A.; Yoshino, H.; Misono, H.; Soda, K. A new antitumor enzyme, L-lysine alpha-oxidase from Trichoderma viride. Purification and enzymological properties. *J. Biol. Chem.* **1980**, *255*, 976–981. [CrossRef]
2. Lukasheva, E.V.; Berezov, T.T. L-Lysine α-Oxidase: Physicochemical and Biological Properties. *Biochemistry* **2002**, *67*, 1152–1158. [CrossRef]
3. Lukasheva, E.V.; Babayeva, G.; Karshieva, S.S.; Zhdanov, D.D.; Pokrovsky, V.S. L-Lysine α-Oxidase: Enzyme with Anticancer Properties. *Pharmaceuticals* **2021**, *14*, 1070. [CrossRef]
4. Amano, M.; Mizuguchi, H.; Sano, T.; Kondo, H.; Shinyashiki, K.; Inagaki, J.; Tamura, T.; Kawaguchi, T.; Kusakabe, H.; Imada, K.; et al. Recombinant expression, molecular characterization and crystal structure of antitumor enzyme, L-lysine α-oxidase from Trichoderma viride. *J. Biochem.* **2015**, *157*, 549–559. [CrossRef]
5. Pokrovsky, V.S.; Treshalina, H.M.; Lukasheva, E.V.; Sedakova, L.A.; Medentzev, A.G.; Arinbasarova, A.Y.; Berezov, T.T. Enzymatic properties and anticancer activity of L-lysine α-oxidase from Trichoderma cf. aureoviride Rifai BKMF-4268D. *Anti.-Cancer Drugs* **2013**, *24*, 846–851. [CrossRef] [PubMed]
6. Kusakabe, H.; Kodama, K.; Kuninaka, A.; Yoshino, H.; Soda, K. Effect ofl-Lysine α-Oxidase on Growth of Mouse Leukemic Cells. *Agric. Biol. Chem.* **1980**, *44*, 387–392. [CrossRef]
7. Griffith, R.S.; Norins, A.L.; Kagan, C. A Multicentered Study of Lysine Therapy in Herpes simplex Infection. *Dermatology* **1978**, *156*, 257–267. [CrossRef]
8. Lukasheva, E.V.; Makletsova, M.G.; Lukashev, A.N.; Babayeva, G.; Arinbasarova, A.Y.; Medentsev, A.G. Fungal Enzyme l-Lysine α-Oxidase Affects the Amino Acid Metabolism in the Brain and Decreases the Polyamine Level. *Pharmaceuticals* **2020**, *13*, 398. [CrossRef]
9. Molina, M.A.; Codony-Servat, J.; Albanell, J.; Rojo, F.; Arribas, J.; Baselga, J. Trastuzumab (herceptin), a humanized anti-Her2 receptor monoclonal antibody, inhibits basal and activated Her2 ectodomain cleavage in breast cancer cells. *Cancer Res.* **2001**, *61*, 4744–4749.
10. Saravanakumar, K.; Sathiyaseelan, A.; Mariadoss, A.V.A.; Jeevithan, E.; Hu, X.; Shin, S.; Wang, M.-H. Dual stimuli-responsive release of aptamer AS1411 decorated erlotinib loaded chitosan nanoparticles for non-small-cell lung carcinoma therapy. *Carbohydr. Polym.* **2020**, *245*, 116407. [CrossRef] [PubMed]
11. Lee, S.J.; Min, H.S.; Ku, S.H.; Son, S.; Kwon, I.C.; Kim, S.H.; Kim, K. Tumor-targeting glycol chitosan nanoparticles as a platform delivery carrier in cancer diagnosis and therapy. *Nanomedicine* **2014**, *9*, 1697–1713. [CrossRef]
12. Nascimento, A.V.; Singh, A.; Bousbaa, H.; Ferreira, D.; Sarmento, B.; Amiji, M.M. Overcoming cisplatin resistance in non-small cell lung cancer with Mad2 silencing siRNA delivered systemically using EGFR-targeted chitosan nanoparticles. *Acta Biomater.* **2016**, *47*, 71–80. [CrossRef]

13. Schellmann, N.; Deckert, P.M.; Bachran, D.; Fuchs, H.; Bachran, C. Targeted enzyme prodrug therapies. *Mini-Rev. Med. Chem.* **2010**, *10*, 887–904. [CrossRef]
14. Appel, E.; Rabinkov, A.; Neeman, M.; Kohen, F.; Mirelman, D. Conjugates of daidzein-alliinase as a targeted pro-drug enzyme system against ovarian carcinoma. *J. Drug Target.* **2010**, *19*, 326–335. [CrossRef]
15. Majd, M.H.; Asgari, D.; Barar, J.; Valizadeh, H.; Kafil, V.; Coukos, G.; Omidi, Y. Specific targeting of cancer cells by multifunctional mitoxantrone-conjugated magnetic nanoparticles. *J. Drug Target.* **2013**, *21*, 328–340. [CrossRef]
16. Kumar, A.; Zhang, X.; Liang, X.-J. Gold nanoparticles: Emerging paradigm for targeted drug delivery system. *Biotechnol. Adv.* **2013**, *31*, 593–606. [CrossRef]
17. Singh, R.; Lillard, J.W., Jr. Nanoparticle-based targeted drug delivery. *Exp. Mol. Pathol.* **2009**, *86*, 215–223. [CrossRef]
18. Baharifar, H.; Khoobi, M.; Bidgoli, S.A.; Amani, A. Preparation of PEG-grafted chitosan/streptokinase nanoparticles to improve biological half-life and reduce immunogenicity of the enzyme. *Int. J. Biol. Macromol.* **2019**, *143*, 181–189. [CrossRef]
19. Singh, A.P.; Biswas, A.; Shukla, A.; Maiti, P. Targeted therapy in chronic diseases using nanomaterial-based drug delivery vehicles. *Signal Transduct. Target. Ther.* **2019**, *4*, 33. [CrossRef]
20. Yuan, S.; Hua, J.; Zhou, Y.; Ding, Y.; Hu, Y. Doxorubicin Loaded Chitosan-W18 O49 Hybrid Nanoparticles for Combined Photothermal-Chemotherapy. *Macromol. Biosci.* **2017**, *17*, 1700033. [CrossRef]
21. Kumar, S.; Jana, A.K.; Dhamija, I.; Maiti, M. Chitosan-assisted immobilization of serratiopeptidase on magnetic nanoparticles, characterization and its target delivery. *J. Drug Target.* **2013**, *22*, 123–137. [CrossRef] [PubMed]
22. Sathiyaseelan, A.; Saravanakumar, K.; Mariadoss, A.V.A.; Wang, M.-H. pH-controlled nucleolin targeted release of dual drug from chitosan-gold based aptamer functionalized nano drug delivery system for improved glioblastoma treatment. *Carbohydr. Polym.* **2021**, *262*, 117907. [CrossRef] [PubMed]
23. Calvo, P.; Remuñán-López, C.; Vila-Jato, J.L.; Alonso, M.J. Novel hydrophilic chitosan-polyethylene oxide nanoparticles as protein carriers. *J. Appl. Polym. Sci.* **1997**, *63*, 125–132. [CrossRef]
24. Scheeren, L.E.; Nogueira, D.R.; Macedo, L.B.; Vinardell, M.P.; Mitjans, M.; Infante, M.R.; Rolim, C.M.B. PEGylated and poloxamer-modified chitosan nanoparticles incorporating a lysine-based surfactant for pH-triggered doxorubicin release. *Colloids Surfaces B Biointerfaces* **2016**, *138*, 117–127. [CrossRef]
25. Dziawer, Ł.; Majkowska-Pilip, A.; Gaweł, D.; Godlewska, M.; Pruszyński, M.; Jastrzębski, J.; Wąs, B.; Bilewicz, A. Trastuzumab-Modified Gold Nanoparticles Labeled with 211At as a Prospective Tool for Local Treatment of HER2-Positive Breast Cancer. *Nanomaterials* **2019**, *9*, 632. [CrossRef]
26. Arinbasarova, A.Y.; Ashin, V.V.; Makrushin, K.V.; Medentsev, A.G.; Lukasheva, E.V.; Berezov, T.T. Isolation and properties of L-lysine-α-oxidase from the fungus Trichoderma cf. aureoviride RIFAI VKM F-4268D. *Microbiology* **2012**, *81*, 549–554. [CrossRef]
27. Yoon, H.Y.; Son, S.; Lee, S.J.; Gil You, D.; Yhee, J.Y.; Park, J.H.; Swierczewska, M.; Lee, S.; Kwon, I.C.; Kim, S.H.; et al. Glycol chitosan nanoparticles as specialized cancer therapeutic vehicles: Sequential delivery of doxorubicin and Bcl-2 siRNA. *Sci. Rep.* **2014**, *4*, 6878. [CrossRef] [PubMed]
28. Saravanakumar, K.; Mariadoss, A.V.A.; Sathiyaseelan, A.; Wang, M.-H. Synthesis and characterization of nano-chitosan capped gold nanoparticles with multifunctional bioactive properties. *Int. J. Biol. Macromol.* **2020**, *165*, 747–757. [CrossRef]
29. Saravanakumar, K.; Sriram, B.; Sathiyaseelan, A.; Hu, X.; Mariadoss, A.V.A.; MubarakAli, D.; Wang, M.-H. Molecular identification, volatile metabolites profiling, and bioactivities of an indigenous endophytic fungus (Diaporthe sp.). *Process Biochem.* **2020**, *102*, 72–81. [CrossRef]
30. Jaiswal, Y.; Tatke, P.; Gabhe, S.; Vaidya, A. Antidiabetic activity of extracts of Anacardium occidentale Linn. leaves on n -streptozotocin diabetic rats. *J. Tradit. Complement. Med.* **2016**, *7*, 421–427. [CrossRef]
31. Sakhtianchi, R.; Atyabi, F.; Yousefpour, P.; Vasheghani-Farahani, E.; Movahedi, A.-A.M.; Dinarvand, R. Targeted delivery of doxorubicin-utilizing chitosan nanoparticles surface-functionalized with anti-Her2 trastuzumab. *Int. J. Nanomed.* **2011**, *6*, 1977–1990. [CrossRef]
32. Niu, S.; Williams, G.R.; Wu, J.; Wu, J.; Zhang, X.; Chen, X.; Li, S.; Jiao, J.; Zhu, L.-M. A chitosan-based cascade-responsive drug delivery system for triple-negative breast cancer therapy. *J. Nanobiotechnology* **2019**, *17*, 95. [CrossRef] [PubMed]
33. Mandal, B.; Mittal, N.K.; Balabathula, P.; Thoma, L.A.; Wood, G.C. Development and in vitro evaluation of core–shell type lipid–polymer hybrid nanoparticles for the delivery of erlotinib in non-small cell lung cancer. *Eur. J. Pharm. Sci.* **2015**, *81*, 162–171. [CrossRef] [PubMed]
34. Forest, V.; Pourchez, J. Preferential binding of positive nanoparticles on cell membranes is due to electrostatic interactions: A too simplistic explanation that does not take into account the nanoparticle protein corona. *Mater. Sci. Eng. C* **2017**, *70*, 889–896. [CrossRef] [PubMed]
35. Hu, D.; Pan, M.; Yang, Y.; Sun, A.; Chen, Y.; Yuan, L.; Huang, K.; Qu, Y.; He, C.; Wei, Q.; et al. Trimodal Sono/Photoinduced Focal Therapy for Localized Prostate Cancer: Single-Drug-Based Nanosensitizer under Dual-Activation. *Adv. Funct. Mater.* **2021**, *31*, 2104473. [CrossRef]
36. Ito, T.; Sun, L.; Bevan, M.A.; Crooks, R.M. Comparison of Nanoparticle Size and Electrophoretic Mobility Measurements Using a Carbon-Nanotube-Based Coulter Counter, Dynamic Light Scattering, Transmission Electron Microscopy, and Phase Analysis Light Scattering. *Langmuir* **2004**, *20*, 6940–6945. [CrossRef] [PubMed]
37. Sathiyaseelan, A.; Saravanakumar, K.; Mariadoss, A.V.A.; Wang, M.-H. Biocompatible fungal chitosan encapsulated phytogenic silver nanoparticles enhanced antidiabetic, antioxidant and antibacterial activity. *Int. J. Biol. Macromol.* **2020**, *153*, 63–71. [CrossRef]

38. Melo, M.N.; Pereira, F.M.; Rocha, M.A.; Ribeiro, J.G.; Diz, F.M.; Monteiro, W.F.; Ligabue, R.A.; Severino, P.; Fricks, A.T. Immobilization and characterization of horseradish peroxidase into chitosan and chitosan/PEG nanoparticles: A comparative study. *Process Biochem.* **2020**, *98*, 160–171. [CrossRef]
39. Barth, A. Infrared spectroscopy of proteins. *Biochim. Biophys. Acta Bioenerg.* **2007**, *1767*, 1073–1101. [CrossRef]
40. Cipolatti, E.P.; Valério, A.; Nicoletti, G.; Theilacker, E.; Araújo, P.H.H.; Sayer, C.; Ninow, J.L.; de Oliveira, D. Immobilization of Candida antarctica lipase B on PEGylated poly(urea-urethane) nanoparticles by step miniemulsion polymerization. *J. Mol. Catal. B Enzym.* **2014**, *109*, 116–121. [CrossRef]
41. Garg, N.K.; Dwivedi, P.; Campbell, C.; Tyagi, R.K. Site specific/targeted delivery of gemcitabine through anisamide anchored chitosan/poly ethylene glycol nanoparticles: An improved understanding of lung cancer therapeutic intervention. *Eur. J. Pharm. Sci.* **2012**, *47*, 1006–1014. [CrossRef]
42. Antoniou, J.; Liu, F.; Majeed, H.; Qi, J.; Yokoyama, W.; Zhong, F. Physicochemical and morphological properties of size-controlled chitosan–tripolyphosphate nanoparticles. *Colloids Surfaces A Physicochem. Eng. Asp.* **2015**, *465*, 137–146. [CrossRef]
43. Niu, S.-J.C.C.-C.; Kuo, C.-F.H.S.-M. Evaluation of chitosan-g-PEG copolymer for cell anti-adhesion application. *J. Med. Biol. Eng.* **2007**, *27*, 41–46.
44. Helmi, O.; Elshishiny, F.; Mamdouh, W. Targeted doxorubicin delivery and release within breast cancer environment using PEGylated chitosan nanoparticles labeled with monoclonal antibodies. *Int. J. Biol. Macromol.* **2021**, *184*, 325–338. [CrossRef] [PubMed]
45. Yuan, Z.; Ye, Y.; Gao, F.; Yuan, H.; Lan, M.; Lou, K.; Wang, W. Chitosan-graft-β-cyclodextrin nanoparticles as a carrier for controlled drug release. *Int. J. Pharm.* **2013**, *446*, 191–198. [CrossRef] [PubMed]
46. Gatenby, R.A.; Gillies, R.J. Why do cancers have high aerobic glycolysis? *Nat. Rev. Cancer* **2004**, *4*, 891–899. [CrossRef] [PubMed]
47. Kesavan, S.; Meena, K.; Sharmili, S.A.; Govindarajan, M.; Alharbi, N.S.; Kadaikunnan, S.; Khaled, J.M.; Alobaidi, A.S.; Alanzi, K.F.; Vaseeharan, B. Ulvan loaded graphene oxide nanoparticle fabricated with chitosan and d-mannose for targeted anticancer drug delivery. *J. Drug Deliv. Sci. Technol.* **2021**, *65*, 102760. [CrossRef]
48. Wu, S.; Xu, C.; Zhu, Y.; Zheng, L.; Zhang, L.; Hu, Y.; Yu, B.; Wang, Y.; Xu, F.-J. Biofilm-Sensitive Photodynamic Nanoparticles for Enhanced Penetration and Antibacterial Efficiency. *Adv. Funct. Mater.* **2021**, *31*, 2103591. [CrossRef]
49. Wang, F.; Li, J.; Tang, X.; Huang, K.; Chen, L. Polyelectrolyte three layer nanoparticles of chitosan/dextran sulfate/chitosan for dual drug delivery. *Colloids Surf. B Biointerfaces* **2020**, *190*, 110925. [CrossRef] [PubMed]
50. Tai, W.; Mahato, R.; Cheng, K. The role of HER2 in cancer therapy and targeted drug delivery. *J. Control. Release* **2010**, *146*, 264–275. [CrossRef] [PubMed]
51. Accardo, A.; Aloj, L.; Aurilio, M.; Morelli, G.; Tesauro, D. Receptor binding peptides for target-selective delivery of nanoparticles encapsulated drugs. *Int. J. Nanomed.* **2014**, *9*, 1537–1557. [CrossRef]
52. Master, A.M.; Gupta, A.S. EGF receptor-targeted nanocarriers for enhanced cancer treatment. *Nanomedicine* **2012**, *7*, 1895–1906. [CrossRef]
53. Ando, T.; Nagumo, M.; Ninomiya, M.; Tanaka, K.; Linhardt, R.J.; Koketsu, M. Synthesis of coumarin derivatives and their cytoprotective effects on t -BHP-induced oxidative damage in HepG2 cells. *Bioorganic Med. Chem. Lett.* **2018**, *28*, 2422–2425. [CrossRef] [PubMed]
54. Pajaniradje, S.; Mohankumar, K.; Pamidimukkala, R.; Subramanian, S.; Rajagopalan, R. Antiproliferative and Apoptotic Effects ofSesbania grandifloraLeaves in Human Cancer Cells. *BioMed Res. Int.* **2014**, *2014*, 474953. [CrossRef]
55. Murad, H.; Hawat, M.; Ekhtiar, A.; Aljapawe, A.; Abbas, A.; Darwish, H.; Sbenati, O.; Ghannam, A. Induction of G1-phase cell cycle arrest and apoptosis pathway in MDA-MB-231 human breast cancer cells by sulfated polysaccharide extracted from Laurencia papillosa. *Cancer Cell Int.* **2016**, *16*, 39. [CrossRef] [PubMed]
56. Kuang, W.; Hu, W.; Ren, H.; Shao, Y.; Liu, B. Plant derived coumestrol phytochemical targets human skin carcinoma cells by inducing mitochondrial-mediated apoptosis, cell cycle arrest, inhibition of cell migration and invasion and modulation of m-TOR/PI3K/AKT signalling pathway. *Saudi J. Biol. Sci.* **2021**, *28*, 2739–2746. [CrossRef] [PubMed]
57. Wu, C.; Wang, C.; Zheng, Y.; Zheng, Y.; Liu, Z.; Xu, K.; Zhong, W. Triple Enzyme-Regulated Molecular Hydrogels for Carrier-Free Delivery of Lonidamine. *Adv. Funct. Mater.* **2021**, *31*, 2104418. [CrossRef]
58. Liao, T.; Liu, C.; Ren, J.; Chen, H.; Kuang, Y.; Jiang, B.; Chen, J.; Sun, Z.; Li, C. A chitosan/mesoporous silica nanoparticle-based anticancer drug delivery system with a "tumor-triggered targeting" property. *Int. J. Biol. Macromol.* **2021**, *183*, 2017–2029. [CrossRef]

MDPI
St. Alban-Anlage 66
4052 Basel
Switzerland
www.mdpi.com

Pharmaceutics Editorial Office
E-mail: pharmaceutics@mdpi.com
www.mdpi.com/journal/pharmaceutics